AF896591

Nanotechnology Based Delivery of Phytoconstituents and Cosmeceuticals

Deep Pooja • Hitesh Kulhari
Editors

Nanotechnology Based Delivery of Phytoconstituents and Cosmeceuticals

Springer

Editors
Deep Pooja
School of Pharmacy
National Forensic Science University
Gandhinagar, Gujarat, India

Hitesh Kulhari
School of Nano Sciences
Central University of Gujarat
Gandhinagar, Gujarat, India

ISBN 978-981-99-5313-4 ISBN 978-981-99-5314-1 (eBook)
https://doi.org/10.1007/978-981-99-5314-1

© The Editor(s) (if applicable) and The Author(s), under exclusive license to Springer Nature Singapore Pte Ltd. 2024
This work is subject to copyright. All rights are solely and exclusively licensed by the Publisher, whether the whole or part of the material is concerned, specifically the rights of translation, reprinting, reuse of illustrations, recitation, broadcasting, reproduction on microfilms or in any other physical way, and transmission or information storage and retrieval, electronic adaptation, computer software, or by similar or dissimilar methodology now known or hereafter developed.
The use of general descriptive names, registered names, trademarks, service marks, etc. in this publication does not imply, even in the absence of a specific statement, that such names are exempt from the relevant protective laws and regulations and therefore free for general use.
The publisher, the authors, and the editors are safe to assume that the advice and information in this book are believed to be true and accurate at the date of publication. Neither the publisher nor the authors or the editors give a warranty, expressed or implied, with respect to the material contained herein or for any errors or omissions that may have been made. The publisher remains neutral with regard to jurisdictional claims in published maps and institutional affiliations.

This Springer imprint is published by the registered company Springer Nature Singapore Pte Ltd.
The registered company address is: 152 Beach Road, #21-01/04 Gateway East, Singapore 189721, Singapore

Paper in this product is recyclable.

Preface

Phytoconstituents (PCs) are the molecules which are obtained from plants. Recently PCs have emerged as alternative medicines to synthetic drugs. The wide therapeutic indices of PCs, i.e., high safety margin make them attractive for treating various diseases including complex diseases like Alzheimer, rheumatoid arthritis, and cancer. However, the clinical application of PCs is often hampered by their poor physicochemical properties like low aqueous solubility, very slow dissolution in physiological fluids, rapid degradation, and poor bioavailability. Recently nanotechnology has emerged as a new tool to design more effective nanocarriers for PCs to overcome the inherent problems of PCs. Various novel nanocarriers have been developed in recent years as they offer unique advantages over conventional formulations. These nanocarriers are made of different biomaterials like polymers, lipids, and proteins. Dendrimer nanostructures have also been attempted to improve the delivery of PCs.

Nanotechnology Based Delivery of Phytoconstituents and Cosmeceuticals explores the importance of nanoscience and nanotechnology in the delivery of phytoconstituents and cosmeceuticals. Chapter 1 introduces the basics of nanoscience and nanotechnology and their applications in various fields. Chapters 2 and 3 include the various types and roles of various phytoconstituents. Chapters 4–9 describe the applications of different types of nanocarriers in the delivery of phytoconstituents. The major carriers include polymeric nanoparticles, lipid-based nanoparticles, supramolecules, metal/metal oxide nanoparticles, protein-based nanoparticles, and dendrimers. The following chapter highlights the applications of nanoscience and nanotechnology in the delivery of cosmeceuticals. One chapter describes the toxicity issues related to nanocarrier systems. The last chapter discusses the regulatory guidelines that need to be considered while designing nanocarriers for clinical applications of phytoconstituents and cosmeceuticals.

Finally, we thank all the authors of the chapters of this book for their generous support and contributions.

Gandhinagar, Gujarat, India — Deep Pooja
Gandhinagar, Gujarat, India — Hitesh Kulhari

Contents

Editors and Contributors

About the Editors

Deep Pooja is an Assistant Professor in the School of Pharmacy at National Forensic Sciences University, Gandhinagar. She has worked as CSIR-Senior Research Fellow at CSIR-Indian Institute of Chemical Technology and received her Ph.D. degree from Osmania University, Hyderabad. Her research interests are focused on the development of nanotechnology-based drug delivery systems. She has more than 65 peer-reviewed international publications. These publications include research articles, reviews, and book chapters. Dr. Pooja has been awarded several awards and fellowships including an Australian Endeavour Research Fellowship by the Department of Education, Australian Government. She has delivered numerous oral and poster presentations at international meetings.

Hitesh Kulhari is an Assistant Professor at the School of Nano Sciences, Central University of Gujarat, Gandhinagar. Previously, he worked as Associate Professor in the Department of Pharmaceutical Technology (Formulations) at the National Institute of Pharmaceutical Sciences and Drug Research, Guwahati. Dr. Kulhari received his PhD degree from RMIT University, Melbourne, in Nanomedicines. His research focus is on the Designing of Targeted Drug Delivery Systems and Pharmaceutical Nanotechnology. He has published 70 journal publications in peer-reviewed international journals, 15 book chapters, and edited 01 book on Pharmaceutical Applications of Dendrimers. Dr. Kulhari is the recipient of several awards and research grants including Prof CNR Rao Research Excellence Award in Material Science in 2014, INSPIRE Faculty Award from DST, New Delhi, in 2015, and Appreciation Award from Cancer Research Foundation, India, in 2020. He has more than 13 years of teaching and research experience in Pharmaceutical Technology and Nanotechnology.

Contributors

Mounisha Bandakinda Department of Pharmacology and Toxicology, National Institute of Pharmaceutical Education and Research (NIPER), Guwahati, Assam, India

Saumyadeep Bora School of Nano Sciences, Central University of Gujarat, Gandhinagar, Gujarat, India

Akanksha Chaturvedi Department of Pharmacy, Banasthali Vidyapith, Jaipur, Rajasthan, India

Sunaina Chaurasiya School of Nano Sciences, Central University of Gujarat, Gandhinagar, Gujarat, India

Jaya Dwivedi Department of Chemistry, Banasthali Vidyapith, Jaipur, Rajasthan, India

Poonam Jain Department of Medical Biomaterials Engineering, College of Biomedical Science, Kangwon National University, Chuncheon, Republic of Korea

Bhavana Jodha School of Life Sciences, Central University of Gujarat, Gandhinagar, India

Hitesh Kulhari School of Nano Sciences, Central University of Gujarat, Gandhinagar, Gujarat, India

Ankit Kumar Department of Pharmacology and Toxicology, National Institute of Pharmaceutical Education and Research (NIPER), Guwahati, Assam, India

Kanakraju Medicherla Department of Human Genetics, College of Science and Technology, Andhra University, Visakhapatnam, Andhra Pradesh, India

Akshada Mhaske Department of Pharmaceutics, National Institute of Pharmaceutical Education and Research-Raebareli, Lucknow, UP, India

Awanish Mishra Department of Pharmacology and Toxicology, National Institute of Pharmaceutical Education and Research (NIPER), Guwahati, Assam, India

Sarvesh Paliwal Department of Pharmacy, Banasthali Vidyapith, Jaipur, Rajasthan, India

Sunita Patel School of Life Sciences, Central University of Gujarat, Gandhinagar, India

Deep Pooja School of Pharmacy, National Forensic Science University, Gandhinagar, Gujarat, India

Sonali Priyadarshini School of Nano Sciences, Central University of Gujarat, Gandhinagar, Gujarat, India

Divya Bharti Rai School of Nano Sciences, Central University of Gujarat, Gandhinagar, Gujarat, India

Swapnil Sharma Department of Pharmacy, Banasthali Vidyapith, Jaipur, Rajasthan, India

Rahul Shukla Department of Pharmaceutics, National Institute of Pharmaceutical Education and Research-Raebareli, Lucknow, UP, India

Shalini Shukla Department of Pharmaceutics, National Institute of Pharmaceutical Education and Research-Raebareli, Lucknow, UP, India

Sanju Kumari Singh School of Life Sciences, Central University of Gujarat, Gandhinagar, India

Raghu Solanki School of Life Sciences, Central University of Gujarat, Gandhinagar, India

Kanika Verma Department of Pharmacy, Banasthali Vidyapith, Jaipur, Rajasthan, India

Hyuk Sang Yoo Department of Medical Biomaterials Engineering, College of Biomedical Science, Kangwon National University, Chuncheon, Republic of Korea

1 Introduction of Nanoscience and Nanotechnology

Saumyadeep Bora, Deep Pooja, and Hitesh Kulhari

1.1 Introduction

In today's fast-growing scientific world, nanoscience and nanotechnology are the two most important pillars on which the development and the growth of the society rely. It encompasses structures, technologies, and systems with unique features and functionalities due to the arrangement of their atoms on the nanoscale, i.e., 1–100 nm. Nanoscience is a discipline of science that studies characteristics of materials at the nanoscale, with a special emphasis on the unique, size-dependent properties of solid-state materials. Nanoscience is the interdisciplinary study of materials at the atomic and molecular sizes. It incorporates physics, chemistry, material science, and biology (Mulvaney 2015). On the other hand, nanotechnology is the field that encompasses the production, design, and use of materials called nanomaterials (NMs), which range in size from 1 to 100 nm (Saifullah et al. 2013). "There's plenty of room at the bottom," a famous line from Nobel laureate Richard Feynman's 1959 speech, is often seen as the beginning of the notions of nanoscience and nanotechnology. In 1974, the term "nanotechnology" was first introduced by Norio Taniguchi to describe how dimensional precision has increased over time, and 1980 onward, there is significant development on the discipline of nanoscience and nanotechnology (Feynman 1959; Taniguchi 1974). However, the applications of nanoscience and nanotechnology have already been reported in history.

Even prior to the advent of nanotechnology, people were unintentionally interacting with and using numerous nanoscale items. Hair coloring in black was popular in Egyptian civilization and was long thought to be hinged on herbal items

S. Bora · H. Kulhari (✉)
School of Nano Sciences, Central University of Gujarat, Gandhinagar, Gujarat, India
e-mail: hitesh.kulhari@cug.ac.in

D. Pooja (✉)
School of Pharmacy, National Forensic Science University, Gandhinagar, Gujarat, India

© The Author(s), under exclusive license to Springer Nature Singapore Pte Ltd. 2024

D. Pooja, H. Kulhari (eds.), *Nanotechnology Based Delivery of Phytoconstituents and Cosmeceuticals*, https://doi.org/10.1007/978-981-99-5314-1_1

like henna (Tolochko 2009). Later on, the studies revealed that hair was colored with a mixture of lime, lead oxide, and water during which PbS nanoparticles (NPs) were formed that provided steady dyeing (Walter et al. 2006). The Lycurgus Cup, dating from the fourth century CE, is arguably the most well-known illustration of early nanotechnology usage. This ancient Roman cup has peculiar optical characteristics; it changes color depending on where the light source is coming from. The cup appears green in the daylight but becomes red when lit from the inside (by a candle) (Barber and Freestone 1990). Recent research of this cup revealed that it includes Au and Ag NPs, ranging in size from 50 to 100 nm, which are the cause of the cup's distinctive coloration due to plasmon resonance where plasmon excitation of electrons takes place (Atwater 2007). The evidences of the employment of nanotechnology techniques, principles and properties were also reported in Ancient India, Mesopotamia, and the Maya times. Thus, it can be concluded that the unintentional usage of nanotechnology in ancient history was not limited to a specific part of the world (Brill and Cahill 1988; Sharon 2019).

Nanoscience and nanotechnology are now developing quickly with new uses in several sectors. For instance, by generating NMs and nanostructures (NSs), and other nanoscale objects, nano-engineering develops multiple prospects for advancement in the sectors of electronics, energy, food/agriculture, environment remediation, health, and other areas. The development of several diverse NPs and nanostructured materials, as well as their use in related applications ranging from developing novel NMs to direct atomic-scale management of matter, has been facilitated by a variety of scientific domains, from organic chemistry to semiconductor physics (Jeevanandam et al. 2018). NMs and NSs are being employed in a variety of applications due to their unique features, comprising catalysis, water treatment, energy storage, medicine, agriculture, and so on. In this book chapter, we have discussed about the different unique properties, classification, synthesis methods, characterization techniques, and applications of NMs and NSs in various fields.

1.2 Unique Properties of Nanomaterials and Nanostructures

When compared to the bulk form, the characteristics of materials at nanoscale level vary significantly. At nanoscale, size-dependent effects are more noticeable than bulk form of materials. Various properties like surface, optical, electrical, mechanical, thermal, and magnetic properties of NMs and NSs can be tweaked by varying the size of NMs (Thomas et al. 2019; Roduner 2006). And owing to these unique properties, NMs and NSs have been used in several fields. The most significant physicochemical characteristics that are altering at the nanoscale are discussed as follows.

1.2.1 Surface Properties

When a bulk material is partitioned into nanoscale components, the overall volume remains constant while the collective surface area increases. Thus, the surface-to-volume (S/V) ratio of the NMs increases as compared to bulk form. As a result, more number of atoms or molecules are present at the surface of NMs, hence enhancing their reactivity and sensing properties. Because of these enhanced surface properties, NMs have wide applications in sensing and catalysis (Tomar et al. 2020).

1.2.2 Optical Properties

For many applications, including photocatalysts and photovoltaics, the optical characteristics of NMs, particularly semiconductor materials, are crucial. The fundamental laws of light and Beer-Lambert law can be used to determine the optical characteristics of NMs. In semiconductor NPs, enhanced wavelength absorption is controlled by a number of variables, including size distribution, shape, size, and the kind of modifiers (Kelly et al. 2003). Quantum confinement causes the increase in bandgap of NMs due to which more energy is required in the absorption that results in blue shift due to the emission of shorter wavelength. Hence, by tuning the size of NMs, bandgap can be tuned accordingly, and thus the same material can be used for the emission of different colors (Piri et al. 2016; Mikhailov et al. 2018).

1.2.3 Electrical Properties

The electrical properties of NMs are influenced by the size, surface area, composition, and surface modification of the NMs (Shin et al. 2015; Henkel et al. 2020). In bulk form, the conduction of electrons is delocalized, but at the nanoscale range, due to the quantum confinement, the energy bands are replaced by discrete energy states which make the conducting materials either semiconductors or insulators. Moreover, other electrical properties like dielectric constant and electrical conductivity of NMs can be tuned through tuning their sizes (Eustis and El-Sayed 2006).

1.2.4 Mechanical Properties

In comparison to bulk materials, NMs have distinct mechanical properties. Owing to their enormous surface area and ease of modification, NMs have improved mechanical properties including elastic modulus, adhesion, hardness, stress, and strain. Due to these enhanced properties, NMs have wide applications in supercapacitors, drug delivery, etc. (Guo et al. 2014).

1.2.5 Thermal Properties

The thermal properties of any material are basically based on its thermal stability, thermal conductivity, heat capacity, and thermoelectric power (Savage and Rao 2006). The NMs possess good thermal properties by virtue of their broad surface area due to which heat transfer takes place straightly on materials' surface. Therefore, nanofluids (NFs) demonstrated improved thermal conductivities as well as thermal stability of NMs as compared to their conventional form. NFs are those fluids in which NMs are dispersed into liquid such as water, ethylene glycol, or oils (Andrievski 2014; Qiu et al. 2020).

1.2.6 Magnetic Properties

The magnetic properties of NMs are firmly dependent upon their size. The magnetic properties vary from strong magnetics, i.e., ferromagnetic to paramagnetic and then to superparamagnetic as size of NMs decreases (Lamouri et al. 2020; Shrimali et al. 2020). The principle behind the enhancement of magnetic properties can be stated that as the size of the material decreases to nanoscale, the S/V ratio increases which results in a different local environment for the surface atoms to strongly interact magnetically with their neighboring atoms, thus having enhanced magnetic properties. In addition to this, the various ways that NMs are synthesized affect the magnetic characteristics of NMs (Lakshmiprasanna et al. 2019; Owens 2015).

1.3 Classification of Nanomaterials and Nanostructures

NMs and NSs can be broadly classified into distinct types depending on their dimensions, materials, origin, porosity, crystallinity, and dispersion.

1.3.1 On the Basis of Dimensions

Based on the dimensions, NMs and NSs are of four types, viz., zero-dimensional (0D), one-dimensional (1D), two-dimensional (2D), and three-dimensional (3D) NMs and NSs (Pokropivny and Skorokhod 2007; Kolahalam et al. 2019).

1.3.1.1 0D NMs and NSs

In 0D NMs and NSs, all three dimensions are confined in nanoscale range, i.e., less than 100 nm range. Quantum dots and fullerenes are some examples of 0D NMs and NSs. These NMs and NSs have huge applications in biomedical as well as electronics because of their optical properties.

1.3.1.2 1D NMs and NSs

In 1D NMs and NSs, two of the dimensions are confined in less than 100 nm range. Some of the examples of 1D NMs and NSs are nanotubes, nanorods, nanowires, etc.

1.3.1.3 2D NMs and NSs

In 2D NMs and NSs, only one dimension is confined in less than the 100 nm range. Nanosheets and nanofilms are some of the examples of 2D NMs and NSs.

1.3.1.4 3D NMs and NSs

In 3D NMs and NSs, none of the three dimensions is confined in less than 100 nm range. Bulk powder and bundles of nanowires and nanotubes are some of the examples of these 3D NMs and NSs.

1.3.2 On the Basis of Materials

NMs and NSs can be further classified into mainly three types: organic, inorganic, and carbon-based NMs depending on their material or elemental composition.

1.3.2.1 Organic NMs

These NMs have distinct functions as a result of the chemical interaction of their main element, i.e., carbon with other elements which impart unique functions and reactivity to the NMs. This class of NMs is made up of polymers, lipids, proteins, and other organic substances (Pan and Zhong 2016). Micelles, liposomes, dendrimers, and protein complexes like ferritin are some of the most well-known examples of this class (Fig. 1.1) (Ng and Zheng 2015). Organic NMs are nontoxic and biodegradable in nature and occasionally even having a hollow core. Since these materials are often formed of noncovalent interaction, that gives them a more pliable character and provides a means of elimination from the body. Due to these characteristics, organic NMs have wide applications in the biomedical field, especially in delivery of drugs (Gujrati et al. 2014).

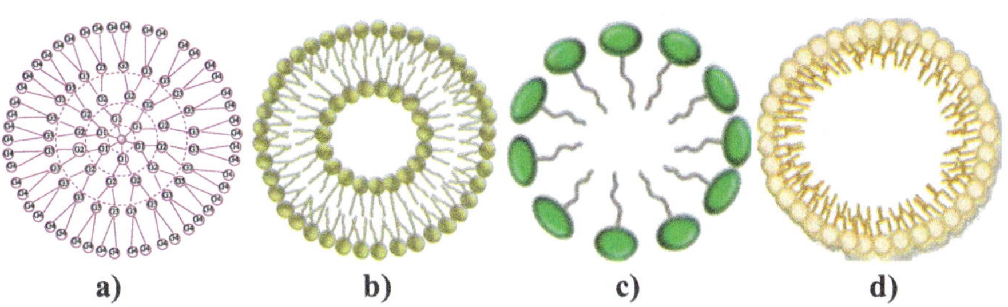

Fig. 1.1 Schematic representation of some organic NMs: (**a**) dendrimer, (**b**) liposome, (**c**) micelle, and (**d**) solid lipid NPs

1.3.2.2 Inorganic NMs

This class of NMs is composed of noncarbon elements like metals, metal oxides, etc. This class includes semiconductor NPs, metal NPs, and ceramic NPs. Semiconductor NPs possess characteristics of both metals and nonmetals, as they are prepared from the semiconductor materials. In comparison to bulk form, semiconductor NPs have distinctive broad bandgaps and thus exhibit a dramatic change in their characteristics with bandgap modification. And due to this, they have applications in photocatalysis, optical, and electrical devices (Khan et al. 2019; Gupta and Tripathi 2012; Sun et al. 2000). Metal NPs are composed of purely metal precursors, and they exhibit excellent optical, thermal, biological, and magnetic properties (Toshima and Yonezawa 1998; Ealias and Saravanakumar 2017). Ceramic NPs are made of metals, metal oxides, metalloids, carbides, phosphates, other inorganic materials, etc. These NPs can be in different forms like polycrystalline, amorphous, porous, or hollow. These NPs are highly stable and have high loading capacity, so they have been utilized in optoelectronics and biomedical applications (Thomas et al. 2015; Moreno-Vega et al. 2012; D'Amato et al. 2013).

1.3.2.3 Carbon-Based NMs

This class of NMs is completely made of carbon. These NMs include carbon quantum dots, fullerenes, graphene, carbon nanotubes (CNTs), nanohorns, activated carbon, carbon black, etc. (Fig. 1.2). Different methods such as laser ablation, arc discharge, and chemical vapour deposition (CVD) are used for the synthesis of carbon-based NMs. These NMs have high strength and excellent thermal, electrical, sorption, and optical properties. Due to these unique characteristics, they are used in various fields like energy storage, drug delivery, optoelectronics, bioimaging, etc. (Mauter and Elimelech 2008; Oh et al. 2010; Liu et al. 2018; Chandra et al. 2011).

1.3.3 On the Basis of Origin

NMs and NSs can be broadly classified into two different types depending on their origin, viz., natural and anthropogenic NMs and NSs.

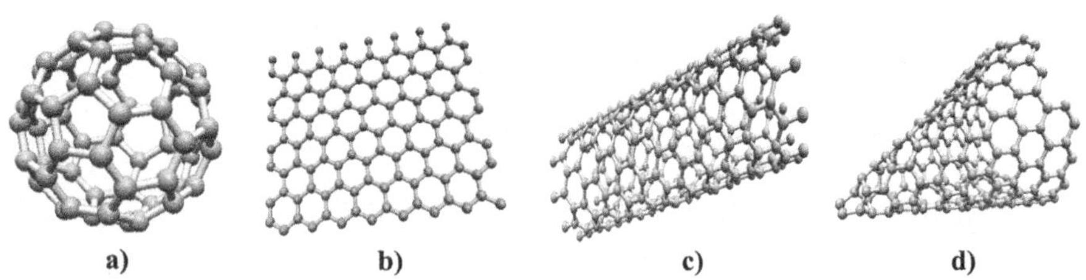

Fig. 1.2 Schematic representation of some carbon-based NMs: (**a**) fullerene, (**b**) graphene, (**c**) CNT and (**d**) nanohorn

1.3.3.1 Natural NMs and NSs

These NMs or NSs are produced by different natural processes without any association of artificial or man-made processes. Some examples of natural NMs and NSs are those present in volcanic ash and clay etc. (Ngô and Van de Voorde 2014). Mainly these NMs or NSs are synthesized or originated during the natural processes like dust storms, forest fires etc. Some microorganisms also have nanoscale size.

1.3.3.2 Anthropogenic NMs and NSs

These NMs and NSs are produced for specific applications depending on their dimensions and other properties. These can be further classified as engineered and bioinspired NMs and NSs. The engineered NMs and NSs have regular shapes due to the specific synthesis process. Examples of engineered NMs and NSs are CNTs, fullerenes etc. (Samyn and Barhoum 2018; Albalawi et al. 2021). Whereas, bioinspired NMs and NSs are the engineered NMs which impersonate the properties of natural NMs. Example of bioinspired NMs is mechanochromic elastomers that mimic the photonic structure of the chameleon iridophore cells (Lee et al. 2017).

1.3.4 On the Basis of Porosity

Depending on the pore size, NMs and NSs can be classified as microporous, mesoporous, and macroporous materials (Barhoum et al. 2017).

1.3.4.1 Microporous NMs

NMs having pore size of less than 2 nm are known as microporous NMs. Examples of microporous NMs are metal-organic frameworks (MOFs) and zeolite.

1.3.4.2 Mesoporous NMs

NMs having pore size of 2–50 nm are known as mesoporous NMs. Activated carbon is an exceptionally good example of mesoporous NMs.

1.3.4.3 Macroporous NMs

NMs having pore size of larger than 50 nm are known as mesoporous NMs.

1.3.5 On the Basis of Crystallinity

NMs and NSs can be classified into amorphous, crystalline, and polycrystalline materials depending on their crystallinity. In amorphous NMs, there is an absence of long-range order of atoms. Whether in crystalline NMs, there is a long-range order of atoms, and in the polycrystalline NMs, there are several regions present in which atoms are arranged in a long-range order (Fig. 1.3) (Illath et al. 2020; Ashraf et al. 2018).

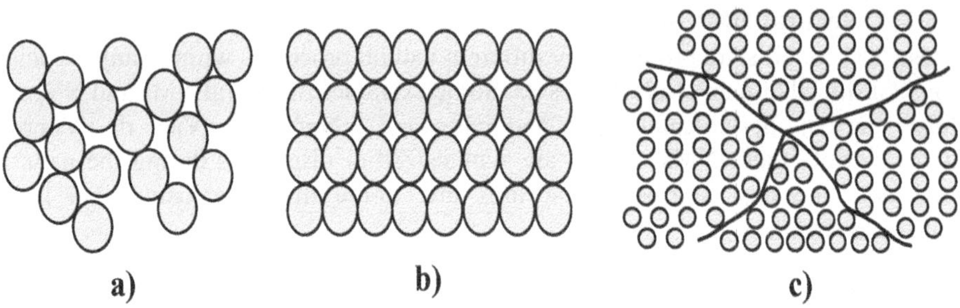

Fig. 1.3 Schematic representation of (**a**) amorphous, (**b**) crystalline, and (**c**) polycrystalline NMs

Table 1.1 Classification of NMs on the basis of PDI value

PDI value	Dispersibility
0.1–0.3	Monodispersed
0.3–0.5	Moderately dispersed
0.5–1	Aggregated

Table 1.2 Classification of NMs on the basis of stability

Zeta potential (mV)	Stability
0–5	Instable
10–30	Low stability
30–40	Moderate stability
40–60	High stability
>60	Excellent stability

1.3.6 On the Basis of Dispersions

Based on how easily they disperse in a solvent and what kind of solvent they dissolve in, NMs can be divided into aggregated and dispersed categories. The polydispersity index (PDI) describes the distribution of particle sizes in NMs. And on the basis of PDI value, NMs can be classified into monodispersed, moderately dispersed, and aggregated NMs (Table 1.1) (Strojan et al. 2017; Kumar and Dixit 2017). Furthermore, the stability of the NM's dispersions can be determined using the zeta potential value. The NM dispersions having higher zeta potential values (positive or negative) are highly stable, whereas NM dispersions having lower zeta potential values are less stable (Table 1.2) (Colvin 2003).

1.4 Synthesis of Nanomaterials and Nanostructures

The key component of nanoscience and nanotechnology is the synthesis of NMs and NSs. Several ways of synthesizing NMs have been developed during the last few decades. Finding out for what reason these NMs are being manufactured is one of the salient goals of the process. In order to properly synthesize the NMs, the researchers must be aware of their uses. NMs will be produced differently for industrial use than for biological or medicinal use. This is because distinct products will be developed for diverse industrial uses of NMs. Improved performance and cheaper cost are two further goals of the researchers who are synthesizing NMs. Many physical and chemical techniques have been employed in recent years to increase the functionality of NMs that exhibit improved characteristics. Top-down and bottom-up approaches are the two main methods of synthesis of NMs and NSs (Fig. 1.4).

1.4.1 Top-Down Approach

In these methods, bulk materials are converted into NMs. Mechanical ball milling, sputtering, electrospinning, lithography, and laser ablation are examples of top-down methods, and these are briefly discussed below.

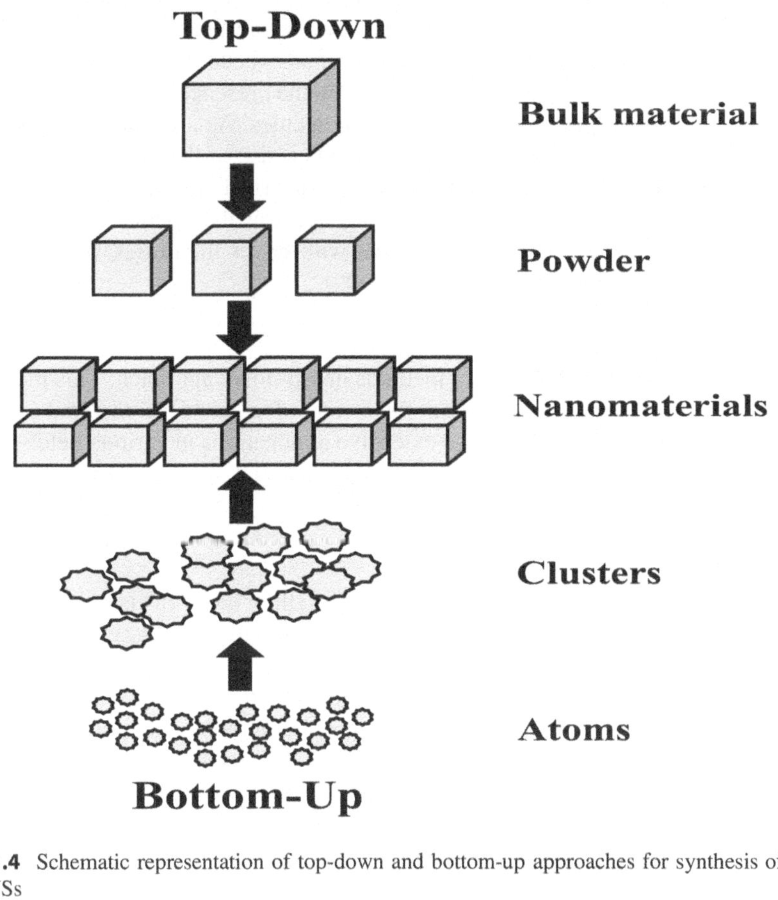

Fig. 1.4 Schematic representation of top-down and bottom-up approaches for synthesis of NMs and NSs

1.4.1.1 Mechanical Ball Milling

As the name suggests, this method consists of balls made up of hardened steel, iron, tungsten carbide or silicon carbide, and a stainless-steel mill chamber, to rotate inside the mill. In the mechanical ball milling method, the bulk powder of interest is taken inside the mill chamber, and with the help of mechanical force (balls), NMs are produced. This approach has gained popularity because of its ease of use, low cost, and eco-friendliness, as well as its potential to obtain extremely high yields. Ball milling has been shown to promote material reactivity and ensure uniform elemental distribution in space (Takacs 2002).

1.4.1.2 Sputtering

In the sputtering method, a solid surface is bombarded with high-energy particles like plasma or gas to produce NMs, mostly for the synthesis of nanofilms. During sputtering, the energetic gaseous ions hit the surface of precursor or source material, and on the basis of the incident energy, physical expulsion of the atomic clusters occurs to form the nanofilms. There are various ways to perform sputtering such as using a DC diode, radiofrequency diode, and magnetron (Ayyub et al. 2001; Son et al. 2017; Wender et al. 2013). Generally, sputtering process takes place in the presence of a sputtering gas (mostly inert gas) in an evacuated chamber. During this process, the gas ions are released at the cathode in the presence of high voltage due to the collision between free electrons and gas molecules. As a result of the collision, atoms are ejected from the surface of the target because of the fast hastening of the positively charged ions toward cathode in the electric field (Muñoz-García et al. 2009). The sputtering approach is intriguing because the sputtered nanomaterial composition is similar to the target material with fewer impurities, and it is more cost-effective than lithography (Nie et al. 2009).

1.4.1.3 Electrospinning

It is one of the easiest NM synthesis methods in top-down approach. This method is mainly used in the synthesis of nanofibers out of a range of materials, mostly polymers. And these nanofibers have extensive applications in various fields including sensing, energy storage, biomedicine, and material science (Ostermann et al. 2011).

1.4.1.4 Lithography

In the lithography technique, a focused beam of light or electrons is used to produce the nanopatterns. Primarily there are two types of nanolithography, viz., masked and maskless nanolithography. By the use of a particular mask or template, nanopatterns are created over a sizable surface area in masked nanolithography. This lithography technique includes nanoimprint lithography, soft lithography, and photolithography (Pimpin and Srituravanich 2012; Szabó et al. 2013; Kuo et al. 2003; Yin et al. 2001). On the contrary, in maskless nanolithography, the nanopatterns are created without using a mask or template. Electron beam lithography, focused ion beam lithography, and scanning probe lithography are examples of maskless nanolithography (Xu and Chen 2020; Matsumoto et al. 2020; Garg et al. 2020).

1.4.1.5 Laser Ablation

This method produces NPs by striking a strong laser beam to the the source or precursor material. As these laser radiations have high intensity, the source or precursor material vaporizes, and results into the formation or generation of NPs. This method is used in the synthesis of various NMs like carbon NMs, oxide nanocomposites, metal nanoparticles, etc. (Amendola and Meneghetti 2009; Zhang et al. 2017; Ismail et al. 2019; Chrzanowska et al. 2015; Duque et al. 2019).

1.4.2 Bottom-Up Approach

In this approach, NMs and NSs are synthesized from very small building blocks like atoms, molecules, or clusters through self-assembly that follows nucleation and growth mechanism. In the bottom-up technique, the starting material is either in a liquid state or a gaseous state. The major advantages of this approach are: first, the production or formation of NPs of various sizes (very small to large scale); second, a uniform and narrow particle size distribution, and third, better control over size and properties of NMs and NSs through optimization of the reaction conditions. Therefore, bottom-up methods are more attractive for both laboratory and industrial scale. There are various methods in bottom-up approaches which include the sol-gel method, hydrothermal method, chemical vapor deposition method, physical vapor deposition, etc.

1.4.2.1 Sol-Gel Method

An important wet chemical approach for the synthesis of NMs is the sol-gel method. A variety of metal-oxide-based (MO) NMs can be synthesized by utilizing sol-gel method. This method involves the modification of the liquid precursor into a sol which then further changes into a group of network-like structure known as gel during the complete synthesis. The common precursors in the synthesis of NMs using this method are metal alkoxides. The various steps involved during this process are hydrolysis, condensation followed by aging and drying, and lastly calcination. There are many conditions affecting the synthesis of NMs such as pH, water content, hydrolysis rate, concentration, temperature, and drying conditions. The production of homogenous material is one of the numerous perquisites of the sol-gel process. Moreover, this has many other benefits such as a facile synthesis process, low temperature, and eco-friendly nature (Danks et al. 2016; Parashar et al. 2020; De Coelho Escobar and Dos Santos 2014).

1.4.2.2 Hydrothermal Method

The hydrothermal process is among the most prevalent and extensively employed methods for producing NMs and NSs. In this technique, at high temperatures and high pressure, a heterogenous reaction is carried out in an aqueous medium in an autoclave to form the NMs (Wu et al. 2011; Cao et al. 2016; Li et al. 2015). Various shaped NSs like nanorods, nanowires, and nanospheres can be synthesized using the hydrothermal method (Dong et al. 2020; Jiang et al. 2018; Chai et al. 2018).

1.4.2.3 Chemical Vapor Deposition (CVD)

In this method, solid-phase NMs are formed by condensing substances that are in vapor phase. The production of carbon-based NMs relies heavily on the CVD method due to high purity and large-scale production. Precursors are deemed appropriate for CVD if they are low-cost, eco-friendly, and have sufficient volatility and high chemical purity. The synthesis of 2D NPs via CVD is a well-known and effective approach for creating high-quality NMs (Jones and Hitchman 2009; Ago 2015; Machac et al. 2020).

1.4.2.4 Physical Vapor Deposition (PVD)

PVD is a dynamic bottom-up technique for synthesizing NMs. It is a vaporization coating method in which atomic-level transfer of materials occurs. There are several steps involved in the PVD method: (a) first, the material to be deposited is vaporized using high temperature in vacuum, (b) from the source to surface, the vapor is transferred to a low-pressure region, and (c) finally, thin films formed on the substrate by condensation of the vapor. Usually, films of thicknesses ranging from a few nm to thousands of nm can be deposited using the PVD method. Moreover, this method can also be used to synthesize freestanding structures, graded composition deposits, and multilayer coatings (Okuyama and Lenggoro 2003; Abegunde et al. 2019).

1.5 Characterization Techniques for Nanomaterials and Nanostructures

In comparison to bulk materials, NMs have a significant surface-to-volume ratio. There are different factors such as concentration of reactant, reaction conditions, time of reaction, temperature, etc. on which the size and shape of the synthesized NMs depends. For development of consistent synthesis of NMs, characterization of the synthesized NMs is very crucial. The characterization includes the analysis of various properties of the synthesized NMs such as size, shape, composition, magnetism, conductivity, and other physical and chemical properties. There are various techniques utilized in the characterization of synthesized NMs.

1.5.1 Microscopic Techniques

1.5.1.1 Scanning Electron Microscopy (SEM)

It's regarded as a robust procedure for the analysis of NMs with a multitude of potential uses. SEM is a multipurpose instrument which is used in determination of surface topography, morphology, and basic composition of the NMs. In SEM, high-energy electron beam is used, and the sample is exposed to this electron beam. Then, the signals achieved from the interaction between electron beam and sample's composition give the all required information about the sample. Yet, compared to

SEM, field-emissive SEM (FESEM) can offer a significantly greater resolution and magnification of NMs (Su 2017; Sant'Anna et al. 2005).

1.5.1.2 Transmission Electron Microscopy (TEM)

The most common method for characterizing NMs in electron microscopy is TEM. The most popular method for analyzing NM size and shape is TEM because it gives the most precise estimate of the homogeneity of the particles while also providing clear pictures of the sample. In TEM, the interaction between electron beam and sample gives the required information about it. When the electron beam interacts with the sample, parts of the electron are transmitted through the sample, while others get scattered elastically or inelastically. The transmitted electrons give the required sample information. As compared to SEM, TEM gives high-resolution images in both dark-field and bright-field conditions and better compositional and crystallographic details about the sample (Su 2017).

1.5.1.3 Atomic Force Microscopy (AFM)

It is a microscopic technique which provides information on the 3D profile of the sample with high resolution. In AFM, a probe is scanned over the sample surface, and the interaction forces between probe and sample are measured. Hence, AFM is a sort of scanning probe microscopy. This probe is a pointed tip attached to the cantilever's end. Depending upon the forces, i.e., attractive, or repulsive between tip and sample surface, the cantilever gets deflected when AFM scans the sample. This deflection is measured on the basis of the reflected laser from the backside of cantilever. Hence, finally forces are measured using this laser variation data and cantilever stiffness value. AFM works on three distinct modes, viz., contact, non-contact, and tapping mode. In case of very subtle samples, noncontact mode is highly preferred as to prevent any damages to the sample during the scanning (Vilalta-Clemente et al. 2008).

1.5.2 Spectroscopic Techniques

1.5.2.1 UV-Visible Spectroscopy (UV-Vis)

This is an elementary and cost-effective spectroscopic method for the analysis of NMs. A UV-Vis instrument measures the difference in light intensity reflected from the sample and reference following the Beer-Lambert law. Since NMs exhibit optical characteristics which depend on size, shape, aggregation rate, concentration, surface charge, etc., UV-Vis becomes a crucial technique for identifying and characterizing the NMs as well as assessing the stability in colloidal solution (Hendel et al. 2014).

1.5.2.2 Fourier Transform Infrared Spectroscopy (FTIR)

It focuses on the interaction of a molecule with IR radiation (4000–400 cm^{-1}) in the electromagnetic spectrum. IR radiation affects the transitions between the vibrational states of the molecule by exciting the vibrations of the covalent bonds within that molecule. Hence, FTIR is used to confirm the molecular structure and interactions and evaluate the presence of the functional groups on the NMs by assessing the vibrational frequencies of the chemical bonds (Manor et al. 2012; Deepty et al. 2019).

1.5.2.3 Nuclear Magnetic Resonance Spectroscopy (NMR)

It's another significant spectroscopic tool used in the characterization of NMs that gives quantitative and structural information about the NMs. This method is based on the idea that when a strong magnetic field is applied to a nuclei with nonzero spin, there is a small energy difference between the spin-up and spin-down states. Electromagnetic radiation with a wavelength of radio wave can be used in studying these states' transitions. The distinct spectral lines of nuclei in various habitats are seen using NMR, which is incredibly sensitive to the magnetic environment of nuclei. When atomic nuclei that have magnetic moments and angular momentum are exposed to a magnetic field, NMR magnetically organizes the materials and interacts with them to generate the chemical information of the materials (Günther 2013).

1.5.2.4 Raman Spectroscopy (RS)

RS is an important spectroscopic technique based on the scattering principle of monochromatic radiation. Here, the incoming monochromatic radiation interacts with sample which produces scattered light whose energy is decreased due to the vibrational modes present in the sample. The light after passing through sample either gets reflected, absorbed, or scattered. Raman effect is closely related to Raman scattering, where change in the wavelength of light occurs when a light beam is deflected by the molecules. Its foundation is the interaction of incoming radiation with vibrating molecules, which results in the inelastic scattering of that radiation. This technique is used to characterize the structure of synthesized NMs which are Raman active. Raman-active molecules are defined as the molecules which show change in polarizability. The change in polarization should occur w.r.t. the vibrational coordinate corresponding to the rotational-vibrational-electronic state for a molecule to exhibit Raman effect. The extent of change in polarization defines the intensity of Raman scattering. The use of surface-enhanced RS (SERS) has a significant impact on the Raman signals in addition to the RS. In tip-enhanced RS (TERS), to obtain surface enhancement in the Raman signals, a less metallic tip is used. In comparison to SERS and TERS, RS gives topological data as well as the structural, chemical, and electrical characteristics of the NMs and NSs (Mulvaney and Keating 2000; Huang et al. 2009; Lin et al. 2020).

1.5.2.5 X-Ray Photoelectron Spectroscopy (XPS)

XPS is a spectroscopic procedure for evaluating the chemical composition of NMs' surfaces, which works on the principle of the photoelectric effect. In XPS, the NMs are analyzed by exposing the target material to an X-ray beam while both counting the number of expelled electrons and measuring the kinetic energy of the target material. XPS is a very effective quantitative method for the determination of oxidation state, electronic structure, and elemental composition of NMs. Thus, it can analyze the core/shell structures, ligand exchange interactions, and surface functionalization of NMs (Fadley 2010).

1.5.3 Thermal Techniques

1.5.3.1 Thermogravimetric Analysis (TGA)

In this method, the change in weight of the substance is recorded as a function of temperature or time. During TGA, the mass of the sample is continually tracked while it is heated or cooled in a certain environment. Hence, various properties of a sample can be studied through TGA like adsorption, desorption, sublimation, vaporization, oxidation, and reduction. Through TGA we can analyze the volatile substances and gaseous products present in the sample as the loss of weight of sample indicates the presence of those substances in a chemical reaction for different NMs. Thus, the thermal stability of the synthesized NMs and the kinetics of the chemical reactions can be determined using the TGA (Wagner 2017; Rami et al. 2020; Saadatkhah et al. 2020).

1.5.3.2 Differential Scanning Calorimetry (DSC)

This is another thermal analytical technique used for the determination of thermal behaviour of NMs and NSs. Herein, the difference between the amount of heat needed to raise the temperature of a sample and a reference is assessed. In order to monitor and keep the temperature difference at zero, sample and reference cells are heated separately. Based on whether a material is undergoing an exothermic or endothermic phase transition, the amount of heat required to keep a constant sample temperature will change as the substance changes. The amount of extra heat emitted or absorbed as a result of sample transitions determines the energy difference necessary to get the sample temperature to that of the reference. DSC is used to evaluate the different physical properties and thermal transitions of NMs, and it gives information about the crystallization and interaction of drugs with NMs (Charsley et al. 2006; Horiuchi 2004; Wang et al. 2014; Illers and Kanig 1982; Pérez-Alonso et al. 2008).

1.5.4 Dynamic Light Scattering (DLS)

It's a commonly employed method for determining particle size of NMs present in a colloidal suspension which are in continuous Brownian motion. The laser beam of the DLS instrument gets scattered by the NMs, and this intensity of scattering as a function of time is measured. Further, this data combined with Stokes-Einstein theory gives the hydrodynamic diameter of the NMs. To prevent the multiple scattering effects in DLS, a low concentration of NMs is required. DLS also gives information about the polydispersity index, size distribution, and aggregation rate of the NMs present in the solution (Kato 2012).

1.5.5 Diffraction Technique

1.5.5.1 X-Ray Diffraction (XRD)

Diffraction methods are employed to determine the average particle size and basic structure of materials. One of the methods that is most frequently used for characterizing NMs is XRD. The XRD patterns give various information about NMs and NSs such as the phase of the NM, the crystal structure of the NM along with crystallinity, the grain size of the NM, the lattice parameters of the NM, etc. The grain size of a sample is determined with the help of Scherrer equation by broadening of the most intense peak obtained in XRD pattern. The XRD method has the benefit of delivering data in statistically relevant, volume-averaged values. Generally, JCPDS database file is used as a reference file for analyzing the obtained XRD pattern to determine the material's composition and phase via assessing the position and intensity of the peaks. However, due to the presence of broad XRD peaks, this technique is inappropriate for analysis of amorphous samples (Upadhyay et al. 2016; Yan et al. 2005).

1.6 Applications of Nanoscience and Nanotechnology

Considering the unique properties of NMs and NSs, they have been used in various fields of science like energy, electronics, agriculture/food, and healthcare (drug and gene delivery), and these are briefly discussed as mentioned below.

1.6.1 Energy

Energy plays a significant part in everyday life and it's regarded as the primary source for human activities. New technology with low energy consumption is greatly needed as a potential means of energy conservation due to the rising energy demand. As their effectiveness and increase in lifespan, Li-ion batteries, LEDs, fuel cells, supercapacitors, and solar cells are employed to preserve energy. Future energy production, transmission, and storage systems are predicted to benefit from the use of nanotechnology since they will be more affordable and effective. In numerous applications of energy savings, including hydrogen production and storage and solar photovoltaic systems, the fabrication of materials and structures at the nanoscale has the potential to produce highly efficient devices with cost-effective and low energy demand. Because of this, the use of nanotechnology in energy sector is a major topic in many areas of science. Hence, energy nanotechnology should be prioritized in order to achieve better efficiency with lower manufacturing costs. Thermoelectric, piezoelectric, catalytic, and photovoltaic are principal NMs that significantly influenced a number of energy applications. Due to their special characteristics, inorganic NMs have better electrical and thermal conductivity, large surface area, and high chemical stability. Nanosized materials demonstrated thermoelectric performance that was two times greater than that of typical materials for the purpose of generating energy. By using an advanced nanostructuring method, the active sites of catalytic materials dramatically enhanced in energy conversion applications. The recent developments in research found that it would be possible to expand the notions of altering the hybrid characteristics at nano-range that may result in the creation of the next generation of materials for energy applications (Christian et al. 2013; Yianoulis and Giannouli 2008; Wang et al. 2020).

1.6.2 Electronics

The goal of nanotechnology in electronics is to create devices that are extremely small. Material synthesis, handling, integration, gadget designing, improved output characterization, and analysis are some of the areas where nanoelectronics has been explored. The many types of materials and technologies covered by this subfield of nanotechnology all have the property of being smaller in size. At this size, materials' characteristics also change. The operation of these devices depends heavily on interatomic interactions and quantum phenomena. Nanoelectronics has enabled the usage of components such as nanotransistors, nano-displays, quantum computers, nanosensors, chips, lasers, batteries, cells, and multiplexers, as well as memory storage, computer processing, energy generation, health diagnostics, and nanorobotics (Lu and Lieber 2006; Hashim 2015; Nsofor 2010; Lundgaard et al. 2008; Zhou et al. 2010; Segal et al. 1998; Hwang et al. 2010; Toshishige 2004; Graham et al. 2005).

1.6.3 Agriculture

Several technical advancements including hybrid varieties, synthetic fertilizers, and pesticides have a long history of benefiting agriculture. In order to address the worldwide concerns of food security and climate change, agricultural specialists are increasingly aware of innovative development, like nanotechnology, which is significantly essential for agronomy. The use of nanotechnology in agriculture has just recently become significant, yet study on the subject began roughly 50 years ago. NMs are utilized for a variety of purposes, including enhancing fertilization, production, lowering the use of pesticides; delivering fertilizer and pesticide; quick, accurate pathogen as well as toxic chemicals sensing in foods; and intelligent devices in food packaging, processing, and managing food safety. The use of NMs in targeted delivery of drugs and genes as well as genetic manipulation of animals and plants can increase agricultural production. Moreover, pollution can also be reduced and degraded by using nanofilters or nanocatalysts. Thus, nanotechnology application in agriculture can result in the controlled and on-demand release of nanofertilizers and nanopesticides; targeted delivery of drug and nutrient in livestock and fisheries, soil and water treatment, and fishery cleaning and maintenance; and nanosensors in evaluating plant health and soil quality (Table 1.3).

Table 1.3 Some recent developments in the field of agriculture using nanotechnology

Areas	Specific type	Nanomaterials	References
Crop production	Plant protection (antimicrobial activity)	Chitosan NPs	Ing et al. (2012)
		Chitosan NPs	Saharan et al. (2013)
		Nanocomposite of pullulan and Ag NPs	Pinto et al. (2013)
	Nanofertilizers	N, K, P encapsulated chitosan NPs	Corradini et al. (2010)
		Urea encapsulated silica NPs	Wanyika et al. (2012)
		ZnO NP-coated urea and monoammonium phosphate	Milani et al. (2012)
	Precision farming	GPS-connected nanosensors	Kalpana-Sastry et al. (2009), Sai Rohith (2015)
Soil improvement	Water/liquid retention	Zeolite and nanoclays	Manjaiah et al. (2018), Sarkar et al. (2014)
Water purification	Wastewater treatment and pollutant removal	TiO_2 nanofilms	McMurray et al. (2006)
		Nanoscale zerovalent iron (nZVI)	Nadagouda and Varma (2009), Crane and Scott (2012)
		CNTs	Theron et al. (2008)
Diagnostic	Nanosensors and diagnostic devices	CNTs and nanofibers	Vamvakaki and Chaniotakis (2007)
	Livestock and fisheries	Nanovaccines	Kalpana-Sastry et al. (2009)
Plant breeding	Genetically modified (GM) plants	Silica NPs	Prasanna (n.d.), Torney et al. (2007)
Food industry	Packaging	Silica NPs	Kalpana-Sastry et al. (2009)
	Processing	Liposome nanovesicles	Wen et al. (2005)
		CNTs	Rivas et al. (2007)

1.6.4 Drug and Gene delivery

Nanotechnology has extensively addressed a number of NMs in the biomedical field during the past decade, enabling the potential for effective distribution, efficient therapy, and an enhanced safety profile. NMs become essential in drug delivery systems because they increase the solubility and stability of pharmaceuticals, regulate release patterns, reduce toxicity, and increase therapeutic benefits. On the otherside, NM-mediated gene delivery provides novel possibilities for the treatment of illnesses by administering fresh genetic material to cells ex vivo and/or in vivo. Different NMs developed for the delivery of drugs and genes are discussed below.

Polymeric and lipid-based NMs and NSs are typically taken into account while developing targeted drug delivery systems (Fig. 1.4). However, these NSs' effectiveness as drug delivery systems vary w.r.t. their size, shape, and other innate chemical/biophysical traits. Recent years have seen significant advancements in the technology used to deliver therapeutic agents or naturally occurring substances to their intended locations for the treatment of different ailments, and some of them are mentioned in Table 1.4.

Table 1.4 Various NMs used as nanocarriers for drug delivery against different diseases

Types of nanocarrier	Nanocarrier	Disease	Drug	Reference
Inorganic NMs	Au NPs	Cancer	Methotrexate (MTX)	Tran et al. (2013)
			Morin	Kondath et al. (2014)
			5-Fluorouracil (5-Fu)	Singh et al. (2018)
			Doxorubicin (DOX)	Wu et al. (2018)
			DOX	Oladipo et al. (2020)
			Hesperidin (HSP)	Sulaiman et al. (2020)
		Cardiovascular	BSA/PVA nanofibers	Ravichandran et al. (2014)
			miR155	Jia et al. (2017)
		Antimicrobial	Levofloxacin (LFX)	Bagga et al. (2016)
			Gentamicin (GTM)	Ahangari et al. (2013)
	Ag NPs	Cancer	TAT	Liu et al. (2012)
			DOX and Alendronate (AND)	Benyettou et al. (2015)
			DOX	Capanema et al. (2019)
			DOX	Zeng et al. (2018)
		Anesthetic	Tetracaine hydrochloride (TCN)	Srivastava et al. (2019)
		Antibiotic	Ciprofloxacin (CFX)	Kooti et al. (2018)
		Malaria	Chloroquine (CLQ) and fosmidomycin (FDM)	Rai et al. (2017)
	TiO_2 NPs	Cancer	DOX	Ren et al. (2013)
			Paclitaxel PTX	Mund et al. (2014)

(continued)

Table 1.4 (continued)

Types of nanocarrier	Nanocarrier	Disease	Drug	Reference
			Erlotinib (ERL) and vorinostat (VNT)	Abdel-Ghany et al. (2020)
			Curcumin (CCN)	Zamani et al. (2018)
	ZnO NPs	Cancer	DOX	Sharma et al. (2016)
			DOX	Liu et al. (2016)
			Quercetin (QRT)	Sadhukhan et al. (2019)
			Taxifolin (TFN)	Sundraraman and Jayakumari (2020)
		Antibiotic	*A. socotrina* extract	Fahimmunisha et al. (2019)
	Mn_3O_4 NPs	Cancer	Gemcitabine (GEM)	Jain et al. (2021)
			5-Fu	Jain et al. (2020)
	Fe_3O_4 NPs	Cancer	CCN	Shafiee et al. (2019)
Lipid-based NMs	SLN	Cancer	GEM	Soni et al. (2016)
			Resveratrol (RSV)	Wang et al. (2017)
			CCN	Wang et al. (2018)
			DOX	Shen et al. (2018)
		Alzheimer	RSV	Loureiro et al. (2017)
			Erythropoietin (ETP)	Dara et al. (2019)
	NLC	Cancer	PTX	Olerile et al. (2017)
			DOX	Li et al. (2018)
			RSV	Poonia et al. (2019)
			CCN	Kamel et al. (2019)
		Alzheimer	Huperzine (HPN)	Tapeinos et al. (2017)

(continued)

Table 1.4 (continued)

Types of nanocarrier	Nanocarrier	Disease	Drug	Reference
	Liposome	Cancer	DOX	Tahover et al. (2015)
			DOX	Du et al. (2017)
			CCN	Jose et al. (2018)
Polymeric NMs	PAMAM	Cancer	DOX	Zhang et al. (2018)
			DOX	Almuqbil et al. (2020)
			DOX	Zhu et al. (2014)
			Cisplatin (CPT)	Kulhari et al. (2015a)
			Piperlongumine (PPL)	Jangid et al. (2022)
			Docetaxel (DCL)	Kulhari et al. (2015b)
			Camptothecin (CTC)	Pooja et al. (2020)
		Hypercholesterolemia	Simvastatin (SMT)	Kulhari et al. (2013)
	PPI	Alzheimer	Maltose-histidine shell	Aso et al. (2019)
	ß cyclodextrins	Cancer	DOX	Hyun et al. (2019)
			DOX	Fan et al. (2019)
			DOX	Yang et al. (2019)
Carbon-based NMs	SWCNT	Cancer	DOX	Heister et al. (2012)
			PTX	Berlin et al. (2010)
			GEM	Razzazan et al. (2016)
		Antibiotic	Azithromycin (AZ)	Darabi et al. (2014)
	MWCNT	Cancer	DOX	Li et al. (2011)
			PTX	Magid and Al-Karam (2021)
			GEM	Singh et al. (2013)
			MTX	Pastorin et al. (2006)

(continued)

Table 1.4 (continued)

Types of nanocarrier	Nanocarrier	Disease	Drug	Reference
	Graphene oxide (GO) nanosheet	Cancer	DOX	Huang et al. (2018)
			5-Fu	Aliabadi et al. (2018)
			PTX	Deng et al. (2018)
			MTX	Karimi Shervedani et al. (2018)
	Nanodiamond	Cancer	DOX	Long et al. (2019)
			PTX	Yu et al. (2019)
			GEM	Lu et al. (2016)
	Fullerene	Cancer	DOX	Grebinyk et al. (2019)
		Covid-19	CLQ	Bagheri Novir and Aram (2020)

The effective transport of genetic materials to their targeted site using NMs attracted much attention. To produce the protein products of the inserted gene, the foreign genetic material often has to be transported to the targeted cells' nuclei. The perfect carrier is a non-immunogenic, benign, and able to produce the gene product without producing toxic effects in the quantities required to fix the flaw and for the timing of transgene expression. It does this by transferring a certain quantity of genetic material into a particular kind of cell. Gene-loaded polymeric nanocarriers have emerged as an innovative, viable approach in cancer therapy. This is because they can enhance the drug's pharmacokinetics as well as act as a supplementary reaction to permeation and retention effects that enhance the drug accretion at the tumor site throughout treating cancer. Because of their simple production and adaptable qualities, biodegradable polymeric nanocarriers like chitosan, dextran, gelatin, and pullulan have gained a leading role in gene therapy. Moreover, various NMs and NSs such as SLNs, NLCs, Au NPs, CNTs, and silica NPs have also emerged as efficient nanocarriers for gene delivery in the treatment of different ailments, mainly cancer (Table 1.5).

Table 1.5 Various NMs used as nanocarriers for gene delivery against different diseases

Types of nanocarrier	Nanocarrier	Disease	Genetic material	References
Inorganic NMs	Au NPs	Cancer	Oligonucleotides	Mendes et al. (2017)
			Aptamer (sgc8c)	Khoshfetrat and Mehrgardi (2017)
			ssDNA	Vinhas et al. (2016)
	Silica NPs	Cancer	sgc8c	Tan et al. (2016)
			siRNA	Hom et al. (2010)
			pDNA	Babaei et al. (2017)
Carbon-based NMs	SWCNT	Cancer	Peptide	Ohta et al. (2016)
			shRNA	Taghavi et al. (2016)
			shRNA	Taghavi et al. (2017)
			siRNA	Bora et al. (2022)
			RTA	Bora et al. (2022)
Polymeric NMs	PAMAM	Cancer	siRNA	Waite and Roth (2009)
			siRNA-VEGFA	Xu et al. (2017)
			DNA	Choi et al. (2005)
			miRNA	Fan et al. (2016)
	Poly (beta-amino easter) (PBAE)	Cancer	pDNA	Choi et al. (2020)
			siRNA	Kozielski et al. (2014)
			miRNA	Lopez-Bertoni et al. (2018)
	Polyethylenimine (PEI)	Cancer	siRNA	Zhupanyn et al. (2019)
	Chitosan NPs	Cancer	siRNA	Zhang et al. (2019)
	Curdlan NPs	Cancer	siRNA	Su et al. (2020)
			siRNA	Erdene-Ochir et al. (2020)

(continued)

Table 1.5 (continued)

Types of nanocarrier	Nanocarrier	Disease	Genetic material	References
Lipid-based NMs	SLN	Cancer	pDNA	Jin and Kim (2014)
			p53	Choi et al. (2008)
	NLC	Amyloidosis	ALN-TTRsc	Fitzgerald et al. (2014)
		Hypercholesterolemia	ALN-PCS02	Fitzgerald et al. (2014)

1.7 Conclusion and Future Perspectives

Nanoscience and nanotechnology have advanced in many ways throughout several branches of science. Innovative hydrogen fuel cells, storage devices and solar cells are being constructed using NMs and NSs to provide clean energy to societies and to lower the dependency on conventional fossil fuels. Moreover, NMs are being utilized in the agriculture/food sector to significantly improve manufacturing, packing, storability, and nutritional bioavailability. The most significant achievements in nanotechnology, however, have been made in the vast area of healthcare, particularly in cancer therapies. This is due to their considerable ability to provide creative ways of overcoming the constraints imposed by existing chemotherapy and radiation techniques. Latest developments in the field of nanoscience and nanotechnology generate a range of NMs and NSs with exceptional characteristics such as their capability to act as nanocarriers and inherent cytotoxic activity that helps in treating different cancers that are now resistant to conventional therapies. Yet, researchers continue to strive for new developments in nanoscience and nanotechnology to improve the comfort and ease of human existence.

Acknowledgment S.B. and H.K. acknowledge the Central University of Gujarat, Gandhinagar for providing the necessary facilities and support. S.B. acknowledges University Grant Commission, New Delhi for a Ph.D. fellowship. D. P. acknowledges the National Forensic Science University, Gandhinagar for providing the necessary facilities and support.

References

Abdel-Ghany S, Raslan S, Tombuloglu H, Shamseddin A, Cevik E, Said OA, Madyan EF, Senel M, Bozkurt A, Rehman S, Sabit H (2020) Vorinostat-loaded titanium oxide nanoparticles (anatase) induce G2/M cell cycle arrest in breast cancer cells via PALB2 upregulation. 3 Biotech 10(9): 1–14. https://doi.org/10.1007/s13205-020-02391-2

Abegunde OO, Akinlabi ET, Oladijo OP, Akinlabi S, Ude AU (2019) Overview of thin film deposition techniques. AIMS Mater Sci 6(2):174–199. https://doi.org/10.3934/MATERSCI.2019.2.174

Ago H (2015) CVD growth of high-quality single-layer graphene. In: Frontiers of graphene and carbon nanotubes. Springer, Tokyo, pp 3–20. https://doi.org/10.1007/978-4-431-55372-4_1

Ahangari A, Salouti M, Heidari Z, Kazemizadeh AR, Safari AA (2013) Development of gentamicin-gold nanospheres for antimicrobial drug delivery to staphylococcal infected foci. Drug Deliv 20(1):34–39. https://doi.org/10.3109/10717544.2012.746402

Albalawi F, Hussein MZ, Fakurazi S, Masarudin MJ (2021) Engineered nanomaterials: the challenges and opportunities for nanomedicines. Int J Nanomedicine 16:161–184. https://doi.org/10.2147/IJN.S288236

Aliabadi M, Yunessnia Lehi A, Shagholani H, Gerayeli A (2018) Planar polymer-graphene oxide nanohybrid as a 5-fluorouacil carrier in pH-responsive controlled release. J Drug Deliv Sci Technol 43:103–106. https://doi.org/10.1016/j.jddst.2017.09.020

Almuqbil RM, Heyder RS, Bielski ER, Durymanov M, Reineke JJ, Da Rocha SRP (2020) Dendrimer conjugation enhances tumor penetration and efficacy of doxorubicin in extracellular matrix-expressing 3D lung cancer models. Mol Pharm 17(5):1648–1662. https://doi.org/10.1021/acs.molpharmaceut.0c00083

Amendola V, Meneghetti M (2009) Laser ablation synthesis in solution and size manipulation of noble metal nanoparticles. Phys Chem Chem Phys 11(20):3805–3821. https://doi.org/10.1039/b900654k

Andrievski RA (2014) Review of thermal stability of nanomaterials. J Mater Sci 49(4):1449–1460. https://doi.org/10.1007/s10853-013-7836-1

Ashraf MA, Peng W, Zare Y, Rhee KY (2018) Effects of size and aggregation/agglomeration of nanoparticles on the interfacial/interphase properties and tensile strength of polymer nanocomposites. Nanoscale Res Lett 13:214. https://doi.org/10.1186/s11671-018-2624-0

Aso E, Martinsson I, Appelhans D, Effenberg C, Benseny-Cases N, Cladera J, Gouras G, Ferrer I, Klementieva O (2019) Poly(propylene imine) dendrimers with histidine-maltose shell as novel type of nanoparticles for synapse and memory protection. Nanomedicine 17:198–209. https://doi.org/10.1016/j.nano.2019.01.010

Atwater HA (2007) The promise of plasmonics. Sci Am 296(4):56–63. http://www.jstor.org/stable/26069230

Ayyub P, Chandra R, Taneja P, Sharma AK, Pinto R (2001) Synthesis of nanocrystalline material by sputtering and laser ablation at low temperatures. Appl Phys A Mater Sci Process 73(1): 67–73. https://doi.org/10.1007/s003390100833

Babaei M, Eshghi H, Abnous K, Rahimizadeh M, Ramezani M (2017) Promising gene delivery system based on polyethylenimine-modified silica nanoparticles. Cancer Gene Ther 24(4): 156–164. https://doi.org/10.1038/cgt.2016.73

Bagga P, Ansari TM, Siddiqui HH, Syed A, Bahkali AH, Rahman A, Sajid Khan M, Khan MS (2016) Bromelain capped gold nanoparticles as the novel drug delivery carriers to aggrandize effect of the antibiotic levofloxacin. EXCLI J 15:772–780

Bagheri Novir S, Aram MR (2020) Quantum mechanical simulation of chloroquine drug interaction with C60 fullerene for treatment of COVID-19. Chem Phys Lett 757:137869. https://doi.org/10.1016/j.cplett.2020.137869

Barber DJ, Freestone IC (1990) An investigation of the origin of the colour of the lycurgus cup by analytical transmission electron microscopy. Archaeometry 32(1):33–45. https://doi.org/10.1111/j.1475-4754.1990.tb01079.x

Barhoum A, Melcher J, Van Assche G, Rahier H, Bechelany M, Fleisch M, Bahnemann D (2017) Synthesis, growth mechanism, and photocatalytic activity of zinc oxide nanostructures: porous microparticles versus nonporous nanoparticles. J Mater Sci 52(5):2746–2762. https://doi.org/10.1007/s10853-016-0567-3

Benyettou F, Rezgui R, Ravaux F, Jaber T, Blumer K, Jouiad M, Motte L, Olsen JC, Platas-Iglesias C, Magzoub M, Trabolsi A (2015) Synthesis of silver nanoparticles for the dual delivery of doxorubicin and alendronate to cancer cells. J Mater Chem B 3(36):7237–7245. https://doi.org/10.1039/c5tb00994d

Berlin JM, Leonard AD, Pham TT, Sano D, Marcano DC, Yan S, Fiorentino S, Milas ZL, Kosynkin DV, Price BK, Lucente-Schultz RM, Wen X, Raso MG, Craig SL, Tran HT, Myers JN, Tour JM (2010) Effective drug delivery, in vitro and in vivo, by carbon-based nanovectors noncovalently loaded with unmodified paclitaxel. ACS Nano 4(8):4621–4636. https://doi.org/10.1021/nn100975c

Bora S, Pooja D, Kulhari H (2022) Applications of surface modified carbon nanotubes in drug delivery. In: Surface modified carbon nanotubes volume 2: industrial applications; acs symposium series, vol 1425. American Chemical Society, Washington, DC, pp 2–19. https://doi.org/10.1021/bk-2022-1425.ch002

Brill RH, Cahill ND (1988) A red opaque glass from sardis and some thoughts on red opaques in general. J Glass Stud 30:16–27. http://www.jstor.org/stable/24190804

Cao S, Zhao C, Han T, Peng L (2016) Hydrothermal synthesis, characterization and gas sensing properties of the WO3 nanofibers. Mater Lett 169:17–20. https://doi.org/10.1016/j.matlet.2016.01.053

Capanema NSV, Carvalho IC, Mansur AAP, Carvalho SM, Lage AP, Mansur HS (2019) Hybrid hydrogel composed of carboxymethylcellulose-silver nanoparticles-doxorubicin for anticancer and antibacterial therapies against melanoma skin cancer cells. ACS Appl Nano Mater 2(11): 7393–7408. https://doi.org/10.1021/acsanm.9b01924

Chai B, Xu M, Yan J, Ren Z (2018) Remarkably enhanced photocatalytic hydrogen evolution over MoS2 nanosheets loaded on uniform CdS nanospheres. Appl Surf Sci 430:523–530. https://doi.org/10.1016/j.apsusc.2017.07.292

Chandra S, Das P, Bag S, Laha D, Pramanik P (2011) Synthesis, functionalization and bioimaging applications of highly fluorescent carbon nanoparticles. Nanoscale 3(4):1533–1540. https://doi.org/10.1039/c0nr00735h

Charsley EL, Laye PG, Palakollu V, Rooney JJ, Joseph B (2006) DSC studies on organic melting point temperature standards. Thermochim Acta 446(1–2):29–32. https://doi.org/10.1016/j.tca.2006.02.035

Choi Y, Thomas T, Kotlyar A, Islam MT, Baker JR (2005) Synthesis and functional evaluation of DNA-assembled polyamidoamine dendrimer clusters for cancer cell-specific targeting. Chem Biol 12(1):35–43. https://doi.org/10.1016/j.chembiol.2004.10.016

Choi SH, Jin SE, Lee MK, Lim SJ, Park JS, Kim BG, Ahn WS, Kim CK (2008) Novel cationic solid lipid nanoparticles enhanced P53 gene transfer to lung cancer cells. Eur J Pharm Biopharm 68(3):545–554. https://doi.org/10.1016/j.ejpb.2007.07.011

Choi J, Rui Y, Kim J, Gorelick N, Wilson DR, Kozielski K, Mangraviti A, Sankey E, Brem H, Tyler B, Green JJ, Jackson EM (2020) Nonviral polymeric nanoparticles for gene therapy in pediatric CNS malignancies. Nanomedicine 23:102115. https://doi.org/10.1016/j.nano.2019.102115

Christian F, Edith S, Adityawarman D, Indarto A (2013) Application of nanotechnologies in the energy sector: a brief and short review. Front Energy 7(1):6–18. https://doi.org/10.1007/s11708-012-0219-5

Chrzanowska J, Hoffman J, Małolepszy A, Mazurkiewicz M, Kowalewski TA, Szymanski Z, Stobinski L (2015) Synthesis of carbon nanotubes by the laser ablation method: effect of laser wavelength. Phys Status Solidi Basic Res 252(8):1860–1867. https://doi.org/10.1002/pssb.201451614

Colvin VL (2003) The potential environmental impact of engineered nanomaterials. Nat Biotechnol 21(10):1166–1171

Corradini E, de Moura MR, Mattoso LHC (2010) A preliminary study of the incorparation of NPK fertilizer into chitosan nanoparticles. Express Polym Lett 4(8):509–515. https://doi.org/10.3144/expresspolymlett.2010.64

Crane RA, Scott TB (2012) Nanoscale zero-valent iron: future prospects for an emerging water treatment technology. J Hazard Mater 211–212:112–125. https://doi.org/10.1016/j.jhazmat.2011.11.073

D'Amato R, Falconieri M, Gagliardi S, Popovici E, Serra E, Terranova G, Borsella E (2013) Synthesis of ceramic nanoparticles by laser pyrolysis: from research to applications. J Anal Appl Pyrolysis 104:461–469. https://doi.org/10.1016/j.jaap.2013.05.026

Danks AE, Hall SR, Schnepp Z (2016) The evolution of "sol-gel" chemistry as a technique for materials synthesis. Mater Horizons 3(2):91–112. https://doi.org/10.1039/c5mh00260e

Dara T, Vatanara A, Sharifzadeh M, Khani S, Vakilinezhad MA, Vakhshiteh F, Nabi Meybodi M, Sadegh Malvajerd S, Hassani S, Mosaddegh MH (2019) Improvement of memory deficits in the rat model of Alzheimer's disease by erythropoietin-loaded solid lipid nanoparticles. Neurobiol Learn Mem 166:107082. https://doi.org/10.1016/j.nlm.2019.107082

Darabi HR, Roozkhosh A, Jafar Tehrani M, Aghapoor K, Sayahi H, Balavar Y, Mohsenzadeh F (2014) Characterization of ester- or thioamide-functionalized single-walled carbon nanotube-azithromycin conjugates. Appl Surf Sci 288:122–129. https://doi.org/10.1016/j.apsusc.2013.09.159

De Coelho Escobar C, Dos Santos JHZ (2014) Effect of the sol-gel route on the textural characteristics of silica imprinted with rhodamine B. J Sep Sci 37(7):868–875. https://doi.org/10.1002/jssc.201301143

Deepty M, Srinivas C, Kumar ER, Mohan NK, Prajapat CL, Rao TVC, Meena SS, Verma AK, Sastry DL (2019) XRD, EDX, FTIR and ESR spectroscopic studies of co-precipitated Mn-substituted Zn–ferrite nanoparticles. Ceram Int 45(6):8037–8044. https://doi.org/10.1016/j.ceramint.2019.01.029

Deng W, Qiu J, Wang S, Yuan Z, Jia Y, Tan H, Lu J, Zheng R (2018) Development of biocompatible and VEGF-targeted paclitaxel nanodrugs on albumin and graphene oxide dual-carrier for photothermal-triggered drug delivery in vitro and in vivo. Int J Nanomedicine 13:439–453. https://doi.org/10.2147/IJN.S150977

Dong Y, Du X-q, Liang P, Man X-l (2020) One-pot solvothermal method to fabricate 1D-VS4 nanowires as anode materials for lithium ion batteries. Inorg Chem Commun 115:107883. https://doi.org/10.1016/j.inoche.2020.107883

Du Y, Liang X, Li Y, Sun T, Jin Z, Xue H, Tian J (2017) Nuclear and fluorescent labeled PD-1-liposome-DOX-64Cu/IRDye800CW allows improved breast tumor targeted imaging and therapy. Mol Pharm 14(11):3978–3986. https://doi.org/10.1021/acs.molpharmaceut.7b00649

Duque JS, Madrigal BM, Riascos H, Avila YP (2019) Colloidal metal oxide nanoparticles prepared by laser ablation technique and their antibacterial test. Colloids Interfaces 3(1):25. https://doi.org/10.3390/colloids3010025

Ealias AM, Saravanakumar MP (2017) A review on the classification, characterisation, synthesis of nanoparticles and their application. IOP Conf Ser Mater Sci Eng 263(3):032019. https://doi.org/10.1088/1757-899X/263/3/032019

Erdene-Ochir T, Ganbold T, Zandan J, Han S, Borjihan G, Baigude H (2020) Alkylation enhances biocompatibility and SiRNA delivery efficiency of cationic curdlan nanoparticles. Int J Biol Macromol 143:118–125. https://doi.org/10.1016/j.ijbiomac.2019.12.048

Eustis S, El-Sayed MA (2006) Why gold nanoparticles are more precious than pretty gold: noble metal surface plasmon resonance and its enhancement of the radiative and nonradiative properties of nanocrystals of different shapes. Chem Soc Rev 35(3):209–217. https://doi.org/10.1039/b514191e

Fadley CS (2010) X-Ray photoelectron spectroscopy: progress and perspectives. J Electron Spectros Relat Phenomena 178–179:2–32. https://doi.org/10.1016/j.elspec.2010.01.006

Fahimmunisha BA, Ishwarya R, AlSalhi MS, Devanesan S, Govindarajan M, Vaseeharan B (2019) Green fabrication, characterization and antibacterial potential of zinc oxide nanoparticles using Aloe socotrina leaf extract: a novel drug delivery approach. J Drug Deliv Sci Technol 2020(55):101465. https://doi.org/10.1016/j.jddst.2019.101465

Fan W, Wang X, Ding B, Cai H, Wang X, Fan Y, Li Y, Liu S, Nie S, Lu Q (2016) Thioaptamer-conjugated CD44-targeted delivery system for the treatment of breast cancer in vitro and in vivo. J Drug Target 24(4):359–371. https://doi.org/10.3109/1061186X.2015.1077850

Fan W, Xu Y, Li Z, Li Q (2019) Folic acid-modified β-cyclodextrin nanoparticles as drug delivery to load DOX for liver cancer therapeutics. Soft Mater 17(4):437–447. https://doi.org/10.1080/1539445X.2019.1624265

Feynman RP (1959) Plenty of room at the bottom. In: American Physical Society annual meeting. https://doi.org/10.1511/2012.96.226

Fitzgerald K, Frank-Kamenetsky M, Shulga-Morskaya S, Liebow A, Bettencourt BR, Sutherland JE, Hutabarat RM, Clausen VA, Karsten V, Cehelsky J, Nochur SV, Kotelianski V, Horton J, Mant T, Chiesa J, Ritter J, Munisamy M, Vaishnaw AK, Gollob JA, Simon A (2014) Effect of an RNA interference drug on the synthesis of proprotein convertase subtilisin/kexin type 9 (PCSK9) and the concentration of serum LDL cholesterol in healthy volunteers: a randomised, single-blind, placebo-controlled, phase 1 trial. Lancet 383(9911):60–68. https://doi.org/10.1016/S0140-6736(13)61914-5

Garg V, Mote R, Fu J (2020) Facile fabrication of functional 3D micro-nano architectures with focused ion beam implantation and selective chemical etching. Appl Surf Sci 526:146644. https://doi.org/10.1016/j.apsusc.2020.146644

Graham AP, Duesberg GS, Seidel RV, Liebau M, Unger E, Pamler W, Kreupl F, Hoenlein W (2005) Carbon nanotubes for microelectronics? Small 1(4):382–390. https://doi.org/10.1002/smll.200500009

Grebinyk A, Prylutska S, Grebinyk S, Prylutskyy Y, Ritter U, Matyshevska O, Dandekar T, Frohme M (2019) Complexation with C60 fullerene increases doxorubicin efficiency against leukemic cells in vitro. Nanoscale Res Lett 14:1–10. https://doi.org/10.1186/s11671-019-2894-1

Gujrati M, Malamas A, Shin T, Jin E, Sun Y, Lu ZR (2014) Multifunctional cationic lipid-based nanoparticles facilitate endosomal escape and reduction-triggered cytosolic SiRNA release. Mol Pharm 11(8):2734–2744. https://doi.org/10.1021/mp400787s

Günther H (2013) NMR spectroscopy: basic principles, concepts and applications in chemistry. Wiley, New York

Guo D, Xie G, Luo J (2014) Mechanical properties of nanoparticles: basics and applications. J Phys D Appl Phys 47(1):013001. https://doi.org/10.1088/0022-3727/47/1/013001

Gupta SM, Tripathi M (2012) An overview of commonly used semiconductor nanoparticles in photocatalysis. High Energy Chem 46(1):1–9. https://doi.org/10.1134/S0018143912010134

Hashim Y (2015) A review on transistors in nano dimensions. Int J Eng Technol Sci 4(1):8–18

Heister E, Neves V, Lamprecht C, Silva SRP, Coley HM, McFadden J (2012) Drug loading, dispersion stability, and therapeutic efficacy in targeted drug delivery with carbon nanotubes. Carbon N Y 50(2):622–632. https://doi.org/10.1016/j.carbon.2011.08.074

Hendel T, Wuithschick M, Kettemann F, Birnbaum A, Rademann K, Polte J (2014) In situ determination of colloidal gold concentrations with UV-Vis spectroscopy: limitations and perspectives. Anal Chem 86(22):11115–11124. https://doi.org/10.1021/ac502053s

Henkel C, Wittmann JE, Träg J, Will J, Stiegler LMS, Strohriegl P, Hirsch A, Unruh T, Zahn D, Halik M, Guldi DM (2020) Mixed organic ligand shells: controlling the nanoparticle surface morphology toward tuning the optoelectronic properties. Small 16(2):e1903729. https://doi.org/10.1002/smll.201903729

Hochella MF, Mogk DW, Ranville J, Allen IC, Luther GW, Marr LC, McGrail BP, Murayama M, Qafoku NP, Rosso KM, Sahai N, Schroeder PA, Vikesland P, Westerhoff P, Yang Y (2019) Natural, incidental, and engineered nanomaterials and their impacts on the earth system. Science 363(6434):eaau8299. https://doi.org/10.1126/science.aau8299

Hom C, Lu J, Liong M, Luo H, Li Z, Zink JI, Tamanoi F (2010) Mesoporous silica nanoparticles facilitate delivery of SiRNA to shutdown signaling pathways in mammalian cells. Small 6(11):1185–1190. https://doi.org/10.1002/smll.200901966

Horiuchi K (2004) DSC studies on structural phase transitions and molecular motions in some A2MCl4 compounds. Phys Status Solidi Appl Res 201(4):723–726. https://doi.org/10.1002/pssa.200306777

Huang M, Yan H, Chen C, Song D, Heinz TF, Hone J (2009) Phonon softening and crystallographic orientation of strained graphene studied by Raman spectroscopy. Proc Natl Acad Sci U S A 106(18):7304–7308. https://doi.org/10.1073/pnas.0811754106

Huang C, Wu J, Jiang W, Liu R, Li Z, Luan Y (2018) Amphiphilic prodrug-decorated graphene oxide as a multi-functional drug delivery system for efficient cancer therapy, vol 89. Elsevier, Amsterdam. https://doi.org/10.1016/j.msec.2018.03.017

Hwang JG, Zahn M, O'Sullivan FM, Pettersson LAA, Hjortstam O, Liu R (2010) Effects of nanoparticle charging on streamer development in transformer oil-based nanofluids. J Appl Phys 107(1):014310. https://doi.org/10.1063/1.3267474

Hyun H, Lee S, Lim W, Jo D, Jung JS, Jo G, Kim SY, Lee DW, Um S, Yang DH, Chun HJ (2019) Engineered beta-cyclodextrin-based carrier for targeted doxorubicin delivery in breast cancer therapy in vivo. J Ind Eng Chem 70:145–151. https://doi.org/10.1016/j.jiec.2018.09.052

Illath K, Narasimahan AK, Nagai M, Wankhar S, Santra TS (2020) Microfluidics-based metallic nanoparticle synthesis and applications. In: Microfluidics and bio-MEMS. Jenny Stanford Publishing, Singapore, pp 429–501. https://doi.org/10.1201/9781003014935-10

Illers KH, Kanig G (1982) Heat of fusion and lamellar structure of polyethylene single crystal mats. Colloid Polym Sci 260(6):564–569. https://doi.org/10.1007/BF01422587

Ing LY, Zin NM, Sarwar A, Katas H (2012) Antifungal activity of chitosan nanoparticles and correlation with their physical properties. Int J Biomater 2012:632698. https://doi.org/10.1155/2012/632698

Ismail RA, Mohsin MH, Ali AK, Hassoon KI, Erten-Ela S (2019) Preparation and characterization of carbon nanotubes by pulsed laser ablation in water for optoelectronic application. Phys E Low Dimens Syst Nanostruct 2020(119):113997. https://doi.org/10.1016/j.physe.2020.113997

Jain P, Patel K, Jangid AK, Guleria A, Patel S, Pooja D, Kulhari H (2020) Modulating the delivery of 5-fluorouracil to human colon cancer cells using multifunctional arginine-coated manganese oxide nanocuboids with MRI properties. ACS Appl Bio Mater 3(10):6852–6864. https://doi.org/10.1021/acsabm.0c00780

Jain P, Patel K, Jangid AK, Guleria A, Patel S, Pooja D, Kulhari H (2021) Biotinylated Mn3O4 nanocuboids for targeted delivery of gemcitabine hydrochloride to breast cancer and MRI applications. Int J Pharm 606:120895. https://doi.org/10.1016/j.ijpharm.2021.120895

Jangid AK, Patel K, Joshi U, Patel S, Singh A, Pooja D, Saharan VA, Kulhari H (2022) PEGylated G4 dendrimers as a promising nanocarrier for piperlongumine delivery: synthesis, characterization, and anticancer activity. Eur Polym J 179:111547. https://doi.org/10.1016/j.eurpolymj.2022.111547

Jeevanandam J, Barhoum A, Chan YS, Dufresne A, Danquah MK (2018) Review on nanoparticles and nanostructured materials: history, sources, toxicity and regulations. Beilstein J Nanotechnol 9(1):1050–1074. https://doi.org/10.3762/bjnano.9.98

Jia H, Chen H, Wei M, Chen X, Zhang Y, Cao L, Yuan P, Wang F, Yang G, Ma J (2017) Gold nanoparticle-based MiR155 antagonist macrophage delivery restores the cardiac function in ovariectomized diabetic mouse model. Int J Nanomedicine 12:4963–4979. https://doi.org/10.2147/IJN.S138400

Jiang Y, Peng ZY, Zhang S, Li F, Liu Z, Zhang J, Liu Y, Wang K (2018) Facile in-situ solvothermal method to synthesize double shell ZnIn2S4 nanosheets/TiO2 hollow nanosphere with enhanced photocatalytic activities. Ceram Int 44(6):6115–6126. https://doi.org/10.1016/j.ceramint.2017.12.244

Jin SE, Kim CK (2014) Charge-mediated topical delivery of plasmid DNA with cationic lipid nanoparticles to the skin. Colloids Surf B Biointerfaces 116:582–590. https://doi.org/10.1016/j.colsurfb.2014.01.053

Jones AC, Hitchman ML (2009) Overview of chemical vapour deposition. In: Chemical vapour deposition: precursors, processes and applications, vol 1. Royal Society of Chemistry, London, pp 1–36

Jose A, Labala S, Ninave KM, Gade SK, Venuganti VVK (2018) effective skin cancer treatment by topical co-delivery of curcumin and STAT3 SiRNA using cationic liposomes. AAPS PharmSciTech 19(1):166–175. https://doi.org/10.1208/s12249-017-0833-y

Kalpana-Sastry R, Rashmi HB, Rao NH, Ilyas SM (2009) Nanotechnology and agriculture in India: the second green revolution. In: OECD conference "potential environ benefits nanotechnol foster safe innovation-led growth", vol 7, pp 15–17

Kamel AE, Fadel M, Louis D (2019) Curcumin-loaded nanostructured lipid carriers prepared using Peceol^{TM} and olive oil in photodynamic therapy: development and application in breast cancer cell line. Int J Nanomedicine 14:5073–5085. https://doi.org/10.2147/IJN.S210484

Karimi Shervedani R, Mirhosseini H, Samiei Foroushani M, Torabi M, Rahsepar FR, Norouzi-Barough L (2018) Immobilization of methotrexate anticancer drug onto the graphene surface and interaction with calf thymus DNA and 4T1 cancer cells. Bioelectrochemistry 119:1–9. https://doi.org/10.1016/j.bioelechem.2017.08.004

Kato H (2012) Size determination of NPs by dynamic light scattering. In: Nanomaterials: processing and characterization with lasers. Wiley, New York

Kelly KL, Coronado E, Zhao LL, Schatz GC (2003) The optical properties of metal nanoparticles: the influence of size, shape, and dielectric environment. J Phys Chem B 107(3):668–677. https://doi.org/10.1021/jp026731y

Khan I, Saeed K, Khan I (2019) Nanoparticles: properties, applications and toxicities. Arab J Chem 12(7):908–931. https://doi.org/10.1016/j.arabjc.2017.05.011

Khoshfetrat SM, Mehrgardi MA (2017) Amplified detection of leukemia cancer cells using an aptamer-conjugated gold-coated magnetic nanoparticles on a nitrogen-doped graphene modified electrode. Bioelectrochemistry 114:24–32. https://doi.org/10.1016/j.bioelechem.2016.12.001

Kolahalam LA, Kasi Viswanath IV, Diwakar BS, Govindh B, Reddy V, Murthy YLN (2019) Review on nanomaterials: synthesis and applications. Mater Today Proc 18:2182–2190. https://doi.org/10.1016/j.matpr.2019.07.371

Kondath S, Raghavan BS, Anantanarayanan R, Rajaram R (2014) Synthesis and characterisation of morin reduced gold nanoparticles and its cytotoxicity in MCF-7 Cells. Chem Biol Interact 224: 78–88. https://doi.org/10.1016/j.cbi.2014.09.025

Kooti M, Sedeh AN, Motamedi H, Rezatofighi SE (2018) Magnetic graphene oxide inlaid with silver nanoparticles as antibacterial and drug delivery composite. Appl Microbiol Biotechnol 102(8):3607–3621. https://doi.org/10.1007/s00253-018-8880-1

Kozielski KL, Tzeng SY, De Mendoza BAH, Green JJ (2014) Bioreducible cationic polymer-based environmentally triggered cytoplasmic SiRNA delivery to primary human brain cancer cells. ACS Nano 8(4):3232–3241. https://doi.org/10.1021/nn500704t

Kulhari H, Kulhari DP, Prajapati SK, Chauhan AS (2013) Pharmacokinetic and pharmacodynamic studies of poly(amidoamine) dendrimer based simvastatin oral formulations for the treatment of hypercholesterolemia. Mol Pharm 10(7):2528–2533. https://doi.org/10.1021/mp300650y

Kulhari H, Pooja D, Singh MK, Chauhan AS (2015a) Optimization of carboxylate-terminated poly (amidoamine) dendrimer-mediated cisplatin formulation. Drug Dev Ind Pharm 41(2):232–238. https://doi.org/10.3109/03639045.2013.858735

Kulhari H, Pooja D, Shrivastava S, Kuncha M, Naidu VGM, Bansal V, Sistla R, Adams DJ (2015b) Trastuzumab-grafted PAMAM dendrimers for the selective delivery of anticancer drugs to HER2-positive breast cancer. Sci Rep 2016(6):1–13. https://doi.org/10.1038/srep23179

Kumar A, Dixit CK (2017) Methods for characterization of nanoparticles. In: Advances in nanomedicine for the delivery of therapeutic nucleic acids. Elsevier, Amsterdam, pp 44–58. https://doi.org/10.1016/B978-0-08-100557-6.00003-1

Kuo CW, Shiu JY, Cho YH, Chen P (2003) Fabrication of large-area periodic nanopillar arrays for nanoimprint lithography using polymer colloid masks. Adv Mater 15(13):1065–1068. https://doi.org/10.1002/adma.200304824

Lakshmiprasanna HR, Jagadeesha Angadi V, Rajesh Babu B, Pasha M, Manjunatha K, Matteppanavar S (2019) Effect of Pr3+-doping on the structural, elastic and magnetic properties of Mn–Zn ferrite nanoparticles prepared by solution combustion synthesis method. Chem Data Collect 24:100273. https://doi.org/10.1016/j.cdc.2019.100273

Lamouri R, Mounkachi O, Salmani E, Hamedoun M, Benyoussef A, Ez-Zahraouy H (2020) Size effect on the magnetic properties of CoFe2O4 nanoparticles: a Monte Carlo study. Ceram Int 46(6):8092–8096. https://doi.org/10.1016/j.ceramint.2019.12.035

Lee GH, Choi TM, Kim B, Han SH, Lee JM, Kim SH (2017) Chameleon-inspired mechanochromic photonic films composed of non-close-packed colloidal arrays. ACS Nano 11(11): 11350–11357. https://doi.org/10.1021/acsnano.7b05885

Li R, Wu R, Zhao L, Hu Z, Guo S, Pan X, Zou H (2011) Folate and iron difunctionalized multiwall carbon nanotubes as dual-targeted drug nanocarrier to cancer cells. Carbon N Y 49(5): 1797–1805. https://doi.org/10.1016/j.carbon.2011.01.003

Li J, Wu Q, Wu J (2015) Synthesis of nanoparticles via solvothermal and hydrothermal methods. In: Aliofkhazraei M (ed) Handbook of nanoparticles. Springer International Publishing, Basel. https://doi.org/10.1007/978-3-319-15338-4_17

Li X, Jia X, Niu H (2018) Nanostructured lipid carriers co-delivering lapachone and doxorubicin for overcoming multidrug resistance in breast cancer therapy. Int J Nanomedicine 13:4107–4119. https://doi.org/10.2147/IJN.S163929

Lin T, Song YL, Liao J, Liu F, Zeng TT (2020) Applications of surface-enhanced raman spectroscopy in detection fields. Nanomedicine 15(30):2971–2989. https://doi.org/10.2217/nnm-2020-0361

Liu J, Zhao Y, Guo Q, Wang Z, Wang H, Yang Y, Huang Y (2012) TAT-modified nanosilver for combating multidrug-resistant cancer. Biomaterials 33(26):6155–6161. https://doi.org/10.1016/j.biomaterials.2012.05.035

Liu J, Ma X, Jin S, Xue X, Zhang C, Wei T, Guo W, Liang XJ (2016) Zinc oxide nanoparticles as adjuvant to facilitate doxorubicin intracellular accumulation and visualize pH-responsive release for overcoming drug resistance. Mol Pharm 13(5):1723–1730. https://doi.org/10.1021/acs.molpharmaceut.6b00311

Liu M, Zhao F, Zhu D, Duan H, Lv Y, Li L, Gan L (2018) Ultramicroporous carbon nanoparticles derived from metal–organic framework nanoparticles for high-performance supercapacitors. Mater Chem Phys 211:234–241. https://doi.org/10.1016/j.matchemphys.2018.02.030

Long W, Ouyang H, Wan W, Yan W, Zhou C, Huang H, Liu M, Zhang X, Feng Y, Wei Y (2019) "Two in one": simultaneous functionalization and DOX loading for fabrication of nanodiamond-based pH responsive drug delivery system. Mater Sci Eng C 2020(108): 110413. https://doi.org/10.1016/j.msec.2019.110413

Lopez-Bertoni H, Kozielski KL, Rui Y, Lal B, Vaughan H, Wilson DR, Mihelson N, Eberhart CG, Laterra J, Green JJ (2018) Bioreducible polymeric nanoparticles containing multiplexed cancer stem cell regulating MiRNAs inhibit glioblastoma growth and prolong survival. Nano Lett 18(7):4086–4094. https://doi.org/10.1021/acs.nanolett.8b00390

Loureiro JA, Andrade S, Duarte A, Neves AR, Queiroz JF, Nunes C, Sevin E, Fenart L, Gosselet F, Coelho MAN, Pereira MC, Latruffe N (2017) Resveratrol and grape extract-loaded solid lipid nanoparticles for the treatment of Alzheimer's disease. Molecules 22(2):1–16. https://doi.org/10.3390/molecules22020277

Lu W, Lieber CM (2006) Semiconductor nanowires. J Phys D Appl Phys 39(21). https://doi.org/10.1088/0022-3727/39/21/R01

Lu M, Wang YK, Zhao J, Lu H, Stenzel MH, Xiao P (2016) PEG grafted-nanodiamonds for the delivery of gemcitabine. Macromol Rapid Commun 37(24):2023–2029. https://doi.org/10.1002/marc.201600344

Lundgaard LE, Hansen W, Ingebrigtsen S (2008) Ageing of mineral oil impregnated cellulose by acid catalysis. IEEE Trans Dielectr Electr Insul 15(2):540–546. https://doi.org/10.1109/TDEI.2008.4483475

Machac P, Cichon S, Lapcak L, Fekete L (2020) Graphene prepared by chemical vapour deposition process. Graphene Technol 5(1–2):9–17. https://doi.org/10.1007/s41127-019-00029-6

Magid M, Al-Karam LQ (2021) Non-covalent functionalization of CNTs with chitosan for drug delivery system. J Phys Conf Ser 2114(1):012038. https://doi.org/10.1088/1742-6596/2114/1/012038

Manjaiah KM, Mukhopadhyay R, Paul R, Datta SC, Kumararaja P, Sarkar B (2018) Clay minerals and zeolites for environmentally sustainable agriculture, vol 2025. Elsevier, Amsterdam. https://doi.org/10.1016/B978-0-12-814617-0.00008-6

Manor J, Feldblum ES, Zanni MT, Arkin IT (2012) Environment polarity in proteins mapped noninvasively by FTIR spectroscopy. J Phys Chem Lett 3(7):939–944. https://doi.org/10.1021/jz300150v

Matsumoto R, Adachi S, Sadki EHS, Yamamoto S, Tanaka H, Takeya H, Takano Y (2020) Maskless patterning of gallium-irradiated superconducting silicon using focused ion beam. ACS Appl Electron Mater 2(3):677–682. https://doi.org/10.1021/acsaelm.9b00781

Mauter MS, Elimelech M (2008) Environmental applications of carbon-based nanomaterials: critical review. Am Chem Soc 42(16):5843–5859

McMurray TA, Dunlop PSM, Byrne JA (2006) The photocatalytic degradation of atrazine on nanoparticulate TiO2 Films. J Photochem Photobiol A Chem 182(1):43–51. https://doi.org/10.1016/j.jphotochem.2006.01.010

Mendes R, Fernandes AR, Baptista PV (2017) Gold nanoparticle approach to the selective delivery of gene silencing in cancer—the case for combined delivery? Genes (Basel) 8(3):94. https://doi.org/10.3390/genes8030094

Mikhailov MM, Yuryev SA, Lovitskiy AA (2018) On the correlation between diffuse reflectance spectra and particle size of BaSO4 powder under heating and modifying with SiO2 nanoparticles. Opt Mater (Amst) 85:226–229. https://doi.org/10.1016/j.optmat.2018.08.059

Milani N, McLaughlin MJ, Stacey SP, Kirby JK, Hettiarachchi GM, Beak DG, Cornelis G (2012) Dissolution kinetics of macronutrient fertilizers coated with manufactured zinc oxide nanoparticles. J Agric Food Chem 60(16):3991–3998. https://doi.org/10.1021/jf205191y

Moreno-Vega AI, Gómez-Quintero T, Nuñez-Anita RE, Acosta-Torres LS, Castaño V (2012) Polymeric and ceramic nanoparticles in biomedical applications. J Nanotechnol 2012:936041. https://doi.org/10.1155/2012/936041

Mulvaney P (2015) Nanoscience vs. nanotechnology. ACS Nano 9(3):2215–2217. https://doi.org/10.1021/acsnano.5b01418

Mulvaney SP, Keating CD (2000) Raman spectroscopy. Anal Chem 72(12):145–158. https://doi.org/10.1021/a10000155

Mund R, Panda N, Nimesh S, Biswas A (2014) Novel titanium oxide nanoparticles for effective delivery of paclitaxel to human breast cancer cells. J Nanopart Res 16(12). https://doi.org/10.1007/s11051-014-2739-x

Muñoz-García J, Vázquez L, Cuerno R, Sánchez-García JA, Castro M, Gago R (2009) Self-organized surface nanopatterning by ion beam sputtering. In: Toward functional nanomaterials. Springer, New York, pp 323–398. https://doi.org/10.1007/978-0-387-77717-7_10

Nadagouda MN, Varma RS (2009) Risk reduction via greener synthesis of noble metal nanostructures and nanocomposites. In: Nanomaterials: risks and benefits. Springer, Dordrecht, pp 209–217. https://doi.org/10.1007/978-1-4020-9491-0_15

Ng KK, Zheng G (2015) Molecular interactions in organic nanoparticles for phototheranostic applications. Chem Rev 115(19):11012–11042. https://doi.org/10.1021/acs.chemrev.5b00140

Ngô C, Van de Voorde MH (2014) Nanomaterials: doing more with less. In: Nanotechnology in a nutshell: from simple to complex systems. Atlantis Press, Paris, pp 55–70. https://doi.org/10.2991/978-94-6239-012-6_4

Nie M, Sun K, Meng DD (2009) Formation of metal nanoparticles by short-distance sputter deposition in a reactive ion etching chamber. J Appl Phys 106(5):054314. https://doi.org/10.1063/1.3211326

Nsofor EC (2010) Recent patents on nanofluids (nanoparticles in liquids) heat transfer. Recent Patents Mech Eng 1(3):190–197. https://doi.org/10.2174/1874477x10801030190

Oh WK, Yoon H, Jang J (2010) Size control of magnetic carbon nanoparticles for drug delivery. Biomaterials 31(6):1342–1348. https://doi.org/10.1016/j.biomaterials.2009.10.018

Ohta T, Hashida Y, Yamashita F, Hashida M (2016) Development of novel drug and gene delivery carriers composed of single-walled carbon nanotubes and designed peptides with PEGylation. J Pharm Sci 105(9):2815–2824. https://doi.org/10.1016/j.xphs.2016.03.031

Okuyama K, Lenggoro WW (2003) Preparation of nanoparticles via spray route. Chem Eng Sci 58(3–6):537–547. https://doi.org/10.1016/S0009-2509(02)00578-X

Oladipo AO, Iku SII, Ntwasa M, Nkambule TTI, Mamba BB, Msagati TAM (2020) Doxorubicin conjugated hydrophilic AuPt bimetallic nanoparticles fabricated from phragmites australis: characterization and cytotoxic activity against human cancer cells. J Drug Deliv Sci Technol 57:101749. https://doi.org/10.1016/j.jddst.2020.101749

Olerile LD, Liu Y, Zhang B, Wang T, Mu S, Zhang J, Selotlegeng L, Zhang N (2017) Near-infrared mediated quantum dots and paclitaxel co-loaded nanostructured lipid carriers for cancer theragnostic. Colloids Surf B Biointerfaces 150:121–130. https://doi.org/10.1016/j.colsurfb.2016.11.032

Ostermann R, Cravillon J, Weidmann C, Wiebcke M, Smarsly BM (2011) Metal-organic framework nanofibers via electrospinning. Chem Commun 47(1):442–444. https://doi.org/10.1039/c0cc02271c

Owens FJ (2015) Properties of magnetic nanoparticles. In: Physics of magnetic nanostructures. Wiley, New York, pp 35–53. https://doi.org/10.1002/9781118989913.ch3

Pan K, Zhong Q (2016) Organic nanoparticles in foods: fabrication, characterization, and utilization. Annu Rev Food Sci Technol 7:245–266. https://doi.org/10.1146/annurev-food-041715-033215

Parashar M, Shukla VK, Singh R (2020) Metal oxides nanoparticles via sol–gel method: a review on synthesis, characterization and applications. J Mater Sci Mater Electron 31(5):3729–3749. https://doi.org/10.1007/s10854-020-02994-8

Pastorin G, Wu W, Wieckowski S, Briand JP, Kostarelos K, Prato M, Bianco A (2006) Double functionalisation of carbon nanotubes for multimodal drug delivery. Chem Commun 11:1182–1184. https://doi.org/10.1039/b516309a

Pérez-Alonso C, Cruz-Olivares J, Barrera-Pichardo JF, Rodríguez-Huezo ME, Báez-González JG, Vernon-Carter EJ (2008) DSC thermo-oxidative stability of red chili oleoresin microencapsulated in blended biopolymers matrices. J Food Eng 85(4):613–624. https://doi.org/10.1016/j.jfoodeng.2007.08.020

Pimpin A, Srituravanich W (2012) Reviews on micro- and nanolithography techniques and their applications. Eng J 16(1):37–55. https://doi.org/10.4186/ej.2012.16.1.37

Pinto RJB, Almeida A, Fernandes SCM, Freire CSR, Silvestre AJD, Neto CP, Trindade T (2013) Antifungal activity of transparent nanocomposite thin films of pullulan and silver against Aspergillus niger. Colloids Surf B Biointerfaces 103:143–148. https://doi.org/10.1016/j.colsurfb.2012.09.045

Piri N, Shams-Nateri A, Mokhtari J (2016) The relationship between refractive index and optical properties of absorbing nanoparticle. Color Res Appl 41(5):477–483. https://doi.org/10.1002/col.21983

Pokropivny VV, Skorokhod VV (2007) Classification of nanostructures by dimensionality and concept of surface forms engineering in nanomaterial science. Mater Sci Eng C 27(5–8 Special Issue):990–993. https://doi.org/10.1016/j.msec.2006.09.023

Pooja D, Srinivasa Reddy T, Kulhari H, Kadari A, Adams DJ, Bansal V, Sistla R (2020) N-acetyl-D-glucosamine-conjugated PAMAM dendrimers as dual receptor-targeting nanocarriers for anticancer drug delivery. Eur J Pharm Biopharm 154:377–386. https://doi.org/10.1016/j.ejpb.2020.07.020

Poonia N, Kaur Narang J, Lather V, Beg S, Sharma T, Singh B, Pandita D (2019) Resveratrol loaded functionalized nanostructured lipid carriers for breast cancer targeting: systematic development, characterization and pharmacokinetic evaluation. Colloids Surf B Biointerfaces 181:756–766. https://doi.org/10.1016/j.colsurfb.2019.06.004

Prasanna, B. M. (n.d.) Nanotechnology in agriculture. In: Parsad R, Gupta VK, Bhar LM, Bhatia VK (eds), Advances in data analytical techniques, vol VI, pp 111–118. Indian Agricultural Statistics Research Institute (I.C.A.R.), New Delhi

Qiu L, Zhu N, Feng Y, Michaelides EE, Żyła G, Jing D, Zhang X, Norris PM, Markides CN, Mahian O (2020) A review of recent advances in thermophysical properties at the nanoscale: from solid state to colloids. Phys Rep 843:1–81. https://doi.org/10.1016/j.physrep.2019.12.001

Rai M, Ingle AP, Paralikar P, Gupta I, Medici S, Santos CA (2017) Recent advances in use of silver nanoparticles as antimalarial agents. Int J Pharm 526(1–2):254–270. https://doi.org/10.1016/j.ijpharm.2017.04.042

Rami JM, Patel CD, Patel CM, Patel MV (2020) Thermogravimetric analysis (TGA) of some synthesized metal oxide nanoparticles. Mater Today Proc 43:655–659. https://doi.org/10.1016/j.matpr.2020.12.554

Ravichandran R, Sridhar R, Venugopal JR, Sundarrajan S, Mukherjee S, Ramakrishna S (2014) Gold nanoparticle loaded hybrid nanofibers for cardiogenic differentiation of stem cells for infarcted myocardium regeneration. Macromol Biosci 14(4):515–525. https://doi.org/10.1002/mabi.201300407

Razzazan A, Atyabi F, Kazemi B, Dinarvand R (2016) In vivo drug delivery of gemcitabine with PEGylated single-walled carbon nanotubes. Mater Sci Eng C 62:614–625. https://doi.org/10.1016/j.msec.2016.01.076

Ren W, Zeng L, Shen Z, Xiang L, Gong A, Zhang J, Mao C, Li A, Paunesku T, Woloschak GE, Hosmane NS, Wu A (2013) Enhanced doxorubicin transport to multidrug resistant breast cancer cells via TiO2 nanocarriers. RSC Adv 3(43):20855–20861. https://doi.org/10.1039/c3ra42863j

Rivas GA, Miscoria SA, Desbrieres J, Barrera GD (2007) New biosensing platforms based on the layer-by-layer self-assembling of polyelectrolytes on nafion/carbon nanotubes-coated glassy carbon electrodes. Talanta 71(1):270–275. https://doi.org/10.1016/j.talanta.2006.03.056

Roduner E (2006) Size matters: why nanomaterials are different. Chem Soc Rev 35(7):583–592. https://doi.org/10.1039/b502142c

Saadatkhah N, Carillo Garcia A, Ackermann S, Leclerc P, Latifi M, Samih S, Patience GS, Chaouki J (2020) Experimental methods in chemical engineering: thermogravimetric analysis—TGA. Can J Chem Eng 98(1):34–43. https://doi.org/10.1002/cjce.23673

Sadhukhan P, Kundu M, Chatterjee S, Ghosh N, Manna P, Das J, Sil PC (2019) Targeted delivery of quercetin via pH-responsive zinc oxide nanoparticles for breast cancer therapy. Mater Sci Eng C 100:129–140. https://doi.org/10.1016/j.msec.2019.02.096

Saharan V, Mehrotra A, Khatik R, Rawal P, Sharma SS, Pal A (2013) Synthesis of chitosan based nanoparticles and their in vitro evaluation against phytopathogenic fungi. Int J Biol Macromol 62:677–683. https://doi.org/10.1016/j.ijbiomac.2013.10.012

Sai Rohith R (2015) Prescision farming using nano based wireless sensor network. Int J Eng Res Gen Sci 3(2):343–350

Saifullah B, Hussein MZ, Hussein-Al-Ali SH, Arulselvan P, Fakurazi S (2013) Antituberculosis nanodelivery system with controlled-release properties based on para-amino salicylate-zinc aluminum-layered double-hydroxide nanocomposites. Drug Des Dev Ther 7:1365–1375. https://doi.org/10.2147/DDDT.S50665

Samyn P, Barhoum A (2018) Engineered nanomaterials for papermaking industry. Elsevier, Amsterdam. https://doi.org/10.1016/B978-0-323-51255-8.00009-4

Sant'Anna C, Campanati L, Gadelha C, Lourenço D, Labati-Terra L, Bittencourt-Silvestre J, Benchimol M, Cunha-e-Silva NL, De Souza W (2005) Improvement on the visualization of cytoskeletal structures of protozoan parasites using high-resolution field emission scanning electron microscopy (FESEM). Histochem Cell Biol 124(1):87–95. https://doi.org/10.1007/s00418-005-0786-1

Sarkar S, Datta SC, Biswas DR (2014) Synthesis and characterization of nanoclay-polymer composites from soil clay with respect to their water-holding capacities and nutrient-release behavior. J Appl Polym Sci 131(6):1–8. https://doi.org/10.1002/app.39951

Savage T, Rao AM (2006) Thermal properties of nanomaterials and nanocomposites. In: Tritt TM (ed) Thermal conductivity. Physics of solids and liquids. Springer, Boston, pp 261–284. https://doi.org/10.1007/0-387-26017-x_12

Segal V, Hjortsberg A, Rabinovich A, Nattrass D, Raj K (1998) AC (60 Hz) and impulse breakdown strength of a colloidal fluid based on transformer oil and magnetite nanoparticles. Conf Rec IEEE Int Symp Electr Insul 2:619–622. https://doi.org/10.1109/elinsl.1998.694869

Shafiee S, Ahangar HA, Saffar A (2019) Taguchi method optimization for synthesis of Fe3O4 @chitosan/Tragacanth Gum nanocomposite as a drug delivery system. Carbohydr Polym 222: 1–9. https://doi.org/10.1016/j.carbpol.2019.114982

Sharma H, Kumar K, Choudhary C, Mishra PK, Vaidya B (2016) Development and characterization of metal oxide nanoparticles for the delivery of anticancer drug. Artif Cells Nanomed Biotechnol 44(2):672–679. https://doi.org/10.3109/21691401.2014.978980

Sharon M (2019) History of nanotechnology. Wiley, New York. https://doi.org/10.1002/9781119460534

Shen MY, Liu TI, Yu TW, Kv R, Chiang WH, Tsai YC, Chen HH, Lin SC, Chiu HC (2018) Hierarchically targetable polysaccharide-coated solid lipid nanoparticles as an oral chemo/thermotherapy delivery system for local treatment of colon cancer. Biomaterials 2019(197): 86–100. https://doi.org/10.1016/j.biomaterials.2019.01.019

Shin SW, Song IH, Um SH (2015) Role of physicochemical properties in nanoparticle toxicity. Nanomaterials 5(3):1351–1365. https://doi.org/10.3390/nano5031351

Shrimali VG, Gadani K, Rajyaguru B, Gohil H, Chudasama DK, Dhruv D, Joshi AD, Pandya DD, Solanki PS, Shah NA (2020) Size dependent dielectric, magnetic, transport and magnetodielectric properties of BiFe0.98Co0.02O3 nanoparticles. J Alloys Compd 817: 152685. https://doi.org/10.1016/j.jallcom.2019.152685

Singh R, Mehra NK, Jain V, Jain NK (2013) Gemcitabine-loaded smart carbon nanotubes for effective targeting to cancer cells. J Drug Target 21(6):581–592. https://doi.org/10.3109/1061186X.2013.778264

Singh P, Pandit S, Mokkapati VRSS, Garg A, Ravikumar V, Mijakovic I (2018) Gold nanoparticles in diagnostics and therapeutics for human cancer. Int J Mol Sci 19(7). https://doi.org/10.3390/ijms19071979

Son HH, Seo GH, Jeong U, Shin DY, Kim SJ (2017) Capillary wicking effect of a Cr-sputtered superhydrophilic surface on enhancement of pool boiling critical heat flux. Int J Heat Mass Transf 113:115–128. https://doi.org/10.1016/j.ijheatmasstransfer.2017.05.055

Soni N, Soni N, Pandey H, Maheshwari R, Kesharwani P, Tekade RK (2016) Augmented delivery of gemcitabine in lung cancer cells exploring mannose anchored solid lipid nanoparticles. J Colloid Interface Sci 481:107–116. https://doi.org/10.1016/j.jcis.2016.07.020

Srivastava A, Thapa U, Saha M, Jalees M (2019) Aggregation behaviour of tetracaine hydrochloride with gemini surfactants and the formation of silver nanoparticles using drug-gemini surfactants mixture. J Mol Liq 276:399–408. https://doi.org/10.1016/j.molliq.2018.12.006

Strojan K, Leonardi A, Bregar VB, Križaj I, Svete J, Pavlin M (2017) Dispersion of nanoparticles in different media importantly determines the composition of their protein corona. PLoS One 12(1):1–21. https://doi.org/10.1371/journal.pone.0169552

Su D (2017) Advanced electron microscopy characterization of nanomaterials for catalysis. Green Energy Environ 2(2):70–83. https://doi.org/10.1016/j.gee.2017.02.001

Su Z, Erdene-Ochir T, Ganbold T, Baigude H (2020) Design of curdlan-based pH-sensitive polymers with endosome buffering functionality for SiRNA delivery. Int J Biol Macromol 146:773–780. https://doi.org/10.1016/j.ijbiomac.2019.10.129

Sulaiman GM, Waheeb HM, Jabir MS, Khazaal SH, Dewir YH, Naidoo Y (2020) Hesperidin loaded on gold nanoparticles as a drug delivery system for a successful biocompatible, anti-cancer, anti-inflammatory and phagocytosis inducer model. Sci Rep 10(1):1–16. https://doi.org/10.1038/s41598-020-66419-6

Sun S, Murray CB, Weller D, Folks L, Moser A (2000) Monodisperse FePt nanoparticles and ferromagnetic FePt nanocrystal superlattices. Science 287(5460):1989–1992. https://doi.org/10.1126/science.287.5460.1989

Sundraraman G, Jayakumari LS (2020) Meticulous taxifolin releasing performance by the zinc oxide nanoparticles: as a short road to drug delivery system for cancer therapeutics. J Clust Sci 31(1):241–255. https://doi.org/10.1007/s10876-019-01642-4

Szabó Z, Volk J, Fülöp E, Deák A, Bársony I (2013) Regular ZnO nanopillar arrays by nanosphere photolithography. Photonics Nanostruct Fundam Appl 11(1):1–7. https://doi.org/10.1016/j.photonics.2012.06.009

Taghavi S, HashemNia A, Mosaffa F, Askarian S, Abnous K, Ramezani M (2016) Preparation and evaluation of polyethylenimine-functionalized carbon nanotubes tagged with 5TR1 aptamer for targeted delivery of Bcl-XL ShRNA into breast cancer cells. Colloids Surf B Biointerfaces 140:28–39. https://doi.org/10.1016/j.colsurfb.2015.12.021

Taghavi S, Nia AH, Abnous K, Ramezani M (2017) Polyethylenimine-functionalized carbon nanotubes tagged with AS1411 aptamer for combination gene and drug delivery into human gastric cancer cells. Int J Pharm 516(1–2):301–312. https://doi.org/10.1016/j.ijpharm.2016.11.027

Tahover E, Patil YP, Gabizon AA (2015) Emerging delivery systems to reduce doxorubicin cardiotoxicity and improve therapeutic index: focus on liposomes. Anti-Cancer Drugs 26(3):241–258. https://doi.org/10.1097/CAD.0000000000000182

Takacs L (2002) Self-sustaining reactions induced by ball milling. Prog Mater Sci 47(4):355–414. https://doi.org/10.1016/S0079-6425(01)00002-0

Tan J, Yang N, Hu Z, Su J, Zhong J, Yang Y, Yu Y, Zhu J, Xue D, Huang Y, Lai Z, Huang Y, Lu X, Zhao Y (2016) Aptamer-functionalized fluorescent silica nanoparticles for highly sensitive detection of leukemia cells. Nanoscale Res Lett 11(1):298. https://doi.org/10.1186/s11671-016-1512-8

Taniguchi N (1974) On the basic concept of nanotechnology. In: Proceedings of the international conference on production engineering, pp 18–23

Tapeinos C, Battaglini M, Ciofani G (2017) Advances in the design of solid lipid nanoparticles and nanostructured lipid carriers for targeting brain diseases. J Control Release 264:306–332. https://doi.org/10.1016/j.jconrel.2017.08.033

Theron J, Walker JA, Cloete TE (2008) Nanotechnology and water treatment: applications and emerging opportunities. Crit Rev Microbiol 34(1):43–69. https://doi.org/10.1080/10408410701710442

Thomas CS, Harshita, Mishra KP, Talegaonkar S (2015) Ceramic nanoparticles: fabrication methods and applications in drug delivery. Curr Pharm Des 21(42):6165–6188. https://doi.org/10.2174/1381612821666151027153246

Thomas S, Sakho EHM, Kalarikkal N, Oluwafemi SO, Wu J (2019) Nanomaterials for solar cell applications. Elsevier, Amsterdam. https://doi.org/10.1016/C2016-0-03432-0

Tolochko NK (2009) The history of nanoscience and nanotechnology, vol 1. https://doi.org/10.1007/978-3-031-02084-1_2

Tomar R, Abdala AA, Chaudhary RG, Singh NB (2020) Photocatalytic degradation of dyes by nanomaterials. Mater Today Proc 29:967–973. https://doi.org/10.1016/j.matpr.2020.04.144

Torney F, Trewyn BG, Lin VSY, Wang K (2007) Mesoporous silica nanoparticles deliver DNA and chemicals into plants. Nat Nanotechnol 2(5):295–300. https://doi.org/10.1038/nnano.2007.108

Toshima N, Yonezawa T (1998) Bimetallic nanoparticles—novel materials for chemical and physical applications. New J Chem 22(11):1179–1201. https://doi.org/10.1039/a805753b

Toshishige Y (2004) Nanoelectronics applications. In: Carbon nanotubes science and applications. Routledge, Milton Park, pp 163–193

Tran NTT, Wang TH, Lin CY, Tai Y (2013) Synthesis of methotrexate-conjugated gold nanoparticles with enhanced cancer therapeutic effect. Biochem Eng J 78:175–180. https://doi.org/10.1016/j.bej.2013.04.017

Upadhyay S, Parekh K, Pandey B (2016) Influence of crystallite size on the magnetic properties of Fe3O4 nanoparticles. J Alloys Compd 678:478–485. https://doi.org/10.1016/j.jallcom.2016.03.279

Vamvakaki V, Chaniotakis NA (2007) Pesticide detection with a liposome-based nano-biosensor. Biosens Bioelectron 22(12):2848–2853. https://doi.org/10.1016/j.bios.2006.11.024

Vilalta-Clemente A, Gloystein K, Frangis N (2008) Principles of atomic force microscopy (AFM). In: Proceedings of physics of advanced materials Winter School, pp 1–8

Vinhas R, Correia C, Ribeiro P, Lourenço A, Botelho de Sousa A, Fernandes AR, Baptista PV (2016) Colorimetric assessment of BCR-ABL1 transcripts in clinical samples via gold nanoprobes. Anal Bioanal Chem 408(19):5277–5284. https://doi.org/10.1007/s00216-016-9622-9

Wagner M (2017) Thermal analysis in practice: fundamental aspects. Carl Hanser Verlag GmbH Co KG, Munich

Waite CL, Roth CM (2009) PAMAM-RGD conjugates enhance SiRNA delivery through a multicellular spheroid model of malignant glioma. Bioconjug Chem 20(10):1908–1916. https://doi.org/10.1021/bc900228m

Walter P, Welcomme E, Hallégot P, Zaluzec NJ, Deeb C, Castaing J, Veyssière P, Bréniaux R, Lévêque JL, Tsoucaris G (2006) Early use of PbS nanotechnology for an ancient hair dyeing formula. Nano Lett 6(10):2215–2219. https://doi.org/10.1021/nl061493u

Wang J, Xie H, Guo Z, Guan L, Li Y (2014) Improved thermal properties of paraffin wax by the addition of TiO2 nanoparticles. Appl Therm Eng 73(2):1541–1547. https://doi.org/10.1016/j.applthermaleng.2014.05.078

Wang W, Zhang L, Chen T, Guo W, Bao X, Wang D, Ren B, Wang H, Li Y, Wang Y, Chen S, Tang B, Yang Q, Chen C (2017) Anticancer effects of resveratrol-loaded solid lipid nanoparticles on human breast cancer cells. Molecules 22(11):1814. https://doi.org/10.3390/molecules22111814

Wang W, Chen T, Xu H, Ren B, Cheng X, Qi R, Liu H, Wang Y, Yan L, Chen S, Yang Q, Chen C (2018) Curcumin-loaded solid lipid nanoparticles enhanced anticancer efficiency in breast cancer. Molecules 23(7):1–13. https://doi.org/10.3390/molecules23071578

Wang H, Liang X, Wang J, Jiao S, Xue D (2020) Multifunctional inorganic nanomaterials for energy applications. Nanoscale 12(1):14–42. https://doi.org/10.1039/c9nr07008g

Wanyika H, Gatebe E, Kioni P, Tang Z, Gao Y (2012) Mesoporous silica nanoparticles carrier for urea: potential applications in agrochemical delivery systems. J Nanosci Nanotechnol 12(3): 2221–2228. https://doi.org/10.1166/jnn.2012.5801

Wen HW, Borejsza-Wysocki W, Decory TR, Baeumner AJ, Durst RA (2005) A novel extraction method for peanut allergenic proteins in chocolate and their detection by a liposome-based lateral flow assay. Eur Food Res Technol 221(3–4):564–569. https://doi.org/10.1007/s00217-005-1202-8

Wender H, Migowski P, Feil AF, Teixeira SR, Dupont J (2013) Sputtering deposition of nanoparticles onto liquid substrates: recent advances and future trends. Coord Chem Rev 257(17–18):2468–2483. https://doi.org/10.1016/j.ccr.2013.01.013

Wu X, Lu GQ, Wang L (2011) Shell-in-shell TiO2 hollow spheres synthesized by one-pot hydrothermal method for dye-sensitized solar cell application. Energy Environ Sci 4(9): 3565–3572. https://doi.org/10.1039/c0ee00727g

Wu D, Wang H, Hou X, Chen H, Ma Y, Hou Y, Hong J, Ding Y (2018) Effects of gold core size on regulating the performance of doxorubicin-conjugated gold nanoparticles. Nano Res 11(6): 3396–3410. https://doi.org/10.1007/s12274-017-1963-y

Xu K, Chen J (2020) High-resolution scanning probe lithography technology: a review. Appl Nanosci 10(4):1013–1022. https://doi.org/10.1007/s13204-019-01229-5

Xu L, Yeudall WA, Yang H (2017) Folic acid-decorated polyamidoamine dendrimer exhibits high tumor uptake and sustained highly localized retention in solid tumors: its utility for local SiRNA delivery. Acta Biomater 57:251–261. https://doi.org/10.1016/j.actbio.2017.04.023

Yan W, Petkov V, Mahurin SM, Overbury SH, Dai S (2005) Powder XRD analysis and catalysis characterization of ultra-small gold nanoparticles deposited on titania-modified SBA-15. Catal Commun 6(6):404–408. https://doi.org/10.1016/j.catcom.2005.04.004

Yang T, Du G, Cui Y, Yu R, Hua C, Tian W, Zhang Y (2019) pH-sensitive doxorubicin-loaded polymeric nanocomplex based on β-cyclodextrin for liver cancer-targeted therapy. Int J Nanomedicine 14:1997–2010. https://doi.org/10.2147/IJN.S193170

Yianoulis P, Giannouli M (2008) Thin solid films and nanomaterials for solar energy conversion and energy saving applications. J Nano Res 2(2):48–60. https://doi.org/10.4028/www.scientific.net/JNanoR.2.49

Yin Y, Gates B, Xia Y (2001) A soft lithographic approach to the fabrication of single crystalline silicon nanostructures with well-defined dimensions and shapes. Mater Res Soc Symp Proc 636(19):1426–1430. https://doi.org/10.1557/proc-636-d4.2.1

Yu Y, Yang X, Liu M, Nishikawa M, Tei T, Miyako E (2019) Amphipathic nanodiamond supraparticles for anticancer drug loading and delivery. ACS Appl Mater Interfaces 11(21): 18978–18987. https://doi.org/10.1021/acsami.9b04792

Zamani M, Rostami M, Aghajanzadeh M, Kheiri Manjili H, Rostamizadeh K, Danafar H (2018) Mesoporous titanium dioxide@ zinc oxide–graphene oxide nanocarriers for colon-specific drug delivery. J Mater Sci 53(3):1634–1645. https://doi.org/10.1007/s10853-017-1673-6

Zeng F, Xu D, Zhan C, Liang C, Zhao W, Zhang J, Feng H, Ma X (2018) Surfactant-free synthesis of graphene oxide coated silver nanoparticles for sers biosensing and intracellular drug delivery. ACS Appl Nano Mater 1(6):2748–2753. https://doi.org/10.1021/acsanm.8b00444

Zhang J, Chaker M, Ma D (2017) Pulsed laser ablation based synthesis of colloidal metal nanoparticles for catalytic applications. J Colloid Interface Sci 489:138–149. https://doi.org/10.1016/j.jcis.2016.07.050

Zhang M, Zhu J, Zheng Y, Guo R, Wang S, Mignani S, Caminade AM, Majoral JP, Shi X (2018) Doxorubicin-conjugated PAMAM dendrimers for pH-responsive drug release and folic acid-targeted cancer therapy. Pharmaceutics 10(3):1–13. https://doi.org/10.3390/pharmaceutics10030162

Zhang W, Xu W, Lan Y, He X, Liu K, Liang Y (2019) Antitumor effect of hyaluronic-acid-modified chitosan nanoparticles loaded with SiRNA for targeted therapy for non-small cell lung cancer. Int J Nanomedicine 14:5287–5301. https://doi.org/10.2147/IJN.S203113

Zhou YX, Wang YS, Tian JH, Sha YC, Jiang XX, Gao SY, Sun QH, Nie Q (2010) Breakdown characteristics in transformer oil modified by nanoparticles. Gaodianya Jishu/High Volt Eng 36: 1155–1159

Zhu J, Zheng L, Wen S, Tang Y, Shen M, Zhang G, Shi X (2014) Targeted cancer theranostics using alpha-tocopheryl succinate-conjugated multifunctional dendrimer-entrapped gold nanoparticles. Biomaterials 35(26):7635–7646. https://doi.org/10.1016/j.biomaterials.2014.05.046

Zhupanyn P, Ewe A, Büch T, Malek A, Rademacher P, Müller C, Reinert A, Jaimes Y, Aigner A (2019) Extracellular vesicle (ECV)-modified polyethylenimine (PEI) complexes for enhanced SiRNA delivery in vitro and in vivo. J Control Release 2020(319):63–76. https://doi.org/10.1016/j.jconrel.2019.12.032

Therapeutic Phytoconstituents-I

2

Sanju Kumari Singh and Sunita Patel

2.1 Introduction

Bioactive chemical substances that occur in plants which are not considered nutrients are called phytoconstituents. They are accountable for protecting the plant from microbial infections, pest infestations, disease attacks, and other forms of predation that may be caused by other species. Some of them are cause for things like color, scent, and other organoleptic qualities (Alamgir 2018). The vast majority of that carbon, nitrogen, and energy is eventually incorporated into molecules which are cosmopolitan throughout cells and are necessary for the efficient operation of cells and body (Taiz and Zeiger 2013; Jones 1953). The term “primary metabolites” refers to these macromolecules, which can include lipids, proteins, nucleic acids, and carbohydrates. Plants produce a vast and varied spectrum of organic chemicals, most of which don’t appear to have any direct role in the life processes. In contrast to animals, the majority of plant species are capable of diverting a large portion of their organic synthesis into the production of secondary metabolites or natural products (Hoffmann 2003; Bennett and Wallsgrove 1994a). Primary metabolites can be found all over the plant kingdom, but secondary metabolites are typically found in a sole species or a tiny handful of species. They are beneficial to plants because they protect them from herbivory. In addition, it aids in the prevention of infection caused by pathogens (Taiz et al. 2015; Hopkins and Hüner 1995). In this chapter, we will discuss many different classes of therapeutic phytoconstituents, as well as the distribution of certain plants across the many families of plants and some of the most important medicinal uses.

Secondary plant metabolites are organized into a variety of classes. Figure 2.1 depicts metabolism and the resultant metabolites, their participation in the synthesis

S. K. Singh · S. Patel (✉)
School of Life Sciences, Central University of Gujarat, Gandhinagar, India
e-mail: sunitap@cug.ac.in

© The Author(s), under exclusive license to Springer Nature Singapore Pte Ltd. 2024

D. Pooja, H. Kulhari (eds.), *Nanotechnology Based Delivery of Phytoconstituents and Cosmeceuticals*, https://doi.org/10.1007/978-981-99-5314-1_2

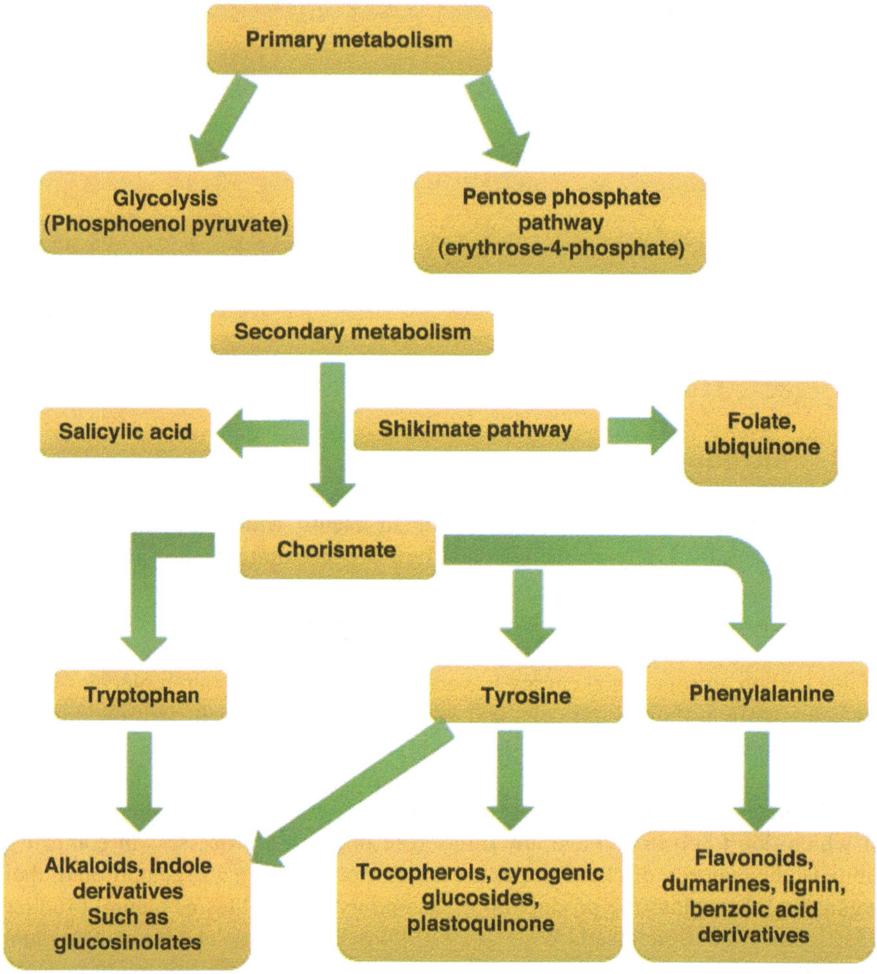

Fig. 2.1 Primary metabolism and secondary metabolism in plants

of other secondary metabolites. The pharmaceutical benefits of plants are due to their unique chemical structure (Bennett and Wallsgrove 1994b). It is possible to gain a better understanding of a plant's potential biomedical value by acquiring a thorough understanding of the chemical components that make up the plant. In 1910, Nobel Prize in Physiology or Medicine awardee Albrecht Kossel is credited with being the very first person to define the concept of secondary metabolite. After a period of 30 years, Czapek referred to them as completed products. According to him, these products are created through a process that he referred to as "secondary modifications," which includes deamination (Bourgaud et al. 2001; Sintupachee et al. 2020). This process originates in the nitrogen metabolism. Around the time that we call the center of the twentieth century, developments in analytical

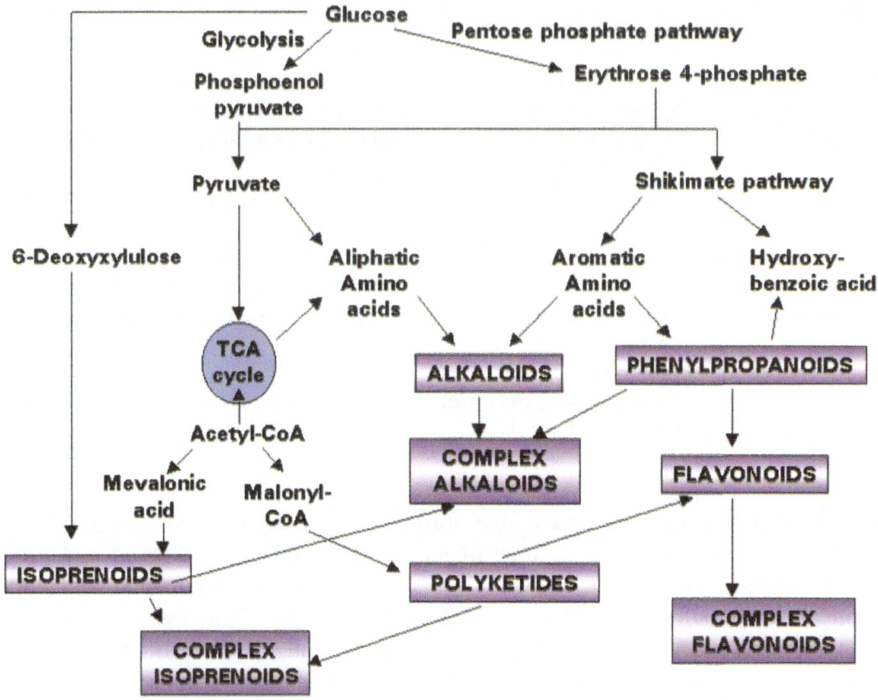

Fig. 2.2 Secondary metabolic gateways leading to the synthesis of secondary metabolites

techniques such as methods of investigation such as chromatography made it possible to recover an increasing number of these molecules (Bennett and Wallsgrove 1994a; Bourgaud et al. 2001). This served as the foundation for the establishment of the field of study known as phytochemistry. It has been evidenced that secondary metabolites possess a wide range of biological implications, which together form the scientific foundation for the use of botanicals in the conventional healers of a diverse range of ancient communities. Because they have been shown to be effective as antibiotics, antifungals, and antivirals, they are in a position to defend plants against disease-causing organisms (Venkataramaiah 2020). In addition to this, they are significant sources of UV-absorbing compounds, protecting the leaves from severe damage caused by exposure to light (Hussein and El-Anssary 2019). There are some herbs and fodder grasses, such as clover and alfalfa, that have been found to interact with animal fertility due to the expression of estrogenic characteristics (Hussein and El-Anssary 2019). Terpenes, phenolic compounds, glycosides, alkaloids, and saponins are major classes of secondary plant metabolites (Boyer and Liu 2004). Figure 2.2 shows how glucose molecule converts into different molecules leading to the synthesis of many metabolites via distinct pathways in the form of a simplified flow chart.

Secondary metabolites are not fundamental to the molecular structure or function of plant cells. They have been employed as folk cures, soaps, medications, essences

Table 2.1 Major classes of phytoconstituents and their applications

Class	Major phytoconstituents	Applications
Polyphenols	Quercetin, stilbenes, apigenin, catechins	Antioxidant, antibacterial, antifungal, anti-inflammatory, neuroprotective
Alkaloids	Codeine, morphine, heroin	Antifungal, predator aversion, aposematic colorations
Glycosides	Flavonoids, coumarin, anthraquinone, digitoxose	Treatment of heart failure, atrial arrhythmia
Phenylpropanoids	Coumarins, lignin	Anticoagulants, antimicrobial, antioxidant, anticancer, antidiabetic, renoprotective
Flavonoids	Flavones, flavonols, anthocyanidin, isoflavonoid	Nutraceutical, pharmaceutical, cosmetic, medicinal
Saponins	Medigenic acid glucoside, diosgenin glycoside	Detergent properties, defense mechanisms, surfactants, medicines, fish poisons

and other therapeutic essentials, dye stuffs, feed stocks for chemical industries (gums, resins, rubber), and a range of substances used to flavor food and drink from antiquity. Table 2.1 shows classes, major phytoconstituents, and their applications broadly (Elshafie et al. 2023). Majorly, they function as therapeutic phytochemicals in various aspects of human and animal's lives. Furthermore, many products of commercial benefit are procured from these phytochemicals on a very large scale utilizing advanced processing mechanisms and equipment (Shaalan et al. 2005; Johnson 2007; Molyneux et al. 2007).

2.2 Terpenes

Terpene comes from the term "turpentine," which originates from the old French word ter(e)bintb, which literally translates to "resin." Terpene is a component of turpentine. They are commonly found in plants and make up the bulk of the constituents that make up the essential oils that plants produce. Terpenes are generally lipophilic polymers of 5-carbon unit, which is derived from the mevalonate pathway. Terpenes are important components of many natural products that offer an organism a range of health benefits, and they play a number of different roles. Terpenes can be derived from a variety of plants, the most common of which are tea, Spanish sage, and citrus (e.g., lemon, mandarin). Terpenes have a broad variety of applications in the medical field; among these applications, antiplasmodial activity stands out due to the fact that its mode of action is strikingly similar to that of the largely prescribed antiplasmodial drug chloroquine. Particularly, monoterpenes have garnered a lot of attention due to the antiviral research conducted on them. Terpenes have the perspective to be used as antitumor and antidiabetic testing agents, which is significant considering the rising rates of cancer patients and diabetics in the modern world. In addition to these properties, terpenes enable a flexible administration route and the reduction of side impacts. Terpenes also have anti-inflammatory properties. In traditional natural medicine, certain terpenes were frequently used as active ingredients. Curcumin is an example of this type of terpene, and it possesses a wide variety of properties, including those that are oxidative, antitumor, astringent, antiprotozoal, abrasive, digestive, and diuretic. Curcumin has recently gained popularity in healthy foods, which has opened the door to a plethora of medical research options. This chapter provides a concise overview of the several terpenes, including their origins, biopharmaceutical attributes, mode of action, and current studies which are underway to design terpenes as a lead molecule for use in contemporary research medicine (Cox-Georgian et al. 2019). They are all derived, in a chemical sense, from isoprene units with five carbons and are assembled in a variety of ways (Sintupachee et al. 2020; Venkataramaiah 2020; Londonkar and Rajani 2021). The number of carbon atoms in a terpene determines its classification. In Table 2.2, classification of terpenes is given. A molecule contains isoprene units, and a prefix in the chemical name indicates the number of these units (Hussein and El-Anssary 2019). Terpene units are classified as hemiterpenes, monoterpenes, sesquiterpenes, diterpines, triterpenes, tetraterpenes, and polyterpenes in order of size (Paduch et al. 2007). Figure 2.3 shows the mevalonic acid pathway which is one of the key processes in the synthesis of secondary metabolites or the potent therapeutic phytochemicals specially terpenes.

2.2.1 Hemiterpenes

They are made up of a single unit of the isoprene molecule. Isoprene is the single known hemiterpene, but hemiterpenoids contain angelic acid, which are derived from *Angelica archangelica*, and isovaleric acid, which originates from *Vaccinium myrtillus*, as well as prenol, which comes from citrus fruit, tomato, and cranberry. Plants contain both of these acids (Builders 2018). Essential oils are the concentrated chemical essence of many plants, including citrus, mint, *Eucalyptus*, and different herbs (sage, thyme, etc.). Essential oils are complex mixes of alcohols, aldehydes, ketones, and terpenoids (essence in perfume). Essential oils have insect-repellent properties (Chung et al. 2013; Adorjan and Buchbauer 2010). Figure 2.4 shows structure of some common hemiterpenes.

2.2.2 Monoterpenes

Monoterpenes have the molecular formula $C_{10}H_{16}$ and are composed of two isoprene units and a ten-carbon chain structure. They are important components of essential oils or volatile oils derived from plants. *Lamiaceae*, *Rutaceae*, *Apiaceae*, and *Pinaceae* are only few of the plant families that feature these organisms. Only a handful of chemicals, including geraniol, are found in all plants, albeit only in minute quantities in their volatile exudations. They are classified into three groups. Monoterpenes have numerous pharmacological applications. Camphor and menthol, for example, are useful as anti-irritants, pain relievers, and anti-itching agents for

Table 2.2 Classification of terpenoids

Number of carbon atoms	Class	Example
5	Hemiterpenoid	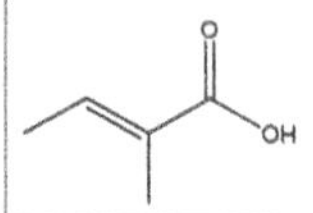 Tiglic acid
10	Monoterpenoid	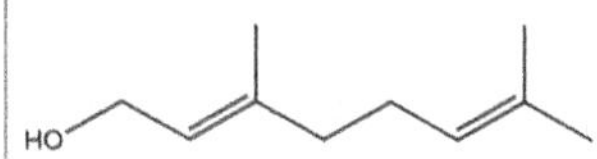 Geraniol
10	Cyclic monoterpenoid	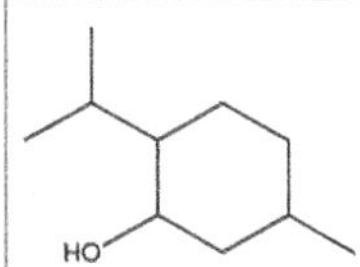 Menthol
15	Sesquiterpenoid	Farnesol 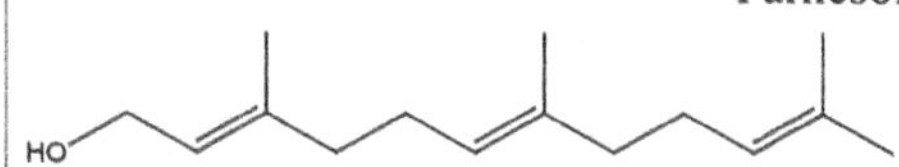
20	Diterpenoid	Phytol 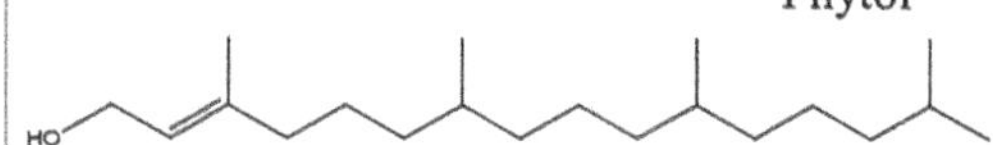
30	Triterpenoid	Squalene 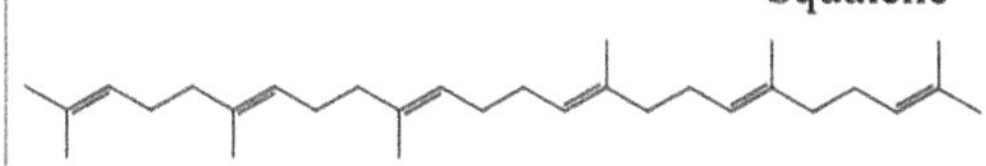
30	Triterpenoid	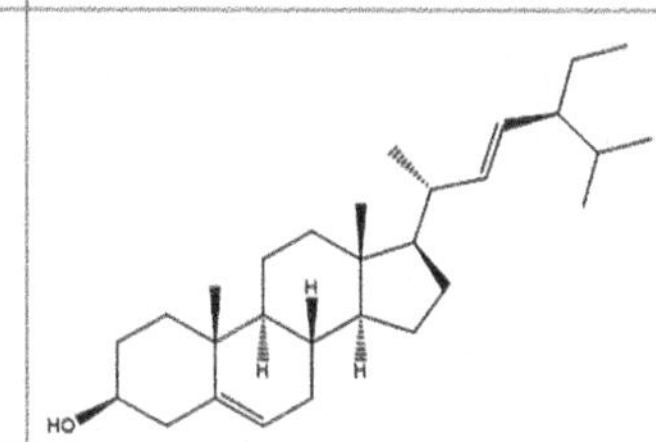 Stigmasterol
40	Tetraterpenoid	Lycopene 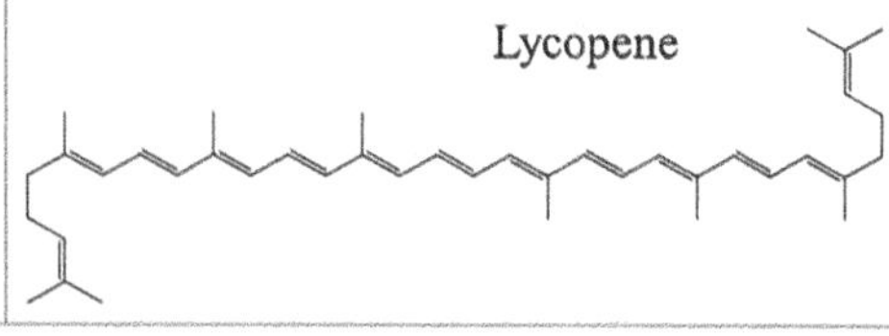

skin infections. They are also used to treat worms. A class of monoterpene glycosides has been shown to have a vasodilatory impact on the heart and femoral artery system (Kochar Kaur et al. 2019; Kandar 2021; Bergman et al. 2019). Terpenes' ability to inhibit the growth of microorganisms is intimately connected to the lipophilic nature of their constituents. Monoterpenes have a predominately

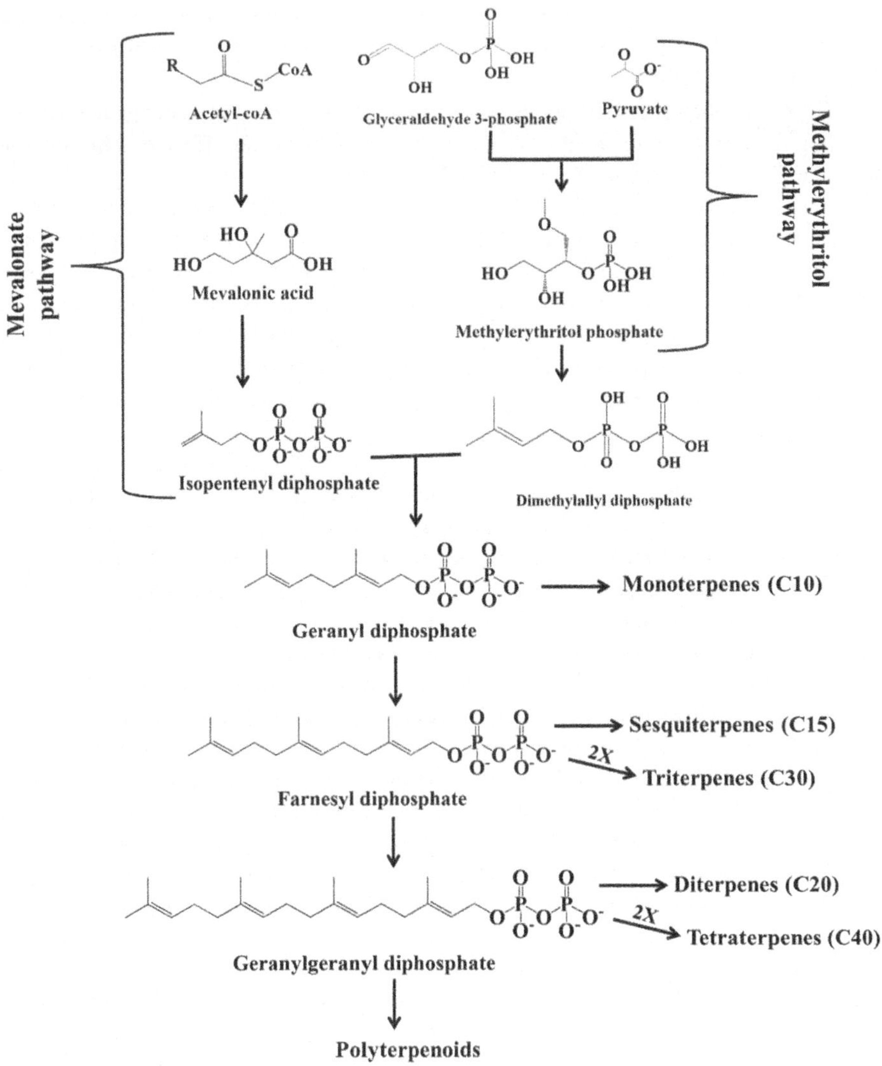

Fig. 2.3 Mevalonic acid pathway

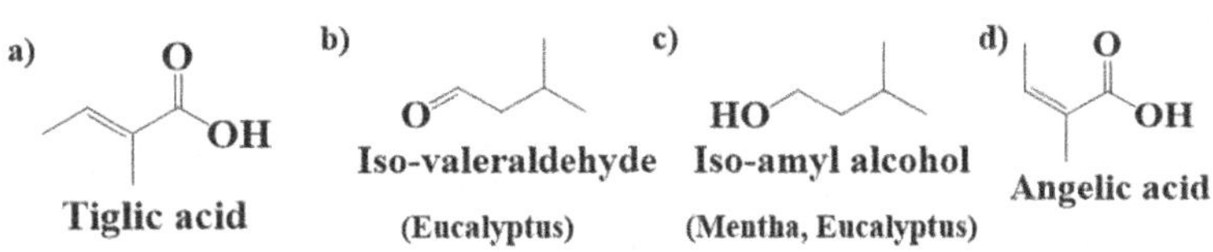

Fig. 2.4 Structure of hemiterpenes (**a**) tiglic acid, (**b**) iso-valeraldehyde, (**c**) iso-amyl alcohol, (**d**) angelic acid

disruptive effect on membrane structures, leading to membrane topological changes, an expansion of membrane permeability, and alterations in the respiratory chain (Trombetta et al. 2005).

The posttranslational isoprenylation of proteins that regulate cell growth is the most significant mechanism that monoterpenes can have an effect on. In general, prenylated proteins regulate cell growth and transformation; consequently, the antitumor activity of monoterpenes may be due to a disturbance in the protein prenylation (particularly in the case of Ras-regulated small GTP-binding proteins); this may be the case because prenylation is regulated by monoterpenes (Crowell 1997). Perillyl alcohol is a monocyclic monoterpene that occurs naturally and has been shown to have both preventative and therapeutic activity against a wide array of tumor types used in preclinical research. It has been demonstrated, using both in vitro and in vivo testing methods, that perillyl alcohol can cause apoptosis to occur in a wide variety of cancer cell types, including those of the pancreas, the mammary gland, the colon, and the liver. As a result of its activity, tumors that had been chemically induced or transplanted into animals completely regressed. Perillyl alcohol was investigated further as a potential carcinogenesis inhibitor after being exposed to ultraviolet light. It was discovered that suntan lotions that contained an increased amount of perillyl alcohol might reduce the risk of developing melanoma. Therefore, it is hypothesized that this monoterpene functions as an effective agent in preventing the development of skin carcinoma (Gupta and Myrdal 2004). Activation of the signaling pathway for transforming growth factor (TGF) is yet another mechanism by which perillyl alcohol exerts its chemotherapeutic effects. The TGF that is produced is in a dormant state and must be activated before it can be used. The activation of TGF is mediated by perillyl alcohol, which in turn leads to an increase in the synthesis of mRNA encoding its receptors. Perillyl alcohol causes activation of the TGF-signaling pathway, which is closely associated with increased synthesis of pro-apoptotic proteins (Bax, Bak, and Bad) but does not influence the expression of p53 or Bcl-2 (Ahn et al. 2003a). In addition, perillyl alcohol downregulates the cell cycle by acting through TGF. This has the effect of altering the production of cyclin and cyclin-dependent kinases, as well as the interactions between the two. As a consequence of this, it causes cells to undergo apoptosis and arrest in the G1 phase (Shi and Gould 2002; Ahn et al. 2003b). The stopping of the production of coenzyme Q by inhibition is an additional mechanism by which perillyl alcohol kills tumor cells (Ahn et al. 2003b; Ren and Gould 1998). CoQ is engaged in the process of transmembrane electron transfer in the plasma membrane, which contributes to the stability of extracellular ascorbate and helps to keep it stable. As a result of this, a decrease in the amount of CoQ found in cell membranes may impair cellular signal transduction and metabolism, and it may also induce tumor cells to undergo the process of apoptosis (Brea-Calvo et al. 2006). It is essential to emphasize the fact that the enantiomeric derivatives of these monoterpenes do not exert any influence on the metabolic processes, proliferation, or differentiation of cancer cells (Gelb et al. 1995). Because of this, the stereoisomeric configuration of natural chemotherapeutic agents needs to be taken into consideration during the preparation process. Essential oils are produced in the majority of plants by glandular trichomes,

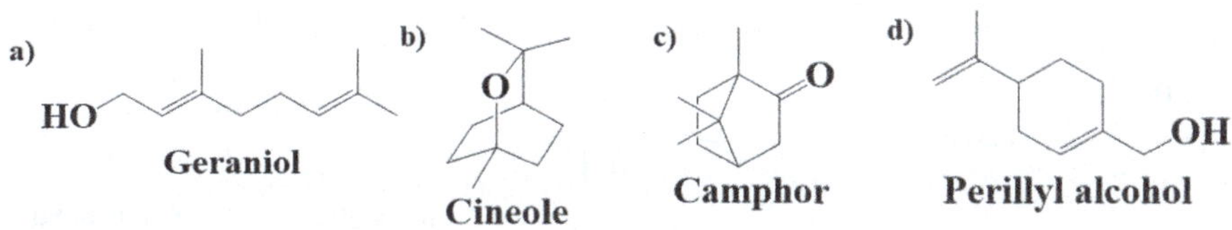

Fig. 2.5 Structure of monoterpenes: (**a**) geraniol, (**b**) cineole, (**c**) camphor, (**d**) perillyl alcohol

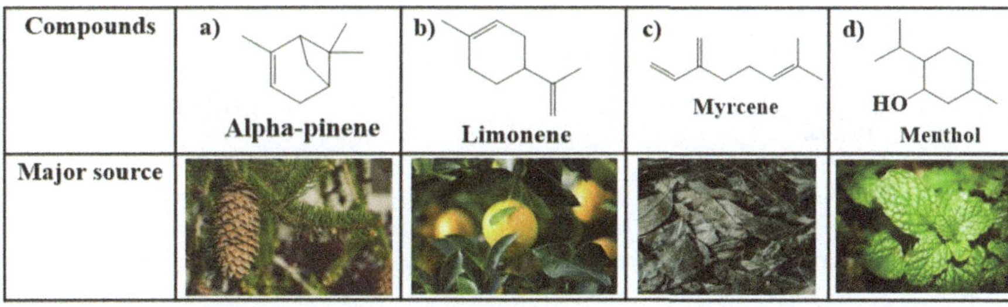

Fig. 2.6 Sources of essential oils with insecticidal properties: (**a**) alpha-pinene, (**b**) limonene, (**c**) myrcene, (**d**) menthol (images adopted from pexels.com)

also known as hairs that are located on the surface of the leaf. Citrus peels also contain glands (Çakmakçı et al. 2020). Figure 2.5 shows monoterpenes like geraniol, camphor, etc. with their structure.

There is a high concentration of pinene and myrcene in the resins of certain conifers. The essential oil of peppermint contains a high concentration of menthol as its primary component (*Mentha piperita*). Pinene, limonene, and menthol are all essential oils that can kill insects (Patel et al. 2007). Figure 2.6 shows monoterpenes along with their major source in pictorial form.

2.2.3 Sesquiterpenes

Sesquiterpenes are large terpene molecules. The structure of sesquiterpenes can be broken down into three distinct isoprene units (C15), and they can either be acyclic (farnesol) or contain rings, in addition to undergoing a variety of other modifications (cadinene). There was evidence of activity against mycobacteria from sesquiterpenes as well as their lactones. In addition, some of the sesquiterpene lactones, like costunolide and parthenolide, among others, displayed activity against mycobacteria (Copp 2003). Phospholipase C was stymied by a number of sesquiterpenes, the most effective of which was petasin (Thomet et al. 2001; Cornwell et al. 1996). Figure 2.7 shows structure of sesquiterpenes like farnesol, cadinene, etc.

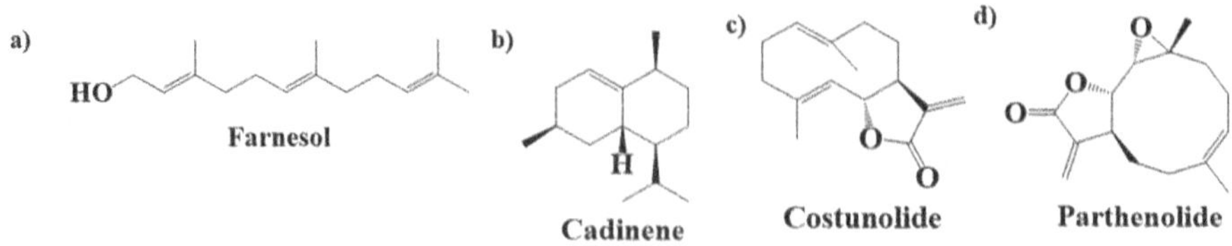

Fig. 2.7 Structure of sesquiterpenes: (**a**) farnesol, (**b**) cadinene, (**c**) costunolide, (**d**) parthenolide

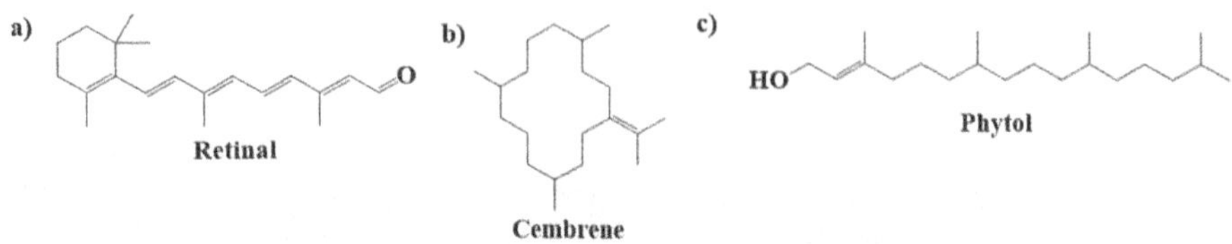

Fig. 2.8 Structure of diterpenes: (**a**) retinal, (**b**) cembrene, (**c**) phytol

2.2.4 Diterpenes

Diterpenes are made up of four separate units of isoprene (C20). The formation of diterpenes involves the condensation of four isoprene units, which can occur either via the mevalonate or the deoxyxylulose phosphate metabolic pathways. This results in a structurally heterogeneous division of C20 natural chemicals that are widely distributed in nature (Lanzotti 2013). Cembrene and retinal are two examples of diterpenes. Retinal is a fundamental chromophore that plays a role in the conversion of light into visual signals. Cembrene is an example of a diterpene. Triterpenes are made up of six different units of isoprene (C30). A significant amount of squalene can be found in shark liver oil. Squalene is a linear triterpene. A large group of biologically active substances is comprised of plant triterpenes and the derivatives of these compounds. Tetraterpenes have a total of eight isoprene units in their structure (C40). Among the tetraterpenes with significant biological relevance are carotenes and lycopene, which is found in tomatoes. Polyterpenes are long chains of many isoprene units strung together in a chain. Polyisoprene with multiple cis-double bonds is what natural rubber is made of (Yadav et al. 2014). Figure 2.8 contains the structure of well-known diterpenes such as retinal, cembrene, and phytol.

2.2.5 Triterpenes

Triterpenes are one of the crucial elements that directly contribute to the stability of cell membrane. Moreover, they are the controllers of permeability as well as the reactions that enzymes carry out. Triterpenes are made up of six different units of isoprene (C30). A significant amount of squalene can be found in shark liver oil. Squalene is a linear triterpene. A large group of biologically active substances is comprised of plant triterpenes and the derivatives of these compounds (de Carvalho

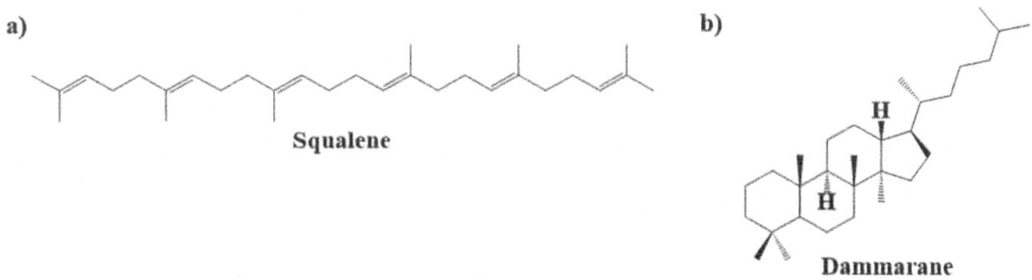

Fig. 2.9 Structure of triterpenes: (**a**) squalene, (**b**) dammarane

and da Fonseca 2006). Additionally, several plant triterpenes demonstrated antitumor activity when tested in vitro. It has been demonstrated that betulinic acid can cause apoptosis to occur in a variety of human tumor cells, including melanoma and glioma cells. Alterations in the mitochondrial permeability transition potential were responsible for triggering apoptosis in the cells. This was accomplished in conjunction with the upregulation of pro-apoptotic proteins (Bax) and the decrease in expression of bcl-2. It was also proven that betulinic acid serves as an inhibitor of topoisomerase I, which is an enzyme in the nucleus that is responsible for catalyzing changes in the topology of DNA. This was one of the more interesting findings of the study (Chowdhury et al. 2002; Fulda et al. 1998). As a direct result, this causes apoptosis to occur in the affected cells. Research has demonstrated that these plant triterpenes impede the formation of leukemia cells. Other fruits also contain natural wax (Cipak et al. 2006). Induction of production of nitric oxide and tumor necrosis factor (TNF) led to a suppression of the growth of numerous transplantable tumors in animals. This was achieved by blocking the production of NO and TNF (Liu 1995; Patocka 2003). These triterpenoids and their derivatives have an effect on the development of tumors at a number of different stages, inhibiting the process of tumor initiation and promotion while also inducing tumor cell differentiation and apoptosis. In addition to this, they are powerful inhibitors of the processes of angiogenesis, invasion, and metastasis that are involved in the spread of tumor cells (Liu 2005). Oleanolic acid and ursolic acid, both members of the triterpene family, exhibited antimycobacterial activity against *M. intracellulare* (Copp 2003) as well as bacteria that are Gram-positive and Gram negative (Mallavadhani et al. 2004). Figure 2.9 shows structure of triterpenes squalene and dammarane. Squalene is a straight chain while dammarane is a 4-hexane ring structure.

2.2.6 Tetraterpenes

Tetraterpene pigments can be found in a wide variety of organisms, including photosynthetic bacteria, certain archaea and fungal species, algae, plants, and even animals. Tetraterpenes are naturally occurring compounds that include pigments. Their backbones are linear C40 hydrocarbons (Cho et al. 2018). Figure 2.10 shows

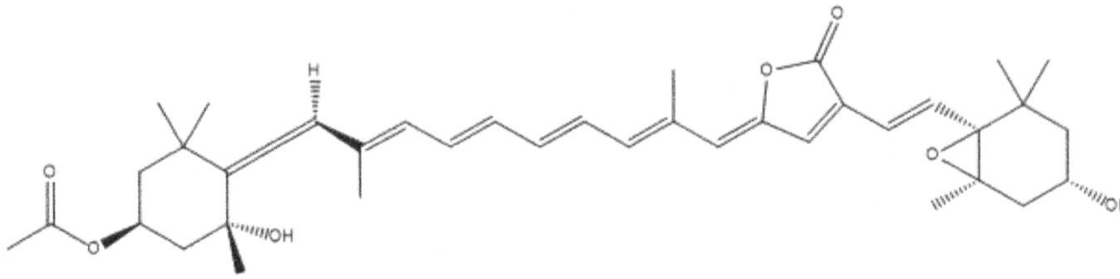

Fig. 2.10 Structure of tetraterpene: peridinin

the structure of peridinin, a widely distributed tetraterpene in all the kingdoms like bacteria, fungi, plant, and animal.

2.2.7 Polyterpenes

The hydrocarbon molecules known as polyterpenes make up a huge class and are found in every known living thing. Their responsibilities have now been enlarged, despite the fact that they were originally defined as having only physiological functions, such as pigments, hormones, protein massive influence molecules, and energy transduction molecules (Satoh et al. 2008; Kiyama 2017). It has been demonstrated that several of the most well-known polyterpenes, such as sterols and hopanoids, can operate as regulators of the properties of membranes. At least some species of the genus *Archaea* are capable of synthesizing each of the four groups of polyterpenes. Even though there isn't a lot of evidence, what there is suggests that each of the four classes of molecules has the potential to act as a membrane regulator in archaea and that these regulators may play a role in how archaea react to various stresses or in how different groups of archaea interact with one another (Cario et al. 2015; Matsuno et al. 2009). Figure 2.11 comprises the structure of polyterpenes like pinene, diadinoxanthin, and dipentene.

2.3 Phenolic Compounds

Polyphenols, which are natural molecules that have significant antioxidant effects and chemical similarities to phenolic substances, can only be synthesized by plants because only plants have this capability. The majority of these molecules or chemical classes can be found in nutrition from fresh produce, green tea, and whole grains (Singla et al. 2019; Rice-Evans et al. 1996). Polyphenols are well-known phenolic systems with at least two phenyl rings and one or more hydroxyl substituents. One-third of the total intake is made up of phenolic acids, while the other two-thirds is made up of flavonoids. The most prevalent types of dietary flavonoids are flavanols, which consist of catechins and proanthocyanidins combined, anthocyanins, and their oxidation products. Fruits and drinks like beer, wine, tea,

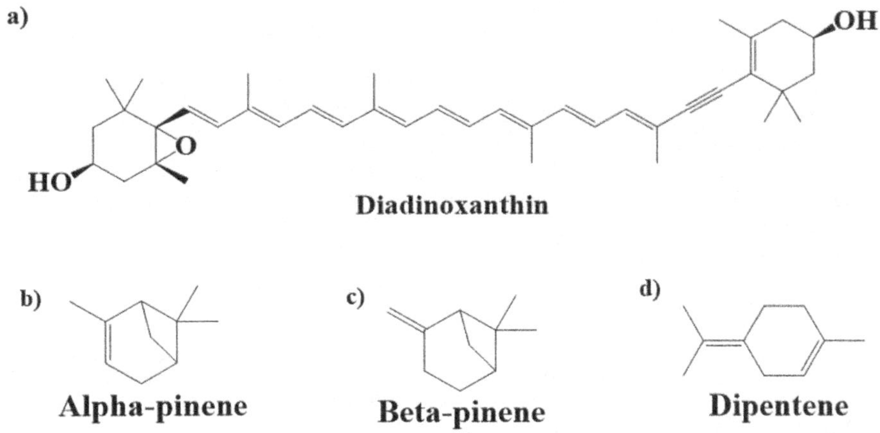

Fig. 2.11 Structure of polyterpenes: (**a**) alpha-pinene, (**b**) beta-pinene, (**c**) dipentene

coffee, and chocolate are the primary food sources of polyphenols, followed by vegetables, dry legumes, and cereals (Singla et al. 2019). The significance of polyphenols in the treatment of cardiovascular disease, osteoporosis, neurodegenerative illness, cancer, and diabetes has been investigated (Rasouli et al. 2017; Attanayake n.d.). Many fruits and vegetables, including kale, onions, and broccoli, contain phenolic acids as important polyphenols. Fruits such as pomegranate juice, tea extracts, and grape extracts were witnessed to diminish atherosclerotic lesions. The preventive impact of polyphenols in health and illness is well-recognized in foods that are rich in antioxidants and phytochemicals when ingested routinely (Polat 2018). Moreover, polyphenols exhibit the pro-oxidant feature, which is detrimental to the cell's metabolic activity. It may also entail preventing cell proliferation and death (Rasouli et al. 2017; Xu 2013).

Due to increased understanding, public interest in natural products has grown substantially over the past several decades. Among numerous bioactive chemicals, polyphenols are considered as an exceptional source of a wide range of compounds with extraordinarily diverse compositions. The biological activity and possible applications of these chemicals have been the subject of a considerable amount of research. Polyphenols are plant-derived secondary metabolites that are widespread. Because of their prevalence in plant-based foods and substantial evidence of a negative association between their consumption and cancer, diabetes, and cardiovascular disease, these chemicals have achieved a prominent position.

Polyphenols are the most common and widely spread bunch of bioactive molecules. Two general classes of polyphenols include flavonoids and phenolic acids. The most prevalent antioxidants in our diet are polyphenols. They inhibit the oxidation of low-density lipoprotein, which is the fundamental mechanism underlying endothelial lesions in atherosclerosis (Abbas et al. 2017). Figure 2.12 shows structure of phenolic compounds like quercetin. Figure 2.13 depicts the

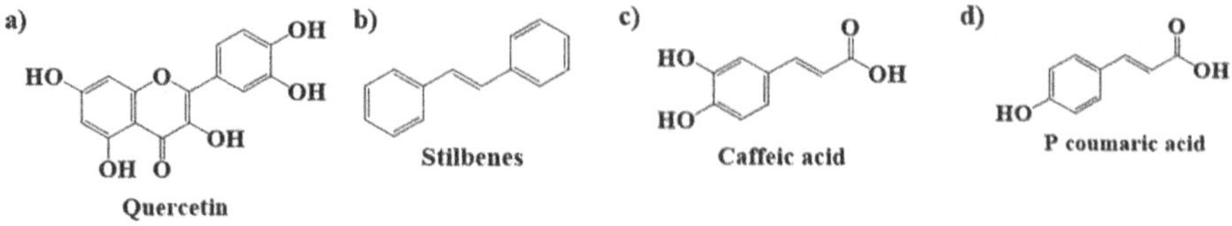

Fig. 2.12 Structure of phenolic compounds: (**a**) quercetin, (**b**) stilbenes, (**c**) caffeic acid, (**d**) P coumaric acid

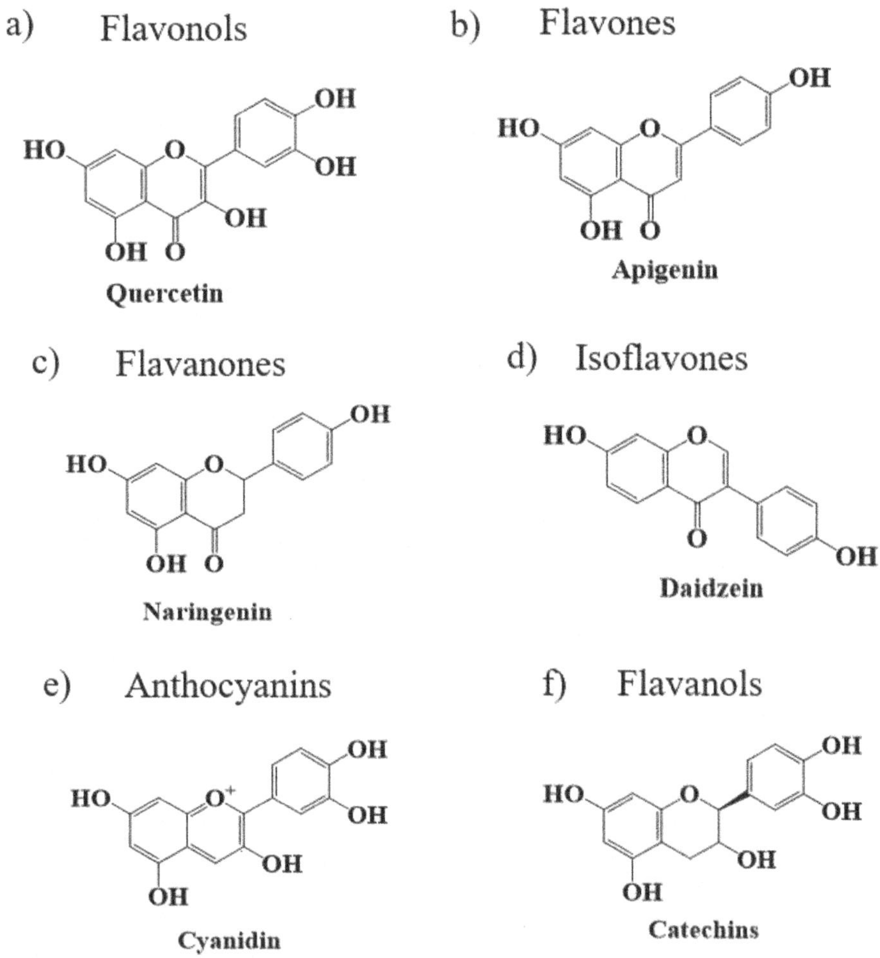

Fig. 2.13 Structure of flavonoids: (**a**) flavonols, (**b**) flavones, (**c**) flavanones, (**d**) isoflavones, (**e**) anthocyanins, (**f**) flavanol

Table 2.3 Main classes of flavonoids

Flavonoids	Dietary sources
Flavonols	Fruits, vegetables, tea, cocoa, pulses, spices, beer
Flavones	Broccoli, onion, cherry, tomato, apple
Flavanol	Tea, red wine, chocolates, grape peels, apple peels, blueberry
Flavanones	Pulses, tomato, aromatic plants
Isoflavonoid	Leguminous plants, soy, nuts
Anthocyanidin	Flower petals, fruits, vegetables, black rice, mixture of grains
Proanthocyanidin	Berries, drupes, wine, pulses, tea, nuts, pommes

Table 2.4 Main classes of phenolic acids

Phenolic acids	Dietary sources
Hydroxybenzoic acid	Nuts, red wine, tea, fruit tea
Hydroxycinnamic acids	Coffee, cherries, pear
Stilbenes	Red wine, white wine, grapes, berry fruits, strawberries
Lignans	Citrus, fruit, olive, oranges, garlic
Hydroxyphenylacetic acid	Olives, wine, beer
Hydroxybenzaldehydes	Olives, wine
Tyrosols	Olives, red wine, beer

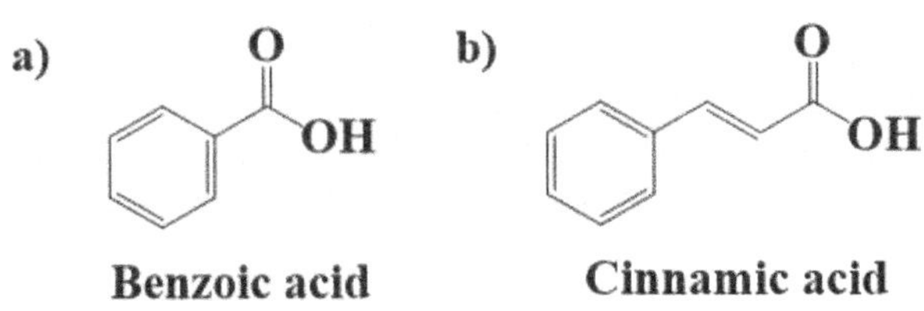

Fig. 2.14 Structure of phenolic acids: (**a**) benzoic acid, (**b**) cinnamic acid

structure of flavonoids. Table 2.3 shows the main classes of flavonoids and their most familiar dietary sources.

Phenolic acids are usually benzoic acid or cinnamic acid derivatives. The flavonoids in fruit peel vary by species and light exposure. Polyphenols have several molecular and biological functions due to their structure. Polyphenols protect hypercholesterolemia, hyperglycemia, hyperlipidemia, and cancer. C1–C6 and C3–C6 backbones are found in cinnamic acid and benzoic acid derivatives, respectively. Free-form phenolic acids can be found in vegetables and fruits. However, the hull, bran, and seed of many plants contain phenolic acids in bound form (Hussain et al. 2005; Rudra et al. 2021). Table 2.4 shows main classes of phenolic acids and their common dietary sources. Figure 2.14 shows structure of two phenolic acids, that is, benzoic acid and cinnamic acid.

2.4 Glycosides

The utility of cardiac glycosides for the treatment of cardiac diseases dates back to ancient time. On the other hand, it has been demonstrated that cardenolides may have anticancer properties, according to preclinical investigations and two phase I studies. One possible mechanism underlying these anticancer effects is an intracellular shift toward higher levels of sodium and calcium at the expense of potassium (Slingerland et al. 2013; Calderón-Montaño et al. 2014; Friend 1962). The production of IL-8 and the pathway involving TNF and NF-κB are both inhibited by intracellular acidification. DNA topoisomerase II inhibition along with the activation of the Src kinase pathway could be another mechanisms for the anticancer activities of these compounds. In order to combat cancer, three distinct cardiac glycosides have been produced up till this day. These cardiac glycosides were put through the first phase of the clinical study in order to ascertain the maximum tolerated dose as well as the dose that caused dose-limiting toxicities. It is imperative that further research be conducted on this distinct class of anticancer drugs in order to determine the probable benefit of these medications in the treatment of cancer (Deepak et al. 1996; Heller 1990). One of the principle cardenolides is digitoxose (Wiesner et al. 1985). Figure 2.15 shows structure of glycosides like coumarin, digitoxose, etc.

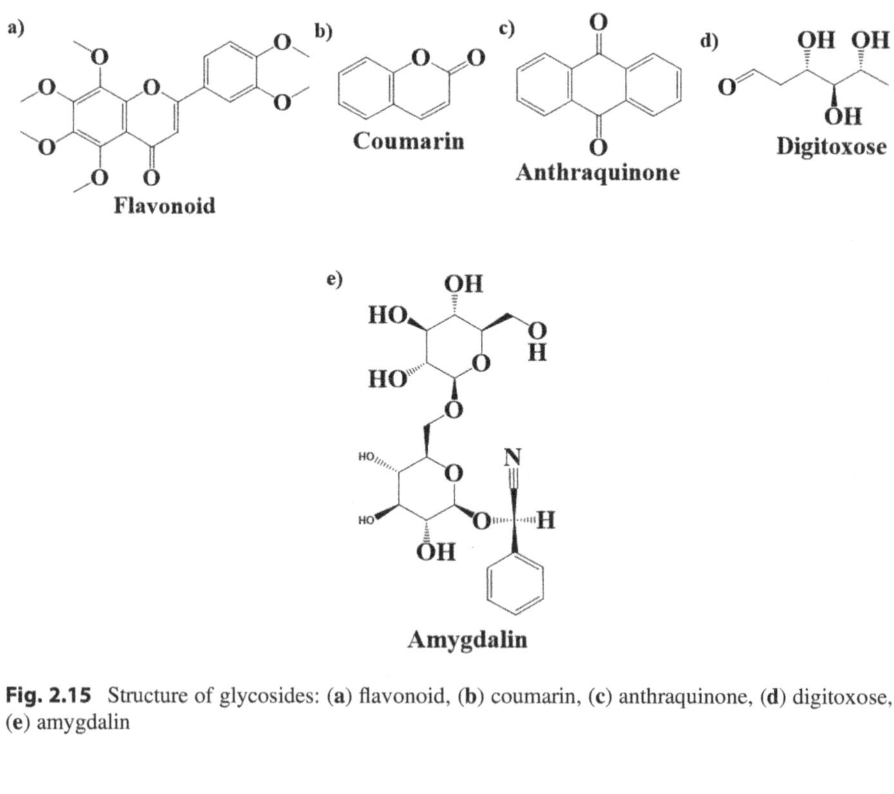

Fig. 2.15 Structure of glycosides: (**a**) flavonoid, (**b**) coumarin, (**c**) anthraquinone, (**d**) digitoxose, (**e**) amygdalin

2.5 Cyanogenic Glycosides

Plants are capable of producing compounds that are able to release poisons such as hydrogen cyanide and prussic acid (HCN). The cyanogenic glycosides are considered to be a part of the natural products that come from plants as well as the products of the secondary metabolism. These compounds have a sugar moiety in addition to an aglycone as one of its components (mostly d-glucose) (Conn 1969, 1981). There are at least 2500 different taxa that contain cyanogenic glycosides, and many of these cyanogenic glycoside-containing taxonomic families include those of the *Fabaceae*, *Rosaceae*, *Linaceae*, and Compositae, among others. The presence of cyanogenic glycosides, often known as CGs, is found in a substantial portion of the plant kingdom (Francisco and Pinotti 2000). There is not one specific mechanism for the genetic control of cyanogenesis; instead, there is heterogeneity among plants in the amount of HCN that is produced (Vetter 2000; Cressey and Reeve 2019). There are two stages involved in the enzymatic degradation of cyanogenic glycosides (Vetter 2000). Figure 2.16 shows the chemical reaction for the production of HCN.

2.6 Alkaloids

Phytochemicals that are members of the alkaloid family are distinguished examples of specialized metabolites that are endowed with an extensive number of biological activities. These naturally occurring nitrogenous compounds serve as a sort of armor for plants, helping them to defend themselves against a wide variety of environmental stresses. Both conventional and alternative medical practices have been able to tap into the possibility that these chemical molecules have as treatments for a broad range of illnesses. This can be accomplished by conducting research (Bhambhani et al. 2021; Khan et al. 2017).

There are three places where alkaloids come from: biogenic amines, the tetrodotoxin class of water-soluble alkaloids, and lipid-soluble alkaloids. But the eggs and larvae of some salamanders and toads are poisonous and unpleasant because they contain tetrodotoxins and bufodienolides. These chemicals are only found in the skin of adult amphibians, but they can be found in other parts of the body as well. Most substances can only be found in the skin of adult amphibians. The defensive value of these diverse metabolites is unmistakably demonstrated by predator aversion, as well

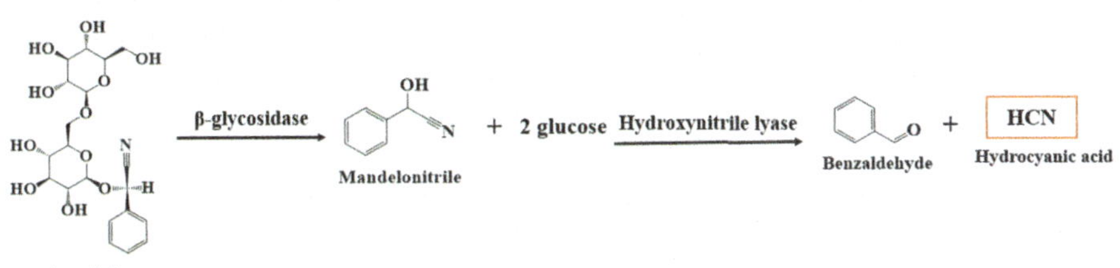

Fig. 2.16 Production of HCN

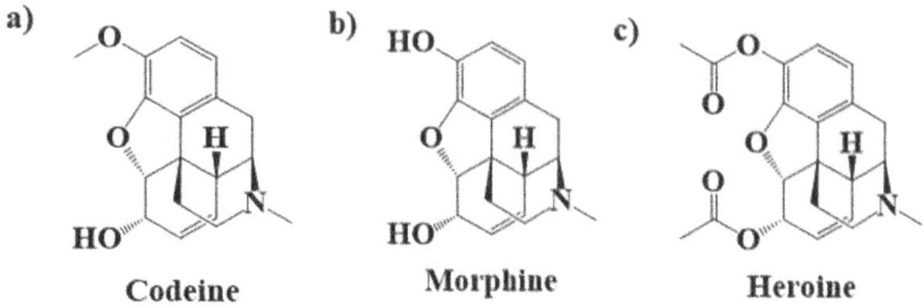

Fig. 2.17 Structure of alkaloids: (**a**) codeine, (**b**) morphine, (**c**) heroin

as a variety of antipredator behaviors and aposematic colorations. This is the case regardless of whether or not these metabolites are generated mainly (like alkaloids) or secondarily (like certain peptides and biogenic amines) for this function. Alkaloids are prime examples of the former (Daly et al. 1987). Opium is a latex gum that contains a mixture of over 20 distinct alkaloids, including morphine, codeine, and papaverine. These alkaloids are responsible for opium's narcotic effects (Shetge et al. 2020). Figure 2.17 shows the structure of alkaloids: codeine, morphine, and heroine.

Codeine and morphine are two types of naturally occurring alkaloids that can be extracted from the seed capsules of the poppy (*Papaver somniferum*). The acetylation of morphine leads to the production of the semisynthetic alkaloid known as heroin. Codeine is frequently utilized in the medical community as both a cough suppressant and a local anesthetic. Morphine's primary function is that of an analgesic, or pain reliever (Ceder and Jones 2001; Maurer et al. 2006).

2.7 Saponins

Saponins are a broad set of molecules that can be found in a large variety of plant-based foods and have a structure that consists of a triterpene or steroid aglycone and one or more sugar chains. Saponins are extensively distributed across the plant kingdom. Saponins have commercial importance because of the growing demand for natural products and their surfactant properties. They are used more and more in the food, personal care products, and pharmaceutical industries. It is necessary to create new methods of processing in order to meet the processing problems provided by the complexities of these products in order to exploit their full potential in the commercial market (Güçlü-Ustündağ and Mazza 2007; Sharma et al. 2021). Figure 2.18 shows the structure of medigenic acid glucoside (triterpenoid saponin) and diosgenin glycoside which is a steroidal saponin.

Figure 2.19 presents a summary of the pathways originating from photosynthesis and proceeding toward formation of several therapeutic phytoconstituents.

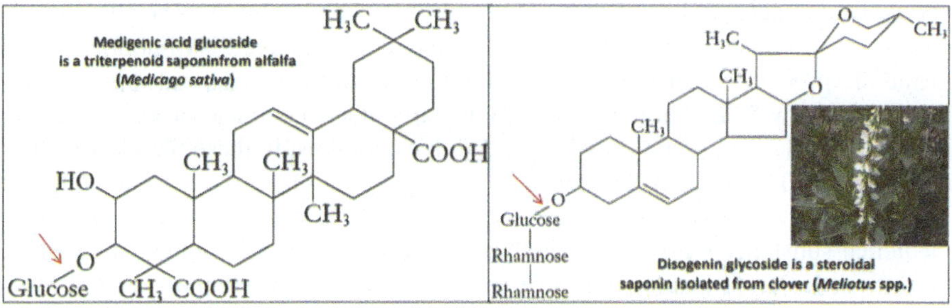

Fig. 2.18 Medigenic acid glucoside and diosgenin glycoside

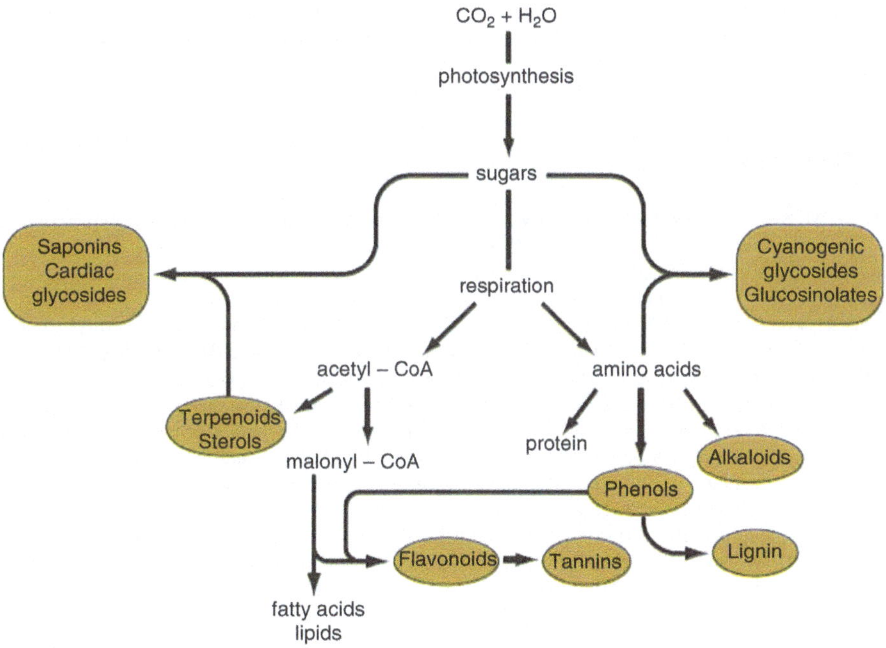

Fig. 2.19 Formation of phytochemicals or secondary metabolites from primary metabolites

2.8 Conclusion

Secondary metabolites, often known as SM or the therapeutic phytochemicals, are substances that are produced by a cell or organism but are not required for the cell's or organism's continued existence. However, secondary metabolites do play a role in the way that the cell or organism engages with its surrounding environment. These chemical plant hormones are typically involved in the process by which plants fight themselves against biotic or abiotic stressors. Secondary metabolites come from a variety of distinct families of metabolites and have the potential to be significantly induced in response to stressors. Primary metabolites are incredibly important to the metabolic process because they play a function in both feeding and reproduction. There are a few SMs that are employed as particularly chemical products, such as medications, flavors, scents, pesticides, and dyes; as a result, these SMs have a high market value. These newly developed technologies will serve to extend and improve the enduring utility of higher plants as renewable chemical sources, especially medicinal substances. It is anticipated that the effective biotechnological manufacture of particular, valuable, and as of yet unidentified plant compounds will result from the continuance and development of activities in this sector.

Acknowledgments Author Sanju Kumari Singh acknowledges and thanks UGC for the PhD fellowship, Dr. Sunita Patel acknowledges the Central University of Gujarat for providing the infrastructure and all necessary facilities. ChemDraw professional 15.1 was used for generating structures of compounds.

References

Abbas M, Saeed F, Anjum FM, Afzaal M, Tufail T, Bashir MS et al (2017) Natural polyphenols: an overview. Int J Food Prop 20(8):1689–1699

Adorjan B, Buchbauer GJF (2010) Biological properties of essential oils: an updated review. Flavour Fragr J 25(6):407–426

Ahn KJ, Lee CK, Choi EK, Griffin R, Song CW, Park HJ (2003a) Cytotoxicity of perillyl alcohol against cancer cells is potentiated by hyperthermia. Int J Radiat Oncol Biol Phys 57(3):813–819

Ahn K-J, Lee CK, Choi EK, Griffin R, Song CW, Park HJ (2003b) Cytotoxicity of perillyl alcohol against cancer cells is potentiated by hyperthermia. Int J Radial Oncol Biol Phys 57(3):813–819

Alamgir A (2018) Phytoconstituents—active and inert constituents, metabolic pathways, chemistry and application of phytoconstituents, primary metabolic products, and bioactive compounds of primary metabolic origin. In: Therapeutic use of medicinal plants and their extracts, vol 2. Springer, Cham, pp 25–164

Attanayake A (n.d.) DIETARY POLYPHENOLS: AN OVERVIEW

Bennett RN, Wallsgrove RM (1994a) Secondary metabolites in plant defence mechanisms. New phytologist 127(4):617–633

Bennett RN, Wallsgrove RM (1994b) Secondary metabolites in plant defence mechanisms. New Phytol 127(4):617–633

Bergman ME, Davis B, Phillips MA (2019) Medically useful plant terpenoids: biosynthesis, occurrence, and mechanism of action. Molecules 24(21):3961

Bhambhani S, Kondhare KR, Giri AP (2021) Diversity in chemical structures and biological properties of plant alkaloids. Molecules 26(11):3374

Bourgaud F, Gravot A, Milesi S, Gontier E (2001) Production of plant secondary metabolites: a historical perspective. Plant Sci 161(5):839–851

Boyer J, Liu RH (2004) Apple phytochemicals and their health benefits. Nutr J 3(1):1–15

Brea-Calvo G, Rodríguez-Hernández A, Fernández-Ayala DJ, Navas P, Sánchez-Alcázar JA (2006) Chemotherapy induces an increase in coenzyme Q10 levels in cancer cell lines. Free Radic Biol Med 40(8):1293–1302

Builders PF (2018) Introductory chapter: introduction to herbal medicine. In: Herbal medicine. IntechOpen, London

Çakmakçı R, Mosber G, Milton AH, Alatürk F, Ali B (2020) The effect of auxin and auxin-producing bacteria on the growth, essential oil yield, and composition in medicinal and aromatic plants. Curr Microbiol 77(4):564–577

Calderón-Montaño JM, Burgos-Morón E, Orta ML, Maldonado-Navas D, García-Domínguez I, López-Lázaro M (2014) Evaluating the cancer therapeutic potential of cardiac glycosides. Biomed Res Int 2014:794930, 1

Cario A, Grossi V, Schaeffer P, Oger PM (2015) Membrane homeoviscous adaptation in the piezo-hyperthermophilic archaeon Thermococcus barophilus. Front Microbiol 6:1152

Ceder G, Jones AW (2001) Concentration ratios of morphine to codeine in blood of impaired drivers as evidence of heroin use and not medication with codeine. Clin Chem 47(11): 1980–1984

Cho KS, Shin M, Kim S, Lee SB (2018) Recent advances in studies on the therapeutic potential of dietary carotenoids in neurodegenerative diseases. Oxidative Med Cell Longev 2018:4120458

Chowdhury AR, Mandal S, Mittra B, Sharma S, Mukhopadhyay S, Majumder HK (2002) Betulinic acid, a potent inhibitor of eukaryotic topoisomerase I: identification of the inhibitory step, the major functional group responsible and development of more potent derivatives. Med Sci Monit 8(7):BR254–BR265

Chung SK, Seo JY, Lim JH, Park HH, Yea MJ, Park HJ (2013) Microencapsulation of essential oil for insect repellent in food packaging system. J Food Sci 78(5):E709–EE14

Cipak L, Grausova L, Miadokova E, Novotny L, Rauko P (2006) Dual activity of triterpenoids: apoptotic versus antidifferentiation effects. Arch Toxicol 80(7):429–435

Conn EE (1969) Cyanogenic glycosides. J Agric Food Chem 17(3):519–526

Conn EE (1981) Cyanogenic glycosides. 7:479–500

Copp BR (2003) Antimycobacterial natural products. Nat Prod Rep 20(6):535–557

Cornwell PA, Barry BW, Bouwstra JA, Gooris GS (1996) Modes of action of terpene penetration enhancers in human skin; differential scanning calorimetry, small-angle X-ray diffraction and enhancer uptake studies. Int J Pharm 127(1):9–26

Cox-Georgian D, Ramadoss N, Dona C, Basu C (2019) Therapeutic and medicinal uses of terpenes. In: Joshee N, Dhekney SA, Parajuli P (eds) Medicinal plants: from farm to pharmacy. Springer International Publishing, Cham, pp 333–359

Cressey P, Reeve J (2019) Metabolism of cyanogenic glycosides: a review. Food Chem Toxicol 125:225–232

Crowell PL (1997) Monoterpenes in breast cancer chemoprevention. Breast Cancer Res Treat 46(2):191–197

Daly JW, Myers CW, Whittaker N (1987) Further classification of skin alkaloids from neotropical poison frogs (Dendrobatidae), with a general survey of toxic/noxious substances in the amphibia. Toxicon 25(10):1023–1095

de Carvalho CC, da Fonseca MM (2006) Biotransformation of terpenes. Biotechnol Adv 24(2): 134–142

Deepak D, Srivastava S, Khare NK, Khare A (1996) Cardiac glycosides Fortschritte der Chemie organischer Naturstoffe = Progress in the chemistry of organic natural products Progres dans la chimie des substances organiques naturelles. 69:71–155

Elshafie HS, Camele I, Mohamed AA (2023) A Comprehensive review on the biological, agricultural and pharmaceutical properties of secondary metabolites based-plant origin. Int J Mol Sci 24(4):3266

Francisco IA, Pinotti MH (2000) Cyanogenic glycosides in plants. Braz Arch Biol Technol 43:487–492

Friend DG (1962) Cardiac glycosides. N Engl J Med 266:402–404; concl, 300

Fulda S, Scaffidi C, Susin SA, Krammer PH, Kroemer G, Peter ME et al (1998) Activation of mitochondria and release of mitochondrial apoptogenic factors by Betulinic acid. J Biol Chem 273(51):33942–33948

Gelb MH, Tamanoi F, Yokoyama K, Ghomashchi F, Esson K, Gould MN (1995) The inhibition of protein prenyltransferases by oxygenated metabolites of limonene and perillyl alcohol. Cancer Lett 91(2):169–175

Güçlü-Ustündağ O, Mazza G (2007) Saponins: properties, applications and processing. Crit Rev Food Sci Nutr 47(3):231–258

Gupta A, Myrdal PB (2004) Development of a perillyl alcohol topical cream formulation. Int J Pharm 269(2):373–383

Heller M (1990) Cardiac glycosides. New/old ideas about old drugs. Biochem Pharmacol 40(5): 919–925

Hoffmann D (2003) Medical herbalism: the science and practice of herbal medicine. Simon and Schuster, New York, NY

Hopkins WG, Hüner NP (1995) Introduction to plant physiology

Hussain T, Gupta S, Adhami VM, Mukhtar H (2005) Green tea constituent epigallocatechin-3-gallate selectively inhibits COX-2 without affecting COX-1 expression in human prostate carcinoma cells. Int J Cancer 113(4):660–669

Hussein RA, El-Anssary AA (2019) Plants secondary metabolites: the key drivers of the pharmacological actions of medicinal plants. Herb Med 1(3)

Johnson IT (2007) Phytochemicals and cancer. Proc Nutr Soc 66(2):207–215

Jones ME (1953) Albrecht Kossel, a biographical sketch. Yale J Biol Med 26(1):80

Kandar CC (2021) Secondary metabolites from plant sources. In: Bioactive natural products for pharmaceutical applications. Springer, Cham, pp 329–377

Khan H, Mubarak MS, Amin S (2017) Antifungal potential of alkaloids as an emerging therapeutic target. Curr Drug Targets 18(16):1825–1835

Kiyama R (2017) Estrogenic terpenes and terpenoids: pathways, functions and applications. Eur J Pharmacol 815:405–415

Kochar Kaur K, Allahbadia G, Singh M (2019) Monoterpenes-a class of terpenoid group of natural products as a source of natural antidiabetic agents in the future—a review. CPQ Nutr 3:1–21

Lanzotti V (2013) Diterpenes for therapeutic use. Nat Prod 1:3173–3191

Liu J (1995) Pharmacology of oleanolic acid and ursolic acid. J Ethnopharmacol 49(2):57–68

Liu J (2005) Oleanolic acid and ursolic acid: research perspectives. J Ethnopharmacol 100(1–2): 92–94

Londonkar R, Rajani KS (2021) PHYTOCHEMICAL STUDIES ON THE LEAF EXTRACTS OF STRELITZIA REGINAE. Plant Arch 21:09725210

Mallavadhani UV, Mahapatra A, Jamil K, Reddy PS (2004) Antimicrobial activity of some pentacyclic triterpenes and their synthesized 3-O-lipophilic chains. Biol Pharm Bull 27(10): 1576–1579

Matsuno Y, Sugai A, Higashibata H, Fukuda W, Ueda K, Uda I et al (2009) Effect of growth temperature and growth phase on the lipid composition of the archaeal membrane from Thermococcus kodakaraensis. Biosci Biotechnol Biochem 73(1):104–108

Maurer HH, Sauer C, Theobald DS (2006) Toxicokinetics of drugs of abuse: current knowledge of the isoenzymes involved in the human metabolism of tetrahydrocannabinol, cocaine, heroin, morphine, and codeine. Ther Drug Monit 28(3):447–453

Molyneux RJ, Lee ST, Gardner DR, Panter KE, James LF (2007) Phytochemicals: the good, the bad and the ugly? Phytochemistry 68(22–24):2973–2985

Paduch R, Kandefer-Szerszeń M, Trytek M, Fiedurek J (2007) Terpenes: substances useful in human healthcare. Arch Immunol Ther Exp 55(5):315

Patel T, Ishiuji Y, Yosipovitch G (2007) Menthol: a refreshing look at this ancient compound. J Am Acad Dermatol 57(5):873–878

Patocka J (2003) Biologically active pentacyclic triterpenes and their current medicine signification. J Appl Biomed 1(1):7–12

Polat M (2018) Yield and some pomological characteristics of organically grown “Alyanak” and “Hasanbey” apricots (Prunus armeniaca L.)

Rasouli H, Farzaei MH, Khodarahmi R (2017) Polyphenols and their benefits: a review. Int J Food Prop 20(sup2):1700–1741

Ren Z, Gould MN (1998) Modulation of small G protein isoprenylation by anticancer monoterpenes in in situ mammary gland epithelial cells. Carcinogenesis 19(5):827–832

Rice-Evans CA, Miller NJ, Paganga G (1996) Structure-antioxidant activity relationships of flavonoids and phenolic acids. Free Radic Biol Med 20(7):933–956

Rudra A, Arvind I, Mehra R (2021) Polyphenols: types, sources and therapeutic applications. Int J Home Sci 7(3):69–75

Satoh T, Kinugawa Y, Tamaki M, Kitajyo Y, Sakai R, Kakuchi T (2008) Synthesis, structure, and characteristics of hyperbranched polyterpene alcohols. Macromolecules 41(14):5265–5271

Shaalan EA-S, Canyon D, Younes MWF, Abdel-Wahab H, Mansour AH (2005) A review of botanical phytochemicals with mosquitocidal potential. Environ Int 31(8):1149–1166

Sharma P, Tyagi A, Bhansali P, Pareek S, Singh V, Ilyas A et al (2021) Saponins: extraction, bio-medicinal properties and way forward to anti-viral representatives. Food Chem Toxicol 150: 112075

Shetge SA, Dzakovich MP, Cooperstone JL, Kleinmeier D, Redan BW (2020) Concentrations of the opium alkaloids morphine, codeine, and thebaine in poppy seeds are reduced after thermal and washing treatments but are not affected when incorporated in a model baked product. J Agric Food Chem 68(18):5241–5248

Shi W, Gould MN (2002) Induction of cytostasis in mammary carcinoma cells treated with the anticancer agent perillyl alcohol. Carcinogenesis 23(1):131–142

Singla RK, Dubey AK, Garg A, Sharma RK, Fiorino M, Ameen SM et al (2019) Natural polyphenols: chemical classification, definition of classes, subcategories, and structures. J AOAC Int 102(5):1397–1400

Sintupachee S, Pollar M, Ruang-on S (2020) Preliminary phytochemical profile analysis of Thespesia populnea (Linn.) Soland ex Correa in Pak Phanang, Nakorn Si Thammarat

Slingerland M, Cerella C, Guchelaar HJ, Diederich M, Gelderblom H (2013) Cardiac glycosides in cancer therapy: from preclinical investigations towards clinical trials. Investig New Drugs 31(4): 1087–1094

Taiz L, Zeiger E (2013) Plant physiology= fisiologia vegetal. JEAPA, Brazil

Taiz L, Zeiger E, Møller IM, Murphy A (2015) Plant physiology and development: Sinauer associates incorporated

Thomet OA, Wiesmann UN, Blaser K, Simon HU (2001) Differential inhibition of inflammatory effector functions by petasin, isopetasin and neopetasin in human eosinophils. Clin Exp Allergy 31(8):1310–1320

Trombetta D, Castelli F, Sarpietro MG, Venuti V, Cristani M, Daniele C et al (2005) Mechanisms of antibacterial action of three monoterpenes. Antimicrob Agents Chemother 49(6):2474–2478

Venkataramaiah C (2020) Chapter-1 phytoconstituents of the plants: the vital carters of the pharmacological deeds, p 1

Vetter J (2000) Plant cyanogenic glycosides. Toxicon 38(1):11–36

Wiesner K, Tsai TY, Jin H (1985) On cardioactive steroids. XVI. Stereoselective β-glycosylation of digitoxose: the synthesis of digitoxin. Helv Chim Acta 68(2):300–314

Xu T (2013) Anti-cancer effects of phenolic-rich extracts of button mushrooms (Agaricus bisporus)

Yadav M, Chatterji S, Gupta SK, Watal G (2014) Preliminary phytochemical screening of six medicinal plants used in traditional medicine. Int J Pharm Pharm Sci 6(5):539–542

Therapeutic Phytoconstituents-II

3

Bhavana Jodha and Sunita Patel

3.1 Introduction

Cancer is one of the most detrimental disorders and an enormous challenge for mankind. It represents one of the most serious healthcare concerns confronting humanity and necessarily requires a proactive cure strategy. Global demographic trends indicate that cancer incidences will rise in the coming decades, with more than 20 million new cancer cases expected each year by 2025 (Soerjomataram and Bray 2021). According to the International Agency for Research on Cancer, one in every five people develops cancer during their lifetime (Ferlay et al. 2021). Astonishingly, one in every eight men and one in every eleven women die from this disease. Among all types of cancer, breast cancer (BC) is the most commonly diagnosed in women across the globe. Women are also more likely to develop colorectal, lung, cervical, and thyroid cancers among other types of cancer. Lung and prostate cancer are the most common types of cancer in men, accounting for nearly one-third of all male cancers (Sung et al. 2021).

Over the last 15 years, advances in molecular and tumor biology have significantly altered the patterns of cancer treatment. Classification of cancer and its treatment previously was exclusively based on organ of origin or histomorphologic characteristics. Cancer treatment mainly includes interventions such as psychosocial support, radiotherapy, chemotherapy, and surgical removal of tumors (Pavet et al. 2011). Chemotherapy is the most common treatment option for cancer. It uses cytotoxic and cytostatic drugs and has proven to be very effective when combined with other therapies (Arruebo et al. 2011). However, most of the chemotherapeutic agents show immunogenicity and severe side effects. Cancer chemotherapeutics

B. Jodha · S. Patel (✉)
School of Life Sciences, Central University of Gujarat, Gandhinagar, India
e-mail: sunitap@cug.ac.in

© The Author(s), under exclusive license to Springer Nature Singapore Pte Ltd. 2024
D. Pooja, H. Kulhari (eds.), *Nanotechnology Based Delivery of Phytoconstituents and Cosmeceuticals*, https://doi.org/10.1007/978-981-99-5314-1_3

currently in use consist primarily of alkylating agents, antimetabolites, antitumor antibiotics, and natural plant-derived anticancer agents (Nygren 2001).

3.2 Pathophysiology of Cancer

Cancer is a population of aberrant cells that divide uncontrollably and have the capacity to infiltrate other tissues. It arises from a single transformed cell and becomes a mass of tissue over a period of time through sequential processes (Sporn 1996). These events are often stimulated or triggered by various factors like genes, exposure to radiation or chemicals, viral infection, chronic inflammation, tobacco smoke, aging, etc. (DePinho 2000).

Cells after transformation become capable of evading apoptosis, suppressing tumor suppressor genes, sustaining angiogenesis, metastasizing, invading distant tissues, and gaining the limitless potential to replicate (Hanahan and Weinberg 2011). These are among the few trademark properties that a normal cell must gain to become cancerous. Figure 3.1 illustrates the stages of carcinogenesis, highlighting the key events in the multistep process that leads to the development of cancer.

Cancerous tumor formation at the initial stage can be prevented by altering transmembrane transport, modifying metabolism, preventing reactive oxygen species (ROS), protecting deoxyribonucleic acid (DNA) structure, regulating DNA metabolism and repair, and managing gene expression, either in the extracellular or intracellular environment (Tabassum and Polyak 2015).

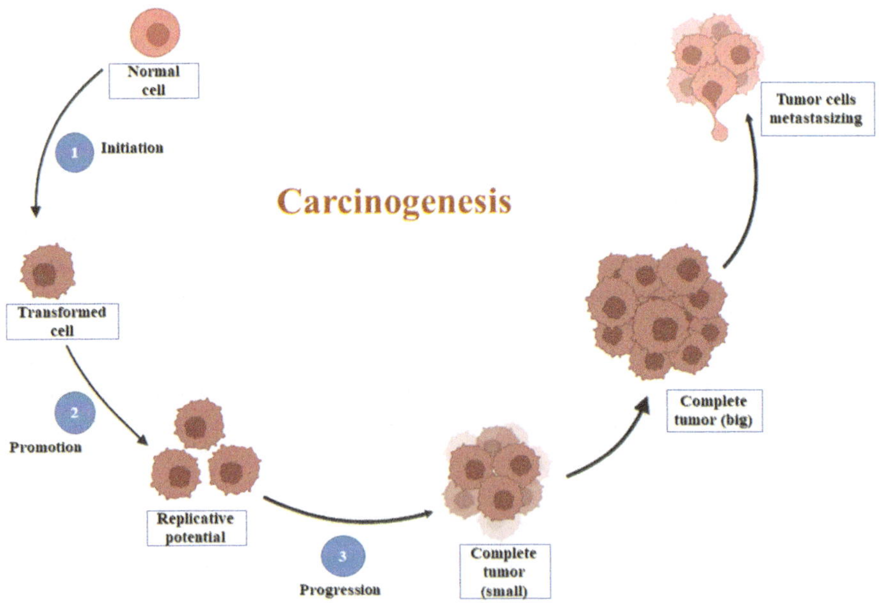

Fig. 3.1 Phases of carcinogenesis: initiation, promotion, progression, and metastasis

The second step of carcinogenesis is the promotion of the tumor, which is followed by tumor progression (Vincent and Gatenby 2008). Both processes can be reduced by reducing genotoxicity, increasing anti-inflammatory action, suppressing proteases, preventing cell proliferation, initiating cell differentiation, modifying apoptosis and other cell signaling pathways, and safeguarding intercellular communications (Compton 2020).

Furthermore, tumor progression can be slowed down by influencing the hormonal status and immune system in numerous ways, as well as by limiting tumor angiogenesis.

3.3 Drugs for Cancer Treatment and Their Limitations

Alkylating drugs, topoisomerase inhibitors, tubulin-binding compounds, and antimetabolites are examples of imperative chemotherapeutics (Nussbaumer et al. 2011). Alkylating drugs that cause DNA damage include carboplatin, cisplatin, oxaliplatin, cyclophosphamide, and melphalan (Ralhan and Kaur 2007). Topoisomerase inhibitors such as doxorubicin and camptothecin prevent DNA replication (Denny 2004). Drugs acting or binding on tubulin disrupt mitotic spindles and halt mitosis. Tubulins are targeted by paclitaxel, docetaxel, eribulin, vinblastine, and vincristine (Galluzzi et al. 2012).

Anticancer medications available at present target aggressively dividing cells. Under normal conditions, some cells in our body also grow rapidly, like cell in the bone marrow and hair follicle cells. As a result, many chemotherapeutic medications also harm normal cells, resulting in major adverse side effects (Shewach and Kuchta 2009). Other harmful effects include myelosuppression and mucositis. Along with this, these drugs even lead to poor appetite, weak immunity, loss of memory, weight gain or loss, and many more side effects. Another constraint is cancer cell resistance to existing treatments. Cancer cells mutate and develop resistance to the medications that are introduced, and hence, efficacy of the available treatment options decreases.

3.4 Anticancer Potential of Phytochemicals: A Novel Approach

Plants and their constituents have been employed in traditional medicine since the dawn of time. Many plants are repositories for novel chemical entities and offer a promising avenue for cancer research (Miller and Snyder 2012). The medicinal actions of plants are primarily attributable to the existence of secondary metabolites (Schmidt et al. 2007). Ongoing research and development are focused on the discovery of new cytotoxic components derived from medicinal plants with anticancerous properties.

Hartwell published the first part of a series of essays in December 1967 discussing the use of plants by mankind to treat cancer (Hartwell 1967; Hartwell 1971). This monumental work published gradually over a period of 5 years lists around 3000 plant species that were considered to have alleged anticancer effects. However, in many cases discussed, the term "cancer" is left ambiguous, or reference is made to ailments like hard swellings, ulcers, warts, corns, lesions, cysts, or tumors, to mention a few (Graham et al. 2000). Hence, for a better understanding, more studies were further conducted to establish the potential anticancerous nature of phytochemicals.

With the discovery and production of the vinca alkaloids, vinblastine, and vincristine, as well as the isolation of the cytotoxic podophyllotoxins, the quest for anticancer medicines from plant sources began to take shape in the 1950s (Singh et al. 2016). As a result, the National Cancer Institute (NCI) of the United States launched an intensive plant collection effort in 1960 (Thomas et al. 2015). This initiative led to the discovery of many plant bioactive agents capable of cytotoxic activity.

In human systems, phytoconstituents and their metabolites found in the root, leaf, flower, stem, and bark serve a variety of therapeutic actions. Alkaloids, flavonoids, stilbenes, polyphenols, terpenoids, glycosides, gums, and oils are among the many helpful and relevant substances obtained from medicinal plants (Shin et al. 2018).

Phytochemicals obtained from plants have exhibited encouraging results in their ability to elicit anticancer effects through diverse mechanisms. Additionally, extensive research has focused on studying the chemopreventive potential of these compounds, aiming to impede or delay cancer development in individuals at high risk or with precancerous conditions. Notably, numerous phytoconstituents have progressed to the stage of clinical trials, which play a vital role in assessing their safety and tolerability (Dhupal and Chowdhury 2020). These trials provide valuable insights into adverse effects, drug interactions, and potential toxicity, ultimately guiding the advancement of medical knowledge. Table 3.1 provides an overview of several anticancer phytochemicals currently under investigation in clinical trials.

Properties such as solubility, lipophilicity, and molecular weight can impact the absorption, distribution, metabolism, and elimination of phytochemicals (Aqil et al. 2013). Understanding these properties is vital for developing nanomedicines that enhance the stability, solubility, and bioavailability of phytochemicals, thus augmenting their therapeutic potential. The biological activity and efficacy of phytochemicals are influenced by their structural characteristics, including functional groups, stereochemistry, and molecular arrangement (Arif et al. 2022). These structural features allow researchers to discern how phytochemicals interact with target cells, biomolecules, or disease pathways, especially in cancer research (see Fig. 3.2 for the structural representation of select anticancer phytochemicals).

3.5 Hallmarks of Cancer and Action of Phytoconstituents

Hanahan and Weinberg's descriptions of cancer hallmarks have been crucial in our knowledge of cancer's common characteristics and rational treatment design. The conceptualization of hallmarks of cancer provides the organizing principle of this complex disorder that helps in rationalizing therapeutic approaches for managing and curing this horrifying disease.

Table 3.1 Anticancer phytoconstituents currently being used in cancer therapy

Phytochemical	Plant source	Molecular targets	Clinical usage	Commercial name	References
Camptothecin	*Camptotheca acuminata*	Topoisomerase I	Ovarian, colorectal, childhood cancer	Topotecan	Hertzberg et al. (1989)
Paclitaxel, docetaxel	*Taxus brevifolia*	Microtubules	Phase I–III clinical trials, breast cancer	Taxol, Abraxane, Taxoprexin	Oudard et al. (2017)
Etoposide	*Sinopodophyllum hexandrum*	Topoisomerase II	Cervical, nasopharyngeal, colon cancer	Vepesid, Etopophos	Cao et al. (2015)
Vinblastine, Vincristine	*Catharanthus roseus*	Tubulin	NSCLC, breast, leukemia	Alkaban AQ, Velban	Martino et al. (2018)

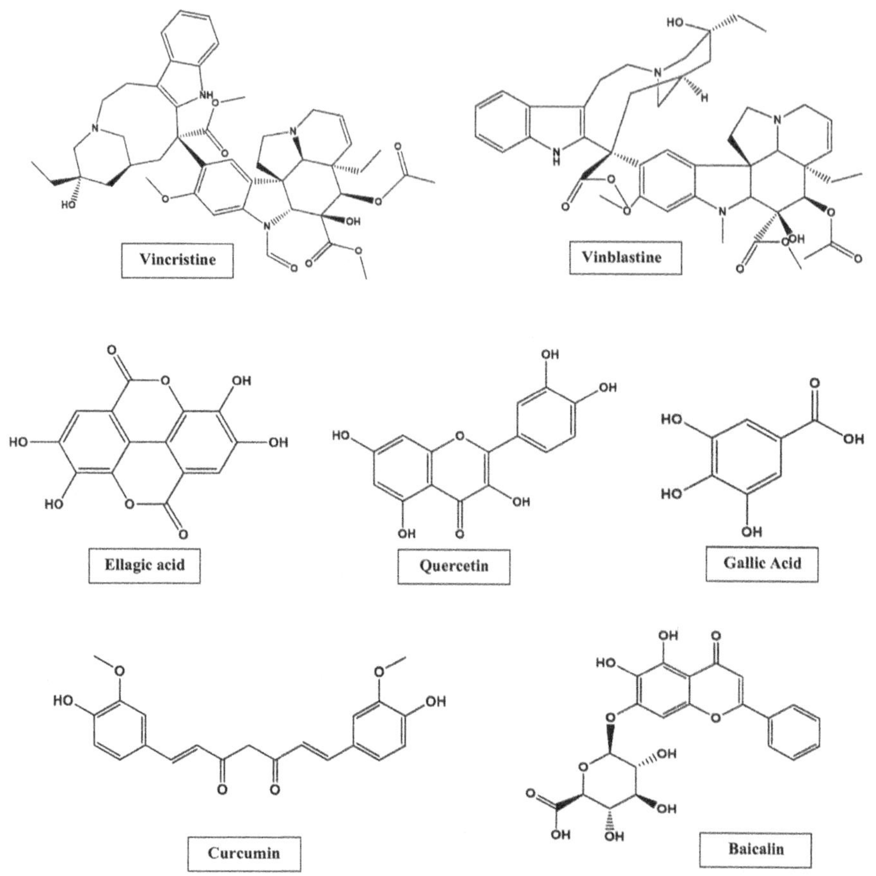

Fig. 3.2 Structures of anticancer phytochemicals

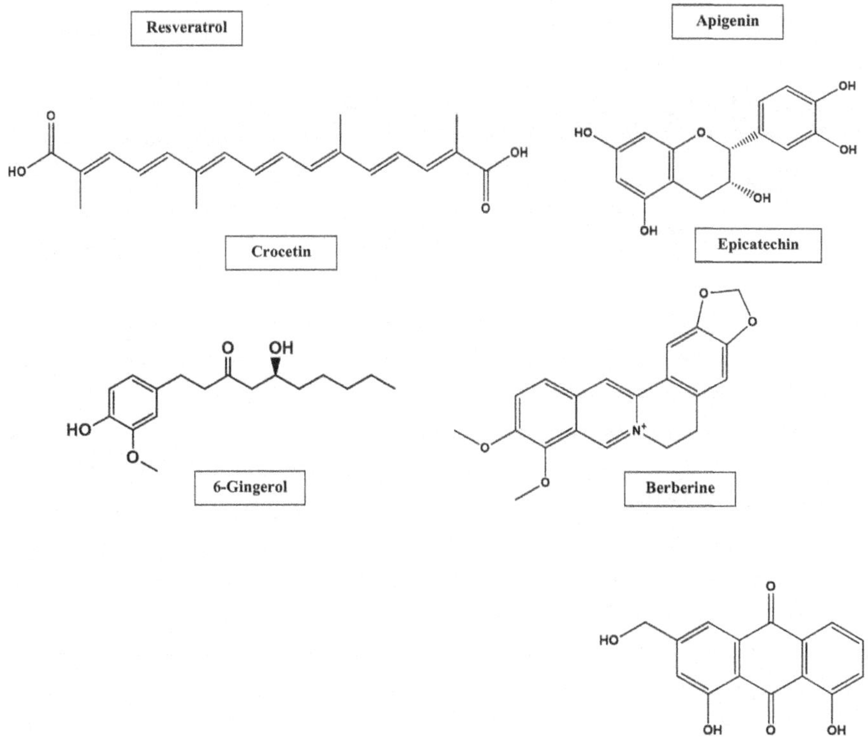

Fig. 3.2 (continued)

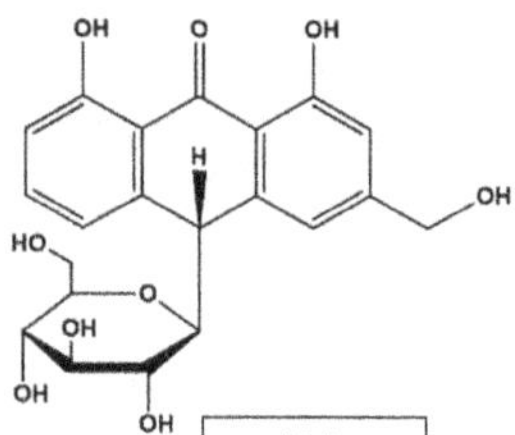

Aloin

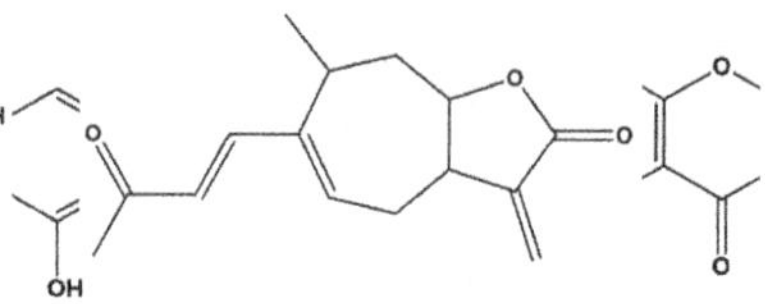

Xanthatin

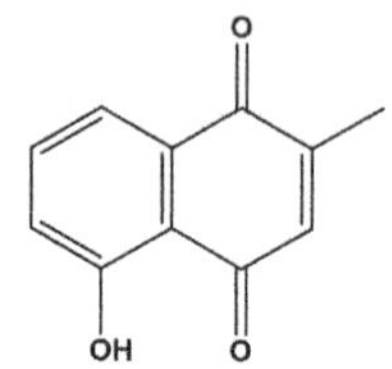

Plumbagin

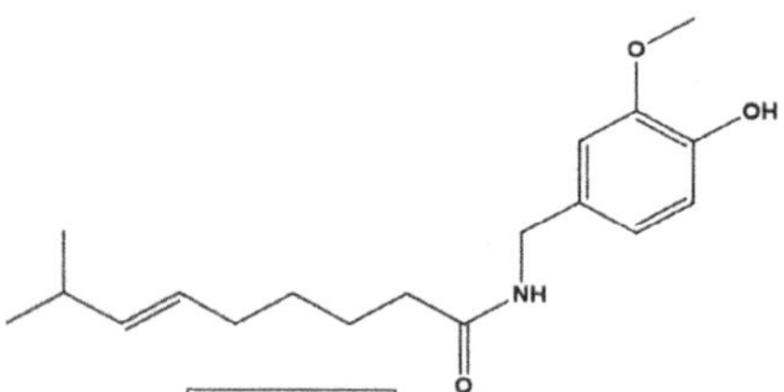

Capsaicin

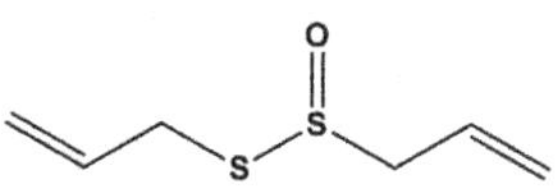

Allicin

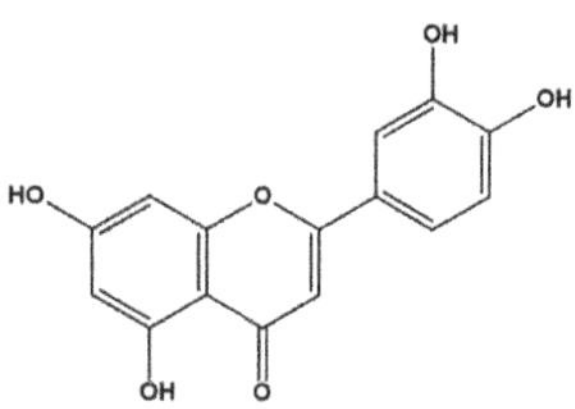

Luteolin

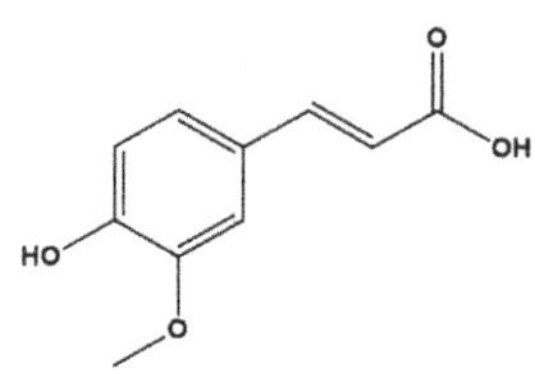

Ferulic Acid

Aloe - emodin

Fig. 3.2 (continued)

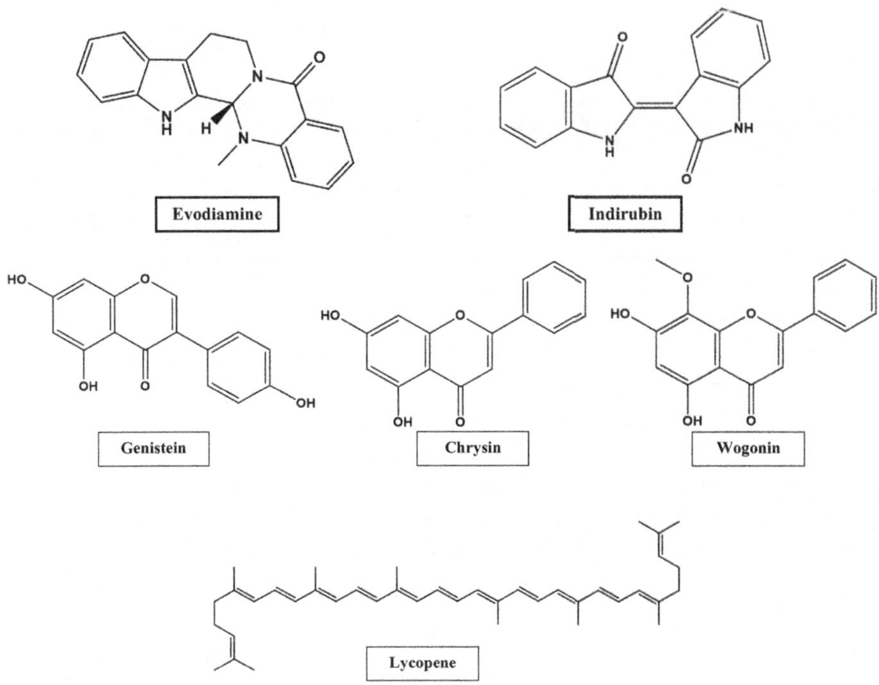

Fig. 3.2 (continued)

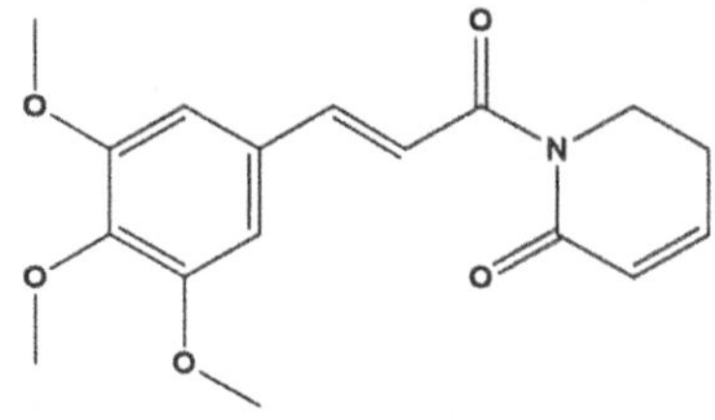

Piperlongumine

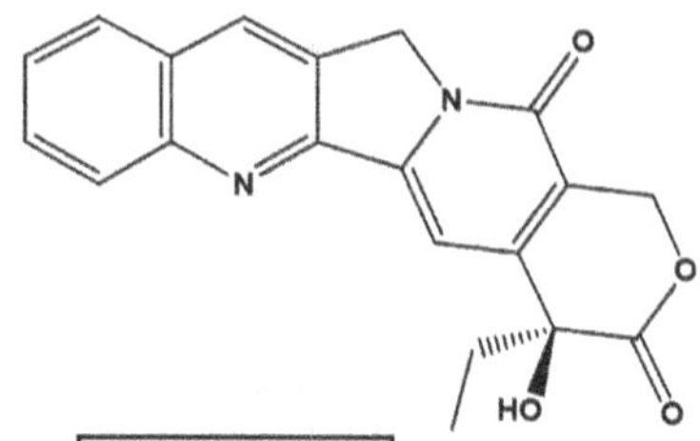

Camptothecin

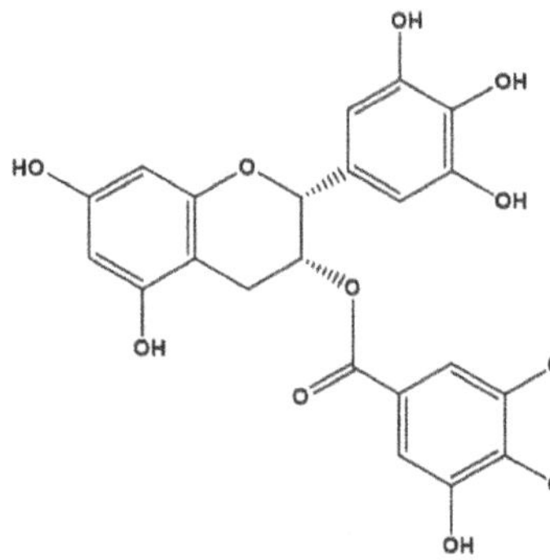

Epigallocatechin gallate (EGCG)

Fig. 3.2 (continued)

Figure 3.3 visually depicts the different hallmarks of cancer along with the names of various phytoconstituents that have the potential to inhibit or reverse these characteristics. Overall, comprehending the hallmarks of cancer and the impact of phytochemicals on these hallmarks facilitates the development of targeted therapies, identifies complementary or alternative treatment options, reveals anticancer mechanisms, enables preventive measures and early interventions, and supports personalized medicine approaches.

3.5.1 Phytoconstituents Targeting Apoptosis

Apoptosis is a well-ordered and coordinated cellular process that happens in both normal and pathological situations (Lowe and Lin 2000). It is also one of the most researched areas in cellular biology. Understanding the underlying mechanism of apoptosis is critical since it plays a critical role in the pathophysiology of many illnesses. Apoptosis is now one of the new ways of identifying and developing innovative anticancer treatments (Kasibhatla and Tseng 2003).

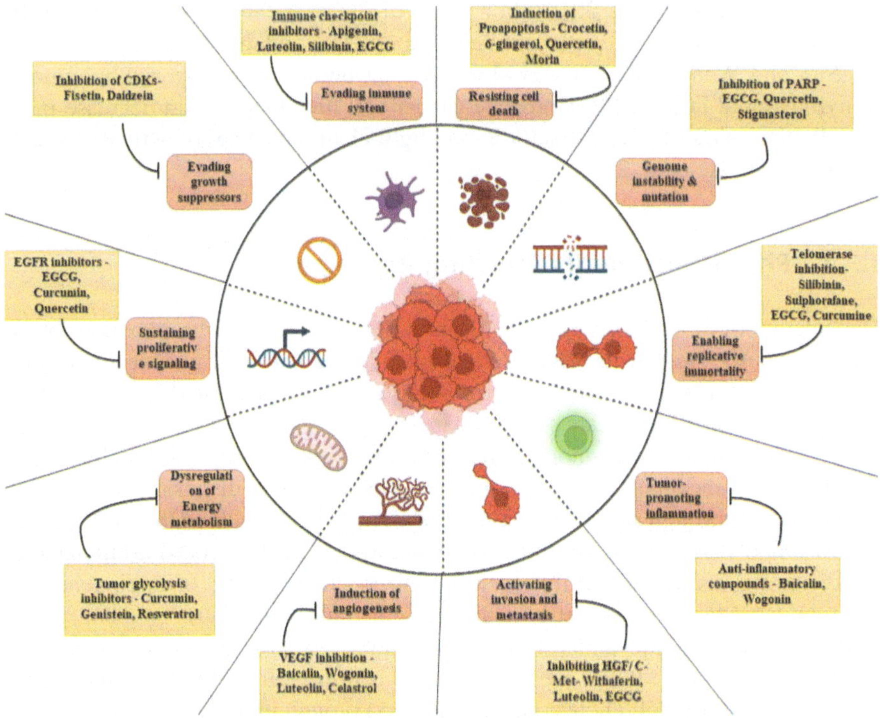

Fig. 3.3 Role of phytoconstituents on hallmarks of cancer

In cells, it is characterized by cellular morphological changes such as reduction in cell size or shrinkage of cytoplasmic matter, blebbing of cellular membrane, chromatin condensation, DNA fragmentation, formation of apoptotic bodies, and so on. It is a complex process that incorporates numerous molecular pathways (Cotter 2009). Defects can develop at any point along these pathways, resulting in the malignant transformation of afflicted cells, tumor spread, and resistance to drugs. A cell can experience apoptosis by one of the three methods: extrinsic pathways, intrinsic pathways, and perforin-granzyme apoptotic pathways (Wong 2011).

It was observed that when saponins from *Astragalus* were administered to HT-29 colon cancer cells, it caused apoptosis via the extrinsic route (Auyeung et al. 2010). Apigenin another phytochemical caused apoptotic cell death in HER2-overexpressing breast cancer cells via the extrinsic route (Seo et al. 2012). Harmine is a β-carboline alkaloid derived from the plant *Peganum harmala* that triggers both intrinsic and extrinsic apoptotic pathways (Hamsa and Kuttan 2011). Curcumin mediated high production of ROS-induced apoptosis in mouse fibroblast-L929 cells, by regulating various apoptotic pathways (Thayyullathil et al. 2008).

Allicin extracted from *Allium sativum* commonly known as garlic plant also shows potential anticancer effects. When administered in vitro in the human gastric cancer cell line SGC-7901, allicin induced apoptosis and decreased overall cancer cell viability (Zhang et al. 2010). In a study of stomach cancer, allicin lowered the activity of telomerase, a key player in cellular division and, hence, triggered apoptosis (Sun and Wang 2003). It was observed that on administering xanthohumol, a prenylflavonoid in the BPH-1 prostate cancer cell line, there was a decrease in the overall expression of nuclear factor kappa-light-chain-enhancer of activated B cells (NF-κB) (Colgate et al. 2007).

3.5.2 Phytoconstituents Affecting Angiogenesis

The creation of new blood capillaries from a pre-existing vascular plexus is known as angiogenesis. Despite its physiological role, dysregulation of this mechanism has been linked to a number of diseases, including cardiovascular diseases, diabetic retinopathy, arthritis, and cancer (Folkman 1984).

Malignant tumors, like normal tissue, require continuous food delivery and removal of intracellular waste to survive. Delivery and disposal are generally accomplished through tumor neovascularization, which involves the development of new blood vessels from existing vasculature (Nishida et al. 2006). Malignancies, unlike normal tissue, utilize this well-organized characteristic of angiogenesis to promote unhindered malignant growth (Carmeliet and Jain 2000).

Phenols and anthocyanins from the apple plant are known to show anti-angiogenic effects in MCF-7 breast carcinoma cell lines. It also reduced the expression of proliferating cell nuclear antigen (PCNA) protein in the same (Yoon and Liu 2007). Another phytochemical, catechin, found in green tea has demonstrated inactivation of NF-κB, the activity of vascular endothelial growth factor (VEGF) promotor, and the production of VEGF in MDA-MB-231 cell lines (Sartippour et al. 2001). Another green tea phytoconstituent (−)-epigallocatechin-3-gallate (EGCG) has been associated with reduced risk of breast cancer, prostate cancer, and liver cancer. It is also observed that EGCG inhibits the expression and/or activity of VEGF in cancers like gastric cancer (Zhu et al. 2007) and head and neck squamous cell carcinoma (Masuda et al. 2002). Curcumin commonly known as turmeric also showed anti-angiogenic effect in MDA-MB-231 cells by suppressing VEGF and basic fibroblast growth factor (bFGF) (Shao et al. 2002). Apigenin extracted from herbs and fruits has been demonstrated to work against cyclooxygenase-2 (COX-2), reduces VEGF, and induces NF-κB inhibition in human pancreatic cancer cells (Melstrom et al. 2011). Another study concerning effect of apigenin on non-small cell lung carcinoma revealed that it suppresses the angiogenesis of tumor via hypoxia-inducible factor 1-alpha (HIF-1α) inhibition (Fu et al. 2022).

In vivo investigations in nude mice xenografted with MCF-7 and MDA-MB-231 tumors duplicated the in vitro effects of genistein, a soy isoflavone on breast cancer and angiogenesis. In both models, genistein slowed tumor development and decreased matrix metalloproteinase-9 (MMP-9) mRNA expression. Transforming growth factor β1 (TGF-β1) and VEGF protein levels were also reduced in the serum and in the tumor cells of mice models (Farina et al. 2006).

3.5.3 Phytoconstituents Targeting Epithelial to Mesenchymal Transition (EMT)

The process through which epithelial cells become mesenchymal cells, gaining fibroblast-like features and exhibiting reduced intercellular adhesion and enhanced motility, is known as epithelial-mesenchymal transition (EMT) (Roche 2018). EMT is crucial in the progression of cancer. EMT suppression or reversal is thus a crucial method of managing many malignancies (Zhang and Weinberg 2018).

Metastasis is a complex, multistep process that consists of a series of interconnected, sequential stages such as invasion, migration, adhesion, infiltration, colonization at a distant place, and the development of new capillaries (Lamouille et al. 2014). EMT-specific junction proteins are E-cadherin and N-cadherin, respectively. Upregulation of E-cadherin and N-cadherin is thus associated with cell invasion suppression and augmentation, respectively (Georgakopoulos-Soares et al. 2020). Hence, at later stages of diagnosed cancer where cells have great migratory potential, EMT is targeted to stall the spread.

A downregulation of mesenchymal factors like MMP-2, MMP-9, and vimentin was observed in emodin-treated colorectal cancer cells. However, upregulation in E-cadherin was observed in the same, showing inhibition of EMT (Pooja and Karunagaran 2014). On treating melanoma cells with fisetin, the activity of N-cadherin, fibronectin, and vimentin decreased; on the other hand, there was an increase in the activity of E-cadherin and desmoglein (Pal et al. 2014). In one study chrysin, a flavone majorly found in honey, passion flowers, etc., also reverted the EMT markers in the case of breast carcinoma cells (Yang et al. 2014). Resveratrol, a versatile phytochemical, has shown anti EMT effect in many cancer cell lines like prostate (Li et al. 2014), lung (Wang et al. 2013), colorectal (Ji et al. 2013), and pancreatic cancer (Li et al. 2013).

Gallic acid (GA) effectively inhibits AGS cell migration by reducing the production of MMP-2/−9 and cytoskeletal F-actin. GA's antimigratory impact could be due to the reduction of NF-κB activity and several proteins involved in metastasis and cytoskeletal rearrangement (Ho et al. 2013).

3.5.4 Phytoconstituents as Anti Inflammatory Agents

Inflammation is another hallmark of cancer. Chronic inflammation can develop if the underlying cause of the inflammation persists or specific control systems responsible for shutting down the process fail (Murakami and Hirano 2012). When these inflammatory reactions become chronic, cell mutation and proliferation can occur, frequently producing an environment favorable for cancer formation (Shacter and Weitzman 2002).

Chronic inflammation causes a variety of problems. ROS production as a result of inflammation causes tissue damage by damaging nucleic acids, proteins, and lipids. Damaged tissue may stimulate progenitor cells for tissue regeneration. ROS from inflammation harms stem cells, and the resulting mutations can accumulate potentially producing cancer stem cells (Coussens and Werb 2002).

There is substantial evidence that biologically active phytochemicals possess anti-inflammatory properties. Among these, baicalein, curcumin, oridonin, and wogonin have been shown to inhibit the proinflammatory activity of IL-6 at the transcriptional level or protein levels (Issa et al. 2006).

Shogaols are formed when gingerols are dehydrated. Shogaols are a key ingredient in dried ginger powder. 6-Shogaol is an anti-inflammatory phytoconstituent (Dugasani et al. 2010). The researchers discovered that 6-Shogaol suppressed the synthesis of tumor necrosis factor alpha (TNF-α), Interleukin-1 (IL-1), IL-6, and Prostaglandin E2 (PGE2), which are all generated by lipopolysaccharide (LPS) activity. It also caused NF-κB activation in BV2 microglia cells (Han et al. 2017). Furthermore, a study found that 6-shogaol inhibited inducible nitric oxide synthase (iNOS), COX-2 gene expression, and NF-κB activation in murine macrophages (Pan et al. 2008).

3.5.5 Phytoconstituents Altering Autophagy

Autophagy is a cellular catabolic self-cannibalism that removes defective intracellular cytoplasmic molecules where cargo-containing autophagosomes fuse with lysosomes to maintain homeostasis (Klionsky 2005). Autophagy maintains a dynamic interconnection between cytoprotective and cytostatic functions throughout the malignant transformation in cancer cells (Levine 2007). It is strictly governed and regulated by a group of genes known as autophagy-related genes (ATGs) and is stimulated in response to a number of stressors, including pathogenic, metabolic, and genotoxic (Mizushima 2007).

To support tumor development and dissemination, autophagy improves cancer cell survival in nutrient-limited and hypoxic environments (Levine 2007). Modulation of the autophagy process has been reported to be a promising method for cancer therapy. Furthermore, autophagic inhibition causes sensitization and increases anticancer activity in tumor cells (Jeda et al. 2022). Natural cancer therapies have been demonstrated to prevent cancer growth by modulating epigenetics via autophagy (Rebecca and Amaravadi 2016). In this regard, a wide spectrum of natural epigenetic modulators derived from various plants have been found to have anticancer properties (Patra et al. 2020).

Curcumin suppressed DNA methyltransferases (DNMT1) and DNMT3B while restoring the activities of miR-143 and miR-145 in prostate cancer cells, inhibiting protective autophagy (Liu et al. 2017). Resveratrol reduced IL-6-promoted cell migration in ovarian cancer by activating ARH-1, an imprinted tumor suppressor that regulates autophagy induction (Ferraresi et al. 2017). Orientin, a phytoconstituent extracted from plant *Ocimum sanctum*, is a flavonoid that carries out the suppression of inflammasome through targeting NF-κB (Xiao et al. 2017). Baicalin, a flavone glycoside derived from *Scutellaria* species, inhibits tumor growth by inducing autophagy in tumor-associated macrophages (Tan et al. 2015). When quercetin was administered to AGS and MKN28 gastric cancer cells, it induced cytoprotective autophagy by increasing the activity of autophagosome and also led to the increase in the expressions of various ATGs, like ATG7 and ATG5. On cellular level it also resulted in the downregulation of the Akt/mTOR pathway (Wang et al. 2011).

Further, information about several phytochemicals and their anticancer effect on different types of cancer is summarized in Table 3.2.

3.6 Seeking Innovative Methods for Using Phytochemicals Against Cancer

3.6.1 Extraction

Pharmaceutical properties and their complexities differ among plant parts. These bioactive phytoconstituents have the potential to be employed in anticancer treatments, but more research is needed. Combinatorial chemistry, isolation tests, and bioassay-guided fractionation may all be used in the purification of active phytochemicals (Jha and Sit 2021). To extract diverse bioactive molecules from a mixture of substances, bioassay-guided fractionation using various analytical techniques could be utilized (Jha and Sit 2021).

3.6.2 Synergism

Herbal medications or extracts themselves consist of various active ingredients that interact with one another and with other pharmaceutical drugs to either improve (synergize) or reduce (antagonize) the therapeutic impact (Liu 2004).

A study on human leukemia cells demonstrated the synergism of quercetin and ellagic acid with resveratrol, which resulted in promoting apoptosis and causing cell cycle arrest (Mertens-Talcott and Percival 2005). In a separate study, the combination of resveratrol and quercetin was responsible for anticancer activity in colon cancer cells via repressing oncogenic microRNA (Del Follo-Martinez et al. 2013). According to the researchers, both resveratrol and quercetin either stabilize each other or interact with different cell signaling cascades.

Sulindac is a nonsteroidal anti-inflammatory medicine utilized as a cancer-preventative agent, but its use is limited due to its harmful side effects (Ishikawa et al. 1997). High doses of this medication suppress COX-1, which might cause gastrointestinal bleeding. It was found that when a combination of EGCG and sulindac was used on PC-9 cells, its apoptosis-generating activity was enhanced (Suganuma et al. 1999).

In vitro studies found that genistein works synergistically with eicosapentaenoic acid to change glucose oxidation, limiting the proliferation of human breast cancer cells (Nakagawa et al. 2000).

Table 3.2 List of anticancer phytoconstituents and their therapeutic actions in different types of cancer

Plant name	Parts of plant used	Bioactive agent	Anticancer activities	Cancer suppressed	References
Curcuma longa	Root and rhizome	Curcumin	– Pro-apoptotic and anti-inflammatory – Scavenges ROS	Breast, colon, bladder, lung, cancer, leukemia	Sultana et al. (2021)
Silybum marianum	Leaves, flowers	Silibinin	– Modulates apoptosis – Antimetastatic – Anti-inflammatory	Gastrointestinal, Bladder, breast, cervical, lung cancer	Fallah et al. (2021)
Piper longum	Fruits, root	Piperlongumine	– Immunomodulatory – Anti-inflammatory – Tubulin inhibitor	Glioma, colorectal, breast, pancreatic, prostate cancer	Henrique et al. (2020), Parama et al. (2021)
Evodia rutaecarpa	Fruit	Evodiamine	– Inhibits cell growth – Induces cell cycle arrest – Antimetastatic	Colorectal, breast, gastric cancer	Panda et al. (2022)
Glycine max	Seed, fruit	Genistein	– Anti-angiogenic – Induces apoptosis – Anti-inflammatory – Antimetastatic	Glioblastoma, breast, colorectal, lung cancer	Tuli et al. (2019)

Scutellaria baicalensis	Root	Wogonin Baicalin	– Induces apoptosis – Inhibits cell proliferation	Cholangiocarcinoma, breast, cervical, hepatic cancer, lymphoma, leukemia	Banik et al. (2022)
Petroselinum crispum *Apium graveolens* *Spinacia oleracea* *Matricaria chamomilla*	Flower, leaves	Apigenin	– Antiproliferative – Inhibits chemoresistance – Inhibits invasion	Pancreatic, colon, prostate, breast, lung cancer	Ahmed et al. (2021), Zhou et al. (2022), Ashrafizadeh et al. (2020)
Berberis vulgaris	Rhizome, barks, leaves, stems	Berberine	– Inhibits cell proliferation – Inhibits telomerase activity – Prevents cell invasion	Liver, cervical, lung cancer, leukemia	Wang et al. (2020)
Passiflora caerulea	Flower	Chrysin	– Pro-apoptotic – Induces cell death – Inhibits tumor angiogenesis	Colorectal, lung, bladder cancer	Mani and Natesan (2018), Lima et al. (2020)
Leea indica	Fruit, root, leaves	Gallic acid	– Anti-inflammatory – Inhibits metastasis	Breast, lung, ovarian cancer, Myeloid leukemia	Subramanian et al. (2015)
Vitis vinifera	Fruit	Resveratrol	– Anti-inflammatory – Induces cell cycle arrest	Colorectal, prostate, lymphoma, glioma	Vervandier-Fasseur and Latruffe (2019)

(continued)

Table 3.2 (continued)

Plant name	Parts of plant used	Bioactive agent	Anticancer activities	Cancer suppressed	References
Fragaria ananassa Malus domestica	Fruit	Fisetin	– Inhibits proliferation and invasion – Induces apoptosis – Prevents cell cycle progression	Lung, colorectal, cervical cancer, skin-malignant melanoma	Kashyap et al. (2019), Imran et al. (2021)
Plumbago zeylanica	Leaves	Plumbagin	– Targets autophagy and apoptosis, – Antimetastatic – Anti-invasion	Colorectal, liver, breast cancer, fibrosarcoma, squamous cell carcinoma	Yin et al. (2020)
Allium cepa Rubus idaeus	Fruits, vegetables, seeds, leaves	Quercetin	– Scavenges ROS – Induces cell cycle arrest	Prostate, pancreatic ductal carcinoma, cholangiocarcinoma, oral cavity cancer	Kashyap et al. (2019), Ulusoy and Sanlier (2020), Dajas (2012)
Peganum harmala	Seed coat	Harmine	– Suppresses cell proliferation	Pancreatic, breast, ovarian, lung cancer	Zhang et al. (2020)
Ocimum sanctum	Leaves	Eugenol	– Induces cell death – Inhibits migration – Induces cell cycle arrest	Breast, liver cancer, fibrosarcoma	Begum et al. (2022), Zari et al. (2021)
Camptotheca acuminata	Stem bark	Camptothecin	– Inhibits DNA topoisomerase I – Induces apoptosis	Colorectal	Behera and Padhi (2020)

Gossypium herbaceum	Root	Gossypol	– Anti-inflammatory – Inhibits cell growth and progression	Breast, colon, pancreatic, prostate cancer	Cao et al. (2021)
Solanum lycopersicum Psidium guajava L	Fruit	Lycopene	– Induces apoptosis – Inhibits cancer cell growth	Gastric, prostate, lung cancer	Bhuvaneswari and Nagini (2005)
Indigofera arrecta	Roots	Indirubin	– Induces cell cycle arrest – Anti-inflammatory – Inhibit tumor angiogenesis	Breast, colorectal cancer	Yang et al. (2022)

3.6.3 Nanotechnology for the Delivery of Phytoconstituents

Current therapy regimens require selective medication, majorly targeting activity of cancer cells while reducing harm to normal cells (Ahmad et al. 2021). With good outcomes, nanotechnology-driven customized pharmaceuticals and drug delivery systems are being created and launched into the market for better cancer treatment and management (Subramanian et al. 2016). Many obstacles in drug delivery to cancer cells can be prevented by using nano-drug carriers, such as enhancing drug solubility and stability, prolonging drug half-lives in the blood, increasing bioavailability, lowering side effects in nontarget organs, and concentrating medications at the disease site (Solanki et al. 2022).

Encapsulation of genistein in inulin-stearic acid (INU-SA) conjugate increased its overall anticancer effectiveness against human colorectal carcinoma cells (Jangid et al. 2022). In another study concerning activity of resveratrol in HCT 116 cells, it was demonstrated that the formed nanoparticles of resveratrol showed better efficacy when compared to administered pure phytochemical (Jangid et al. 2021).

3.7 Potential Challenges with Phytochemical-Based Anticancer Therapy

Despite widespread recognition and acceptance, as evidenced by a huge number of scientific publications, developing anticancerous phytochemicals as effective chemoprevention and/or chemotherapeutic drugs is far more difficult than it appears. There is still a lack of preclinical and clinical data available on the antitumor functions generated by some phytoconstituents, which continues to remain a problem and a limitation of the present literature in order to gain a better understanding of the molecular mechanisms behind these effects.

Bioavailability, solubility, stability, and immunogenicity still remain major concerns for several plant-based compounds. Hence, it is of utmost priority to address these issues for further paving the way for novel therapeutic strategies for generating better cancer regimens.

3.8 Conclusion and Future Perspective

Cancer is a highly complex disease with multiple complications, and standard therapies have significant side effects. Henceforth, a newer approach toward cancer drug discovery is required. Historically, traditional plants were thought to offer an unlimited source of novel biochemical for the production of new medications and drugs for several ailments. As a result, exploring these agents for their anticancer effects would lay a possible research ground for potential effective therapeutic options, which may be used to narrow down pharmaceutical options and gain more promising results in generating potential anticancer drugs.

In this regard, a number of phytochemicals derived from plants have been found and are now being employed in cancer therapy. For instance, vinca alkaloids and paclitaxel have been shown to have anticancer benefits in clinical trials and have been licensed for clinical use. In addition, certain innovative methods like synergism and nano-drug delivery system would create a whole new avenue for better delivery and efficacy of the plant-derived compounds. The research works summarized here provide strong support for the significance of phytoconstituents as anticancer therapeutics and suggest that their antiproliferative and anti-apoptotic characteristics hold considerable potential in cancer prevention and treatment. Furthermore, a complete understanding of possibilities of anticancer phytochemicals will help in changing the current reality of cancer management. With the discovery of novel therapeutic options that are less toxic and more efficient, the scientific community might be able to develop medications to cure cancer and prevent its associated morbidities.

Acknowledgments Author Bhavana Jodha thanks UGC for the Non-NET fellowship, and Dr. Sunita Patel acknowledges the Central University of Gujarat for providing the infrastructure and all necessary facilities.

References

Ahmad R, Srivastava S, Ghosh S, Khare SKJC, Biointerfaces SB (2021) Phytochemical delivery through nanocarriers: a review. Colloids Surf B 197:111389

Ahmed SA, Parama D, Daimari E, Girisa S, Banik K, Harsha C et al (2021) Rationalizing the therapeutic potential of apigenin against cancer. Life Sci 267:118814

Aqil F, Munagala R, Jeyabalan J, Vadhanam MV (2013) Bioavailability of phytochemicals and its enhancement by drug delivery systems. Cancer Lett 334(1):133–141

Arif JM, Kandimalla R, Aqil F (2022) Role of phytochemicals and structural analogs in cancer chemoprevention and therapeutics. Front Pharmacol 13:865619

Arruebo M, Vilaboa N, Sáez-Gutierrez B, Lambea J, Tres A, Valladares M et al (2011) Assessment of the evolution of cancer treatment therapies. Cancers (Basel) 3(3):3279–3330

Ashrafizadeh M, Bakhoda MR, Bahmanpour Z, Ilkhani K, Zarrabi A, Makvandi P et al (2020) Apigenin as tumor suppressor in cancers: biotherapeutic activity, nanodelivery, and mechanisms with emphasis on pancreatic cancer. Front Chem 8:829

Auyeung KK-W, Mok N-L, Wong C-M, Cho C-H, Ko JK (2010) Astragalus saponins modulate mTOR and ERK signaling to promote apoptosis through the extrinsic pathway in HT-29 colon cancer cells. Int J Mol Med 26(3):341–349

Banik K, Khatoon E, Harsha C, Rana V, Parama D, Thakur KK et al (2022) Wogonin and its analogs for the prevention and treatment of cancer: a systematic review. Phytother Res 36:1854

Begum SN, Ray AS, Rahaman CH (2022) A comprehensive and systematic review on potential anticancer activities of eugenol: from pre-clinical evidence to molecular mechanisms of action. Phytomedicine 107:154456

Behera A, Padhi S (2020) Passive and active targeting strategies for the delivery of the camptothecin anticancer drug: a review. Environ Chem Lett 18(5):1557–1567

Bhuvaneswari V, Nagini S (2005) Lycopene: a review of its potential as an anticancer agent. Curr Med Chem Anticancer Agents 5(6):627–635

Cao B, Chen H, Gao Y, Niu C, Zhang Y, Li L (2015) CIP-36, a novel topoisomerase II-targeting agent, induces the apoptosis of multidrug-resistant cancer cells in vitro. Int J Mol Med 35(3): 771–776

Cao H, Sethumadhavan K, Cao F, Wang TT (2021) Gossypol decreased cell viability and down-regulated the expression of a number of genes in human colon cancer cells. Sci Rep 11(1):1–16
Carmeliet P, Jain RK (2000) Angiogenesis in cancer and other diseases. Nature 407(6801):249–257
Colgate EC, Miranda CL, Stevens JF, Bray TM, Ho E (2007) Xanthohumol, a prenylflavonoid derived from hops induces apoptosis and inhibits NF-kappaB activation in prostate epithelial cells. Cancer Lett 246(1–2):201–209
Compton C (2020) Cancer initiation, promotion, and progression and the acquisition of key behavioral traits. Springer, pp 25–48
Cotter TG (2009) Apoptosis and cancer: the genesis of a research field. Nat Rev Cancer 9(7): 501–507
Coussens LM, Werb Z (2002) Inflammation and cancer. Nature 420(6917):860–867
Dajas F (2012) Life or death: neuroprotective and anticancer effects of quercetin. J Ethnopharmacol 143(2):383–396
Del Follo-Martinez A, Banerjee N, Li X, Safe S, Mertens-Talcott S (2013) Resveratrol and quercetin in combination have anticancer activity in colon cancer cells and repress oncogenic microRNA-27a. Nutr Cancer 65(3):494–504
Denny WA (2004) Emerging DNA topoisomerase inhibitors as anticancer drugs. Expert Opin Emerg Drugs 9(1):105–133
DePinho RA (2000) The age of cancer. Nature 408(6809):248–254
Dhupal M, Chowdhury D (2020) Phytochemical-based nanomedicine for advanced cancer theranostics: perspectives on clinical trials to clinical use. Int J Nanomedicine 15:9125–9157
Dugasani S, Pichika MR, Nadarajah VD, Balijepalli MK, Tandra S, Korlakunta JN (2010) Comparative antioxidant and anti-inflammatory effects of [6]-gingerol,[8]-gingerol, [10]-gingerol and [6]-shogaol. J Ethnopharm 127(2):515–520
Fallah M, Davoodvandi A, Nikmanzar S, Aghili S, Mirazimi SMA, Aschner M et al (2021) Silymarin (milk thistle extract) as a therapeutic agent in gastrointestinal cancer. Biomed Pharm 142:112024
Farina HG, Pomies M, Alonso DF, Gomez DE (2006) Antitumor and antiangiogenic activity of soy isoflavone genistein in mouse models of melanoma and breast cancer. Oncol Rep 16(4): 885–891
Ferlay J, Colombet M, Soerjomataram I, Parkin DM, Piñeros M, Znaor A et al (2021) Cancer statistics for the year 2020: an overview. Int J Cancer 149(4):778–789
Ferraresi A, Phadngam S, Morani F, Galetto A, Alabiso O, Chiorino G et al (2017) Resveratrol inhibits IL-6-induced ovarian cancer cell migration through epigenetic up-regulation of autophagy. Mol Carcinog 56(3):1164–1181
Folkman J (1984) Angiogenesis:412–428
Fu J, Zeng W, Chen M, Huang L, Li S, Li Z et al (2022) Apigenin suppresses tumor angiogenesis and growth via inhibiting HIF-1α expression in non-small cell lung carcinoma. Chem Biol Interact 361:109966
Galluzzi L, Senovilla L, Zitvogel L, Kroemer G (2012) The secret ally: immunostimulation by anticancer drugs. Nat Rev Drug Discov 11(3):215–233
Georgakopoulos-Soares I, Chartoumpekis DV, Kyriazopoulou V, Zaravinos A (2020) EMT factors and metabolic pathways in cancer. Front Oncol 10:499
Graham J, Quinn M, Fabricant D, Farnsworth NR (2000) Plants used against cancer–an extension of the work of Jonathan Hartwell. J Ethnopharmacol 73(3):347–377
Hamsa TP, Kuttan G (2011) Harmine activates intrinsic and extrinsic pathways of apoptosis in B16F-10 melanoma. Chin Med 6(1):1–8
Han Q, Yuan Q, Meng X, Huo J, Bao Y, Xie G (2017) 6-Shogaol attenuates LPS-induced inflammation in BV2 microglia cells by activating PPAR-γ. Oncotarget 8(26):42001
Hanahan D, Weinberg RA (2011) Hallmarks of cancer: the next generation. Cell 144(5):646–674
Hartwell JL (1967) A survey for plants used against cancer. Lloydia 30:379–436
Hartwell JL (1971) Plants used against cancer. A survey. Lloydia 34(1):103–160

Henrique T, Zanon CF, Girol AP, Stefanini ACB, Contessoto NSA, da Silveira NJ et al (2020) Biological and physical approaches on the role of piplartine (piperlongumine) in cancer. Sci Rep 10(1):1–14

Hertzberg RP, Caranfa MJ, Hecht SM (1989) On the mechanism of topoisomerase I inhibition by camptothecin: evidence for binding to an enzyme-DNA complex. Biochemistry 28(11): 4629–4638

Ho H-H, Chang C-S, Ho W-C, Liao S-Y, Lin W-L, Wang C-J et al (2013) Gallic acid inhibits gastric cancer cells metastasis and invasive growth via increased expression of RhoB, downregulation of AKT/small GTPase signals and inhibition of NF-κB activity. Toxicol Appl Pharmacol 266(1):76–85

Imran M, Saeed F, Gilani SA, Shariati MA, Imran A, Afzaal M et al (2021) Fisetin: an anticancer perspective. Food Sci Nutr 9(1):3–16

Ishikawa H, Akedo I, Suzuki T, Narahara H, Otani T (1997) Adverse effects of sulindac used for prevention of colorectal cancer. J Natl Cancer Inst 89(18):1381

Issa AY, Volate SR, Wargovich MJ (2006) The role of phytochemicals in inhibition of cancer and inflammation: new directions and perspectives. J Food Compost Anal 19(5):405–419

Jangid AK, Patel K, Jain P, Patel S, Medicherla K, Pooja D et al (2021) Carrier-free resveratrol nanoparticles: formulation development, in-vitro anticancer activity, and oral bioavailability evaluation. Mat Lett 302:130340

Jangid AK, Solanki R, Patel S, Pooja D, Kulhari H (2022) Genistein encapsulated inulin-stearic acid bioconjugate nanoparticles: formulation development, characterization and anticancer activity. Int J Biol Macromol 206:213–221

Jeda AS, Ghabeshi S, Jazaeri EO, Araiinejad M, Sheikholeslami F, Abdoli M et al (2022) Autophagy modulation and cancer combination therapy: a smart approach in cancer therapy. Cancer Treat Res Commun 1:100512

Jha AK, Sit N (2021) Extraction of bioactive compounds from plant materials using combination of various novel methods: a review. Trends Food Sci Technol 119:579

Ji Q, Liu X, Fu X, Zhang L, Sui H, Zhou L et al (2013) Resveratrol inhibits invasion and metastasis of colorectal cancer cells via MALAT1 mediated Wnt/β-catenin signal pathway. PloS One 8(11):e78700

Kashyap D, Garg VK, Tuli HS, Yerer MB, Sak K, Sharma AK et al (2019) Fisetin and quercetin: promising flavonoids with chemopreventive potential. Biomolecules 9(5):174

Kasibhatla S, Tseng B (2003) Why target apoptosis in cancer treatment? Mol Cancer Ther 2(6): 573–580

Klionsky DJ (2005) Autophagy. Curr Biol 15(8):R282–R2R3

Lamouille S, Xu J, Derynck R (2014) Molecular mechanisms of epithelial–mesenchymal transition. Nat Rev Mol Cell Biol 15(3):178–196

Levine B (2007) Autophagy and cancer. Nature 446(7137):745–747

Li W, Ma J, Ma Q, Li B, Han L, Liu J et al (2013) Resveratrol inhibits the epithelial-mesenchymal transition of pancreatic cancer cells via suppression of the PI-3K/Akt/NF-κB pathway. Curr Med Chem 20(33):4185–4194

Li J, Chong T, Wang Z, Chen H, Li H, Cao J et al (2014) A novel anti-cancer effect of resveratrol: reversal of epithelial-mesenchymal transition in prostate cancer cells. Mol Med Rep 10(4): 1717–1724

Lima APB, Almeida TC, Barros TMB, Rocha LCM, Garcia CCM, da Silva GN (2020) Toxicogenetic and antiproliferative effects of chrysin in urinary bladder cancer cells. Mutagenesis 35(5):361–371

Liu RH (2004) Potential synergy of phytochemicals in cancer prevention: mechanism of action. J Nutr 134(12):3479S–3485S

Liu J, Li M, Wang Y, Luo J (2017) Curcumin sensitizes prostate cancer cells to radiation partly via epigenetic activation of miR-143 and miR-143 mediated autophagy inhibition. J Drug Target 25(7):645–652

Lowe SW, Lin AW (2000) Apoptosis in cancer. Carcinogenesis 21(3):485–495

Mani R, Natesan V (2018) Chrysin: sources, beneficial pharmacological activities, and molecular mechanism of action. Phytochemistry 145:187–196

Martino E, Casamassima G, Castiglione S, Cellupica E, Pantalone S, Papagni F et al (2018) Vinca alkaloids and analogues as anti-cancer agents: looking back, peering ahead. Bioorg Med Chem Lett 28(17):2816–2826

Masuda M, Suzui M, Lim JT, Deguchi A, Soh JW, Weinstein IB et al (2002) Epigallocatechin-3-gallate decreases VEGF production in head and neck and breast carcinoma cells by inhibiting EGFR-related pathways of signal transduction. J Exp Ther Oncol 2(6):350–359

Melstrom LG, Salabat MR, Ding X-Z, Strouch MJ, Grippo PJ, Mirzoeva S et al (2011) Apigenin down-regulates the hypoxia response genes: HIF-1α, GLUT-1, and VEGF in human pancreatic cancer cells. J Surg Res 167(2):173–181

Mertens-Talcott SU, Percival SS (2005) Ellagic acid and quercetin interact synergistically with resveratrol in the induction of apoptosis and cause transient cell cycle arrest in human leukemia cells. Cancer Lett 218(2):141–151

Miller PE, Snyder DC (2012) Phytochemicals and cancer risk: a review of the epidemiological evidence. Nutr Clin Pract 27(5):599–612

Mizushima N (2007) Autophagy: process and function. Genes Dev 21(22):2861–2873

Murakami M, Hirano T (2012) The molecular mechanisms of chronic inflammation development. Front Immunol 3:323

Nakagawa H, Yamamoto D, Kiyozuka Y, Tsuta K, Uemura Y, Hioki K et al (2000) Effects of genistein and synergistic action in combination with eicosapentaenoic acid on the growth of breast cancer cell lines. J Cancer Res Clin Oncol 126(8):448–454

Nishida N, Yano H, Nishida T, Kamura T, Kojiro M (2006) Angiogenesis in cancer. Vasc Health Risk Manag 2(3):213

Nussbaumer S, Bonnabry P, Veuthey J-L, Fleury-Souverain S (2011) Analysis of anticancer drugs: a review. Talanta 85(5):2265–2289

Nygren P (2001) What is cancer chemotherapy? Acta Oncol 40(2–3):166–174

Oudard S, Fizazi K, Sengeløv L, Daugaard G, Saad F, Hansen S et al (2017) Cabazitaxel versus docetaxel as first-line therapy for patients with metastatic castration-resistant prostate cancer: a randomized phase III trial—FIRSTANA. J Clin Oncol 35(28):3189–3197

Pal HC, Sharma S, Strickland LR, Katiyar SK, Ballestas ME, Athar M et al (2014) Fisetin inhibits human melanoma cell invasion through promotion of mesenchymal to epithelial transition and by targeting MAPK and NFκB signaling pathways. PLoS One 9(1):e86338

Pan MH, Hsieh MC, Hsu PC, Ho SY, Lai CS, Wu H et al (2008) 6-Shogaol suppressed lipopolysaccharide-induced up-expression of iNOS and COX-2 in murine macrophages. Mol Nutr Food Res 52(12):1467–1477

Panda M, Tripathi SK, Zengin G, Biswal BK (2022) Evodiamine as an anticancer agent: a comprehensive review on its therapeutic application, pharmacokinetic, toxicity, and metabolism in various cancers. Cell Biol Toxicol 39:1–31

Parama D, Rana V, Girisa S, Verma E, Daimary UD, Thakur KK et al (2021) The promising potential of piperlongumine as an emerging therapeutics for cancer. Explor Target Antitumor Ther 2(4):323

Patra S, Mishra SR, Behera BP, Mahapatra KK, Panigrahi DP, Bhol CS et al (2020) Autophagy-modulating phytochemicals in cancer therapeutics: current evidences and future perspectives. In: Seminars in cancer biology. Elsevier, London

Pavet V, Portal M, Moulin J, Herbrecht R, Gronemeyer H (2011) Towards novel paradigms for cancer therapy. Oncogene 30(1):1–20

Pooja T, Karunagaran D (2014) Emodin suppresses Wnt signaling in human colorectal cancer cells SW480 and SW620. Eur J Pharmacol 742:55–64

Ralhan R, Kaur J (2007) Alkylating agents and cancer therapy. Expert Opin Ther Pat 17(9): 1061–1075

Rebecca V, Amaravadi RK (2016) Emerging strategies to effectively target autophagy in cancer. Oncogene 35(1):1–11

Roche J (2018) The epithelial-to-mesenchymal transition in cancer. Cancers 10:52
Sartippour MR, Heber D, Ma J, Lu Q, Liang Go V, Nguyen M et al (2001) Green tea and its catechins inhibit breast cancer xenografts. Nutr Cancer 40(2):149–156
Schmidt BM, Ribnicky DM, Lipsky PE, Raskin I (2007) Revisiting the ancient concept of botanical therapeutics. Nat Chem Biol 3(7):360–366
Seo H-S, Choi H-S, Kim S-R, Choi YK, Woo S-M, Shin I et al (2012) Apigenin induces apoptosis via extrinsic pathway, inducing p53 and inhibiting STAT3 and NFκB signaling in HER2-overexpressing breast cancer cells. Mol Cell Biochem 366(1):319–334
Shacter E, Weitzman SA (2002) Chronic inflammation and cancer. Oncology (Williston Park, NY) 16(2):217–226, 29; discussion 30
Shao ZM, Shen ZZ, Liu CH, Sartippour MR, Go VL, Heber D et al (2002) Curcumin exerts multiple suppressive effects on human breast carcinoma cells. Int J Cancer 98(2):234–240
Shewach DS, Kuchta RD (2009) Introduction to cancer chemotherapeutics. Chem Rev 109(7): 2859–2861
Shin S-A, Moon SY, Kim W-Y, Paek S-M, Park HH, Lee CS (2018) Structure-based classification and anti-cancer effects of plant metabolites. Int J Mol Sci 19(9):2651
Singh S, Sharma B, Kanwar SS, Kumar A (2016) Lead phytochemicals for anticancer drug development. Front Plant Sci 7:1667
Soerjomataram I, Bray F (2021) Planning for tomorrow: Global cancer incidence and the role of prevention 2020–2070. Nat Rev Clin Oncol 18(10):663–672
Solanki R, Jodha B, Prabina KE, Aggarwal N, Patel S (2022) Recent advances in phytochemical based nano-drug delivery systems to combat breast cancer: a review. J Drug Deliv Sci Technol 103832:103832
Sporn MB (1996) The war on cancer. Lancet 347(9012):1377–1381
Subramanian A, John A, Vellayappan M, Balaji A, Jaganathan S, Supriyanto E et al (2015) Gallic acid: prospects and molecular mechanisms of its anticancer activity. Rsc Adv 5(45): 35608–35621
Subramanian A, Jaganathan S, Manikandan A, Pandiaraj K, Gomathi N, Supriyanto E (2016) Recent trends in nano-based drug delivery systems for efficient delivery of phytochemicals in chemotherapy. RSC Adv 6(54):48294–48314
Suganuma M, Okabe S, Kai Y, Sueoka N, Sueoka E, Fujiki H (1999) Synergistic effects of (−)-epigallocatechin gallate with (−)-epicatechin, sulindac, or tamoxifen on cancer-preventive activity in the human lung cancercell line PC-9. Cancer Res 59(1):44–47
Sultana S, Munir N, Mahmood Z, Riaz M, Akram M, Rebezov M et al (2021) Molecular targets for the management of cancer using Curcuma longa Linn. Phytoconstituents: a review. Biomed Pharm 135:111078
Sun L, Wang X (2003) Effects of allicin on both telomerase activity and apoptosis in gastric cancer SGC-7901 cells. World J Gastroenterol 9(9):1930
Sung H, Ferlay J, Siegel RL, Laversanne M, Soerjomataram I, Jemal A et al (2021) Global cancer statistics 2020: GLOBOCAN estimates of incidence and mortality worldwide for 36 cancers in 185 countries. CA Cancer J Clin 71(3):209–249
Tabassum DP, Polyak K (2015) Tumorigenesis: it takes a village. Nat Rev Cancer 15(8):473–483
Tan H, Wang N, Man K, Tsao S, Che C, Feng Y et al (2015) Autophagy-induced RelB/p52 activation mediates tumour-associated macrophage repolarisation and suppression of hepatocellular carcinoma by natural compound baicalin. Cell Death Dis 6(10):e1942
Thayyullathil F, Chathoth S, Hago A, Patel M, Galadari S (2008) Rapid reactive oxygen species (ROS) generation induced by curcumin leads to caspase-dependent and-independent apoptosis in L929 cells. Free Radic Biol Med 45(10):1403–1412
Thomas R, Butler E, Macchi F, Williams M (2015) Phytochemicals in cancer prevention and management. Br J Med Pract 8(2):1–8
Tuli HS, Tuorkey MJ, Thakral F, Sak K, Kumar M, Sharma AK et al (2019) Molecular mechanisms of action of genistein in cancer: recent advances. Front Pharm 10:1336

Ulusoy HG, Sanlier N (2020) A minireview of quercetin: from its metabolism to possible mechanisms of its biological activities. Crit Rev Food Sci Nutr 60(19):3290–3303

Vervandier-Fasseur D, Latruffe N (2019) The potential use of resveratrol for cancer prevention. Molecules 24(24):4506

Vincent TL, Gatenby RA (2008) An evolutionary model for initiation, promotion, and progression in carcinogenesis. Int J Oncol 32(4):729–737

Wang K, Liu R, Li J, Mao J, Lei Y, Wu J et al (2011) Quercetin induces protective autophagy in gastric cancer cells: involvement of Akt-mTOR-and hypoxia-induced factor 1α-mediated signaling. Autophagy 7(9):966–978

Wang H, Zhang H, Tang L, Chen H, Wu C, Zhao M et al (2013) Resveratrol inhibits TGF-β1-induced epithelial-to-mesenchymal transition and suppresses lung cancer invasion and metastasis. Toxicology 303:139–146

Wang Y, Liu Y, Du X, Ma H, Yao J (2020) The anti-cancer mechanisms of berberine: a review. Cancer Manag Res 12:695

Wong RS (2011) Apoptosis in cancer: from pathogenesis to treatment. J Exp Clin Cancer Re 30(1): 1–14

Xiao Q, Qu Z, Zhao Y, Yang L, Gao P (2017) Orientin ameliorates LPS-induced inflammatory responses through the inhibitory of the NF-κB pathway and NLRP3 inflammasome. Evid Based Complement Altern Med 2017:1

Yang B, Huang J, Xiang T, Yin X, Luo X, Huang J et al (2014) Chrysin inhibits metastatic potential of human triple-negative breast cancer cells by modulating matrix metalloproteinase-10, epithelial to mesenchymal transition, and PI3K/Akt signaling pathway. J Appl Toxicol 34(1):105–112

Yang L, Li X, Huang W, Rao X, Lai Y (2022) Pharmacological properties of indirubin and its derivatives. Biomed Pharmacother 151:113112

Yin Z, Zhang J, Chen L, Guo Q, Yang B, Zhang W et al (2020) Anticancer effects and mechanisms of action of plumbagin: review of research advances. Biomed Res Int 2020:6940953

Yoon H, Liu RH (2007) Effect of selected phytochemicals and apple extracts on NF-κB activation in human breast cancer MCF-7 cells. J Agric Food Chem 55(8):3167–3173

Zari AT, Zari TA, Hakeem KR (2021) Anticancer properties of eugenol: a review. Molecules 26(23):7407

Zhang Y, Weinberg RA (2018) Epithelial-to-mesenchymal transition in cancer: complexity and opportunities. Front Med 12(4):361–373

Zhang W, Ha M, Gong Y, Xu Y, Dong N, Yuan Y (2010) Allicin induces apoptosis in gastric cancer cells through activation of both extrinsic and intrinsic pathways. Oncol Rep 24(6): 1585–1592

Zhang L, Li D, Yu S (2020) Pharmacological effects of harmine and its derivatives: a review. Arch Pharm Res 43(12):1259–1275

Zhou Y, Yu Y, Lv H, Zhang H, Liang T, Zhou G et al (2022) Apigenin in cancer therapy: from mechanism of action to nano-therapeutic agent. Food Chem Toxicol 1:113385

Zhu B-H, Zhan W-H, Li Z-R, Wang Z, He Y-L, Peng J-S et al (2007) (−)-Epigallocatechin-3-gallate inhibits growth of gastric cancer by reducing VEGF production and angiogenesis. World J Gastroenterol 13(8):1162

Polymeric Nanocarriers for the Delivery of Phytoconstituents

4

Kanika Verma, Akanksha Chaturvedi, Sarvesh Paliwal, Jaya Dwivedi, and Swapnil Sharma

4.1 Introduction

It is well known that herbal medicines have been recognized globally because of their enormous benefits. People are shifting to herbal products to ensure safety with better efficacy. Hence, the demand for phytoconstituents is also increasing to treat both mild colds and life-threatening cancer. The use of phytomedicine is popularising in the field of skin, metabolic, locomotor, neurodegenerative, or cancer diseases (Rajagopal et al. 2022). But still, along with uncountable benefits, it also has some limitations such as reduced bioavailability of some bioactive, possible toxicity, and incomplete data of clinical trials restricting their usage. Therefore, nanotechnology plays a crucial role in breaking all the limitations associated with herbal drug formulations prepared using conventional methods.

Nanotechnology is spreading its characteristic features across several fields of health and medical sciences. Nanotechnology overcame the drawbacks of conventional formulations by enhancing bioactive bioavailability, biocompatibility, and therapeutic efficacy (Kesharwani 2018). Nanocarriers have gained a lot of attention from formulators as they can easily penetrate the cell membrane of the target cell and can prolong the effect of a drug without inducing any severe side effects. This allows easy incorporation of phytoconstituents into a nanocarrier that could target the target site without any difficulty. Nanocarriers are mainly of four types: biological, polymeric, inorganic, and organic (dendrimers, liposomes, and micelles) (Mansoori et al. 2017).

K. Verma · A. Chaturvedi · S. Paliwal · S. Sharma (✉)
Department of Pharmacy, Banasthali Vidyapith, Jaipur, Rajasthan, India

J. Dwivedi
Department of Chemistry, Banasthali Vidyapith, Jaipur, Rajasthan, India

© The Author(s), under exclusive license to Springer Nature Singapore Pte Ltd. 2024

D. Pooja, H. Kulhari (eds.), *Nanotechnology Based Delivery of Phytoconstituents and Cosmeceuticals*, https://doi.org/10.1007/978-981-99-5314-1_4

Polymeric nanocarriers have shown significant potential in terms of biodegradability, biocompatibility, and sustained effect of the drug. Polymeric nanocarriers are made up of repeating units of monomers that dissociate in the human body to release drugs in a sustained manner. It enhances the water solubility and increases its permeability to the cell by providing an alternative pathway of cell penetration (De et al. 2022). The nanocarriers have peculiar characters and are influenced by physicochemical properties to alter the pharmacokinetic parameters of a drug and the biodegradability of a polymer. Ideally, biodegradable polymers are used to design nanocarriers, but for some conditions such as dental restoration, nonbiodegradable polymers could also be used. Polymeric nanocarriers offer a controlled release of a compound by applying two mechanisms: endogenous (within the body), that is, pH, temperature, and enzymatic degradation of a polymer, and exogenous (influenced by external environment), that is depends upon light, magnetic, and ultrasonic effects (Catalin Balaure and Mihai Grumezescu 2015).

A combination of polymeric nanocarriers and phytoconstituents has set a remarkable impression in the research field. It combines the benefits of phytoconstituents and nanocarriers and masks its limitations to present an effective nanoformulation that possesses the potential to treat chronic diseases. Many phytoconstituents like polyphenols, viz., naringenin, apigenin, resveratrol, curcumin, etc. are formulated as polymeric nanocarriers and tested in vitro or in vivo (Krsti'c et al. 2021). In the present chapter, we are going to discuss polymeric nanocarriers and different methods of preparation. The chapter will also focus on the types of polymers and the influence of physicochemical properties on these polymers. Preclinical investigations of phytoconstituent-loaded polymeric nanocarriers are also discussed in this chapter along with their future perspectives.

4.2 Polymeric Nanocarriers

Polymeric nanocarrier is a promising candidate to deliver drugs at the target site. Its use is more prevalent in the areas of medicine and biology because of its attractive features such as significant pharmacokinetics, therapeutic efficacy, and biopharmaceutics (Mogosanu et al. 2016; Dang and Guan 2020). They have limited the drawbacks associated with conventional approaches, such as usage of high dosages, low bioavailability, poor biocompatibility, high risk of toxicity, the short half-life of drugs, difficulty to incorporate water-insoluble drugs, and usage of nondegradable compounds (Roy et al. 2017; Tiwari et al. 2012).

Polymeric nanocarriers are composed of repeating units of monomers linked with a wide range of functional groups (De et al. 2022). The use of polymers allows a nanocarrier to be biodegradable and biocompatible. Proper selection of polymer potentiates the therapeutic effect of the drug at the target site and ensures its sustained release. This also reduces the drug dosage making it a convenient option for patients and ensures the controlled release of drugs. Specifically, stimuli-responsive polymers play a crucial role in targeted drug delivery (Roy et al. 2017; Tiwari et al. 2012). The polymers are responsible to maintain the pharmacokinetic properties of a compound and are influenced by changing its physicochemical features like structure, stability, molecular weight, and size of a polymer. For instance, reducing its size can help a polymer to cross even small capillaries of blood vessels and mucous layers of target cells (Kamaly et al. 2016). The polymers are efficient in encapsulating both hydrophilic and lipophilic molecules under the same unit. Polymeric nanocarriers gained enormous attention to develop anticancer, antimicrobial, antidiabetic, and ocular drugs that assure all the characteristics required to potentiate the efficacy of a compound. In 1976, Langer and Folkman developed the first polymeric nanocarrier to target cancer cells and also ensured the sustained release of a drug. Earlier in the development of polymeric nanocarriers, nondegradable polymers such as polystyrene, poly(methyl methacrylate) (PMMA), polyacrylates, and polyacrylamides with rapid clearance and better therapeutic effect but chronic toxicity were used. In addition, nondegradable polymers were also associated with inflammatory responses. Hence, later nondegradable polymers were replaced with biodegradable polymers which possess impressive physicochemical properties and are temperature and pH-dependent (Anju et al. 2020).

While designing a nanocarrier, it is essential to assure no interaction of the outer surface of polymers with proteins present at the target site (Chen et al. 2008). After ingestion of polymeric nanocarrier, intracellular and extracellular depolymerase play a key role to degrade the polymer. Extracellular polymerase (released from cells) depolymerizes polymeric chains and produces oligomers, dimers, and monomers. On the other hand, intracellular depolymerase penetrates the cell membrane of the target site to degrade the polymer and release the drug. Another way of degradation includes hydrolysis of the polymer (poly(lactic-co-glycolide) (PLGA) degrades after hydrolysis of ester bonds). The degradation time decides the fate of drug release and ensures the half-life of the compound (Makadia and Siegel 2011). Moreover, polymeric nanocarriers surpass the benefits of other nanocarriers like micelles, inorganic nanosystems, and dendrimers (van Vlerken et al. 2007). They ensure safety and better tolerability along with biocompatibility. Some of the biodegradable polymers are even FDA and European Medicines Agency (EMA) approved (Palma et al. 2018). The nanocarriers are prepared by different methods depending on the type of drug to be incorporated. Polymeric nanocarriers like micelles, polymersomes, polyplex, nanocapsules, and nanogels are prepared for targeted drug delivery (Fig. 4.1). In this chapter, we will be covering all the possible aspects of polymeric nanocarriers.

4.3 Types of Polymers

Polymers used in a nanosystem are prepared by polymerizing the monomers and then organized and arranged in a manner to develop a nanocarrier. Their size ranges between 10 and 100 nm. Polymeric nanocarriers have been used in several ailments like cancer, diabetes, inflammation, and ophthalmic, cerebral, and infectious diseases. Nowadays, biodegradable polymers are more emphasized as nondegradable polymers take a longer duration of time to degrade. Conversely, biodegradable

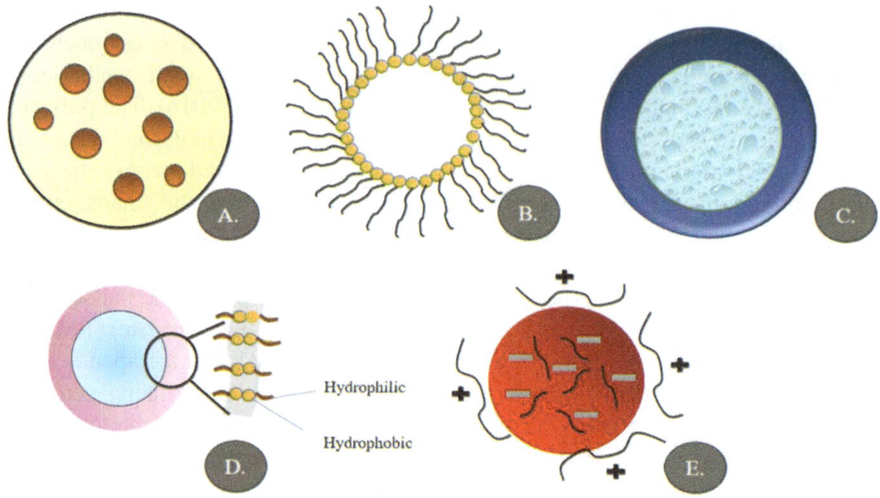

Fig. 4.1 Schematic representation of polymeric nanocarriers. (**a**) Polymeric nanoparticle, (**b**) polymeric micelle, (**c**) nanocapsule, (**d**) polyplex, (**e**) polymersomes

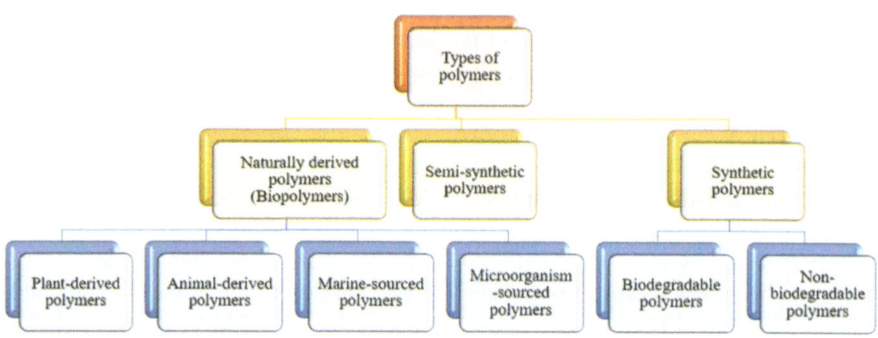

Fig. 4.2 Types of polymers

polymers depend on certain physicochemical parameters such as structure, size, and molecular weight and a few environmental factors like pH and temperature. These polymers can be used as plasma expanders and surgical materials or may be incorporated for bone replacement, and delivery systems for vaccines, genes, proteins, and peptides to formulate a better treatment against untreatable diseases (De et al. 2022).

Polymers are broadly classified under three categories. These include natural, semisynthetic, and synthetic polymers (Figs. 4.2 and 4.3). All these categories are discussed in detail in this section.

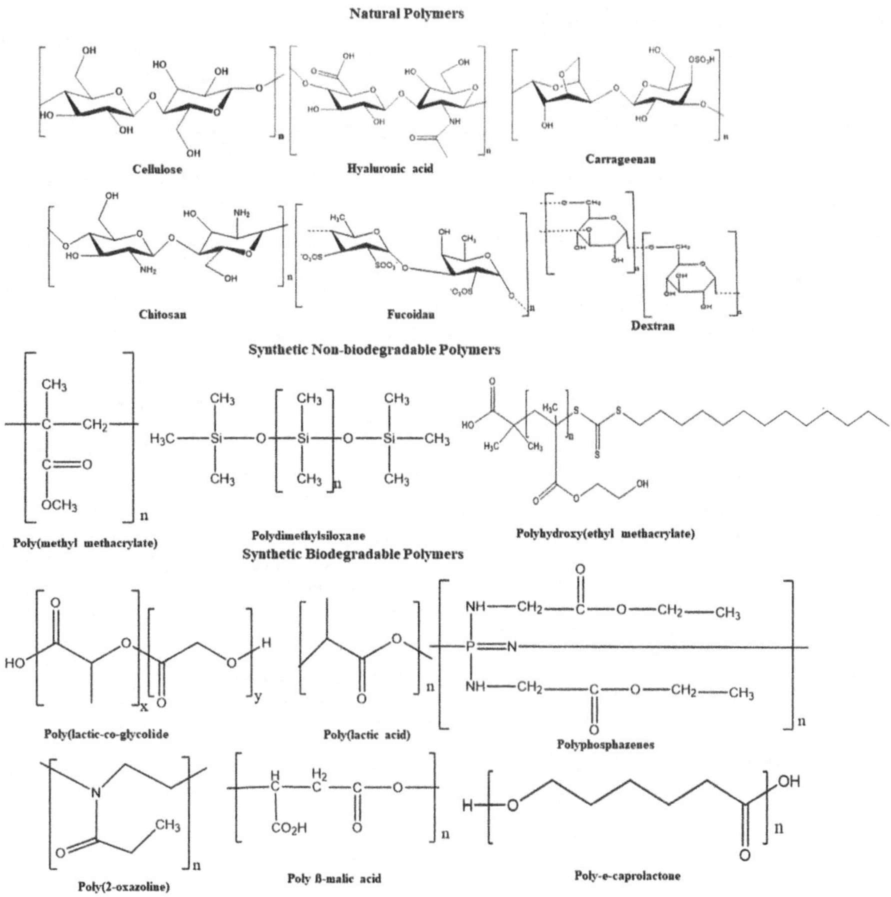

Fig. 4.3 Chemical structures of common polymers used in the manufacturing of polymeric nanocarriers

4.3.1 Natural Polymers

Natural polymers are also referred to as biopolymers, as they are biodegradable and biocompatible in nature. This class possesses different categories of proteins and polysaccharides such as chitosan, cellulose, starch, gelatin, collagen, albumin, soy protein, zein, hyaluronic acid, mauran, carrageenan, alginate, etc. They are generally used in gene therapy, cell-based transplantation, and tissue engineering. Because of their biocompatibility, nontoxicity, biodegradability, and abundance in the environment, natural polymers are preferred. But their immunogenic characteristic limits their usage (Eroglu et al. 2017; Gálisová et al. 2020; Karlsson et al. 2018). Natural polymers could be obtained from plants, animals, marine sources, and microbes.

4.3.1.1 Plant-Derived Polymers

These are abundant in nature with less immunogenic properties than animal-based polymers. This class involves a variety of polymers (Table 4.1; Fig. 4.2).

Cellulose

Cellulose is a homopolymer of 1,4-β-glycoside-linked D-glucopyranose which is highly hydrophilic in nature that resists phagocytosis of the polymer (Fig. 4.3). It is not immunogenic and, thus, safe to use. It resists phagocytosis and is the most suitable polymer for incorporating hydrophobic drugs. It prolongs its residence time in the bloodstream and accumulates easily at the target site to show therapeutic action (Meyabadi et al. 2014; Halib et al. 2017). Even though cellulose possesses

Table 4.1 Types of polymeric nanocarriers for targeted drug delivery based on polymer

Classification	Name of polymers	Applications	Reference
Naturally derived polymers			
Plant-based	Soy protein	Gene delivery	MaHam et al. (2009); Cavalu et al. (2018)
Animal-based	Gelatin, collagen, hyaluronic acid	Controlled release of drugs, gene delivery, formation of nanoparticles, film coating, and binding agents	
Marine-sourced	Guar gum, alginate, cyclodextrins, chitosan	Film coating and binding agents and facilitates the sustained release of drugs	
Synthetic polymers			
Biodegradable			
Polyesters	Poly (glycolic acid), poly (lactic acid), poly(dioxanes)	Dialysis of membrane and protein delivery	Rai et al. (2021)
Cellulose	Ethyl cellulose, hydroxy propyl methyl cellulose, carboxymethylcellulose, cellulose acetate	Emulsifying, binding, coating, and disintegrating agents in tablets and capsules	Arif et al. (2022)
Polyamides	Poly(iminocarbonates), polyamino acids	Preferred for sustained and controlled release of drugs, catheters, and sutures for angioplasty	Avcu et al. (2022)
Polyanhydrides	Poly(sebacic) acid, poly (adipic) acid	Used to coat drugs for controlled release and used in medical implants	Kenry and Liu (2018)
Nonbiodegradable polymers			
Acrylic polymers	Polyhydroxy (ethyl acrylates), polymethacrylates	Thermos-gelling agents that show high H-bond interactions and show changes with a change in temperature in hydrophobic molecules	Priya James et al. (2014)
Silicones	Polydimethylsiloxane, colloidal silica	Used in transdermal delivery of drugs, and implants, and used in therapeutic devices	Mashak and Rahimi (2009)

these advantages that could help to prepare a sustained release drug delivery nanocarrier, it is noteworthy to have a sound knowledge regarding the manufacturing, capital, and processing while using cellulose to develop a cost-effective product. In addition, selecting a particular type of cellulose (such as bacterial or plant-based) is an important step in designing a nanoparticle (Masek and Kosmalska 2022).

Zein

It is isolated from the cytoplasm of corn cell endosperm that is insoluble in water but solubilizes in the presence of alkali, anionic detergents, urea, and alcohol. Zein has a low molecular weight (around 20 kDa) and hence is a promising agent in developing a nanocarrier. However, its insolubility in water often limits the encapsulation of hydrophilic components into the protein for oral drug delivery systems (DeFrates et al. 2018; Gagliardi et al. 2019, 2021; Elzoghby et al. 2017).

Starch

Starch is a polysaccharide that is generally obtained from tuberous plants such as potatoes and cereals like wheat, rice, corn, and beans. It is metabolized by glucosidases and amylases into small glucose units that serve as an energy reserve for different plant parts (Alcázar-Alay and Meireles 2015). It is used as a polymer in a nanosystem to enhance the bioavailability, biocompatibility, drug stability, and solubility and to minimize toxicity (George et al. 2019). Because of these advantages, starch is an alternative to developing polymeric nanocarriers. In spite of that, inorganic materials are preferred over starch in the development of a nanocarrier in order to obtain a cost-effective end product. This is due to the time and capital investment required to develop and calibrate a protocol that could incorporate starch as a polymer to design a nanocarrier (Abe et al. 2021).

Soy Protein

Soy proteins are cost-friendly and abundant agents found in nature. It is composed of polar and nonpolar groups along with bonding with some amino acids such as aspartate, leucine, and glutamate. It provides significant efficiency to entrap hydrophobic compounds. It also provides desired solubility characteristics to administer the drug from different routes (DeFrates et al. 2018). Chien and Shah (2012) prepared scaffolds for tissue regeneration using soy protein, and it suggested better cell proliferation in a duration of 15 days after treating it with an enzyme transglutaminase. However, the enzyme crosslinker could not boost the mechanical strength of the scaffold, and the cell proliferation at different concentrations of soy protein (3% and 5%) was also uncertain in the presence of transglutaminase. Hence, cell studies are essential to justify the biocompatibility of soy protein in tissue regeneration.

4.3.1.2 Animal-Derived Polymers

These are obtained from animals and are formulated to develop a nanocarrier. These types of polymers are discussed in this section (Table 4.1).

Albumin

It is a negatively charged blood protein with a molecular weight of around 69kDA. Albumin is extracted from human plasma and then fractionated with chilled alcohol that is later cooled down to avoid any microbial growth. Along with human serum albumin, bovine serum albumin is also used in developing a nanosystem. Albumin provides longer t1/2 and serves better pharmacokinetic properties (Tullis 1977). Yet the method of manufacturing largely contributes to the biological activity of a nanosystem.

Hyaluronic Acid

It is a mucopolysaccharide that is made up of D-glucuronic acid and N-acetylglucosamine that are linked alternatively with β-1,4- and β-1,3-glycosidic bonds (Fig. 4.3). These are isolated from intracellular domains and extracellular matrix of all living organisms. Hyaluronic acid is metabolized by the enzyme hyaluronidase and rapidly clears from the body (Tripodo et al. 2015). Even after the clearance of hyaluronic acid, the carboxyl and hydroxyl groups linked with amino acids promotes the blood residence time of the polymer. Therefore, because of their long-acting and target-specific properties, these are preferable candidates in pharmaceutical and biomedical preparations (Taghipour-Sabzevar et al. 2019). Hsiao et al. (2019) developed an antibiotic system using hyaluronic acid incorporating epigallocatechin that could combat antibiotic resistance. Although this controlled released system proved itself effective by boosting the healing of the skin and suppressing the oxidative stress in a duration of 10 days, it is less efficient in wound healing compared to collagenase-induced models.

Gelatin

Gelatin is a heterogeneous mixture isolated from animal collagen. It is obtained by the process of partial hydrolysis that forms two types of gelatin, i.e., type A and type B (Karlsson et al. 2018). Gelatin is a better option to design polymeric nanocarriers as it is cheap, easy to source, and easily modifiable to take targeting molecules to the desired site (Elzoghby et al. 2017). But still, along with many advantages, it also carries some limitations such as the induction of communicable infections due to contamination. This can be prevented by developing recombinant human gelatin. It can be adapted by formulators if the cost of the polymer could be deduced.

4.3.1.3 Marine Organisms: Originated Polymers

Marine organisms are another source to obtain naturally derived polymers that could be renewable, nontoxic, and stable. Some of the marine-sourced polymers are alginate, fucoidan, carrageenan, and chitosan (Table 4.1).

Carrageenan

It is a sulfated polysaccharide obtained from red algae called *Rhodophyta*. The sulfated polysaccharide is composed of D-galactose and 3,6-anhydrogalactose linked by α-1,3- and β-1,4-glycosidic linkages arranged alternatively (Fig. 4.3). It is reported that the polymer prolongs the drug release, and hence, it is the optimal choice when targeting epithelial or mucosal tissues. Carrageenan is of three types, i.e., λ, ι, and κ type to which λ type is the most sulfated one (Shukla et al. 2016). Contrarily, κ type of the polymer suggested better loading capacity of a drug and caused the sustained release. This was explained by Sathuvan et al. (2017) while encapsulating curcumin with κ-type of polymer. Still, establishing a correlation between the chemical structure and biological activity is necessary in order to obtain desired and sustained effect (Mokhtari et al. 2021).

Agarose

Agar is constituted of two components: agarose and agaropectin. Agarose is obtained from red seaweed and contains a dominant part of agar and other parts of agaropectin. Agar is made up of linear polymers arranged by repeating units of monomeric agarobiose. On the other hand, agaropectin is a heterogeneous compound made up of D-galactose and 3,6-anhydro-L-galactopyranose (Chopin et al. 2014). Kolanthai et al. (2017) designed agarose incorporated polymer that showed satisfactory results in drug delivery, but immunogenicity and variation while manufacturing are certain limitations that are still needed to be improved.

Chitosan

It is a marine-sourced polysaccharide obtained from shrimps' or crabs' shells. It is an FDA-approved polymer because of its benefits that start from reduced toxicity, and biocompatibility, to non-immunogenicity (George et al. 2019). It is a positively charged polysaccharide that could easily bind with negative compounds and attain stability to produce a controlled release of a drug. It is comprised of N-acetyl-glucosamine and glucosamine (Fig. 4.3) and is optimal to be involved in the formation of a nanostructure because its bio-adhesive properties allow it to penetrate small junctions of endothelial cells (Safdar et al. 2019). Still, it is challenging to incorporate drug molecules of alkaline nature along with reduced mechanical stress of the polymer. The method of manufacturing can hinder the physicochemical properties of a nanocarrier (Garg et al. 2019).

Alginate

It is also obtained from seaweeds that produce polysaccharides made up of β-L-guluronic and α-D-mannuronic acids. It is cost-effective and can easily be incorporated with phytoconstituents making it an ideal polymer to develop nanocarriers (Joye and McClements 2014). A limitation of this polymer is that it could get contaminated with pollutants that could develop an immune response in the human body (Lapasin 2012). In addition, their bulkier chemical structure makes the manufacturing procedure more complicated and hence restricted to imply targeted drug delivery.

Fucoidan

It is a sulfated polysaccharide obtained from brown seaweed composed of sulfated ester groups and L-fucopyranose units (Fig. 4.3). It is reported to possess ideal properties to design anticancer drugs using the polymer (Wu et al. 2016). However, extraction of fucoidan by employing different manufacturing techniques such as ultrasound-assisted extraction or ultrafiltration can largely influence the quality and biological activity of the polymer. In addition, it also alters the physicochemical properties of the polymer leading to instability (Haggag et al. 2023).

4.3.1.4 Microorganism-Sourced Polymers

From the microbes, many polymers were outsourced that are exploited to design nanocarriers. Some of them will be discussed in this section.

Dextran

It is extracted from enzymes of the bacteria *Streptococcus mutans* that convert sucrose into dextran (Fig. 4.3). It could also be developed by chemical modifications in levoglucosan (ring-opening polymerization). It is a water-soluble and high molecular weight polymer that could hydrolyze under extreme pH conditions (highly acidic or alkaline environment) (Dias et al. 2011). Because of its biodegradability and biocompatibility, Jamwal et al. 2019 designed glucose oxidase immobilized-acryloyl crosslinked dextran dialdehyde nanoparticles that showed effective release of insulin. Nonetheless, it is still not used extensively in clinical trials due to the presence of some other synthetic polymers in the market.

Mauran

It is an exopolysaccharide obtained from the bacteria *Halomonas maura.* It is made up of repeating units of glucose, mannose, glucuronic acid, and galactose (Arias et al. 2003). It could bear extreme temperatures, pH, salt concentrations, and freeze-thawing. Its unique thixotropic properties steal the attention of researchers (Llamas et al. 2006). It contains high concentrations of sulfate and uronic acid that exhibit antiproliferative and immunomodulatory activities (Arias et al. 2003). Antigenicity and variation while manufacturing could be the possible drawbacks that limit its usage (Dadwal et al. 2014).

4.3.2 Semisynthetic Polymers

The semisynthetic polymers are designed by combining natural and synthetic molecules. These are produced by chemically modifying the structure of either natural or synthetic molecules and mimic components of human tissue.

4.3.3 Synthetic Polymers

Synthetic polymers gained a lot of attention because of their unique characteristics including flexibility, stability, reduced immunogenicity, high-temperature tolerability, and biodegradability. Some of the known synthetic degradable polymers are poly(lactones), poly(α-hydroxy acids), poly(alkylcyanoacrylates), and polyhydroxyalkanoates (Abasian et al. 2020) (Table 4.1). Synthetic polymers have been divided into biodegradable and nonbiodegradable polymers. Nondegradable polymers are preferred in genetic engineering to design nanoformulations that could exist for a longer duration of time. For instance, Zato et al. developed nanoformulations intending to restore dental health. It was prepared by adding 2-hydroxyethyl methacrylate and trimethylolpropane trimethacrylate that preserves cetylpyridinium chloride or fibroblast growth factor-2 (FGF-2) during dental restoration (Imazato et al. 2017). Polymers such as polyhydro(ethyl methacrylate) (Fig. 4.3), polymethacrylates, polydimethylsiloxane (Fig. 4.3), poly(methyl methacrylate) (Fig. 4.3), and colloidal silica acrylic polymers are certain examples of nonbiodegradable polymers. These polymers possess some unavoidable adverse effects which range from heat sensitivity, the hindered release of drug with time sustainably, uneven drug distribution, to chances of microbial growth (Kluin et al. 2013).

On the other hand, biodegradable polymers are used to exhibit the release of drug-influenced by pH response, surrounding nature, or in vivo degradation to remove drugs and polymers from the body easily (Rahman and Hasan 2019). Some of the biodegradable polymers are discussed in this section.

4.3.3.1 Poly(Dioxanones) (PDX)

The polymer is composed of both ether and ester groups that provide some unique characteristics making it different from other polymers. The presence of structural specificity and bulk material characteristics makes it an ideal candidate to process as a nanoformulation. The copolymerization of polymers ensures the sustained release of a drug. Hence, it is frequently used in tissue engineering to develop cartilage, vascular tissues, urologic parts, and scaffolds in bone. It could also be incorporated in nano-sizes to manufacture a nanocarrier. For instance, Goonoo et al. (2015) reported preparations of a microparticle using this polymer loaded with bovine serum protein. Poly(dioxanones) include variety of polymers such as P(HDX-co-DX) [poly(5-hydroxymethyl-1,4-dioxan-2-one-co-DX)], chitosan-g-PDX copolymers, PBDX [poly(5-benzyloxymethyl-1,4-dioxan-2-one], and poly (x-pentadecalactone-co-dioxanone)[poly(PDL-co-DX)] copolymers.

4.3.3.2 Polyesters

It is constituted of ester groups that are synthesized by esterification condensation of acids and alcohols. The degradation rate of the polymer is ensured by setting its crystallinity, copolymer composition, stereochemistry, and molecular weight of the monomers. The polymer category includes poly-β-malic acid (PMLA), poly (caprolactones)s (PCL), poly(dioxanone)s, poly(3-hydroxybutyrate), poly(lactide)s (PLA), poly(glycolide)s (PGA), and PLGA (Rowe et al. 2016; Södergård and Stolt 2002).

4.3.3.3 Poly(Lactic-Co-Glycolide)

PLGA is a naturally derived polymer chemically modified by ring-opening polymerization that carries specific features such as better crystallinity, optimum molecular weight, and biodegradable and maintained a balance between hydrophilic and lipophilic molecules (Sj 2016). The polymer is composed of poly(glycolic acid) (PGA) and PLA (Fig. 4.3). PGA provides rapid degradation and better incorporation of hydrophilic molecules along with enhanced crystallinity compared to PLA. Hence, the ratio maintained by these two copolymers to attain optimal degradation and other characteristics needed to ensure the sustained release of the drug (Sun et al. 2017). Alteration in the ratios can limit the biological activity of the polymer by degrading it in acidic conditions due to hydrolysis and decreases the immunogenic features present in it while storing the nanoparticle.

4.3.3.4 Poly(Lactic Acid)

Polylactide (PLA) is an aliphatic polyester extracted from natural sources such as sugarcane, cornstarch, and tapioca starch. It is chemically modified by converting 2-hydroxypropanoic acid (lactic acid) by ring-opening polymerization or polycondensation (Fig. 4.3). It is biocompatible and biodegradable which increases its demand to synthesize nanoformulation. A combination of PLA-polyethylene glycol (PEG) can potentiate water solubility and systemic circulation of a drug. Another combination with PLA is polyaspartic acid (PA) which is less toxic to the environment and hence preferred as an alternative (Sin and Tueen 2019; Li and Yuan 2007; Tudorachi et al. 2013). Although its magnificent characteristics increase its usage for the preparation of a nanoparticle, its slow degradation rate, brittleness, and hydrophobic nature can counter as a challenge while manufacturing the nanoformulation alongside difficulty in altering the chemical structure of PLA to design in a desired form (Casalini et al. 2019).

4.3.3.5 Poly(2-Oxazolines)

It is prepared by ring-opening polymerization that produces better characteristics such as dispersity, architecture, and molar mass of the polymer (Fig. 4.3). It is composed of temperature-sensitive, lipophilic, and lipophobic groups that made it a versatile polymer. An example of this class of polymer is PEG which exhibits nontoxicity in vitro and in vivo and surface-coating properties. But still, the use is limited due to certain limitations such as its polyether group tends to undergo oxidation resulting in degradation of the polymer. This limits the use of polymer for long-term release. But at the same time, a part of polymer poly(2-alkyl/aryl-2-oxazoline)s (PAO_x) shows better stability, targeting, and drug loading capacity (Bauer et al. 2012).

4.3.3.6 Phosphate-Linked Polymers

The polymer gained a lot of attention in the field of biomedical research as the macromolecules such as RNA and DNA possess phosphate bonds that made it suitable to design a formulation using phosphorus-based polymers that could easily target genetic material of the body. A combination of the ester group along with phosphate group can serve more biodegradability and better efficacy to handle water molecules. Poly(phosphate ester) is an example of such a category. Other phosphate-based polymers are polyphosphazenes (Fig. 4.3) and polyphosphonates. Chemical modification in the structure of the polymer is essential in order to incorporate hydrophobic or hydrophilic drug molecules (Monge and David 2014).

4.3.3.7 Poly-β-Malic Acid

PMLA is a polyester chemically prepared by adding aspartic acid and malic acid that undergoes polycondensation and ring-opening polymerization. It could incorporate water-soluble molecules easily (Fig. 4.3). PMLA is made up of repeating units of L-mall that are linked with ester bonds and arranged with hydroxyl and carbonyl groups. This synthesizes malic acid. It is used to carry hydrophilic molecules (Lanz Landazuri et al. 2014). PMLA bears negative charge that makes the nanocarrier difficult to penetrate the cell and, hence, reduces its cellular uptake (Zhou et al. 2015).

4.3.3.8 Polyanhydrides (PAs)

They are biocompatible, cost-effective, nontoxic, hydrophobic, not easily degradable, and can easily be excreted from the body. Its solubility is attained in organic solvents that reduce the chances of toxicity. Some PA-based compounds such as Septacin (Abbott Laboratories, North Chicago, Illinois, and Gliadel (MGI Pharma Inc., Bloomington, Minnesota)) are under clinical trials. Copolymers of PAs are poly (sebacic acid) and poly(adipic acid) (Jain et al. 2008). However, they are unstable at extreme temperatures as they degrade by hydrolytic cleavage and form polymeric rings and cyclic dimers (Hill 1930).

4.3.3.9 Poly-ε-Caprolactone (PCL)

It is an aliphatic polyester prepared by ring-opening polymerization of ε-caprolactone (Fig. 4.3). It is semicrystalline, biodegradable, biocompatible, and nontoxic, with a low melting point (59–64 °C) and glass-transition temperature (60 ° C). It is implied in tissue engineering, biomedical research, and controlled release of the drug (Schnell et al. 2007). It has the slowest degradation rate (3–4 years) and, hence, can only be used for long-term drug delivery systems such as bone tissue engineering and implants that could treat for years (Arakawa and DeForest 2017).

4.4 Method of Preparation of Polymeric Nanocarriers

Polymeric nanocarriers are designed to encapsulate both hydrophilic and lipophilic molecules under the same nanocarrier. Molecules like proteins, high molecular weight DNA, and salts can be encapsulated in polymeric nanocarriers (Cosco et al. 2014, 2015; Lombardo et al. 2018). This encapsulation ensures the sustained release of a drug molecule that binds at the target site. Polymeric nanocarriers can be prepared either by encapsulation, entrapment, or binding with polymeric nanoparticles such as nanocapsules, polymeric micelles, nanosphere, or drug conjugate (Dhathathreyan et al. 2017; Patra et al. 2018; Catalin Balaure and Mihai Grumezescu 2015). With the help of these methods of preparations, drugs can be released either by diffusing through polymers, eroding the polymeric matrix, hydrostatic swelling, or a combination of all these methods (Crucho and Barros 2017).

Drugs are dispensed in nanocarriers via these methods. Some of the methods of preparation are discussed in this section.

4.4.1 Emulsification and Solvent Extraction/Evaporation Technique

The method is suitable for the small-scale production of polymeric nanocarriers. It is based on the principle of dissolution of polymers in such a manner that forms an organic solvent layer around the water phase. A solution of drug/polymer is mixed in a volatile organic solvent and then added to the water phase to form an oil-in-water (O/W) emulsion. This emulsion is stabilized using a surfactant (Figs. 4.4 and 4.5). Commonly used organic solvents are ethyl acetate, chloroform, and dichloromethane (Guo et al. 2015). The organic solvent is evaporated under reduced pressure at room temperature after ultrasonication, or homogenization. To eliminate organic solvents, spray drying can also opt for large-scale production of polymeric nanocarriers (Nava-Arzaluz et al. 2012). The spray drying technique is a method of choice for drying heat-sensitive components (Ozeki and Tagami 2014). These precipitate out the nanostructures that are collected by ultracentrifugation (Nava-Arzaluz et al. 2012). The nanostructures are then washed with distilled water to remove toxic residues and residual solvents. Depending upon the nature of the drug/polymer, i.e., either hydrophilic or lipophilic, oil-in-water (o/w) or water-in-oil/in-water (w/o/w) emulsions can be formed (Cosco et al. 2015). For emulsification of hydrophobic drugs, o/w, and hydrophilic drugs, w/o/w emulsions are preferred.

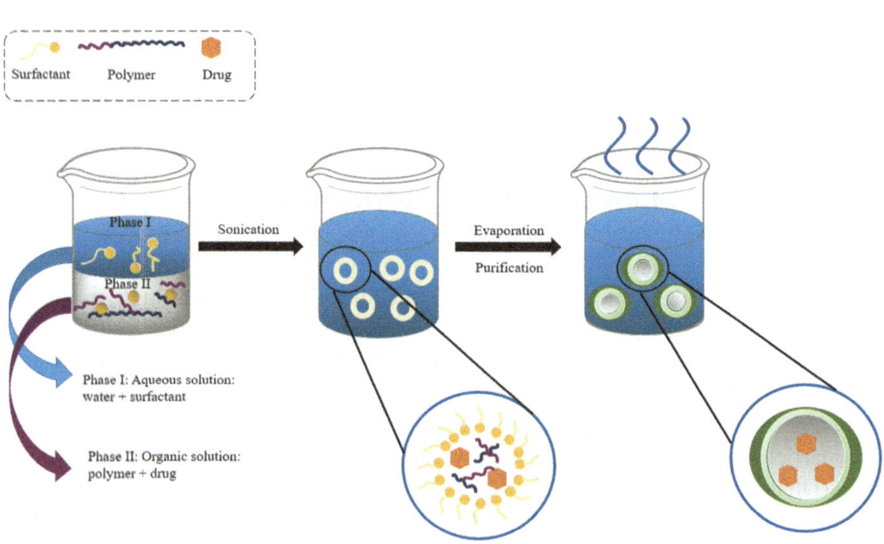

Fig. 4.4 Schematic representation of solvent evaporation method

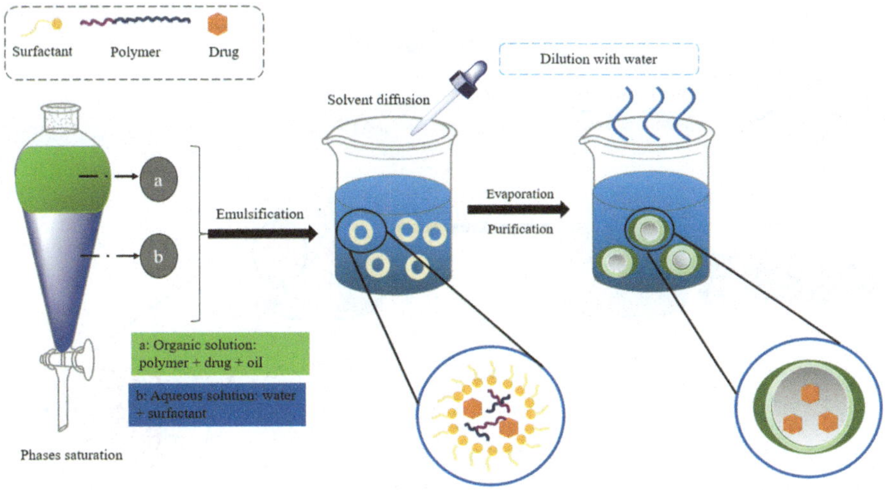

Fig. 4.5 Schematic representation of solvent diffusion/emulsification method

4.4.2 Emulsion Diffusion Technique

The technique is slightly different from the previous method that is discussed above. In this method, a partially water-miscible solvent is used as the organic phase. To this solvent, drug, and polymer are dissolved using a hydrophobic stabilizer that is added to the water phase. This prepares nanostructures, and depending upon the ratio of oil to polymer, nanoparticles or nanocapsules can be formed (Kumar et al. 2012; Piñón-Segundo et al. 2018) (Fig. 4.5).

4.4.3 Nanoprecipitation Method

In this method, the polymer and drug are dissolved in a water-miscible solvent that is added dropwise to the aqueous solution leading to the formation of nanostructures (Rivas et al. 2017) (Fig. 4.6). These nanostructures entrap the drug. Later, the undesired solvent can be removed by evaporation. This technique is suitable to encapsulate lipophilic drug molecules. Encapsulation of cucurbitacin with PLGA to prepare a polymeric nanocarrier prevented the loss of the drug during emulsification along with an increase in drug entrapment in the polymer (Alshamsan 2014).

4.4.4 Supercritical Antisolvent Method

In this method, spray a polymeric solution at high pressure of CO_2 to form nanoparticles. The nanoparticles are formed after the immediate diffusion of tiny droplets of polymeric liquid. The supercritical antisolvent method is one of the

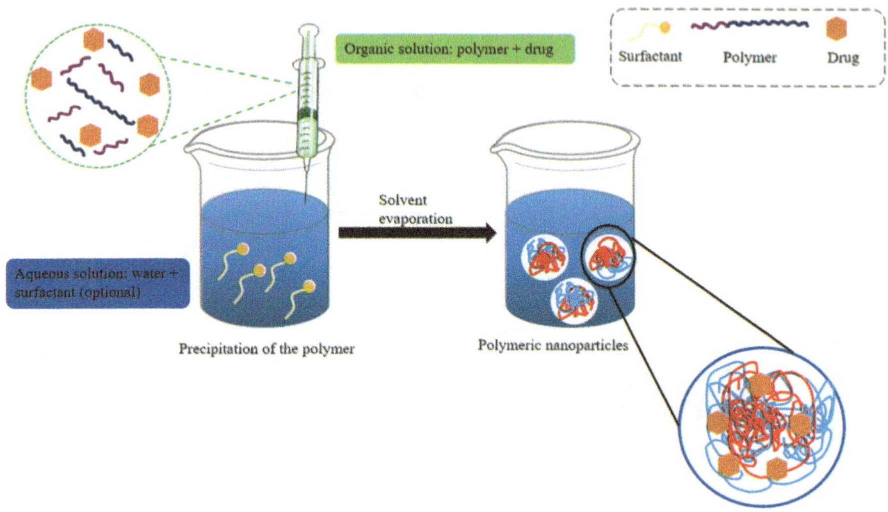

Fig. 4.6 Schematic representation of the nanoprecipitation method

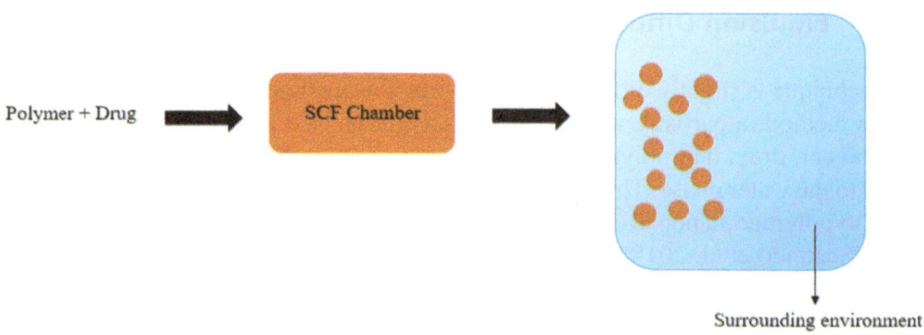

Fig. 4.7 Schematic representation of a supercritical fluid method

approaches adopted by supercritical fluid technology (SCF). Other SCF technologies include aerosol solvent extraction systems, gas antisolvent systems, precipitation using compressed antisolvent, solution-enhanced dispersion by supercritical fluids, etc. (Sekhon 2010; Kankala et al. 2017) (Fig. 4.7).

4.4.4.1 Salting Out

Salting-out technique involves the incorporation of a high concentration of salts into the polymeric solution. Salts are added at a high amount to an aqueous solution in which water-miscible solvent along with drug and polymer is included (Fig. 4.8). This forms an appearance of a coacervate. The coacervate formation is influenced by pH and temperature. Salts like calcium chloride, saccharides, magnesium chloride, and magnesium acetate are used in the salting-out technique (Masood 2016).

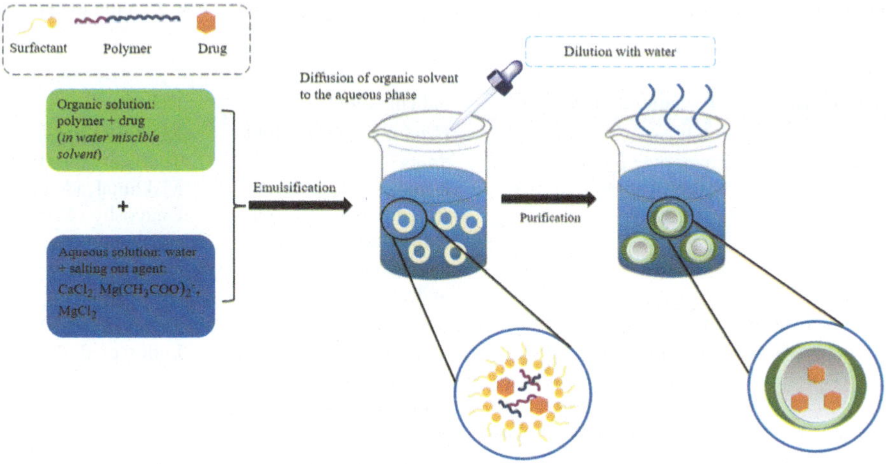

Fig. 4.8 Schematic representation of salting out method

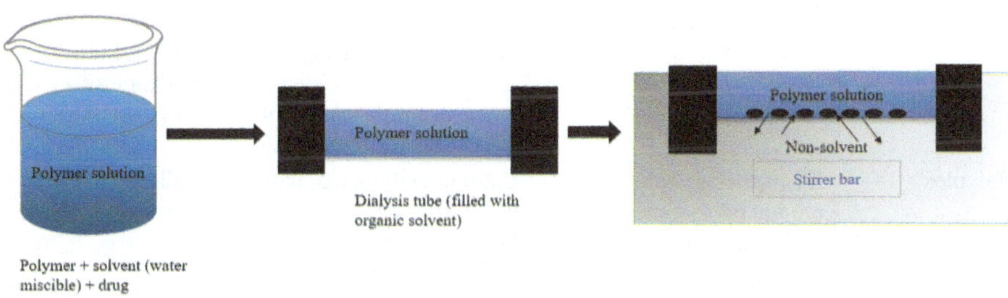

Fig. 4.9 Schematic representation of the dialysis method

4.4.4.2 Emulsion Polymerization

In the method, instead of using polymers, monomers are added to microreactors to initiate radical polymerization. The oil droplets constituting monomers are dispersed in the water phase. The process omits many steps required for the preparation of a normal emulsion. This technique regulates the particle size of formed polymers that removes the need for surfactants and emulsifiers. It also offers biocompatibility (Cho et al. 2019).

4.4.4.3 Dialysis

The formation of nanostructures is quite similar to the nanoprecipitation method. In this method, the water-miscible solvent is placed in a dialysis tube and kept in an aqueous solution (Fig. 4.9). After placing the dialysis tube into the aqueous solution, the aqueous phase displaces the organic phase. This triggers the formation of nanoparticles loaded with drugs. Generally, dimethyl formamide is preferred as a water-miscible solvent to which drug and polymer are added (Nah et al. 1998).

Table 4.2 List of some polymeric nanocarriers for targeted drug delivery based on the method of preparation

Polymeric nanocarrier	Preparation technique	Salient features	References
Polymeric micelles	Solvent evaporation, nanoprecipitation, dialysis, freeze-drying	Used for poorly soluble drugs It makes a core-shell structure that incorporates both hydrophilic and hydrophobic molecules under one unit	Xu et al. (2013), Teotia et al. (2015), Makhmalzade and Chavoshy (2018)
Polymersomes	Rehydration, electroporation, polymerization-induced self-assembly, direct injection, emulsion phase transfer, microfluidics	Integrate lipophilic molecules within its membrane and hydrophilic part in its aqueous interior They can possess high molecular weight copolymers	Fisher et al. (2010), Lefley et al. (2020)
Nanocapsules	Nanoprecipitation	Enhance oral bioavailability, mask the bitter taste of a drug, reduce systemic toxicity of the drug, promoting its selectivity for target cells	Tiarks et al. (2001), Lu et al. (2016), Meier (2000), Whelan (2001)
Polyplex	By pipetting, solutions are mixed to generate electrostatic interactions	Cationic polymeric chains boost drug delivery of bioactive compounds This is done by preventing the entry of phytoconstituents in endocytic vesicles of cytoplasm	Millili et al. (2010), Debus et al. (2012)

Various types of polymeric nanocarriers could be formulated using these methods. Nanocarriers such as polymeric micelles, nanospheres, nanocapsules, polymersomes, polyplexes, etc. (Fig. 4.1) are formulated to load phytoconstituents that deliver to the target site to show antifungal, anticancer, anti-inflammatory, antidiabetic, or anti-leishmanial activities. These types of nanocarriers are briefly discussed in Table 4.2.

4.5 Physicochemical Properties of Polymeric Nanocarriers

After the preparation of nanosystems, it is essential to embrace desired physicochemical properties to show significant activity. If the characterization of a nanoformulation is not done accordingly, it could not be translated properly even if it carries undeniable clinical benefits. Parameters such as molecular weight, size, shape, surface properties, stability, etc. are some physicochemical properties that can influence the release of drugs or the erosion of the polymers (George et al. 2019). For instance, Kamaly et al. (2016) reported that low molecular weight polymers degrade more rapidly as compared to high molecular weight polymers. This degradation affects the release of encapsulated drugs incorporated in a polymeric matrix.

Thermodynamically unstable nanoparticles are stabilized by enhancing their physicochemical characteristics such as by enhancing the resistance to prevent aggregation of molecules to prolong the shelf life of the nanosystem (Madkour et al. 2019). Other than inducing resistance, the addition of surfactants can also reduce the interfacial tension between solid-liquid phases, stabilizing the nanosystem (Heinz et al. 2017). In the following section, discussion on few physicochemical parameters is discussed.

4.5.1 Particle Size

The particle size of nanosystems such as nanomicelles influences the delivery of the drugs to the target site. It is essential to design nanocarriers of a particular size that could rapidly wash out from the blood capillaries and renal filtration or could stuck in these capillaries or renal filtration (De et al. 2022). Change in particle size of a polymer to develop a nanosystem can lead to enhanced pharmacokinetic properties of nanoparticles, distorted binding to the target tissue site, and undesirable interaction with a substrate and/or receptor. Laser scattering techniques such as static or dynamic light scattering or laser diffraction techniques could be used to determine the mean size and diameter of a particle. Other techniques include electron microscopy, particle tracking analysis, ultracentrifugation, field-flow fractionation, centrifugal particle sedimentation, and tunable resistive pulse sensing for mean particle size determination (Halamoda-Kenzaoui et al. 2019). De et al. (2022) reviewed the permeation of nanomicelle of different sizes, i.e., 30, 50, 70, and 100 nm. It was observed that all different-sized polymers could permeate highly permeable tumors, but nanomicelle of size 30 nm could even penetrate poorly permeable nanomicelle. Nanoparticles with particle sizes less than 100 nm could easily escape blood vessels by endothelial fenestrations. Contrarily, microparticles are ingested by Kupffer cells located in the liver of the body, whereas nanoparticles of size more than 500 nm can be uptake from the caveolae-mediated pathway, and particles less than 200 nm can be ingested by the clathrin-mediated pathway (Di Marzio et al. 2016).

4.5.2 Particle Shape

The particle shape of a nanocarrier influences the pharmacological properties of a drug. The spherical shape of nanoparticles is preferred globally because of isosymmetry. However, other shapes such as rod-shaped, star-shaped, and worm-like particles, etc. are being explored by different researchers to enhance the better activity of a nanosystem (Banik et al. 2016).

Stylianopoulos and Jain (2015) reported that spherical-shaped particles could enhance the uptake of drugs despite the way of binding to the cell surface. On the other hand, it was proved that long filamentous rod-shaped particles could exhibit better drug delivery to the target site and they could show better therapeutic efficacy. In the study, spherical, short, and long rod-shaped particles were examined. Zhang et al. (2019) compared sphere-shaped particles to cylindrical-shaped particles and concluded that cylindrical-shaped particles exhibited better tissue permeability, enhanced cellular intake, and rapid body clearance. Contrarily, sphere-shaped particles showed better binding to the tumor site along with prolonged blood circulation duration. Yang et al. (2016) demonstrated the effectiveness of star-shaped particles to deliver hydrophobic drugs to the target site. The drug delivery was stimuli-responsive and responds to an acidic environment and controls its release in a cytoplasmic environment.

4.5.3 Architecture of Nanocarrier

It is significant to arrange polymers and nanocarriers in such a manner that could translate into a functional nanoformulation with better pharmacological properties (Patra et al. 2018; Qiu and Bae 2006). For instance, two sugar-based polymeric micelles, i.e., dynamic and unimolecular, exhibit different properties. The unimolecular micelles were prepared by amphiphilic macromolecules bound covalently to each other, whereas dynamic micelles were prepared by self-assembly of amphiphilic polymers. Tao et al. (2016) noted that unimolecular micelles bear more loading capacity of a drug than dynamic micelles.

4.5.4 Particle Surface

Particle surface also plays a key role to penetrate biological barriers. They must possess defined biocompatibility. It must also carry significant solubility as well as target binding characteristics. Nanoparticles are designed in a manner by which their hydrophilic surface interacts with the aqueous phase of a biological system. For the development of aqueous phase interaction, neutral hydrophilic polymers are mostly preferred as they reduce the chances of phagocytosis. Polymers such as polyethylene oxide and polyethylene glycol are gold standards for designing nanoparticles (Hu et al. 2018). A comparison between PEGylated and non-PEGylated formulations was done by Gref et al. (1994), and it was observed that PEGylated nanoparticles showed prolonged plasma half-life along with decreased liver uptake.

The polymeric nanocarriers also influence the biological behavior of the compound. For instance, cationic or anionic nanoparticles can easily escape opsonization which ultimately avoids phagocytosis of a nanoparticle. These types of nanoparticles attain more stability than neutral ones. Farshbaf et al. (2018) reported that cationic nanoparticles possess strong interaction with genetic materials that are negatively charged. This made it a potential candidate for drug delivery at the target site.

Cationic nanoparticles allow genetic materials to penetrate the cell membrane effectively and check endocytosis to allow uptake through the cell. Apart from using amphiphilic polymers, pH-responsive polymers can also be used to trigger drug release to the target site. This concludes that particle surface affects the targeting of specific receptors or proteins that make a nanosystem target specific. Therefore, polymeric nanocarriers have the potential to load anticancer drugs in them (Banerjee et al. 2016).

4.5.5 Stability

The stability of a nanocarrier is ensured by focusing on the chain length and charge of the polymer. The type of molecular bonding such as covalent, H-bond, or electrostatic bond also affects the stability of polymeric nanocarriers. Therefore, modifications in molecular interactions or particle surface influence the release of drugs from polymeric nanocarriers. An example of such polymeric nanocarriers that ensure stability was explained by Palanikumar et al. (2020). According to their study, a drug loaded in PLGA polymer covalently linked with bovine serum albumin showed biocompatibility as well as pH sensitivity. The covalent bonding of polymer reduced interactions with macrophages and proteins that potentiate target recognition. It is significant to select the proper environment for drug release, polymer composition, and drug encapsulation techniques to design an effective polymeric nanocarrier.

Maintenance of temperature is also essential to attain stability. For instance, nanocarriers made up of polymers PCL and poly(D, L-lactide) (PLA50) could be easily stored at 4 °C and room temperature for up to 1 year without losing their stability, whereas nanocarriers made up of polymers PLA37.5GA25 can only be stable at 4 °C (Lemoine et al. 1996). Therefore, it is noteworthy to maintain stability to ensure better efficacy of a drug against the disease.

4.6 Preclinical and Clinical Evidence of Phytoconstituent-Incorporated Nanocarriers for Targeted Drug Delivery

A proper treatment against metabolic or severe diseases such as diabetes, cancer, and cardiovascular or neurodegenerative disorders is still being investigated. Researchers are discovering new small molecules that contain reduced or no adverse reactions, exhibit drug-drug or drug-food interactions, and possess the potency to eliminate heavy doses. To this, several synthetic compounds are being designed that could eliminate the limitations of conventional medicines up to a certain extent. Thus, while searching for a perfect molecule, naturally occurring phytoconstituents steal the attention of researchers that have the potential to treat disease up to large extent without causing any severe side effects.

The plant-derived bioactive compounds exhibit significant activity against a particular disease. The plant-derived secondary metabolites show their effect by acting on various mechanisms. But many of the phytoconstituents are not bioavailable and biocompatible. Some bioactive compounds also possess less half-life and, therefore, are unable to show sustained effect. Hence, these compounds cannot be formulated even if they bear significant properties. Therefore, to overcome the limitations of designing a formulation using phytoconstituents, nanocarriers are often chosen as an alternative. The nanoformulation of a plant metabolite could overcome all the drawbacks associated with the conventional method of preparations. In this section, we will be discussing some of these phytoconstituents that are designed as polymeric nanocarriers with special emphasis on diseased states.

4.6.1 Polymeric Nanocarriers Loaded with Phytoconstituents for Diabetes

It is well known that diabetes is a condition in metabolic syndrome that results because of high blood glucose levels. It is also known to cause retinopathy, cardiomyopathy, coronary artery disease, nephropathy, neuropathy, peripheral arterial disease, and stroke along with high blood glucose levels (Qaseem et al. 2012; Dewanjee et al. 2018; Dewanjee and Bhattacharjee 2018; Bhattacharjee et al. 2016; Khanra et al. 2017). Several antihyperglycemic drugs are considered either alone or in combination to treat the hyperglycemic state of a patient. However, unwanted side effects of these drugs disappoint the researchers, and clinicians try to avoid prescribing these medications either because of their adverse reactions or certain limitations.

Hence, phytoconstituents such as curcumin, apigenin, resveratrol, naringenin, etc. are designed in a formulation to achieve a potent effect against diabetes with significant bioavailability and minimal side effects. But while developing a formulation using conventional techniques, it was noted that water solubility of phytoconstituents such as quercetin, apigenin, curcumin, resveratrol, luteolin, etc. was diminished along with reduced absorption and half-life. The formulation had a high metabolism and rapid excretion. Thus, polymeric nanocarriers could ensure the efficient delivery of phytoconstituents to lower blood glucose levels. Several preclinical and clinical shreds of evidence support polymeric nanocarriers loaded with phytoconstituents to possess better pharmacokinetic properties (Dewanjee et al. 2020) as mentioned in Table 4.3.

4.6.2 Polymeric Nanocarriers Loaded with Phytoconstituents for Neurodegenerative Disorders

Neurodegenerative disease results from a deficiency in neuronal subtypes causing deformities in the brain signals. Alzheimer's disease (AD) and Parkinson's disease (PD) are the most renowned diseases categorized under neurodegenerative disorders

Table 4.3 Beneficial preclinical and clinical evidence of phytoconstituent-loaded polymeric nanocarrier for delivery of an antidiabetic compound

Phytoconstituent loaded polymeric nanocarrier	Limitation of conventionally developed formulation	Preclinical/clinical evidence	Reference
Curcumin-encapsulated polyvinyl alcohol nanoparticles	Poor water solubility, chemical stability, penetration, absorption, and half-life and increased metabolism, and excretion	A better therapeutic effect over free curcumin on cataract-developed rats	Grama et al. (2013), Suresh and Nangia (2018), Ernest et al. (2018), Metzler et al. (2013)
Curcumin-loaded polylactic acid (PLA)-polyethylene glycol (PEG) polymeric nanoparticles	Poor water solubility, chemical stability, penetration, absorption, and half-life and increased metabolism, and excretion	Effective in reducing blood glucose levels, and diabetes-induced hepatotoxicity in streptozotocin-induced rats	El-Naggar et al. (2019), Suresh and Nangia (2018), Ernest et al. (2018), Metzler et al. (2013)
Naringenin-loaded core-shell polymeric nanoparticles	Poor water solubility, absorption, and oral bioavailability, and increased hepatic transformation	Reversed hyperglycemia, oxidative stress, and dyslipidemia in diabetic rats	Maity et al. (2017), Taghipour et al. (2019), Gera et al. (2017), Song et al. (2015)
Glycyrrhizin-entrapped chitosan-gum Arabic nanoparticles	Poor absorption, hydrolysis of the drug by gastric fluid and GI flora, oral bioavailability, and rapid metabolism	25% pure glycyrrhizin entrapped nanoformulation showed a significant reduction in hyperglycemic state at low doses in diabetic rats	Rani et al. (2018), Jin et al. (2012)
Resveratrol-assembled PLGA nanoparticles	Poor water solubility, plasma concentration, systemic distribution, oral bioavailability, and rapid metabolism and elimination	Showed better efficacy in terms of solubility, absorption, encapsulation, bioavailability, and sustained release	Wan et al. (2018), Taghipour et al. (2019), Peñalva et al. (2018), Chimento et al. (2019)
Curcumin-encapsulated poly (lactic-co-glycolic acid) (PLGA) nanoparticles	Poor water solubility, chemical stability, penetration, absorption, and half-life and increased metabolism, and excretion	Enhanced oral bioavailability and biological half-life by improving water solubility, residence time, permeation, and triggering drug release	Xie et al. (2011), Suresh and Nangia (2018), Ernest et al. (2018), Metzler et al. (2013)

(Bensadoun et al. 2003; Kishima et al. 2004; McArthur 2004; Menei et al. 2005; Flachenecker 2006; Kabanov and Gendelman 2007). To this AD is characterized by dementia followed by learning difficulty, hypokinesia, unbending nature, bradykinesia, and resting quake (Popovic and Brundin 2006; Singh et al. 2007). It is caused because of degeneration of nigrostriatal dopaminergic neurons and plaque formation in the brain due to the accumulation of β-amyloid protein and phosphorylation of tau protein (Breijyeh and Karaman 2020). The exact reason for occurrence is still unknown, but age is considered as a parameter for induction of the disease. Parkinson's disease is another neurodegenerative disease that comes after AD caused because of reduced production of dopamine in the patient. Reduced production is a consequence of the degeneration of dopaminergic neurons. It is characterized by catalepsy, stiffness, the slow movement of the body, and tremor. PD could be induced by multiple system atrophy, supranuclear palsy, drug-induced Parkinsonism, or genetic factors (Tysnes and Storstein 2017).

Drugs like galantamine, rivastigmine, donepezil, etc. are used to treat symptoms of AD, as the exact reason for occurrence is unknown. On the other hand, for the treatment of PD, it is essential to elevate levels of dopamine in the patient, and hence drugs like levodopa and carbidopa are considered first-line treatment. However, it is noteworthy that the drugs also exhibit certain unavoidable side effects due to which an alternative is essential to search. Therefore, researchers are focusing on herbal products that possess minimal side effects, and to enhance their effectiveness, nanocarriers have opted. Phytoconstituents such as curcumin, andrographolide, etc. are a few examples of drug-loaded polymeric nanocarriers (Barbara et al. 2017; Guccione et al. 2017; Mathew et al. 2012). Some of the examples of polymeric nanocarriers assembled with phytoconstituents are mentioned in Table 4.4.

4.6.3 Polymeric Nanocarriers Loaded with Phytoconstituents for the Treatment of Cancer

Cancer is a disease that reported around 0.6 million deaths in the year 2020 (Wong et al. 2020). After cardiac diseases, cancer is the second leading disease-causing death globally. To treat cancer, several conventional methods such as chemotherapy, immunotherapy, radiotherapy, surgical resection, or a combination of any of them are applied (DeVita and Chu 2008). But it is also well known that to treat cancer, the drugs or therapy used also possess menacing side effects such as bone marrow depression, alopecia, peripheral neuropathy, and other systemic toxicities (Tripathi 2008). Chemotherapy also carries low selectivity. Hence, phytoconstituents are preferred due to their solid pieces of evidence to cure cancer. To enhance their efficacy, the phytoconstituents are incorporated into polymeric nanocarriers to attain the optimal effect. Some of the phytoconstituent-loaded polymeric nanocarriers are discussed in Table 4.5.

Table 4.4 Beneficial preclinical and clinical evidence of phytoconstituent-loaded polymeric nanocarrier for delivery of neuroprotective agents

Phytoconstituent-encapsulated polymeric nanocarrier	In vitro/in vivo models	Preclinical/clinical evidence	Reference
Curcumin-loaded Polylactide-co-glycolic-acid (PLGA) nanoparticles	Hippocampal cell cultures	A significant reduction in the aggregation of β-amyloid with enhanced penetration capacity in the blood-brain barrier	Barbara et al. (2017)
Andrographolide-loaded human serum albumin and polyethylcyanocrylate nanoparticles	Human cerebral microvascular endothelial cell line (hCMEC/D3)	Enhanced permeability in the blood-brain barrier with elevated bioavailability	Guccione et al. (2017)
Curcumin-loaded PLGA nanoparticles	GI-glioma cells	Antioxidant activity of curcumin was observed along with suppression of beta-amyloid when Tet-1 peptide was targeted	Mathew et al. (2012)
Resveratrol-loaded polymeric nanocarrier	In vivo findings on Tg2576 mice and in vitro results in PC12 cells	Reduced neuronal apoptosis caused due to beta-amyloid was observed and in vitro findings suggested regulation of silent information regulator 1 (SIRT1)-rho-associated kinase 1 (ROCK1) signaling pathway	La Barbera et al. (2022)

4.6.4 Polymeric Nanocarriers Loaded with Phytoconstituents for Ocular Drug Delivery

The human eye is the most delicate and complex organ of the body. It is made up of physiological and anatomical barriers that protect the organ. The eye is constituted of the retina, optic nerves, cornea, iris, ciliary muscles, vitreous and aqueous humor, choroid, conjunctiva, and lens (Patel et al. 2013; Boddu et al. 2013). Because of its complexity and delicacy, common diseases could be observed in the patients such as dry eye disease, cataract, diabetic retinopathy, glaucoma, age-related macular degeneration, allergic uveitis, and allergic conjunctivitis (Patel et al. 2013).

To treat eye diseases, it is essential to develop a suitable drug that could easily penetrate the physiological barriers of the eye. The topical route of drug delivery is an effective measure to design a formulation to penetrate the eye. The topical route of drug delivery involves solutions, suspensions, and ointments that are preferred to design ophthalmic formulations. As it is mentioned previously, the eye has a complex structure, and it is important to design a formulation that could maintain the isotonicity of the organ along with mitigating the disease. Hence, it is essential to develop a bioactive-containing nanoformulation that produces no side effects such as eye irritation along with better ocular penetrability. Here are a few examples of

Table 4.5 Beneficial preclinical and clinical evidence of phytoconstituent-loaded polymeric nanocarrier for delivery of an anticancer compound

Phytoconstituent-encapsulated polymeric nanocarrier	In vitro cell lines/ in vivo models	Pre-clinical/clinical evidence	Reference
Apigenin-loaded poly (lactic-co-glycolide acid) (PLGA) nanoparticles	Ultraviolet B-induced mouse skin cancer model	Reduction in tissue damage followed by suppression in the frequency of chromosomal aberration and potentiation of mitochondrial aberration	Mirzoeva et al. (2018)
Epigallocatechin-3-gallate-assembled polylactic acid and polyethylene glycol nanoparticles	22Rv1 prostate carcinoma cell lines	Potent activity over free epigallocatechin	Siddiqui et al. (2010)
Curcumin-encapsulated chitosan nanoparticles	MCF-7 breast cancer cell lines	The enhanced release rate of curcumin was reported along with improved cell viability studies	Esfandiarpour-Boroujeni et al. (2017)
Curcumin-loaded gold-polyvinylpyrrolidone nanoparticles	Progesterone/estrogen-negative cells	MTT assay confirmed efficient anticancer activity at lower doses in cells	Mahalunkar et al. (2019)
PLGA nanoparticles loaded with nimbolide	MCF-7 and MDA-MB-231 breast cancer cell lines and AsPC-1 pancreatic cancer cell lines	Enhanced solubility and the cytotoxic effect	Patra et al. (2019)
Paclitaxel-entrapped PLGA nanoparticles	Caco-2 cell lines	The permeability of paclitaxel was increased by fivefold compared to free paclitaxel	Roger et al. (2012)

polymeric nanocarriers loaded with bioactive resulting in better ocular activity in Table 4.6.

Thus, with the increasing prevalence of various diseases, the designing of such multifunctional polymers using recent advances in nanotechnology is anticipated to revolutionize the field of nanomedicine.

Table 4.6 Beneficial preclinical and clinical evidence of phytoconstituent-loaded polymeric nanocarrier for delivery of compound in eye

Phytoconstituent-encapsulated polymeric nanocarrier	In vitro/ in vivo models	Pre-clinical/clinical evidence	Reference
Resveratrol-loaded poly (lactic-co-glycolic-acid) nanoparticles	ARPE-19 cells	Nanoparticles showed a reduction in the expression of VEGF and hence result in a decrease in neovascular age-related macular degeneration	Bhatt et al. (2020)
Naringenin-loaded cyclodextrin-chitosan nanoparticles	Ocular irritation test in rabbits	The formulation does not cause any eye irritation in the rabbit's eye	Zhang et al. (2016)
Myricetin-encapsulated PVCL-PVA-PEG nanoparticles	In vivo corneal permeation in rabbits	Stability and ocular bioavailability were attained at a larger extent than free myricetin	Hou et al. (2019)
Kaempferol-entrapped Polyvinylpyrrolidone (PVP) nanocarrier	Cytotoxicity study in rabbits	Better ocular tolerance was attained without any signs of redness, inflammation, or irritation in the eye	Zhang et al. (2020)

4.7 Conclusion and Future Perspectives

In conclusion, the development of polymeric nanocarriers plays a significant role to overcome drawbacks during the incorporation of phytoconstituents via traditional methods. Phytoconstituents like curcumin, resveratrol, apigenin, etc. are easily incorporated and showed efficient effects during preclinical or clinical studies. Several research had taken place to exploit nanocarriers to develop anticancer, hypoglycemic, neuroprotective, and eye-protective medicines. Thus, depending on a drug for a specific pathological condition, a suitable technique can be selected to design the nanosystem. The technique persists several advantages that accelerate its usage worldwide, but simultaneously, its drawbacks hold its process of designing. For instance, aggregation of monomers or chances of toxicity because of toxic degradation are a few limitations that limit its use. The residual components bound to the polymer may also contribute to the toxic effects of a polymer. Another disadvantage is its complex process of manufacturing. The complexity of manufacturing limits its trials on a large group of people to access its pharmacological significance. The selection of polymers (more precisely biodegradable polymers) is also an important parameter while designing a polymeric nanocarrier. There is also a need to ensure certain shapes, sizes, and other physicochemical properties to attain a stable nanocarrier. It is also noteworthy that a nanosystem must permeate the biological barriers whenever necessary and must show sustained effect. Another complication observed during the manufacturing of polymeric nanoparticles is difficulty in drug loading and, hence, adversity in scaling. A formulator must also carry a sound knowledge regarding the chemical structure of a polymer and could modify its chemical structure to maintain its flexibility and strength. Hence, further continuous research on phytoconstituent-loaded nanocarriers in both clinical and preclinical studies focusing on physical stability and therapeutic efficacy will improve the diagnosis, prevention, and treatment of diseases.

References

Abasian P et al (2020) Polymeric nanocarriers in targeted drug delivery systems: a review. Poly Adv Technol 31(12):2939–2954

Abe MM, Martins JR, Sanvezzo PB, Macedo JV, Branciforti MC, Halley P, Brienzo M (2021) Advantages and disadvantages of bioplastics production from starch and lignocellulosic components. Polymers 13(15):2484

Alcázar-Alay SC, Meireles MA (2015) Physicochemical properties, modifications and applications of starches from different botanical sources. Food Sci Technol 35:215–236

Alshamsan A (2014) Nanoprecipitation is more efficient than emulsion solvent evaporation method to encapsulate cucurbitacin I in PLGA nanoparticles. Saudi Pharm J 22:219–222

Anju S, Prajitha N, Sukanya VS, Mohanan PV (2020) Complicity of degradable polymers in health-care applications. Mater Today Chem 16:100236

Arakawa CK, DeForest CA (2017) Polymer design and development. In: Biology and engineering of stem cell niches. Academic Press, London, pp 295–314

Arias S, Del Moral A, Ferrer MR, Tallon R, Quesada E, Bejar V (2003) Mauran, an exopolysaccharide produced by the halophilic bacterium Halomonas maura, with a novel composition and interesting properties for biotechnology. Extremophiles 7:319–326

Arif ZU et al (2022) Recent advances in 3D-printed polylactide and polycaprolactone-based biomaterials for tissue engineering applications. Int J Biol Macromol 218:930–968

Avcu E et al (2022) Biodegradable polymer matrix composites containing graphene-related materials for antibacterial applications: a critical review. Acta Biomater 151:1–44

Banerjee A, Qi J, Gogoi R, Wong J, Mitragotri S (2016) Role of nanoparticle size, shape and surface chemistry in oral drug delivery. J Control Release 12:33

Banik BL, Fattahi P, Brown JL (2016) Polymeric nanoparticles: the future of nanomedicine. Wiley Interdiscip Rev Nanomed Nanobiotechnol 8:271–299

Barbara R, Belletti D, Pederzoli F, Masoni M, Keller J, Ballestrazzi A, Vandelli MA, Tosi G, Grabrucker AM (2017) Novel curcumin loaded nanoparticles engineered for blood-brain barrier crossing and able to disrupt Abeta aggregates. Int J Pharm 526(1–2):413–424

Bauer M, Lautenschlaeger C, Kempe K, Tauhardt L, Schubert US, Fischer DJ (2012) Poly(2-ethyl-2-oxazoline) as alternative for the stealth polymer poly (ethylene glycol): comparison of in vitro cytotoxicity and hemocompatibility. J Vis Exp 12(7):986–998

Bensadoun J-C, Almeida LPD, Fine EG, Tseng JL, Déglon N, Aebischer P (2003) Comparative study of GDNF delivery systems for the CNS: polymer rods, encapsulated cells, and lentiviral vectors. J Control Release 87:107–115

Bhatt P, Fnu G, Bhatia D, Shahid A, Sutariya V (2020) Nanodelivery of resveratrol-loaded PLGA nanoparticles for age-related macular degeneration. AAPS PharmSciTech 21:1–9

Bhattacharjee N, Barma S, Konwar N, Dewanjee S, Manna P (2016) Mechanistic insight of diabetic nephropathy and its pharmacotherapeutic targets: an update. Eur J Pharmacol 791:8–24

Boddu SHS, Menees AL, Ray A, Mitra AK (2013) A brief overview of ocular anatomy and physiology. In: Mitra AK (ed) Treatise on ocular drug delivery. Bentham Science Publishers, Sharjah, pp 3–19

Breijyeh Z, Karaman R (2020) Comprehensive review on Alzheimer's disease: causes and treatment. Molecules 25(24):5789

Casalini T, Rossi F, Castrovinci A, Perale G (2019) A perspective on polylactic acid-based polymers use for nanoparticles synthesis and applications. Front Bioeng Biotechnol 7:259

Catalin Balaure P, Mihai Grumezescu AJC (2015) Smart synthetic polymer nanocarriers for controlled and site-specific drug delivery. Curr Trends Med Chem 15(15):1424–1490
Cavalu S et al (2018) Novel formulation based on chitosan-arabic gum nanoparticles entrapping propolis extract production, physico-chemical and structural characterization. Rev Chim 69 (12):3756–3760
Chen H, Yuan L, Song W, Wu Z et al (2008) Biocompatible polymer materials: role of protein–surface interactions. Prog Polym Sci 33(11):1059–1087
Chien KB, Shah RN (2012) Novel soy protein scaffolds for tissue regeneration: material characterization and interaction with human mesenchymal stem cells. Acta Biomater 8(2):694–703
Chimento A, De Amicis F, Sirianni R, Sinicropi MS, Puoci F, Casaburi I, Saturnino C, Pezzi V (2019) Progress to improve oral bioavailability and beneficial effects of resveratrol. Int J Mol Sci 20:1381
Cho Y-S, Ji S, Kim YS (2019) Synthesis of polymeric nanoparticles by emulsion polymerization for particle self-assembly applications. J Nanosci Nanotechnol 19(10):6398–6407
Chopin N, Guillory X, Weiss P, Bideau JL, Colliec-Jouault S (2014) Design polysaccharides of marine origin: chemical modifications to reach advanced versatile compounds. Curr Org Chem 18(7):867–895
Cosco D, Federico C, Maiuolo J, Bulotta S, Molinaro R, Paolino D et al (2014) Physicochemical features and transfection properties of chitosan/ poloxamer 188/poly (D, L-lactide-co-glycolide) nanoplexes. Int J Nanomedicine 9:2359–2372. https://doi.org/10.2147/IJN.S58362
Cosco D, Paolino D, De Angelis F, Cilurzo F, Celia C, Di Marzio L et al (2015) Aqueous-core PEG-coated PLA nanocapsules for an efficient entrapment of water soluble anticancer drugs and a smart therapeutic response. Eur J Pharm Biopharm 89:30–39
Crucho CIC, Barros MT (2017) Polymeric nanoparticles: a study on the preparation variables and characterization methods. Mater Sci Eng C 80:771–784
Dadwal M, Solan D, Pradesh H (2014) Polymeric nanoparticles as promising novel carriers for drug delivery: an overview. J Adv Pharm Educ Res 4(1):1
Dang Y, Guan J (2020) Nanoparticle-based drug delivery systems for cancer therapy. Smart Mater Med 1:10–19
De R, Mahata MK, Kim K (2022) Structure-based varieties of polymeric nanocarriers and influences of their physicochemical properties on drug delivery profiles. Adv Sci 9(10):2105373
Debus H, Beck-Broichsitter M, Kissel T (2012) Optimized preparation of pDNA/poly(ethylene imine) polyplexes using a microfluidic system. Lab Chip 12(14):2498–2506
DeFrates KG, Moore R, Borgesi J, Lin G, Mulderig T, Beachley V, Hu X (2018) Protein-based fiber materials in medicine: a review. Nanomaterials 8(7):457
DeVita VT Jr, Chu E (2008) A history of cancer chemotherapy. Cancer Res 68(21):8643–8653
Dewanjee S, Bhattacharjee N (2018) MicroRNA: a new generation therapeutic target in diabetic nephropathy. Biochem Pharmacol 155:32–47
Dewanjee S, Das S, Das AK, Bhattacharjee N, Dihingia A, Dua TK, Kalita J, Manna P (2018) Molecular mechanism of diabetic neuropathy and its pharmacotherapeutic targets. Eur J Pharmacol 833:472–523
Dewanjee S et al (2020) Plant-based antidiabetic nanoformulations: the emerging paradigm for effective therapy. Int J Mol Sci 21(6):2217
Dhathathreyan K, Rajalakshmi N, Balaji R (2017) Nanomaterials for fuel cell technology. Nanotechnol energy sustainability. Part III: energy conversion and harvesting, chapter 24 Germany: Wiley-VCH Verlag GmbH & co. KGaA. p 569–596
Di Marzio L, Ventura CA, Cosco D, Paolino D, Di Stefano A, Stancanelli R et al (2016) Nanotherapeutics for anti-inflammatory delivery. J Drug Deliv Sci Technol 32:174–191
Dias A, Hussain A, Marcos A, Roque A (2011) A biotechnological perspective on the application of iron oxide magnetic colloids modified with polysaccharides. Biotechnol Adv 29(1):142–155
El-Naggar ME, Al-Joufi F, Anwar M, Attia MF, El-Bana MA (2019) Curcumin-loaded PLA-PEG copolymer nanoparticles for treatment of liver inflammation in streptozotocin-induced diabetic rats. Colloids Surf B Biointerfaces 177:389–398

Elzoghby A, Freag M, Mamdouh H, Elkhodairy K (2017) Zein-based nanocarriers as potential natural alternatives for drug and gene delivery: focus on cancer therapy. Curr Pharm Des 23: 5261–5271

Ernest U, Chen HY, Xu MJ, Taghipour Y, Asad M, Rahimi R, Murtaza G (2018) Anti-cancerous potential of polyphenol-loaded polymeric nanotherapeutics. Molecules 23:2787

Eroglu N, Emekci M, Athanassiou CG (2017) Applications of natural zeolites on agriculture and food production. J Sci Food Agric 97(11):3487–3499

Esfandiarpour-Boroujeni S, Bagheri-Khoulenjani S, Mirzadeh H, Amanpour S (2017) Fabrication and study of curcumin-loaded nanoparticles based on folate-chitosan for breast cancer therapy application. Carbohydr Polym 168:14–21

Farshbaf M, Davaran S, Zarebkohan A, Annabi N, Akbarzadeh A, Salehi R (2018) Significant role of cationic polymers in drug delivery systems. Artif Cells Nanomed Biotechnol 46:1872–1891

Fisher O, Khademhosseini A, Peppas N (2010) Drug delivery: nanoscale devices. Encyclopedia of materials: science and technology. Elsevier, Amsterdam

Flachenecker P (2006) Epidemiology of neuroimmunological diseases. J Neurol 253(Suppl. 5):V2–V8

Gagliardi A, Bonacci S, Paolino D, Celia C, Procopio A, Fresta M, Cosco D (2019) Paclitaxel-loaded sodium deoxycholate-stabilized zein nanoparticles: characterization and in vitro cytotoxicity. Heliyon 5(9)

Gagliardi A, Giuliano E, Venkateswararao E, Fresta M, Bulotta S, Awasthi V, Cosco D (2021) Biodegradable polymeric nanoparticles for drug delivery to solid tumors. Front Pharmacol 12:601626

Gálisová A, Jirátová M, Rabyk M, Sticová E, Hájek M, Hrubý M, Jirák D (2020) Glycogen as an advantageous polymer carrier in cancer theranostics: straightforward in vivo evidence. Sci Rep 10(1):10411

Garg U, Chauhan S, Nagaich U, Jain N (2019) Current advances in chitosan nanoparticles based drug delivery and targeting. Adv Pharm Bull 9(2):195

George A, Shah PA, Shrivastav PS (2019) Natural biodegradable polymers based nano-formulations for drug delivery: a review. Int J Pharm 561:244–264

Gera S, Talluri S, Rangaraj N, Sampathi S (2017) Formulation and evaluation of naringenin nanosuspensions for bioavailability enhancement. AAPS PharmSciTech 8:3151–3162

Goonoo N, Jeetah R, Bhaw-Luximon A, Jhurry DJ (2015) Polydioxanone-based biomaterials for tissue engineering and drug/gene delivery applications. Eur J Pharm Biopharm 97:371–391

Grama CN, Suryanarayana P, Patil MA, Raghu G, Balakrishna N, Kumar MR, Reddy GB (2013) Efficacy of biodegradable curcumin nanoparticles in delaying cataract in diabetic rat model. PLoS One 8:e78217

Gref R, Minamitake Y, Peracchia MT, Trubetskoy V, Torchilin V, Langer R (1994) Biodegradable long-circulating polymeric nanospheres. Science 263:1600–1603

Guccione C, Oufir M, Piazzini V, Eigenmann DE, Jahne EA, Zabela V, Faleschini MT, Bergonzi MC, Smiesko M, Hamburger M (2017) Andrographolide-loaded nanoparticles for brain delivery: formulation, characterisation and in vitro permeability using hCMEC/D3 cell line. Eur J Pharm Biopharm 119:253–263

Guo W, Quan P, Fang L, Cun D, Yang M (2015) Sustained release donepezil loaded PLGA microspheres for injection: preparation, in vitro and in vivo study. Asian J Pharm Sci 10:405–414

Halib N, Perrone F, Cemazar M, Dapas B, Farra R, Abrami M, Chiarappa G, Forte G, Zanconati F, Pozzato G, Murena L (2017) Potential applications of nanocellulose-containing materials in the biomedical field. Materials 10(8):977

Haggag YA, Abd Elrahman AA, Ulber R, Zayed A (2023) Fucoidan in pharmaceutical formulations: a comprehensive review for smart drug delivery systems. Mar Drugs 21(2):112

Halamoda-Kenzaoui B, Holzwarth U, Roebben G, Bogni A, BremerHoffmann S (2019) Mapping of the available standards against the regulatory needs for nanomedicines. Wiley Interdiscip Rev Nanomed Nanobiotechnol 11:e1531

Heinz H, Pramanik C, Heinz O, Ding Y, Mishra RK, Marchon D et al (2017) Nanoparticle decoration with surfactants: molecular interactions, assembly, and applications. Surf Sci Rep 14:33–39

Hill JW (1930) Studies on polymerization and ring formation. VI. Adipic anhydride. J Am Chem Soc 52(10):4110–4114

Hou Y, Zhang F, Lan J, Sun F, Li J, Li M, Song K, Wu X (2019) Ultra-small micelles based on polyoxyl 15 hydroxystearate for ocular delivery of myricetin: optimization, in vitro, and in vivo evaluation. Drug Deliv 26:158–167

Hsiao MY, Lin AC, Liao WH, Tyng G, Chia H, Wen S, Feng H (2019) Drug-loaded hyaluronic acid hydrogel as a sustained-release regimen with dual effects in early intervention of tendinopathy. Sci Rep 9:4784

Hu J, Sheng Y, Shi J, Yu B, Yu Z, Liao G (2018) Long circulating polymeric nanoparticles for gene/drug delivery. Curr Drug Metabol 19:723–738

Imazato S et al (2017) Non-biodegradable polymer particles for drug delivery: a new technology for 'bio-active' restorative materials. Dent Mater J 36(5):524–532

Jain JP, Chitkara D, Kumar NJ (2008) Polyanhydrides as localized drug delivery carrier: an update. Expert Opin Drug Deliv 5(8):889–907

Jamwal S, Ram B, Ranote S, Dharela R, Chauhan GS (2019) New glucose oxidase-immobilized stimuli-responsive dextran nanoparticles for insulin delivery. Int J Biol Macromol 123:968–978

Jin S, Fu S, Han J, Jin S, Lv Q, Lu Y, Qi J, Wu W, Yuan H (2012) Improvement of oral bioavailability of glycyrrhizin by sodium deoxycholate/phospholipid-mixed nanomicelles. J Drug Target 20:615–622

Joye IJ, McClements DJ (2014) Biopolymer-based nanoparticles and microparticles: fabrication, characterization, and application. Curr Opin Colloid Interface Sci 19:417–427

Kabanov AV, Gendelman HE (2007) Nanomedicine in the diagnosis and therapy of neurodegenerative disorders. Prog Polym Sci 32:1054–1082

Kamaly N, Yameen B, Wu J, Farokhzad OC (2016) Degradable controlled-release polymers and polymeric nanoparticles: mechanisms of controlling drug release. Chem Rev 116:2602–2663

Kankala RK, Zhang YS, Wang SB, Lee CH, Chen AZ (2017) Supercritical fluid technology: an emphasis on drug delivery and related biomedical applications. Adv Healthc Mater 6(16):1700433

Karlsson J, Vaughan HJ, Green JJ (2018) Biodegradable polymeric nanoparticles for therapeutic cancer treatments. Ann Rev Chem Biomol Eng 9:105–127

Kenry, Liu B (2018) Recent advances in biodegradable conducting polymers and their biomedical applications. Biomacromolecules 19(6):1783–1803

Kesharwani P (2018) Nanotechnology-based approaches for anti-diabetic drugs delivery. Diabetes Res Clin Pract 136:52–77

Khanra R, Bhattacharjee N, Dua TK, Nandy A, Saha A, Kalita J, Manna P, Dewanjee S (2017) Taraxerol, a pentacyclic triterpenoid, from Abroma augusta leaf attenuates diabetic nephropathy in type 2 diabetic rats. Biomed Pharmacother 94:726–741

Kishima H, Poyot T, Bloch J, Dauguet J, Conde F, Dolle F, Hinnen F, Pralong W, Palfi S, Deglon N et al (2004) Encapsulated GDNF-producing C2C12 cells for Parkinson's disease: a pre-clinical study in chronic MPTP-treated baboons. Neurobiol Dis 16:428–439

Kluin OS, Van der Mei HC, Busscher HJ, Neut D (2013) Biodegradable vs non-biodegradable antibiotic delivery devices in the treatment of osteomyelitis. Expert Opin Drug Deliv 10(3):341–351

Kolanthai E, Sindu PA, Arul KT, Chandra VS, Manikandan E, Kalkura SN (2017) Agarose encapsulated mesoporous carbonated hydroxyapatite nanocomposites powder for drug delivery. J Photochem Photobiol B 166:220–231

Krsti'c L, González-García MJ, Diebold Y (2021) Ocular delivery of polyphenols: meeting the unmet needs. Molecules 26:370

Kumar S, Dilbaghi N, Saharan R, Bhanjana GJB (2012) Nanotechnology as emerging tool for enhancing solubility of poorly water-soluble drugs. Bionanoscience 2(4):227–250

La Barbera L et al (2022) Functionalization strategies of polymeric nanoparticles for drug delivery in Alzheimer's disease: current trends and future perspectives. Front Neurosci 16:939855

Langer R, Folkman J (1976) Polymers for the sustained release of proteins and other macromolecules. Nature 263:797–800

Lanz-Landázuri A, Portilla-Arias J, Martinez de Ilarduya A, García-Alvarez M, Holler E, Ljubimova J, Muñoz-Guerra S (2014) Nanoparticles of esterified polymalic acid for controlled anticancer drug release. Macromol Biosci 14(9):1325–1336

Lapasin R (2012) Rheology of industrial polysaccharides: theory and applications. Springer Science & Business Media, Boston, MA

Lefley J, Waldron C, Becer CR (2020) Macromolecular design and preparation of polymersomes. Polym Chem 11(45):7124–7136

Lemoine D, Francois C, Kedzierewicz F, Preat V, Hoffman M, Maincent P (1996) Stability study of nanoparticles of poly(ε-caprolactone), poly(d,l-lactide) and poly(d,l-lactide-co-glycolide). Biomaterials 17(22):2191–2197

Li X, Yuan X (2007) Poly(ethylene glycol)-poly(lactic acid) copolymers for drug carriers. Prog Chem 19(6):973–981

Llamas I, Moral AD, Martínez-Checa F, Arco Y, Arias S, Quesada E (2006) Halomonas maura is a physiologically versatile bacterium of both ecological and biotechnological interest. Antonie Van Leeuwenhoek 89:395–403

Lombardo GE, Maggisano V, Celano M, Cosco D, Mignogna C, Baldan F et al (2018) Anti-htert sirna-loaded nanoparticles block the growth of anaplastic thyroid cancer xenograft. Mol Cancer Ther 17:1187–1195

Lu H, Wang J, Wang T, Zhong J, Bao Y, Hao H (2016) Recent progress on nanostructures for drug delivery applications. J Nanomater 2016:1

Madkour M, Bumajdad A, Al-Sagheer F (2019) To what extent do polymeric stabilizers affect nanoparticles characteristics? Adv Colloid Interf Sci 11:121–136

Mahalunkar S, Yadav AS, Gorain M, Pawar V, Braathen R, Weiss S, Bogen B, Gosavi SW, Kundu GC (2019) Functional design of pH-responsive folate-targeted polymer-coated gold nanoparticles for drug delivery and in vivo therapy in breast cancer. Int J Nanomedicine 14: 8285–8302

MaHam A et al (2009) Protein-based nanomedicine platforms for drug delivery. Small 5(15): 1706–1721

Maity S, Mukhopadhyay P, Kundu PP, Chakraborti AS (2017) Alginate coated chitosan core-shell nanoparticles for efficient oral delivery of naringenin in diabetic animals-an in vitro and in vivo approach. Carbohydr Polym 170:124–132

Makadia HK, Siegel SJ (2011) Poly lactic-co-glycolic acid (PLGA) as biodegradable controlled drug delivery carrier. Polymer 3(3):1377–1397

Makhmalzade BS, Chavoshy FJ (2018) Polymeric micelles as cutaneous drug delivery system in normal skin and dermatological disorders. J Adv Pharm Technol Res 9(1):2

Mansoori B, Mohammadi A, Davudian S, Shirjang S, Baradaran B (2017) The different mechanisms of cancer drug resistance: a brief review. Adv Pharm Bull 7:339. https://doi.org/10.15171/apb.2017.041

Masek A, Kosmalska A (2022) Technological limitations in obtaining and using cellulose biocomposites. Front Bioeng Biotechnol 10:912052

Mashak A, Rahimi A (2009) Silicone polymers in controlled drug delivery systems: a review. Iran Polymer J 18:279–295

Masood F (2016) Polymeric nanoparticles for targeted drug delivery system for cancer therapy. Mater Sci Eng C 60:569–578

Mathew A, Fukuda T, Nagaoka Y, Hasumura T, Morimoto H, Yoshida Y, Maekawa T, Venugopal K, Kumar DS (2012) Curcumin loaded-PLGA nanoparticles conjugated with Tet-1 peptide for potential use in Alzheimer's disease. PLoS One 7:e32616

Meyabadi TF, Dadashian F, Sadeghi GM, Asl HE (2014) Spherical cellulose nanoparticles preparation from waste cotton using a green method. Powder Technol 261:232–240

McArthur JC (2004) HIV dementia: an evolving disease. J Neuroimmunol 157:3–10
Meier W (2000) Polymer nanocapsules. Chem Soc Rev 29:295–303
Menei P, Montero-Menei C, Venier MC, Benoit JP (2005) Drug delivery into the brain using poly (lactide-co-glycolide) microspheres. Expert Opin Drug Deliv 2:363–376
Metzler M, Pfeiffer E, Schulz SI, Dempe JS (2013) Curcumin uptake and metabolism. Biofactors 39:14–20
Millili PG, Yin DH, Fan H, Naik UP, Sullivan MO (2010) Formulation of a peptide nucleic acid based nucleic acid delivery construct. Bioconjug Chem 21(3):445–455
Mirzoeva S et al (2018) Apigenin inhibits UVB-induced skin carcinogenesis: the role of thrombospondin-1 as an anti-inflammatory factor. Neoplasia 20(9):930–942
Mogosanu GD, Grumezescu AM, Bejenaru C, Bejenaru LE (2016) Polymeric protective agents for nanoparticles in drug delivery and targeting. Int J Pharm 510:419–429
Mokhtari H, Tavakoli S, Safarpour F, Kharaziha M, Bakhsheshi-Rad HR, Ramakrishna S, Berto F (2021) Recent advances in chemically-modified and hybrid carrageenan-based platforms for drug delivery, wound healing, and tissue engineering. Polymers 13(11):1744
Monge S, David G (2014) Use of phosphorus-containing polymers for the removal of metal Ions from wastewater. In: Phosphorus-based polymers: from synthesis to applications, vol 11. Royal Society of Chemistry, Cambridge, pp 225–251
Nah J-W, Paek Y-W, Jeong Y-I et al (1998) Clonazepam release from poly (DL-lactide-co-glycolide) nanoparticles prepared by dialysis method. Arch Pharm Res 21(4):418–422
Nava-Arzaluz GM, Piñón-Segundo E, Ganem-Rondero A, LechugaBallesteros D (2012) Single emulsion-solvent evaporation technique and modifications for the preparation of pharmaceutical polymeric nanoparticles. Recent Pat Drug Deliv Formul 6(3):209–223
Ozeki T, Tagami T (2014) Drug/polymer nanoparticles prepared using unique spray nozzles and recent progress of inhaled formulation. Asian J Pharm Sci 9:236–243
Palanikumar L, Al-Hosani S, Kalmouni M et al (2020) pH-responsive high stability polymeric nanoparticles for targeted delivery of anticancer therapeutics. Comm Biol 3:95
Palma E, Pasqua A, Gagliardi A, Britti D, Fresta M, Cosco D (2018) Antileishmanial activity of amphotericin B-loaded-PLGA nanoparticles: an overview. Materials 11:E1167
Patel A, Cholkar K, Agrahari V, Mitra AK (2013) Ocular drug delivery systems: an overview. World J Pharm 2:47–64
Patra JK, Das G, Fraceto LF et al (2018) Nano-based drug delivery systems: recent developments and future. Prospects 16(1):71
Patra A, Satpathy S, Hussain MD (2019) Nanodelivery and anticancer effect of a limonoid, nimbolide, in breast and pancreatic cancer cells. Int J Nanomedicine 14:8095–8104
Peñalva R, Morales J, González-Navarro CJ, Larrañeta E, Quincoces G, Peñuelas I, Irache JM (2018) Increased oral bioavailability of resveratrol by its encapsulation in casein nanoparticles. Int J Mol Sci 2018(19):2816
Piñón-Segundo E, Llera-Rojas VG, Leyva-Gómez G, Urbán-Morlán Z, Mendoza-Muñoz N, Quintanar-Guerrero D (2018) The emulsification diffusion method to obtain polymeric nanoparticles: two decades of research. Nanoscale fabrication, optimization, scale-up and biological aspects of pharmaceutical nanotechnology. Elsevier, Amsterdam, pp 51–83
Popovic N, Brundin P (2006) Therapeutic potential of controlled drug delivery systems in neurodegenerative diseases. Int J Pharm 314:120–126
Priya James H et al (2014) Smart polymers for the controlled delivery of drugs—a concise overview. Acta Pharm Sin B 4(2):120–127
Qaseem A, Humphrey LL, Sweet DE, Starkey M, Shekelle P (2012) Oral pharmacologic treatment of type 2 diabetes mellitus: a clinical practice guideline from the American College of Physicians. Ann Intern Med 156:218–231
Qiu LY, Bae YH (2006) Polymer architecture and drug delivery. Pharm Res 23(1):1–30
Rahman M, Hasan MR (2019) Synthetic biopolymers. In: Mazumder MAJ, Sheardown H, Al-Ahmed A (eds) Functional biopolymers. Springer International Publishing, Cham, pp 1–43

Rai P et al (2021) Recent advances in the sustainable design and applications of biodegradable polymers. Bioresour Technol 325:124739

Rajagopal M, Paul AK, Lee M-T, Joykin AR, Por C-S, Mahboob T, Salibay CC, Torres MS, Guiang MMM, Rahmatullah M et al (2022) Phytochemicals and nano-phytopharmaceuticals use in skin, urogenital and locomotor disorders: are we there? Plan Theory 11:1265

Rani R, Dahiya S, Dhingra D, Dilbaghi N, Kim KH, Kumar S (2018) Improvement of antihyperglycaemic activity of nano-thymoquinone in rat model of type-2 diabetes. Chem Biol Interact 295:119–132

Rivas CJM, Tarhini M, Badri W, Miladi K, Greige-Gerges H, Nazari QA et al (2017) Nanoprecipitation process: from encapsulation to drug delivery. Int J Pharm 532:66–81

Roger E, Kalscheuer S, Kirtane A, Guru BR, Grill AE, Whittum-Hudson J, Panyam J (2012) Folic acid functionalized nanoparticles for enhanced oral drug delivery. Mol Pharm 9:2103–2211

Rowe MD, Eyiler E, Walters KB (2016) Hydrolytic degradation of bio-based polyesters: effect of pH and time. Polym Test 52:192–199

Roy N, Agrawal M, Chaudhary S, Tirkey V, Dhwaj A, Mishra N (2017) Review article on permeation enhancers: a major breakthrough in drug delivery technology. Int J Pharm Sci Res 8:1001

Safdar R, Omar AA, Arunagiri A, Regupathi I, Thanabalan M (2019) Potential of chitosan and its derivatives for controlled drug release applications—a review. J Drug Del Sci Technol 49:642–659

Sathuvan M, Thangam R, Gajendiran M et al (2017) κ-Carrageenan: an effective drug carrier to deliver curcumin in cancer cells and to induce apoptosis. Carbohydr Polym 160:184–193

Schnell E, Klinkhammer K, Balzer S et al (2007) Guidance of glial cell migration and axonal growth on electrospun nanofibers of polyε-caprolactone and a collagen/poly-ε-caprolactone blend. Biomaterials 28(19):3012–3025

Sekhon BS (2010) Supercritical fluid technology: an overview of pharmaceutical applications. Int J Pharm Tech Res 2(1):810–826

Shukla PS, Borza T, Critchley AT, Prithiviraj B (2016) Carrageenans from red seaweeds as promoters of growth and elicitors of defense response in plants. Front Mar Sci 3:81

Siddiqui IA, Adhami VM, Ahmad N, Mukhtar H (2010) Nanochemoprevention: sustained release of bioactive food components for cancer prevention. Nutr Cancer 62:883–890

Sin L, Tueen B (2019) Polylactic acid: a practical guide for the processing, manufacturing, and applications of PLA. William Andrew, Oxford

Singh N, Pillay V, Choonara YE (2007) Advances in the treatment of Parkinson's disease. Prog Neurobiol 81:29–44

Sj A (2016) Types of biodegradable polymers. Introduction to bioplastics engineering. Elsevier, Amsterdam, pp 81–151

Södergård A, Stolt MJ (2002) Properties of lactic acid based polymers and their correlation with composition. Prog Polym Sci 27(6):1123–1163

Song I, Cha J, Choi M (2015) Enhanced oral bioavailability of naringenin administered in a mixed micelle formulation with pluronic F127 and tween 80 in rats. J Pharm Investig 45:633–640

Stylianopoulos T, Jain RK (2015) Design considerations for nanotherapeutics in oncology. Nanomedicine 11:1893–1907

Sun X, Xu C, Wu G, Ye Q, Wang CJP (2017) Poly(lactic-co-glycolic acid): applications and future prospects for periodontal tissue. Regeneration 9(6):189

Suresh K, Nangia A (2018) Curcumin: pharmaceutical solids as a platform to improve solubility and bioavailability. Cryst Eng Comm 24:3277–3296

Taghipour YD, Hajialyani M, Naseri R, Hesari M, Mohammadi P, Stefanucci A, Mollica A, Farzaei MH, Abdollahi M (2019) Nanoformulations of natural products for management of metabolic syndrome. Int J Nanomedicine 14:5303–5321

Taghipour-Sabzevar V, Sharifi T, Moghaddam MM (2019) Polymeric nanoparticles as carrier for targeted and controlled delivery of anticancer agents. Ther Deliv 10:527–550

Tao L, Chan JW, Uhrich KE (2016) Drug loading and release kinetics in polymeric micelles: comparing dynamic versus unimolecular sugar-based micelles for controlled release. J Bioact Compat Polym 31(3):227–241
Teotia A, Sami H, Kumar A (2015) Thermo-responsive polymers: structure and design of smart materials. In: Switchable and responsive surfaces and materials for biomedical applications. Elsevier, Amsterdam, pp 3–43
Tiarks F, Landfester K, Antonietti M (2001) Preparation of polymeric nanocapsules by miniemulsion polymerization. Langmuir 17(3):908–918
Tiwari G, Tiwari R, Sriwastawa B et al (2012) Drug delivery systems: an updated review. Int J Pharm Invest 2(1):2
Tripathi KD (2008) Chemotherapy of neoplastic diseases. In: Essentials of medical pharmacology. Jaypee Brothers Medical Publishers, New Delhi
Tripodo G, Trapani A, Torre ML, Giammona G, Trapani G, Mandracchia D (2015) Hyaluronic acid and its derivatives in drug delivery and imaging: recent advances and challenges. Eur J Pharm Biopharm 97:400–416
Tudorachi N, Lipsa R, Vasile C, Mustata FJ (2013) Poly(lactic acid)-co-aspartic acid copolymers: possible utilization in drug delivery systems. J Polym Environ 21(4):1064–1107
Tullis JL (1977) Albumin: 2. Guidelines for clinical use. JAMA 237(5):460–463
Tysnes OB, Storstein A (2017) Epidemiology of Parkinson's disease. J Neural Transm (Vienna) 124:901–905
van Vlerken LE, Vyas TK, Amiji MM (2007) Poly(ethylene glycol)- modified nanocarriers for tumor-targeted and intracellular delivery. Pharm Res (N. Y.) 24:1405–1414
Wan S, Zhang L, Quan Y, Wei K (2018) Resveratrol-loaded PLGA nanoparticles: enhanced stability, solubility and bioactivity of resveratrol for non-alcoholic fatty liver disease therapy. R Soc Open Sci 5:181457
Whelan J (2001) Nanocapsules for controlled drug delivery. Drug Discov Today 6:1183
Wong KH, Lu A, Chen X, Yang Z (2020) Natural ingredient-based polymeric nanoparticles for cancer treatment. Molecules 25:3620
Wu L, Sun J, Su X, Yu Q, Yu Q, Zhang P (2016) A review about the development of fucoidan in antitumor activity: progress and challenges. Carbohydr Polym 154:96–111
Xie X, Tao Q, Zou Y, Zhang F, Guo M, Wang Y, Wang H, Zhou Q, Yu S (2011) PLGA nanoparticles improve the oral bioavailability of curcumin in rats: characterizations and mechanisms. J Agric Food Chem 59:9280–9289
Xu W, Ling P, Zhang T (2013) Polymeric micelles, a promising drug delivery system to enhance bioavailability of poorly water-soluble drugs. J Drug Delivery 2013:1–15
Yang B, Han X, Ji B, Lu R (2016) Competition between tumor and mononuclear phagocyte system causing the low tumor distribution of nanoparticles and strategies to improve tumor accumulation. Curr Drug Deliv 13:1261–1274
Zhang P, Liu X, Hu W, Bai Y, Zhang L (2016) Preparation and evaluation of naringenin-loaded sulfobutylether-βcyclodextrin/chitosan nanoparticles for ocular drug delivery. Carbohydr Polym 149:224–230
Zhang Z, Liu C, Li C, Wu W, Jiang X (2019) Shape effects of cylindrical versus spherical unimolecular polymer nanomaterials on in vitro and in vivo behaviors. Research
Zhang F, Li R, Yan M, Li Q, Li Y, Wu X (2020) Ultra-small nanocomplexes based on polyvinylpyrrolidone K-17PF: a potential nanoplatform for the ocular delivery of kaempferol. Eur J Pharm Sci 147:105289
Zhou Q, Yang T, Qiao Y, Guo S, Zhu L, Wu H (2015) Preparation of poly (β-L-malic acid)-based charge-conversional nanoconjugates for tumor-specific uptake and cellular delivery. Int J Nanomedicine 10:1941

5 Lipid-Based Nanocarriers for the Delivery of Phytoconstituents

Sonali Priyadarshini, Saumyadeep Bora, and Hitesh Kulhari

5.1 Introduction

Lipid-based nanoparticles (LBNPs) are particles composed of lipids; thus, LBNPs are organic NPs. LBNPs are generally spherical in shape having diameter ranging from 1 to 100 nm (Khan et al. 2019). LBNPs are ionizable lipid-based spherical vesicles that are negatively charged at physiological pH and neutral at low pH (Yang et al. 2022). LBNPs consist of either a solid lipid core or a matrix containing lipid molecules (Huang et al. 2008). The external core of lipid-based NPs is stabilized by surfactants which are also known as surface active agents or emulsifying agents and are used to reduce the surface tension or interfacial tension respectively. These molecules are categorized into two types as ionic: (a) a surfactant which carries an ionic group in the form of negative charge known as anionic surfactant (sodium stearate, sodium cholate, sodium taurodeoxycholate, sodium oleate, sodium taurocholate, sodium glycocholate) and (b) a surfactant which carries an ionic group in the form of positive charge known as cationic surfactant (cetrimonium bromide, benzethonium chloride, triethylamine hydrochloride, dimethyldioctadecylammonium chloride) and non-ionic surfactant (Tweens, Spans, Tyloxapol, Poloxamers, Brij 78, Tego care 450, Solutol HS15) (Cortés et al. 2021). Several criteria are taken into consideration when choosing surfactants for manufacturing of lipid based NPs, including the hydrophilic-lipophilic balance (HLB) scale, desired mode of administration, involvement in in vivo lipid breakdown, impact on particle size, and requirement of lipid modification or not (Shrestha et al. 2014a). There are various types or forms of LBNPs: solid lipid NPs (SLN), emulsion, nanostructured lipid carrier (NLC), liposome, niosome, exosome, etc. (Thi et al. 2021). These NPs have tremendous applications in biomedical fields due to

S. Priyadarshini · S. Bora · H. Kulhari (✉)
School of Nano Sciences, Central University of Gujarat, Gandhinagar, Gujarat, India
e-mail: hitesh.kulhari@cug.ac.in

© The Author(s), under exclusive license to Springer Nature Singapore Pte Ltd. 2024

D. Pooja, H. Kulhari (eds.), *Nanotechnology Based Delivery of Phytoconstituents and Cosmeceuticals*, https://doi.org/10.1007/978-981-99-5314-1_5

their low toxicity, biodegradability, and biocompatibility. The safety, stability, drug-encapsulating capacity, pharmacokinetics, therapeutic effect, and biodistribution of lipid-based nanocarriers can be regulated through structural optimization, surface modification, and material combination (Din et al. 2017). LBNPs have thermal stability, high drug-loading potential, low manufacturing costs, and high yield and are easy to prepare (García-Pinel et al. 2019). Several methods are used for preparation of LBNPs such as solvent displacement method, emulsification, solvent evaporation, high-pressure homogenization, spray drying, etc. (Sathali et al. 2012; Mehnert and Mäder 2012). Another preparation technique is based on supercritical fluid technology. In this method, the extraction of LBNPs occurs above the critical temperature (31 °C) and pressure (74 bar) for CO_2 respectively. Using this technique, dry-powdered LBNPs are produced (Mehnert and Mäder 2012). The rationale for investigating lipid-based systems:

- Ability of lipidic excipients to be flexible
- Flexibility when generating a formulation
- Low hazard profile
- Enhanced oral bioavailability with minimal changes to the plasma profile
- Greater penetration when applying topically
- Possibility of commercialization
- Those have good features that are highly marketable for innovative products
- Enhanced capacity to deal with the crucial issues of technology transfer and manufacturing scale-up (Cerpnjak et al. 2013; Pouton and Porter 2008)

Solubility of the formulation is a major challenge for scientists. As a result, lipid-based drug delivery systems (LBDDS) have become extremely important in recent years owing to their capability to enhance solubility and subsequently bioavailability (Shrestha et al. 2014b). Along with improving storage and delivery, lipid-based systems also prevent decomposition, degradation, and API oxidation (Rawat et al. 2006). The ability of LBDDS to cross the blood-brain barrier (BBB), GIT, blood vessels, and gut is one of their main benefits (Müller et al. 2002a).

5.2 Advantages and Disadvantages of LBNPs

As a drug delivery carrier, LBNPs have numerous advantages such as simple processes for large-scale production (Luo et al. 2006), high biocompatibility, biodegradable nature of NPs, etc. (Silva et al. 2011). It enhances drug solubility (Wang et al. 2016). It protects the drug from external environmental situations (Yoon et al. 2013). It has good stability during storage period (Hamishehkar et al. 2015). It can also be manufactured on a solvent-free process (Bertoni et al. 2022) and is easy to sterilize. It is simpler to validate and obtain regulatory approval (Mishra et al. 2018).

LBNPs also have few disadvantages; however, those are negligible. Drug loading capacity is limited in comparison to other NPs (Wissing et al. 2004). The aqueous dispersion has a variety of colloidal form. The complexity of the physical state of the lipids causes stability issues during storage (Martins et al. 2007). The drug expulsion may happen during administration (Jenning et al. 2000). Dilution of the sample or removal of water can drastically affect the equilibrium between the various colloidal forms and the physical state of the lipid (Mehnert and Mäder 2001). Some of the LBNPs are not stable and require a high amount of surfactant to stabilize them (Göppert and Müller 2003).

5.3 Routes of Administration of LBNPs for Phytoconstituent Delivery

LBNPs can be administered in the human body through different routes such as topical, oral, ocular, parenteral, pulmonary, etc. Various types of drug delivery systems have been designed to deliver the phytoconstituents to different tissues or organs in the human body. The specific advantages and disadvantages of these delivery systems are listed in Table 5.1, while examples of different LBNPs for the delivery of phytoconstituents through various routes of administration are listed in Table 5.2.

5.3.1 LBNPs for Topical DDS

Skin disorders are extremely common all over the world. The most significant constraints in treating these disorders are low therapeutic efficacy due to poor skin penetration or skin permeation of drugs from the most prevalent formulations (Ghasemiyeh and Mohammadi-Samani 2018). LBNPs have been used to increase the skin permeation or penetration (Patzelt et al. 2017). Liposome formulations have been attempted through topical route for the delivery of curcumin to enhance its delivery to brain (Chen and Chiang 2020). Liposomes were used as a nanocarrier for delivering of resveratrol, quercetin, arbutin, pterostilbene, etc. through topical route (Gugleva et al. 2021). Transferosomes were formulated for the delivery of mulberry leaves, epigallocatechin-3-gallate (EGCG), etc. (Nangare et al. 2021; Avadhani et al. 2017). Ethosomes were prepared for the delivery of phytochemicals such as paeonol, ligustrazine phosphate, apigenin, etc. (Ma et al. 2018; Shi et al. 2012; Shen et al. 2014). Niosomes were found very good carriers for transdermal drug delivery of *Annona squamosa* (Abd-Elghany and Mohamad 2020). Nanoemulsions were attempted to encapsulate *Olea europaea* oil and hesperidin for dermal delivery (Battaglia and Ugazio 2019; Durán et al. 2019). SLNs were loaded with hydroquinone for its topical delivery (Ghanbarzadeh et al. 2015).

5.3.2 LBNPs for Oral DDS

Oral drug administration is the most prevalent way of DDS. The main drawbacks of oral drug delivery are the high hepatic first-pass effect and low bioavailability of

Table 5.1 Advantages and disadvantages of administration of lipid-based NPs through various drug delivery systems

Drug delivery system for LBNPs	Advantages	Disadvantages
Topical	LBNPs have higher skin penetration in topical drug delivery system (Pardeike et al. 2009). It is biocompatible and biodegradable in nature (Naseri et al. 2015). Formed clumps and films hydrate the skin (Jain et al. 2014). The process of production is simple and scalable (Gönüllü et al. 2015). It extended skin deposition and improved drug solubility (Ferreira et al. 2016). In drug delivery system, it avoids systemic absorption and side effects (Lauterbach and Müller-Goymann 2015). It may be possible to target particular follicles (Lauterbach and Müller-Goymann 2015)	LBNPs have several limitations in transdermal drug delivery Large quantities of drug loss obtain (Beloqui et al. 2016) Not having a strong controlled medication release (Lauterbach and Müller-Goymann 2015)
Oral	LBNPs minimize the hepatic first pass metabolism (Sangsen et al. 2015). Drug-loaded LBNPs have greater oral bioavailability than pure drugs (Müller et al. 2006). LBNPs show little variation in oral absorption than pure drug (Garg et al. 2017). Drug release can be modulated and controlled by using LBNPs (Nunes et al. 2017). Longer duration and a quicker onset of action (Cirri et al. 2017; Gonçalves et al. 2016)	High levels of water are present in lipid dispersions (Beloqui et al. 2014). There is a chance to expulsion of drug during storage (Beloqui et al. 2014). Hydrophilic drug-loading capacity is limited (Zhang et al. 2016). Polymorphic changes happen (Pandita et al. 2014). The increase of particle size may occur while being stored (Rao and Prestidge 2016). Lipid dispersions are gel in nature (Tran et al. 2014)
Ocular	Encapsulation efficiency is high (Attama et al. 2008). Pharmacokinetic properties are appropriate (Chetoni et al. 2016). It enhances the drug corneal permeability (Sánchez-López et al. 2017a) and increases the bioavailability and distribution of ocular drugs (Araújo et al. 2009). Due to intimate contact with negatively charged mucus, positively charged lipid nanoparticles have a longer ocular retention time (Başaran et al. 2010). Maintain the adequate levels of drugs in the retina, aqueous humor, and vitreous humor (Battaglia et al. 2016). LBNPs are biocompatible and control the drug release (Andrade et al. 2016; Sánchez-López et al. 2017b)	There hasn't been enough research done to determine how hazardous lipid nanoparticles are to retinal cells (Seyfoddin et al. 2010). These formulations have not recently undergone extensive clinical trials, with the majority of studies focusing solely on in vitro or preclinical evaluation (Sánchez-López et al. 2017a). Although lipid nanoparticles are biocompatible, effects of other critical parameters on retinal toxicity need to be investigated. These parameters may include particle size, charge, exposure period, and drug concentration

(continued)

Table 5.1 (continued)

Drug delivery system for LBNPs	Advantages	Disadvantages
Parenteral	Scaling-up potential. Physical stability is high in comparison to poorly stable drugs. An increase of three to five times in drug plasma peaks and a decrease in the cytotoxicity (Wissing et al. 2004). Lower clearance rate and less distribution volume (Ajorlou and Khosroushahi 2017). Less side effects. LBNPS have excellent potential as vaccine adjuvant (Joshi and Müller 2009). Takes a long time for drug circulation. It increases the bioavailability of drug (Hosseini et al. 2016). Improved the drug retention and permeability in tumor tissues. Controlled and sustained drug release (Bhise et al. 2017)	Erosion mechanism used for drug burst release. Drug expulsion obtained (Wissing et al. 2004). Lack of wide clinical studies (Wong et al. 2007). Systemic administration of cytotoxic drugs requires reticuloendothelial system (RES) clearance (Selvamuthukumar and Velmurugan 2012). Particularly when combined with SLN that contains Compritol, lipid buildup in the liver and spleen may result in clinical changes (Prasad and Chauhan 2014). Hydrophilic medicines have a low drug payload (Attama 2011). EPR is a fairly heterogeneous phenomenon that might differ greatly between an animal model and a human or across patients (Ghasemiyeh and Mohammadi-Samani 2018)
Pulmonary	Improved biopharmaceutical characteristics (Cipolla et al. 2014). Achieved a high local concentration. Avoided the hepatic first-pass metabolism (Patil-Gadhe et al. 2014). Obtained higher bioavailability (Hidalgo et al. 2015). Increased the duration of drug residence in the lungs (Patlolla et al. 2010). Incorporating the chemotherapeutic drugs into lipid nanoparticles has the potential to treat lung cancer (Kaur et al. 2016). Prevents the adverse medication effects (Paranjpe and Müller-Goymann 2014). Has excellent storage stability and low toxicity (Pardeike et al. 2011; Nordin et al. 2022). Patient compliance is improved. Dosing intervals are long (Moreno-Sastre et al. 2016). LBNPs can be modified with mucoadhesiveness properties (Makled et al. 2017). Prevented peptide and protein degradation by pulmonary systemic medication administration (Weber et al. 2014)	There are no data on human safety. There were changes in drug release profiles due to lipase degradation in some lipid matrix compositions (Cipolla et al. 2014). Due to smaller particle size, they should be enclosed in lipid microparticles to prevent deep lung delivery (Patil-Gadhe et al. 2014). Drug clearance obtained by macrophages (Hidalgo et al. 2015). Some amount of drug is lost during nebulization (Weber et al. 2014). The release of drugs from these nanocarriers may produce toxic consequences (Xiang et al. 2007). Some fundamental processes are observed, such as agglomeration, clotting, and fragmentation of LBNPs during nebulization (Paranjpe et al. 2014)
Brain	Significant increase of drug uptake in the brain. The drug retention time	

(continued)

Table 5.1 (continued)

Drug delivery system for LBNPs	Advantages	Disadvantages
	increases in the brain. Lipid nanoparticle with polysorbate 80 increases bioavailability (Blasi et al. 2007). BBB disruption can be achieved by coating. Suppression of the efflux system (Tosi et al. 2016). It is possible to encapsulate both lipophilic and hydrophilic drugs (Kaur et al. 2008). It has the ability to administer drug specifically to the brain using a variety of methods. Has potential for SLN apolipoprotein E to target specific brain regions. It has low cytotoxicity, biocompatibility, and biodegradability (Dal Magro et al. 2017)	Rapid removal of the IV-administered drug-loaded SLNs from the bloodstream (Dal Magro et al. 2017)

hydrophobic molecules (Ghasemiyeh and Mohammadi-Samani 2018). To overcome these drawbacks, various types of LBNPs are formulated. SLNs have been attempted through oral route for the delivery of puerarin, triptolide, cantharidin, resveratrol, etc. (Kuo et al. 2008; Luo et al. 2013; Zhang et al. 2013; Dang and Zhu 2013; Neves et al. 2013). Thymoquinone (TQ)-loaded SLNs were formulated for better oral bioavailability (Alam et al. 2018). Similarly, NLCs improved the oral bioavailability of tripterine, silymarin, curcumin, etc. (Shangguan et al. 2014; Yuan et al. 2013). Quercetin was loaded in cationic NLCs to study biodistribution after oral delivery (Liu et al. 2014). Liposomes were loaded with phytoconstituents such as 3-EGCG, curcumin, brucine, etc. (Karewicz et al. 2013; Luo et al. 2014; Li et al. 2013). On the other hand, phytosomes were used to encapsulate silybin, quercetin, kaempferol, and 3-EGCG to enhance absorption, oral bioavailability, and bioefficacy (Gabetta et al. 1988; Song et al. 2008; Singh et al. 2012; Chen et al. 2010; Pietta et al. 1998).

5.3.3 LBNPs for Ocular DDS

Eyes are very sophisticated organ. Due to physiological and anatomical features of the eye, the drug administration through this route is challenging. Major barriers in ocular route administration are tear drainage, corneal epithelium, conjunctival blood flow, and blood ocular barrier. LBNPs were considered as a novel drug delivery system to overcome these challenges (Ghasemiyeh and Mohammadi-Samani 2018). Charged solid SLNs have been attempted to deliver tetrandrine for ocular delivery (Li et al. 2014). LBNPs are capable of passing through blood ocular barrier, to protect the phytochemicals from lacrimal enzymes, to provide sustained drug

Table 5.2 Routes of administration of LBNPs for phytoconstituent delivery

Sr no.	LBNPs	Phytoconstituent	Administration routes	References
1	Liposome	Curcumin	Transdermal/ topical	Chen and Chiang (2020)
	Liposome	Resveratrol, quercetin, arbutin, pterostilbene		Gugleva et al. (2021)
	Transferosome	Mulberry leaves		Nangare et al. (2021)
	Transferosome	Epigallocatechin-3-gallate (EGCG)		Avadhani et al. (2017)
	Ethosome	Paeonol		Ma et al. 2018
	Ethosomes	Ligustrazine phosphate, apigenin		Shi et al. (2012), Shen et al. (2014)
	Niosome	*Annona squamosa*		Abd-Elghany and Mohamad (2020)
	Nanoemulsion	Arbutin		Battaglia and Ugazio (2019)
	NLCs	*Olea europaea* oil, hesperidin		Durán et al. (2019)
	SLNs	Hydroquinone		Ghanbarzadeh et al. (2015)
2	SLNs	Puerarin, triptolide, cantharidin, resveratrol	Oral	Kuo et al. (2008), Luo et al. (2013), Zhang et al. (2013), Dang and Zhu (2013), Neves et al. (2013)
	SLNs	Thymoquinone (TQ)		Alam et al. (2018)
	NLCs	Silymarin, tripterine, curcumin		Shangguan et al. (2014); Yuan et al. (2013)
	Niosome	*Fumaria officinalis*		
	Liposome	3-EGCG, curcumin, brucine		Karewicz et al. (2013), Luo et al. (2014), Li et al. (2013)
	Phytosome	Quercetin, kaempferol, silybin, 3-EGCG, quercetin		Gabetta et al. (1988), Song et al. (2008), Singh et al. (2012), Chen et al. (2010), Pietta et al. (1998)
	Cationic NLCs	Quercetin		Liu et al. (2014)
3	Charged SLNs	Tetrandrine	Ocular	Li et al. (2014)
4	Solid lipid dual-drug NPs	Curcumin	Intraperitoneal route	He et al. (2016)
	NLCs	Curcumin	Intraperitoneal	Chen et al. (2016)
	NLCs	Curcumin	Intranasal	Madane and Mahajan (2016)
	SLNs	Baicalein	Intravenous	Chen et al. (2018)
	Liposomes	Curcumin, NGF	Intravenous	Kuo et al. (2017)

(continued)

Table 5.2 (continued)

Sr no.	LBNPs	Phytoconstituent	Administration routes	References
	Niosome	*Carum carvi*	Parenteral	Barani et al. (2019)
	Ethosome	Thymoquinone (TQ)	Intravenous	Nasri et al. (2020)
	Low-density lipoprotein mimics nanostructured lipid carrier	Curcumin	Intravenous	Meng et al. (2015)
5	Solid lipid microparticles	Quercetin	Pulmonary	Mehta et al. (2018)
	Lipid microparticle	Quercetin		Mehta et al. (2018)
	Nanoemulsion	Quercetin		Arbain et al. (2018)
	Liposome	Curcumin		Adel et al. (2021)

release, and to deliver drugs for a long period (Ghasemiyeh and Mohammadi-Samani 2018).

5.3.4 LBNPs for Parenteral DDS

Nanomedicine and nanotechnology have a significant impact on parenteral drug delivery. Drug-loaded LBNPs can be administered via intravenous infusion, subcutaneous injection, intramuscular injection, etc. It is possible for LBNPs to release drugs through enzymatic diffusion or degradation, which could lead to a sustained release of the drug (Ghasemiyeh and Mohammadi-Samani 2018). Solid lipid dual-drug NPs have been endeavored through intraperitoneal route for the delivery of curcumin (He et al. 2016). Curcumin-loaded NLCs have been used for intraperitoneal and intranasal route administration (Chen et al. 2016; Madane and Mahajan 2016). SLNs prepared for delivering baicalein through intravenous (IV) route (Chen et al. 2018). Liposomes were formulated to deliver curcumin and nerve growth factor (NGF) through IV pathway (Kuo et al. 2017). *Carum carvi*-loaded niosome targeted to tumor cell line through parenteral administration (Barani et al. 2019). Thymoquinone (TQ)-loaded ethosome was introduced through IV route for breast cancer treatment (Nasri et al. 2020).

5.3.5 LBNPS for Pulmonary DDS

Drug delivery using pulmonary route is a reasonably innovative tactic with numerous benefits. It is a nonintrusive method of administering medications both locally and systemically. Drug dosage may be reduced using this direct delivery system, which would reduce drug side effects. High drug permeability could be ensured by the pulmonary system's large surface area and thin alveolar epithelium (Ghasemiyeh and Mohammadi-Samani 2018). Various types of LBNPS are used to deliver quercetin for pulmonary DDS such as solid lipid microparticles, lipid microparticle, nanoemulsion, etc. (Mehta et al. 2018; Arbain et al. 2018). Liposomes were prepared to deliver the curcumin through pulmonary route (Adel et al. 2021).

5.3.6 LBNPs for Brain DDS

Delivering drugs to the brain is one of the pharmaceutical sciences' most critical issues due to the blood-brain barrier (BBB). NPs can be used in brain DDS because of their small size and high drug encapsulation efficiency and can escape RES uptake. Two major barriers in brain DDS are constrained drug penetration across the BBB and efflux of the drugs. The primary colloidal DDSs that have been utilized to overcome these hurdles are SLNs and NLCs (Ghasemiyeh and Mohammadi-Samani 2018).

5.4 Various Types/Forms of LBNPs

A variety of LBNPs have been developed including liposome, solid lipid nanoparticles (SLN), niosome, exosome, ethosome, transferosome, nanostructure lipid carriers (NLC), nanoemulsions, etc. (Fig. 5.1) (Nordin et al. 2022). Drugs that are poorly water-soluble and permeable can be made more bioavailable through the use of such nanostructures, which can also increase the transport of bioactive chemicals (Teixeira et al. 2017).

5.4.1 Liposome

Liposome is the traditional type of LBNPs, discovered by Alec D Bangham in the 1960s (Shah et al. 2020). According to one definition, liposomes are "closed, continuous bilayered structures mainly composed of lipid molecules" (Mozafari et al. 2002). Liposomes are composed of phospholipids having a polar head group with a long hydrophobic tail. The arrangement of phospholipid molecules results in a bilayer structure, with the heads of one layer contacting the media on the outside and the heads of the other layer encircling an internal aqueous phase (Briuglia et al. 2015). Some examples of natural phospholipids are phosphatidylinositol, phosphatidylglycerol, phosphatidylethanolamine, phosphatidylserine, phosphatidylcholine, etc. (Allen and Cullis 2013). Liposomes can entrap both hydrophilic and hydrophobic molecules (Brandelli 2012). Liposomes are spherical structures made up of two layers of lipid molecules with their nonpolar groups facing one another and can spontaneously form bilayers in aqueous condition (Esposto et al. 2021; Liu et al.

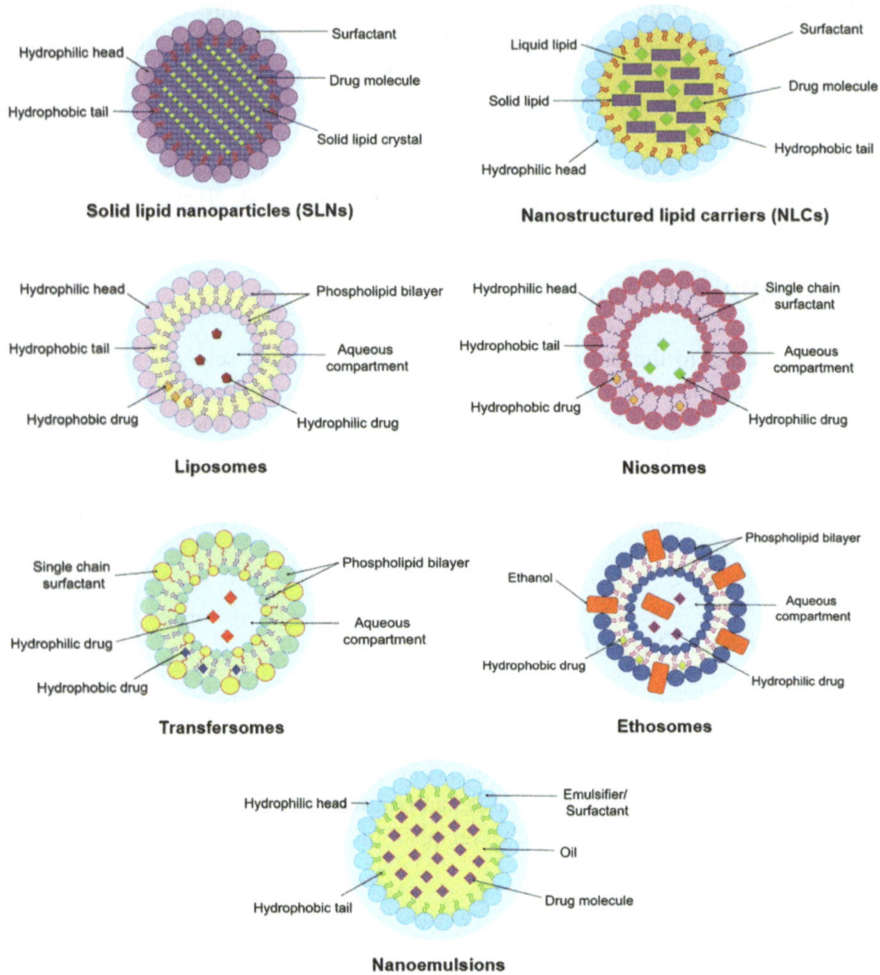

Fig. 5.1 Structural diagrams of various types of lipid-based nanoparticles. (Adapted from Nordin et al. 2022)

2020). On the basis of size, lamellarity, and vesicularity, liposomes are further classified (Fig. 5.2) (Mozafari et al. 2008), such as:

- Unilamellar vesicles (UVL) are type of liposomes consist of one single lipidic bilayer and can be small (SUVL ~10–100 nm) or large (LUVL ~above 100 nm).
- Multilamellar vesicles (MLV) are composed of multiple concentric bilayers (0.5–10 nm).
- Multivesicular vesicles (MVV) are made up of a large number of tiny nonconcentric vesicles encased in a single lipid bilayer.
- Double bilayer vesicles (DBV) are made up of two bilayer membranes (Fernandes et al. 2021).

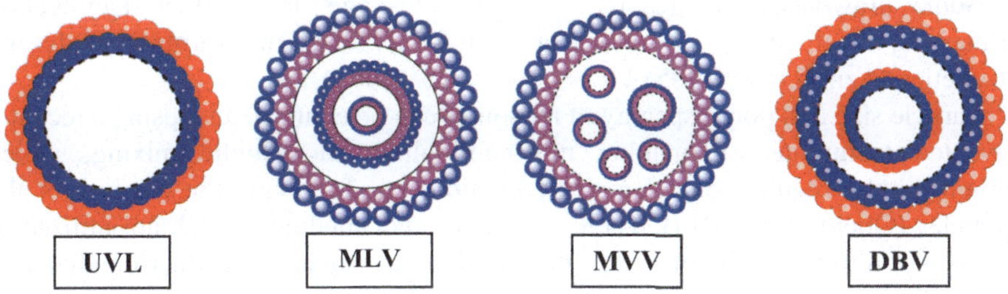

Fig. 5.2 Different types of liposome vesicles

Advanced liposomes can be designed as per the requirements, e.g., cationic liposomes, immunoliposomes, pH-sensitive liposomes, and long-circulating liposomes (Sriraman and Torchilin 2014).

Several methods are used for preparation of liposome such as solvent dispersion method, thin film hydration, mechanical dispersion, detergent removal method, supercritical fluids technology, and high-pressure homogenization technique (Esposto et al. 2021; Pinilla et al. 2021).

Liposomes possess excellent drug-protection and targeting ability (Chacko et al. 2020). Because of these properties, liposome formulations have higher efficiencies than free drugs, enhance the solubility for hydrophobic drug (Shah et al. 2020), and improved bioavailability and biodistribution (Mozafari et al. 2002). Poor hydrophilic drug encapsulation and poor storage stability caused by drug leakage are drawbacks of liposomes (Opatha et al. 2020).

5.4.2 Niosomes

Niosome is an alternative approach to liposome which is prepared by using cholesterol and nonionic surfactants in aqueous condition (Chacko et al. 2020). Thus, it is also called nonionic surfactant vesicles (Baillie et al. 1985). Handjani-Vila and colleagues made the first observation of niosome formation (Azmin et al. 1985). They are likely similar to liposomes, but niosomes have nonionic surfactants that form the bilayers (Baillie et al. 1985). Generally, the size of niosome is in the range of 10–1000 nm and shape is spherical (Thi et al. 2021).

Niosomes have many benefits, including affordability, high stability, accessibility to a variety of nonionic surfactants, biocompatibility, biodegradability, and favorable storage conditions (Sahin 2007). One of the advantages of niosome is improved dispersion for substances with solubility problems. It enhances the penetration of the ingredients and complies with the skin's surface and releases the drug over time. It has greater stability over an extended period as compared to liposomes (Kumar 2019). It is easy to formulate and scale up (Müller et al. 2002a). Like liposomes, niosomes cannot permeate deeper skin layers (Rogerson et al. 1988). It is non-immunogenic and osmotically stable (Kumar et al. 2022).

Some drawbacks of niosomes are drug leakage and particle aggregation (Bhardwaj et al. 2020). Niosomes are commonly used in pharmaceuticals and cosmetics (Azmin et al. 1985).

Particle size and polydispersity of niosomes can be controlled by using a recently invented technique for making niosome called microfluidic mixing, which eliminates the requirement of a size reduction step following the creation of the particles (Obeid et al. 2017). There are some other methods which are utilized to formulate niosome, namely, thin-film hydration, sonication, microfluidization, reverse phase evaporation, etc. (Kumar et al. 2022).

By altering the composition of the surfactants, the physical properties of the niosomes like stability, membrane fluidity, and liquidity can be improved (Kauslya et al. 2021).

5.4.3 Exosomes

Exosomes are nanovesicles with sizes ranging from 30 to 100 nm and densities ranging from 1.13 to 1.19 g/mL. It gets released by practically all cell types and, upon invagination and budding of the limiting membrane of late endosomes during endosome maturation, evolves into intraluminal vesicles (ILVs) of multivesicular bodies (MVBs). Early in the 1980s, exosomes were first discovered in reticulocyte culture media (Johnstone et al. 1987; Harding et al. 1983, 1984). Johnstone first used the name "exosomes" in 1987 to describe these vesicles, which are expelled from cells by exocytosis (Johnstone et al. 1987). These are essentially cytoplasm that has been covered in a lipid bilayer and has transmembrane protein domains exposed on the outside. Exosomes can include a variety of biomolecules, including carbohydrates, proteins, lipids, and a nucleic acid signature that identifies the source of origin. New purification techniques that result in extremely pure exosome preparations have made it possible to employ proteomic and molecular methods to analyze the molecular composition of exosomes. Several techniques, including fluorescence-activated cell sorting (FACS), Western blot, mass spectrometry, and ELISA, have been employed to determine whether exosome formulations from various biological sources include cellular proteins (Chahar et al. 2015).

Some methods of exosome preparation are ultracentrifugation, density gradient ultrafiltration, chromatography, exoQuick polymer, microfluidic method, precipitation technique, immuno-affinity, etc. Exosomes are an excellent drug delivery nanocarrier due to their various benefits, which include strong biocompatibility, permeability, low toxicity, and low immunogenicity (Butreddy et al. 2021).

Exosomes have been linked to both normal and abnormal manifestations of physiology, including immune response, lactation, the onset and progression of liver disease, neuronal function, viral infections, cancer, and neurodegenerative diseases (Admyre et al. 2007; Masyuk et al. 2013; Vella et al. 2008; Bard et al. 2004).

Exosomes have the advantages of minimal immunogenicity, high blood stability, and direct drug delivery to cells as a natural carrier. Exosomes are able to travel across cells, making them ideal to allow cells to share materials and information (Xi et al. 2021).

Exosomes still have several drawbacks in a drug delivery technology, including low yield, difficult manufacture, restricted drug loading, impaired activity of specific medications, and poor clinic targeting efficiency (Moon and Chang 2022).

5.4.4 Transferosomes

The Latin word *transferre* (which means "to carry across") and the Greek word *soma*, which means "a body," are the origins of the word "transferosome" (Kaur 2014). Transferosomes are vesicles composed of phospholipids, 10–25% surfactants, 3–10% organic solvents for increasing the flexibility, and hydrating medium containing of saline phosphate buffer (Kumar et al. 2013). A transferosome carrier is a synthetic vesicle that mimics the properties of a cell vesicle, or a cell engaged in exocytosis. These nanocarriers are also appropriate for controlled and possibly targeted drug delivery. They have high flexibility and automated membranes that make the vesicle very deformable. Transferosome vehicles can efficiently cross microporous barriers, even if the pores are much smaller than the vesicles themselves (Solanki et al. 2016).

The procedures of preparation of transferosomes are simple and easy to scale up. Film hydration is the major process used to manufacture transferosomes. In addition, methods that are used to prepare transferosomes include modified handshaking method, reverse-phase evaporation method, ethanol injection method, and vortexing–sonication method (Das et al. 2022).

In order to accommodate drug molecules varying in solubility, transferosomes have an infrastructure made up of both hydrophilic and hydrophobic components. These can swell and fit through a small opening without suffering much damage. High deformability of this system allows intact vesicles to penetrate more effectively. Transferosomes have been explored to deliver both low and high molecular weight drugs. They are biodegradable, biocompatible, have high entrapment efficiency, and fortify the incorporated drug from metabolic degradation (Kumar et al. 2012). Both systemic and topical medication administration can be accomplished using them. They serve as a storehouse, gradually releasing their contents (Kumar et al. 2013).

Transferosomes are chemically unstable due to their propensity for oxidative destruction. The purity of natural phospholipids is another obstacle in using transferosomes as DDS. Moreover, transferosome formulations are also expensive (Chaurasiya et al. 2019).

5.4.5 Ethosome

As an additional lipid nanocarrier, ethosomes were synthesized by Touitou et al. in 1997 (Bendas and Tadros 2007). Ethosomes are phospholipid nanoparticles that contain a high percentage of ethanol (20–45%). The addition of ethanol increases the permeability and elasticity of the ethosomes that make them useful for transdermal drug and cosmetic delivery via the pores of the stratum corneum, the skin's outermost layer (Touitou et al. 2000; Natsheh et al. 2019). Because of the high concentration of ethanol, which is known for disrupting the organization of the skin's lipid bilayer, the ethosomes are unique. Furthermore, because of the high ethanol content, the lipid membrane is packed less tightly than typical vesicles, yet having equal stability, allowing for a more pliable shape and improved drug distribution ability in the stratum corneum lipids (Verma and Pathak 2010). Ethosomes are prepared by several methods such as cold homogenization method, hot homogenization method, injection method, mechanical dispersion method, etc. (Shabreen and Sangeetha 2020).

Some advantages of ethosomes are passive, noninvasive, and the use of nontoxic raw materials. These are mostly used in cosmetics and topical pharmaceutical formulations. Ethosomes can deliver a variety of molecules with varying physiochemical properties, including peptides, hydrophilic and lipophilic molecules, etc. These are used as gel or cream form and have high patient compliance (Satyam et al. 2010).

Ethosomes have also some disadvantages such as formation of clusters and aggregation, lose their content when they move from the organic to the aqueous layer, and very poor yield. Some individuals may have skin irritation or dermatitis as a result of ethanol, penetration enhancers, or other excipients. The administration of ethosomes did not result in rapid bolus-type drug input. They were made to deliver drugs gradually over time (Aggarwal and Nautiyal 2016).

Applications of ethosomes have been investigated in the delivery of antibiotics, herpetic infection, antiviral agents, anti-arthritis drug, pilosebaceous targeting agent, and anticancer drugs (Sivakranth et al. 2012; Mohite Mukesh et al. 2021).

5.4.6 Solid Lipid Nanoparticles (SLNs)

Professors R.H. Müller and M. Gasco presented SLN as a substitute and superior lipid-based carrier system in the midd 1990s (Üner and Yener 2007). These NPs generally have a spherical form and a diameter between 50 and 1000 nm. A SLN is defined as a lipid monolayer enclosing a solid lipid core (Ganesan and Narayanasamy 2017). The most common lipidic materials used in the production of SLNs are triglycerides, fatty acids, complex glyceride mixtures, and waxes (Ganesan and Narayanasamy 2017; Barroso et al. 2021). High thermal resistance lipids with a melting point exceeding 40 °C should be used to create the nanoparticles, to ensure that the lipids are solid at both body and room temperatures (Teixeira et al. 2017). SLN can solubilize lipophilic compounds because they have a solid lipid core matrix. The lipid core of SLNs is stabilized using surfactants (Patel et al. 2015). There are three different drug ingestion models for SLNs: solid solution, drug-enriched shell, and drug-enriched core model (Ramteke et al. 2012).

SLNs are prepared by various methods like sonication, double-emulsion method, spray drying, solvent emulsification/evaporation method, supercritical fluid method, using membrane contractor, solvent injection technique, high pressure homogenization, etc. (Patel et al. 2015).

Significant benefits of SLNs include protection of drug, drug targeting and controlled drug release, improved drug stability, low cost, easy to prepare, nontoxicity, biodegradability, biocompatibility, and cost-effectiveness (Patidar et al. 2010). Bioavailability of entrapped bioactive compounds has been increased with the help of SLNs. Both lipophilic and hydrophilic drugs can be incorporated. A large amount of drug can be payload. SLN can be prepared without organic solvents. These are easy to scale up and sterilize (Fonseca-Santos et al. 2015; Wilczewska et al. 2012; Agrawal et al. 2018). The toxicity of SLNs to human granulocytes is remarkably low. All these excellent advantages make them a strong candidate for the DDSs. Because of their remarkable benefits, these structures have been utilized in medicine, with applications in cosmetics and pharmaceuticals. The applications of SLNs in the food sector also show great promise as carrier for ingredients found in food items, such as carotenoids, vitamins, antioxidants, and phytosterols (Pinilla et al. 2021). Further, SLNs are appropriate for IV injection in addition to parenteral, pulmonary, and cutaneous usage due to their small size required for delivering drugs to a specific tissue or organ (Üner and Yener 2007). It can be degraded by pancreatic lipase or lipolytic enzyme, depending upon the lipid and surfactant composition. The degradation of lipids also depends upon the fatty acid chain length. The degradation is slowed as the length increases, but the breakdown of surfactants is dependent on steric stabilization (Olbrich and Müller 1999).

Some disadvantages of SLNs are low drug loading of hydrophilic molecules, drug expulsion, particle growth, unexpected gelation tendencies, etc. The expulsion of the drug is generated by crystallization process during the storage (García-Pinel et al. 2019).

5.4.7 Nanostructured Lipid Carriers (NLCs)

NLCs are the second generation of lipid nanoparticles created to address the shortcomings of SLNs. Generally, the sizes of the NLCs are 50–1000 nm and are comprised of a lipid matrix with a unique nanostructure. These NLCs are synthesized from mixture of solid lipids and liquid lipids to improve the loading capacity of drugs (Teixeira et al. 2017; Katouzian et al. 2017). The main components used to synthesize the NLCs are solid lipids, liquid lipids, surfactant, water, and the bioactive compounds (Gordillo-Galeano and Mora-Huertas 2018). Compared to SLNs, the NLCs prevent coalescing of particle with the solid matrix due to the presence of liquid lipids. This allows the encapsulation of bioactive compounds that are better soluble in liquid lipids (Pinilla et al. 2021). The resulting lipid particle matrix has a lower melting point than the original solid lipid, but it remains solid at body temperature (da Silva Santos et al. 2019). Different types of NLCs are produced depending on the method of production and the composition of the lipid blend (Patidar et al. 2010). Basically, there are three categories of NLCs: I, II, and III. NLCs of type I have significant flaws. Type II NLCs may load the drug in both phases and have a very high ratio of liquid lipid to oil. Amorphous lipids found in Type III NLCs hinder the drug from being ejected (Müller et al. 2002b). In an intriguing study, Precirol ATO 5 (solid lipid) and Transcutol RHP (liquid lipid) were used to optimize the NLCs along with different combinations of the surfactants. The results revealed that the optimized combination enhanced the pharmaceutical and physical characteristics of NLCs, including particle size, drug loading effectiveness, drug release rate, solubility, and stability. Optimizing the lipid monomer composition can also enhance the carrier's pharmacokinetic and physical characteristics (Mura et al. 2021).

Several techniques are adopted for synthesis of NLCs; however, high-pressure homogenization (hot or cold) is the most used method. In addition, micro-emulsion method is regularly used for preparation of NLCs (Fang et al. 2013).

As mentioned, NLCs have advantages over SLNs, such as increased encapsulation effectiveness, increased solubility, improved bioavailability, etc. It has lower water content and increases the permeability and half-life (Pinilla et al. 2021).

NLCs also have some disadvantages like requirement of high temperature for operating and not stable over a long time (Fang et al. 2013).

5.4.8 Nanoemulsion (NE)

NEs are composed of two immiscible liquids that are equilibrated into a monophase using surfactants or a surfactant–co-surfactant combination and have size range of 50–200 nm. NEs are an isotropic and thermodynamically stable system. The nanoemulsions are made up of oils, surfactants, or a combination of the two and an aqueous phase. The oil phase is used to dissolve hydrophobic drugs (Souto et al. 2011). NEs are stabilized by surfactants and are heterogeneous dispersions of the compositions "oil in water" (O/W) or "water in oil" (W/O). Double NEs have also been developed, either as W/O/W or O/W/O. Because it helps to lower interfacial tension, an emulsifier is essential for the creation of small droplets. Additionally, it helps to stabilize NEs using repelling electrostatic interactions and steric hindrance (Lai et al. 2013).

NEs have gained recognition over the past few decades, because of their astonishing qualities like transparent occurrence, high surface area, tunable rheology, and dynamic stability. NEs are biodegradable, biocompatible, and easy to prepare. Some other benefits of NEs are lack of flocculation, sedimentation, and creaming. NEs are manufactured with least amount of surfactants, making it less harmful to biological membranes. The nanosized droplets make them resistant toward destabilizing phenomenon. NEs are used to replace the liposomes and other lipid vesicles as these are more stable (Mishra et al. 2018). NEs have been used in a variety of industries, including food, cosmetics, pharmaceuticals, and material synthesis due to their various advantages (Lai et al. 2013). NEs have the benefits of low toxicity, no irritation, and long-term stability in addition to being usable through a variety of routes of administration, particularly in dermal and transdermal drug delivery (Thiagarajan 2011).

It is a nonequilibrium system. Therefore, high-energy techniques like homogenization and ultrasonication are required to obtain or getting a nonequilibrium state in which large droplets formed into submicron size (Sainsbury et al. 2014). The system can remain dispersed without separation because the small droplet size also inhibits flocculation of the drops. The small droplets are nondeformable that prevents the surface fluctuations. For NEs, the range of surfactant concentrations is 5–10%, which is less as compared to microemulsion (Walker et al. 2015; Tadros et al. 2004).

NEs have several factors which limit their applications—expensive to produce because of the need of expensive equipment. The process of creating submicron droplets and the function of surfactants and co-surfactants are not well understood. There is a lack of knowledge regarding the mechanisms that influence and cause issues during the production of nanoemulsions (Tadros et al. 2004). As less amount of surfactant is used, this may cause stability to be temporary. To successfully introduce this system to the market, more research about the mechanisms that cause droplets to form using different techniques is still necessary (Mishra et al. 2018).

5.5 Lipid-Based Nanoparticles for the Delivery of Phytoconstituent

Phytochemicals (PCs) are organic bioactive substances that are naturally present in plant-derived foods like vegetables, fruits, spices, grains, and nuts (Xiao and Bai 2019). Numerous phytochemicals from traditional medicine have been used to maintain health and prevent diseases (Surh 2003; Basnet and Skalko-Basnet 2011). Numerous PCs are reported to have anticancer properties throughout the last decades based on research from cell culture and some animal studies, but some human clinical trials have yielded mixed results (Khushnud and Mousa 2013; Priyadarsini and Nagini 2012). High doses of PCs are impractical for human studies and may be the cause of the inconsistent results. The clinical use of PCs is restricted due to their poor aqueous solubility, stability, low bioavailability, and high rate of digestion and metabolism by enzymes in the liver, the digestive tract, and other tissues (Khushnud and Mousa 2013; Wang et al. 2013).

The most commonly used PCs are silymarin, curcumin, catechin extract, oleuropein, lapachone, *Brucea javanica* oil, resveratrol, naringin, tanshinone, *Nigella sativa*, diosmin, genistein, berberine, brucine, sesamol, tretinoin, myricetin, thymoquinone, celastrol, mangiferin, tea tree oil, glycyrrhetinic acid, colchicine, epigallocatechin-3-gallate (EGCG), etc. (Chuan et al. 2015).

LBNPs have been used as carriers for the delivery of PCs. Numerous PCs have been loaded in biocompatible and biodegradable LBNPs. These NPs can improve their absorption and bioavailability, protect them from degradation by enzymes, increase their stability, prolong their circulation time, and exhibit high differential uptake efficiency in target cells or tissues (Wang et al. 2014). LBNPs such as SLN, NLC, niosome, NE, microemulsion, exosome, liposome, etc. have been used as a carrier (Nordin et al. 2022).

5.5.1 Applications of LBNPs to Deliver Phytoconstituents for the Treatment of Cancer

Cancer is a global illness with a high morbidity and fatality rate. According to the WHO's GCO Program (Global Cancer Observatory), in 2030, there will be more than 20 million people with cancer and 8 million deaths from this disease. Numerous studies have centered on the research and development of innovative technologies and medications from natural products, leading to the development of natural bioactive products with strong anticancer activity (Thangaraj et al. 2022). There are more than a hundred distinct types of cancer, and they vary considerably in terms of size, metastasis, and location of the cancer. Skin cancer, colon and rectal cancer, breast cancer, liver cancer, pancreatic cancer, lung cancer, and prostate cancer are the most prevalent cancers (Chuan et al. 2015). LBNPs increase the delivery of PCs to the tumor tissues and cancer cells and thus improve the effectiveness of the treatment (Table 5.3).

5.5.1.1 Colorectal Cancer

Because of its high death rate, colorectal cancer is a severe public health concern (Bray et al. 2018; Ferlay et al. 2019). In the case of metastatic colon cancer, LBNPs represent a potential approach to improve the current therapy. Low et al. developed a system based on Pickering emulsions (PE), which consist of a magnetic cellulose nanocrystal loaded with curcumin capable of controlling the release of the drug. The growth of HCT 116 human colon cancer cells was slowed by this system in both monolayer and multicellular spheroids (Low et al. 2019).

5.5.1.2 Pancreatic Cancer

With a 5-year survival rate of 10%, pancreatic ductal adenocarcinoma (PDAC) is one of the most fatal gastrointestinal malignancies. In the United States, PDAC is the third biggest cause of cancer-related fatalities (Siegel et al. 2018). Most PDAC patients are ineligible for surgery because of late diagnosis, early metastases, and severe local tissue invasion (Liu et al. 2022). Further, drug distribution to the pancreas also remains very low. Nanotechnology provides some therapeutic methods to raise these patients' chances of recovery. Chirio et al. prepared curcumin-loaded lipid NPs using the microemulsion cold dilution technique. This system increased the inhibition of cell growth in PANC-1 cells (Chirio et al. 2018). Ranjan et al. prepared curcumin-loaded liposomes for pancreatic cancer treatment. The developed liposomal formulation reduced the tumor growth, decreased the cell viability, and reduced the angiogenesis of Mia PaCa human pancreatic cancer cells (Ranjan et al. 2013).

Table 5.3 Lipid-based nanoparticles for the delivery of phytoconstituents against various diseases

Sr no	Lipid nanoparticles	Phytoconstituents	Diseases	Remark	References
1	SLN	Silymarin	Lung and breast cancer cells	Activated apoptosis and cytotoxicity in breast and lung cancer cells	Sezer (2021)
	Nanoemulsion		Liver cancer	Exhibited cytotoxic effect. Promoted intracellular release of ROS. Improved free drug pharmacokinetics. Higher C_{max}, shorter T_{max}, and larger AUC	Ahmad et al. (2018)
	NLCs		Depression	Similar to the antidepressant effect of fluoxetine in rats. Significantly higher brain concentration	Ashraf et al. (2019)
2	Pickering emulsions	Curcumin	Colorectal cancer	Controlled drug release, decreased growth of HCT116 cells	Low et al. (2019)
	MEs (stearoyl chitosan-coated lipid NPs)		Pancreatic cancer	Increase in inhibition of the cancer cell growth and deep penetration of drug	Chirio et al. (2018)
	MEs, cationic liposome		Liver cancer	Increase in both cellular internalization and cytotoxicity to cancer cell	Chang et al. (2018)
	MEs, NEs		Glioblastoma multiform	Enhanced brain uptake, able to boost the effect of the drug against GBM	Kumar et al. (2016)
	NLCs		Prostate cancer	Sustained drug release; decrease in cell viability, increased anticancer activities	Aditya et al. (2013)
	SLNs		Breast cancer	Intracellular uptake, curcumin bioavailability, and anticancer activity increased	Sun et al. (2013)
	SLNs		Liver cancer	Sustained release, cell viability decreased, apoptosis increased	Chuan et al. (2015)
	NLC		Alzheimer	Shown no toxic effects on BCECs	Meng et al. (2015)
	SLNs		Lung cancer	Enhanced stability, anticancer activity	Chuan et al. (2015)

(continued)

Table 5.3 (continued)

Sr no	Lipid nanoparticles	Phytoconstituents	Diseases	Remark	References
	Liposomes		Pancreatic cancer	Cell viability decreased, tumor growth decreased, angiogenesis reduced	Ranjan et al. (2013)
	NLC		Brain cancer	Enhances the inhibition effect, ROS level increased	Chen et al. (2016)
	NLC		CNS histone hyperacetylation	Decreased histone acetylation in CNS	Chen et al. (2016)
3	NEs	Catechin extract	Prostate cancer	Reducing the IC50 of drug against PC-3 cells, allowing a greater tumor volume reduction	Tsai and Chen (2016)
4	Liposome	Oleuropein	Prostate cancer	Improved 22Rv1 cell apoptosis, oleuropein bioavailability, and in vivo survival	Nassir et al. (2019)
5	Liposome	Brucea javanica oil	Liver cancer	High cytotoxic effect in vitro, dose-dependent reduction in intrahepatic metastasis with improved survival rates in vivo	Yue et al. (2015)
6	SLN	Resveratrol	Liver cancer	As compared to free drug, it has lower expression of pro-inflammatory cytokines. HepG2 cell cytotoxicity has increased. Lower hepatic nodule formation. Higher accumulation in tumor tissue	Rahman et al. (2020)
	SLNs		Brain cancer	Higher concentration of resveratrol in the brain	Jose et al. (2014)
	SLNs		Breast cancer	As compared to free resveratrol, it showed superior ability in inhibiting the cell proliferation, exhibited much stronger inhibitory effects on the invasion and migration of cells	Chirio et al. (2018)
7	NLCs	Naringin, coix seed oil	Liver cancer	Shows the highest cytotoxicity effect and the lower IC50	Zhu et al. (2020)
8	ME	Tanshinone	Liver cancer	Induces necrotic effects in H22 cells, able to upregulate Bax; enhances the tumor inhibition and decreases tumor weight	Ma et al. (2013)

9	NE	Nigella sativa	Liver cancer	Induce cytotoxicity in HepG2 while being nontoxic to normal human liver, increased levels of internal (ROS)	Tabassum and Ahmad (2018)
10	Phytosome	Diosmin	Liver cancer	Enhances the intestinal absorption of drug and has the ability to penetrate intestinal membrane	Freag et al. (2013)
11	Phytosome	Genistein	Liver cancer	Enhances the drug accumulation in the liver	Komeil et al. (2021)
	Cationic NE		Skin disease	Increased the retention of isoflavone in epidermis	
	Liposome		Ovarian and prostate carcinomas	Cellular delivery is high. Cytotoxicity increased, apoptosis increased	Chuan et al. (2015)
	Lipidic micelles and NE		Hepatic and colon carcinoma	Cytotoxicity increased, anticancer activities increased	Chuan et al. (2015)
12	SLNs	Sesamol	Skin cancer	Enhanced bioavailability in the skin	Geetha et al. (2015)
13	NLCs	Sesamol	Breast cancer	Apoptosis increased, increment of aqueous solubility improved by 1000 folds, reduced the viability of cancer cell	Chuan et al. (2015)
	Liposome		Breast cancer	Cellular uptake increased, antioxidant activities improved	
	Nanoliposomes		C6 glioma cell	C6 glioma cells were more necrotic	
14	Nanoliposomes, NE	Resveratrol	Aging	When compared to pure resveratrol, it exhibited a strong radical scavenging effect	Vaiserman et al. (2020)
	SLNs	Genistein		Improve the oral bioavailability	
	SLNs	EGCG		Increased the stability, improve the bioavailability after oral delivery, higher antioxidant properties	
15	SLNs	Thymoquinone	Depression	After administration, more TQ reached the target	Alam et al. (2018)

(continued)

Table 5.3 (continued)

Sr no	Lipid nanoparticles	Phytoconstituents	Diseases	Remark	References
16	Nanoliposomes	Epigallocatechin-3-gallate (EGCG)	Breast cancer	Increases EGCG stability and increased proapoptotic, antiproliferative, and antiangiogenic efficacy	de Pace et al. (2013)
	SLNs		Prostate cancer	Improve the stability and anticancer activity of EGCG, shown sustained drug release	Radhakrishnan et al. (2016)
17	liposome	Quercetin	Breast cancer	High cellular uptake and high antioxidant activities	Chuan et al. (2015)
	Nanomicelles		Lung cancer	High quercetin stability, less viability of cancer cells, and reduction of the tumor size	Tan et al. (2012)
18	Niosome	Celastrol	Psoriasis	Water solubility and skin penetration of celastrol increased its anti-psoriasis effectiveness in mice	Meng et al. (2019)
19	NE	Mangiferin	Psoriasis	Reduces edema by ~20-fold	Singh et al. (2021)
	ME	Tea tree oil		Potential in enhanced drug uptake	Sonia and Anupama (2011)
	NE	Glycyrrhetinic acid		Improved formulation stability and transdermal impact	Puglia et al. (2010)
	Liposome	Colchicine		Sustained delivery	Singh et al. (2009)
20	SLNs	Quercetin	Acne vulgaris	Increase the skin penetration through the stratum corneum	Chutoprapat et al. (2022)
	NLCs	Lycopene		High drug loading capacity with enhanced stability	
	NLCs	Resveratrol		Higher stability of RSV	
	NLCs	Eucalyptus essential oils or rosemary essential oils		Higher antimicrobial and wound-healing properties	

21	Liposomes, niosomes, SLNs, NLCs	Curcumin	Skin disease	Improved solubility, stability, and therapeutic efficacy of curcumin	Rajagopal et al. (2022)
22	SLNs	Thymoquinone	Psoriasis	Less skin irritation score	Nordin et al. (2022)
23	NLCs	Hup A	Alzheimer's disease	Showed controlled drug release	Yang et al. (2010)
	MEs, SLNs, NLCs			Safe for skin use	Patel et al. (2013)
	SLNs	Quercetin		Exhibited better memory retention	Dhawan et al. (2011)
	SLNs	Sesamol		Effectively restore cognitive deficits, assistance in reducing nitrosative stress and cytokine release	Sachdeva et al. (2015)
	SLNs	Chrysin		Protect against neuronal damage	Vedagiri and Thangarajan (2016)
	SLNs	Piperine		Reduced the value of superoxide dismutase (SOD) and immobility values while raising acetylcholinesterase values	Yusuf et al. (2013)
24	Ethosome	Ligustrazine phosphate		Enhanced skin permeation in vitro and bioactivity in vivo	Shi et al. (2012)

5.5.1.3 Liver Cancer

Liver cancer is one of the most common and deadly diseases in people. Curcumin, silymarin, gallic acid, and EGCG are examples of phenolic PCs that are secondary plant metabolites. These molecules have been proven to be efficient and can enhance the cell signaling pathways that control the growth, inflammation, invasion, and apoptosis of liver cancer cells (Banik and Patra 2022). Ahmad et al. prepared NE formulations that were used to encapsulate silymarin. In comparison to the untreated control, the drug-loaded NE showed cytotoxic effects as well as apoptotic signs in HepG2 human liver cancer cells (Ahmad et al. 2018). Glycyrrhetinic acid-modified curcumin-loaded cationic liposome was formulated by Chang M. et.al. This targeted nanoformulation showed increase of both cellular internalization and cytotoxicity to cancer cell (Chang et al. 2018). Yue et al. prepared *Brucea javanica* oil-loaded liposomes for targeting in HepG2 cell line-bearing mice. They observed a significant increase in cytotoxicity in vitro, reduced intrahepatic metastasis in a dose-dependent manner, and increased longevity of tumor-bearing mice (Yue et al. 2015). Rahman et al. formulated cationic SLNs incorporated with resveratrol for treating liver cancer. There was a significant decrease in cell viability of HepG2 cells, formation of lower hepatic nodules, and lower expression of proinflammatory cytokines than free drug (Rahman et al. 2020). Zhu et al. formulated naringin-entrapped NLCs composed of coix seed oil. When tested on HepG2 cells, NLCs displayed higher cytotoxicity effect and lower IC_{50} compared to the free drug and increased the inhibition of tumor cell (Zhu et al. 2020). Ma and colleagues created and evaluated efficacy of tanshinone-entrapped microemulsion. They observed its ability to induce necrotic effects in H22 cells and to downregulate Bcl-2 (Ma et al. 2013). Tabassum et al. incorporated the *Nigella sativa* in nanoemulsions. HepG2 cells were susceptible to cytotoxic effect of nanoemulsions, whereas normal human liver (WRL-68) cells were not affected (Tabassum and Ahmad 2018). Freag et al. showed the ability of phytosome to enhance the intestinal absorption of diosmin. This NP system was capable of penetrating intestinal membranes and reaching to plasma (Freag et al. 2013). Komeil et al. prepared genistein-loaded phytosome which was able to improve the drug accumulation in the liver, blood, and intestinal serum lipoproteins in healthy rats (Komeil et al. 2021).

5.5.1.4 Prostate Cancer

Prostate cancer is the second most common malignant tumor worldwide. Additionally, it ranks as the fifth most common reason for cancer-related deaths in males. At this time, the nanotechnology is being tested as a new therapeutic approach for prostate cancer treatment. Aditya et al. formulated curcumin-loaded NLCs. The prostate tumor growth was reduced by 50% when nano-curcumin was used instead of free curcumin (Aditya et al. 2013). Catechin extract-loaded NE reduced the IC_{50} of drug against PC-3 human prostate cancer cells, allowing a greater tumor volume reduction. Considering liposomes, 22Rv1 PrC cells were treated with PEG folate-targeted oleuropein liposomes. The level of 22Rv1 apoptosis and the bioavailability of oleuropein were both increased by these nanoplatforms (Tsai and Chen 2016; Nassir et al. 2019). Epigallocatechin-3-gallate (EGCG)-encapsulated SLNs have improved the stability and anticancer activity of EGCG against human prostate cancer cells (Radhakrishnan et al. 2016).

5.5.1.5 Breast Cancer

Breast cancer is reported as having the highest mortality rate among women in less developed nations. There are many different drugs used to treat cancer, but due to the emergence of resistance and the finiteness of bioavailability, their ability to reduce tumor cells is constrained. Silymarin-loaded SLNs showed to have cytotoxic, antiproliferative, and apoptosis-promoting effects on breast and lung cancer cells (Sezer 2021). Sun et al. showed that curcumin-loaded SLNs had increased in vivo bioavailability and sustained in vitro anticancer efficacy and and cellular uptake (Sun et al. 2013). Rezaei-Sadabady et al. reported that quercetin-loaded liposome nanoparticles show high cellular uptake and high antioxidant activities of quercetin (Rezaei-Sadabady et al. 2016).

5.5.1.6 Lung Cancer

Lung cancer is the top cause of mortality for men in both developed and developing countries. Wang et al. formulated curcumin-loaded SLN for the treatment of lung cancer. A high stability of curcumin, greater distribution of curcumin increases in tumor cell, inhibition of cancer cell, and increase in apoptosis of A549 lung cancer cell were observed (Wang et al. 2012a). In another study, compared to free quercetin, nanomicellar quercetin significantly reduced tumor growth. It exhibited high stability, less viability of cancer cells, and reduction in the tumor size (Tan et al. 2012). On lung cancer cells, silymarin-loaded nanoformulation exhibited cytotoxic, antiproliferative, and apoptosis-promoting effects (Sezer 2021).

5.5.1.7 Brain Cancer

The most aggressive and prevalent kind of malignant brain tumor, glioblastoma multiforme (GBM), has an extremely poor 5-year survival rate (Sasmita et al. 2018). Current therapies still have poor survival rates because they rely on surgical resection, temozolomide (TMZ), and radiation. There are some drawbacks of current GBM treatments, including the difficulty of completely removing all tumors, the BBB which prevents drugs from crossing into the brain, and the presence of protrusions in healthy brain tissues. It has been shown that some NPs can pass the BBB (Kumar et al. 2016; Shinde and Devarajan 2017). Curcumin-loaded lipid nanocarrier was created by Kumar et al. to increase the drug's bioavailability, stability, and cytotoxicity against malignant glioma cells (Kumar et al. 2016). Jose et al. prepared resveratrol-loaded SLN for the treatment of brain cancer (Jose et al. 2014). Wang G. employed quercetin nanoliposomes to cause type III programmed cell death in C6 glioma cells (Wang et al. 2012b). Chen Y prepared curcumin-loaded NLCs which enhanced the bioavailability of drug and increased the ROS level to the tumor cells (Chen et al. 2016).

5.5.2 LBNPs to Deliver Phytoconstituents for the Treatment of Neurological Disorders

Cognitive or motor impairments are common symptoms of neurodegenerative diseases like Alzheimer's disease, amyotrophic lateral sclerosis, and Parkinson's disease (Mehdi et al. 2022).

The most common form of dementia, Alzheimer disease (AD), has approximately 24 million cases globally (Ferri et al. 2005). Memory loss, weakened physical and cognitive abilities, and AD are its defining characteristics. Due to the loss of neuronal cells, it eventually results in patient death. The illness may present as the familiar form of AD, which is typically brought on by genetic factors, or as sporadic AD brought on by environmental factors. The limbic system, which is crucial for controlling emotion, instinctive behavior, learning, and short-term memory, and the neocortex, which processes sensory information primarily, are the two main brain regions that are affected by AD. Currently, all the available AD treatments can only reduce symptoms at the moment; they do not have the ability to reverse the disease or completely cure it (Tapeinos et al. 2017). In this section, we'll concentrate on the studies that use various lipid NPs, which have been used in a number of reports on studies for treating AD. Huperzine A (HupA) is one of the most used phytochemicals for AD. HupA is a cholinergic neurotransmitter that may benefit Alzheimer's patients (Qian and Ke 2014). In the past 7 years, two studies using SLNs and NLCs that contained encapsulated HupA have been presented. The authors formulated cetyl palmitate (CP)-based NLCs loaded with HupA and investigated both their structural and physicochemical features (Yang et al. 2010). In order to find the best method for delivering HupA transdermally, the authors encapsulated it in three different formulations such as microemulsion (MEMs), SLNs, and NLCs. According to the results of the in vivo study, the MEMs had the highest cumulative level of drug permeation, followed by the NLCs and SLNs (Patel et al. 2013). Based on Compritol, quercetin-loaded SLNs were developed with the goal of creating a system that could cross the BBB after IV administration (Dhawan et al. 2011). SLN loaded with piperine (PPR) was prepared to reduce the value of superoxide dismutase (SOD) and the value in surceases while raising acetylcholinesterase values (Yusuf et al. 2013). Sesamol-loaded SLNs were used for AD to restore cognitive deficits and assist in reducing nitrosative stress and cytokine release (Sachdeva et al. 2015). Chrysin-loaded SLNs were able to protect against neuronal damage (Vedagiri and Thangarajan 2016).

5.5.3 LBNPs to Deliver Phytoconstituents for the Treatment of Psoriasis

Psoriasis affects between 2% and 5% of people worldwide. Psoriasis is a chronic inflammatory autoimmune disorder. It is quickly accumulated by skin surface cells, resulting in itchy, unpleasant red areas. Although there is no permanent cure for psoriasis, its symptoms can be reduced by quitting smoking, moisturizing, and stress management. Plaque, guttate, pustular, and erythrodermic are subtypes of psoriasis. Psoriasis is characterized largely by aberrant keratinocyte differentiation and epidermal hyper-proliferation, and it is linked to cardiovascular and metabolic problems (Pradhan et al. 2018; Kim et al. 2017; Raychaudhuri et al. 2014). It is an autoimmune disease that can be acute or chronic and is controlled by T cells. However, psoriasis has a week-long maturation cycle in which cells amass on the skin's surface to create red lesions rather than shed (Sala et al. 2016). Approximately 80% of psoriasis vulgaris patients are treated topically. To overcome from the conventional topical medication, various kinds of nanocarriers are used including LBNPs such as liposome, niosome, microemulsion, nanoemulsion, nanostructure lipid carrier, etc. (Suresh et al. 2013). Hal created a glycyrrhetinic acid-loaded NE that improved formulation stability and transdermal action (Puglia et al. 2010). Khokhra and Diwan formulated tea tree oil-loaded nanoemulsion that has high potential to enhance drug uptake (Sonia and Anupama 2011). Meng et al. prepared celastrol-loaded niosome which showed better drug penetration via the skin and higher bioavailability of poorly absorbed drug (Meng et al. 2019).

5.5.4 LBNPs to Deliver Phytoconstituents for the Treatment of Skin Diseases

Dermatological disorders are common throughout the world and are considered one of the most significant worldwide burdens among numerous illnesses (Hay et al. 2014). Burns, wounds, and acne may create trauma and extra psychological pressures, in addition to probable pain or other aggravations produced by the condition itself (Barankin and DeKoven 2002; Hazarika and Archana 2016). Atopic dermatitis, alopecia, hirsutism, hyperhidrosis, hidradenitis suppurative, vitiligo, psoriasis, and melanoma are all examples of dermatological disorders (Mian et al. 2019). The majority of dermatological disorders affect the skin's outermost layer that is typically dry and dense and serves as an operative blockade against the rapid passage of any outside substances, including chemicals and infectious agent. To exhibit their therapeutic effects, medications for these disorders must penetrate the horny layer and reach the source of skin infection. A therapeutic agent's low molecular weight (20–300 kDa) improves its ability to penetrate the stratum corneum or horny layer (Essendoubi et al. 2016). It is also further enhanced by using ointment, gel, cream, etc. Through drug-loaded NPs, nanotechnology is an essential technique for the delivery of therapeutic drugs for both transdermal and topical applications (Roberts et al. 2017). Mogana Rajagopal et al. demonstrated the impact of curcumin-loaded LBNPs for various skin diseases (Rajagopal et al. 2022). Quercetin-loaded SLNs were used to treat acne vulgaris and increase the skin penetration through the stratum corneum. Lycopene-loaded NLCs produced a biphasic release curve and enhanced lycopene stability (Chutoprapat et al. 2022). Chen et al. developed resveratrol-loaded NLCs for acne vulgaris, which demonstrated gradual degradation over 24 h, whereas resveratrol solution shown fast disintegration in the first 8 h (Chen et al. 2017). Vijayan et al. developed and characterized a SLN-loaded neem oil for topical acne therapy (Vijayan et al. 2013). Saporito et al. demonstrated the use of eucalyptus oil-loaded lipid nanoparticles for wound healing (Saporito et al. 2018).

5.5.5 LBNPs to Deliver Phytoconstituents for the Treatment of Aging

Human aging is closely linked to a variety of pathological changes, such as cancer, heart problems, metabolic disorders like type II diabetes, and neurodegenerative illnesses like AD. Aging is a degenerative process that is conserved in all living things. It is distinguished by a steady degeneration of cellular components and functions, which almost always terminates in death (Li et al. 2021; Ghosh and De 2017). About 90% of deaths in developed countries are caused by aging, which accounts for about 100,000 cases daily and roughly two-thirds of deaths worldwide. According to UN, the percentage of the world's population older than 60 years old is anticipated to almost treble by the year 2050, from 962 million to 2.1 billion (Caruso and Puca 2021). Therefore, it is crucial to learn about the molecular mechanisms underlying the aging process and to look for therapeutic interventions that can lengthen life span and improve health span (Keshavarz et al. 2023). Because the proteostasis network deteriorates with age, proteotoxicity-related diseases such as Parkinson's disease, Alzheimer's disease, Huntington's disease, and amyotrophic lateral sclerosis appear (Ruz et al. 2020). Meanwhile, the search for bioactive compounds with antioxidant properties has intensified. PCs can interact with biochemical systems, which have a variety of positive health effects, including anti-aging, anti-inflammatory, antihypertensive, antidiabetic, anticancer, and antineurodegenerative qualities (Arora 2020).

When compared to pure resveratrol, resveratrol loaded in nanoliposome demonstrated a more pronounced radical scavenging effect. Patients with Parkinson's disease who used the vitamin E-loaded resveratrol NEs showed high ROS scavenging efficiency. Pure genistein shows endocrine-disrupting and toxic effects by accumulation of high doses. To overcome these limitations, LBNPs are used. Additionally, genistein-loaded SLNs were found to have increased oral bioavailability when compared to its suspension formulations or bulk powders. EGCG-loaded SLNs were more stable and had greater oral delivery potential than pure EGCG (Vaiserman et al. 2020).

5.6 Conclusion and Future Perspective

The development of lipid-based nanocarriers for the treatment of different ailments is a cutting-edge, optimistic, and quickly developing scientific topic. LBNPs are very advantageous toward drug delivery systems for various diseases. Generally, LBNPs are biocompatible, biodegradable, highly stable, nontoxic, and easy to prepare and scale up at industrial scale. Synthetic drugs are toxic for the human body. There are also chances of death during high amounts of accumulation of drug in cells. These nanostructures have shown promise in the delivery of phytoconstituents to treat ailments including cancer, neurological issues, skin disorders, etc. In the in vitro studies and some preclinical studies, results are encouraging. However, further preclinical and clinical research is required to validate LBNPs for phytoconstituent delivery.

Acknowledgment The authors acknowledge the Central University of Gujarat, Gandhinagar for providing the necessary facilities and support.

References

Abd-Elghany AA, Mohamad EA (2020) Ex-vivo transdermal delivery of Annona squamosa entrapped in niosomes by electroporation. J Radiat Res Appl Sci 13(1):164–173

Adel IM, ElMeligy MF, Abdelrahim ME, Maged A, Abdelkhalek AA, Abdelmoteleb AM, Elkasabgy NA (2021) Design and characterization of spray-dried proliposomes for the pulmonary delivery of curcumin. Int J Nanomedicine 16:2667

Aditya NP, Shim M, Lee I, Lee Y, Im MH, Ko S (2013) Curcumin and genistein coloaded nanostructured lipid carriers: in vitro digestion and antiprostate cancer activity. J Agric Food Chem 61(8):1878–1883

Admyre C, Johansson SM, Qazi KR, Filén JJ, Lahesmaa R, Norman M, Neve EP, Scheynius A, Gabrielsson S (2007) Exosomes with immune modulatory features are present in human breast milk. J Immunol 179(3):1969–1978. https://doi.org/10.4049/jimmunol.179.3.1969

Aggarwal D, Nautiyal U (2016) Ethosomes: a review. Int J Pharm Med Res 4(4):354–363

Agrawal M, Saraf S, Saraf S, Antimisiaris SG, Hamano N, Li SD, Chougule M, Shoyele SA, Gupta U, Ajazuddin, Alexander A (2018) Recent advancements in the field of nanotechnology for the delivery of anti-Alzheimer drug in the brain region. Expert Opin Drug Deliv 15(6): 589–617. https://doi.org/10.1080/17425247.2018.1471058

Ahmad U, Akhtar J, Singh SP, Ahmad FJ, Siddiqui S (2018) Silymarin nanoemulsion against human hepatocellular carcinoma: development and optimization. Artif Cells Nanomed Biotechnol 46(2):231–241

Ajorlou E, Khosroushahi AY (2017) Trends on polymer- and lipid-based nanostructures for parenteral drug delivery to tumors. Cancer Chemother Pharmacol 79(2):251–265. https://doi.org/10.1007/s00280-016-3168-6

Alam M, Najmi AK, Ahmad I, Ahmad FJ, Akhtar MJ, Imam SS, Akhtar M (2018) Formulation and evaluation of nano lipid formulation containing CNS acting drug: molecular docking, in-vitro assessment and bioactivity detail in rats. Artif Cells Nanomed Biotechnol 46(Suppl 2):46–57. https://doi.org/10.1080/21691401.2018.1451873

Allen TM, Cullis PR (2013) Liposomal drug delivery systems: from concept to clinical applications. Adv Drug Deliv Rev 65(1):36–48. https://doi.org/10.1016/j.addr.2012.09.037

Andrade LM, Rocha KA, De Sá FA, Marreto RN, Lima EM, Gratieri T, Taveira SF (2016) Voriconazole-loaded nanostructured lipid carriers for ocular drug delivery. Cornea 35(6): 866–871

Araújo J, Gonzalez E, Egea MA, Garcia ML, Souto EB (2009) Nanomedicines for ocular NSAIDs: safety on drug delivery. Nanomedicine 5(4):394–401. https://doi.org/10.1016/j.nano.2009.02.003

Arbain NH, Salim N, Wui WT, Basri M, Rahman MBA (2018) Optimization of quercetin loaded palm oil ester based nanoemulsion formulation for pulmonary delivery. J Oleo Sci 67(8): 933–940. https://doi.org/10.5650/jos.ess17253

Arora S (2020) Phytochemicals: benefits, concerns and challenges. In: Advancement in functional food ingredients. Jaya Publishing House, Delhi

Ashraf A, Mahmoud PA, Reda H, Mansour S, Helal MH, Michel HE, Nasr M (2019) Silymarin and silymarin nanoparticles guard against chronic unpredictable mild stress induced depressive-like behavior in mice: involvement of neurogenesis and NLRP3 inflammasome. J Psychopharmacol 33(5):615–631

Attama AA (2011) SLN, NLC, LDC: state of the art in drug and active delivery. Recent Pat Drug Deliv Formul 5(3):178–187. https://doi.org/10.2174/187221111797200524

Attama AA, Reichl S, Müller-Goymann CC (2008) Diclofenac sodium delivery to the eye: in vitro evaluation of novel solid lipid nanoparticle formulation using human cornea construct. Int J Pharm 355(1–2):307–313. https://doi.org/10.1016/j.ijpharm.2007.12.007

Avadhani KS, Manikkath J, Tiwari M, Chandrasekhar M, Godavarthi A, Vidya SM, Hariharapura RC, Kalthur G, Udupa N, Mutalik S (2017) Skin delivery of epigallocatechin-3-gallate (EGCG) and hyaluronic acid loaded nano-transfersomes for antioxidant and anti-aging effects in UV radiation induced skin damage. Drug Deliv 24(1):61–74. https://doi.org/10.1080/10717544.2016.1228718

Azmin MN, Florence AT, Handjani-Vila RM, Stuart JF, Vanlerberghe G, Whittaker JS (1985) The effect of non-ionic surfactant vesicle (niosome) entrapment on the absorption and distribution of methotrexate in mice. J Pharm Pharmacol 37(4):237–242. https://doi.org/10.1111/j.2042-7158.1985.tb05051.x

Baillie AJ, Florence AT, Hume LR, Muirhead GT, Rogerson A (1985) The preparation and properties of niosomes—non-ionic surfactant vesicles. J Pharm Pharmacol 37(12):863–868. https://doi.org/10.1111/j.2042-7158.1985.tb04990.x

Banik D, Patra P (2022) Possibility of liver cancer treatment by nanoformulation of phenolic phytochemicals. Curr Nanomed 12(2):108–125

Barani M, Mirzaei M, Torkzadeh-Mahani M, Adeli-Sardou M (2019) Evaluation of carum-loaded niosomes on breast cancer cells: physicochemical properties, in vitro cytotoxicity, flow cytometric, DNA fragmentation and cell migration assay. Sci Rep 9(1):7139. https://doi.org/10.1038/s41598-019-43755-w

Barankin B, DeKoven J (2002) Psychosocial effect of common skin diseases. Can Fam Physician 48(4):712–716

Bard MP, Hegmans JP, Hemmes A, Luider TM, Willemsen R, Severijnen LA, van Meerbeeck JP, Burgers SA, Hoogsteden HC, Lambrecht BN (2004) Proteomic analysis of exosomes isolated from human malignant pleural effusions. Am J Respir Cell Mol Biol 31(1):114–121. https://doi.org/10.1165/rcmb.2003-0238OC

Barroso LC, Viegas C, Vieira J, Ferreira-Pêgo C, Costa J, Fonte P (2021) Lipid-based carriers for food ingredients delivery. J Food Eng 295:110451

Başaran E, Demirel M, Sirmagül B, Yazan Y (2010) Cyclosporine-A incorporated cationic solid lipid nanoparticles for ocular delivery. J Microencapsul 27(1):37–47. https://doi.org/10.3109/02652040902846883

Basnet P, Skalko-Basnet N (2011) Curcumin: an anti-inflammatory molecule from a curry spice on the path to cancer treatment. Molecules (Basel, Switzerland) 16(6):4567–4598. https://doi.org/10.3390/molecules16064567

Battaglia L, Ugazio E (2019) Lipid nano- and microparticles: an overview of patent-related research. J Nanomater 2019:2834941

Battaglia L, Serpe L, Foglietta F, Muntoni E, Gallarate M, Del Pozo Rodriguez A, Solinis MA (2016) Application of lipid nanoparticles to ocular drug delivery. Expert Opin Drug Deliv 13(12):1743–1757. https://doi.org/10.1080/17425247.2016.1201059

Beloqui A, Solinís MÁ, Delgado A, Evora C, Isla A, Rodríguez-Gascón A (2014) Fate of nanostructured lipid carriers (NLCs) following the oral route: design, pharmacokinetics and biodistribution. J Microencapsul 31(1):1–8. https://doi.org/10.3109/02652048.2013.788090

Beloqui A, Solinís MÁ, Rodríguez-Gascón A, Almeida AJ, Préat V (2016) Nanostructured lipid carriers: promising drug delivery systems for future clinics. Nanomedicine 12(1):143–161. https://doi.org/10.1016/j.nano.2015.09.004

Bendas ER, Tadros MI (2007) Enhanced transdermal delivery of salbutamol sulfate via ethosomes. AAPS PharmSciTech 8(4):E107. https://doi.org/10.1208/pt0804107

Bertoni S, Albertini B, Ronowicz-Pilarczyk J, Calonghi N, Passerini N (2022) Solvent-free fabrication of biphasic lipid-based microparticles with tunable structure. Pharmaceutics 14(1): 54. https://doi.org/10.3390/pharmaceutics14010054

Bhardwaj P, Tripathi P, Gupta RK, Pandey SS (2020) Niosomes: a review on niosomal research in the last decade. J Drug Deliv Sci Technol 56:101581

Bhise K, Kashaw SK, Sau S, Iyer AK (2017) Nanostructured lipid carriers employing polyphenols as promising anticancer agents: quality by design (QbD) approach. Int J Pharm 526(1–2): 506–515. https://doi.org/10.1016/j.ijpharm.2017.04.078

Blasi P, Giovagnoli S, Schoubben A, Ricci M, Rossi C (2007) Solid lipid nanoparticles for targeted brain drug delivery. Adv Drug Deliv Rev 59(6):454–477. https://doi.org/10.1016/j.addr.2007.04.011

Brandelli A (2012) Nanostructures as promising tools for delivery of antimicrobial peptides. Mini Rev Med Chem 12(8):731–741. https://doi.org/10.2174/138955712801264774

Bray F, Ferlay J, Soerjomataram I, Siegel RL, Torre LA, Jemal A (2018) Global cancer statistics 2018: GLOBOCAN estimates of incidence and mortality worldwide for 36 cancers in 185 countries. CA Cancer J Clin 68(6):394–424. https://doi.org/10.3322/caac.21492

Briuglia ML, Rotella C, McFarlane A, Lamprou DA (2015) Influence of cholesterol on liposome stability and on in vitro drug release. Drug Deliv Transl Res 5(3):231–242. https://doi.org/10.1007/s13346-015-0220-8

Butreddy A, Kommineni N, Dudhipala N (2021) Exosomes as naturally occurring vehicles for delivery of biopharmaceuticals: insights from drug delivery to clinical perspectives. Nanomaterials 11(6):1481

Caruso C, Puca AA (2021) Special issue "Centenarians—a model to study the molecular basis of lifespan and healthspan". Int J Mol Sci 22(4):2044. https://doi.org/10.3390/ijms22042044

Cerpnjak K, Zvonar A, Gašperlin M, Vrečer F (2013) Lipid-based systems as a promising approach for enhancing the bioavailability of poorly water-soluble drugs. Acta Pharm 63(4):427–445. https://doi.org/10.2478/acph-2013-0040

Chacko IA, Ghate VM, Dsouza L, Lewis SA (2020) Lipid vesicles: a versatile drug delivery platform for dermal and transdermal applications. Colloids Surf B Biointerfaces 195:111262. https://doi.org/10.1016/j.colsurfb.2020.111262

Chahar HS, Bao X, Casola A (2015) Exosomes and their role in the life cycle and pathogenesis of RNA viruses. Viruses 7(6):3204–3225. https://doi.org/10.3390/v7062770

Chang M, Wu M, Li H (2018) Antitumor activities of novel glycyrrhetinic acid-modified curcumin-loaded cationic liposomes in vitro and in H22 tumor-bearing mice. Drug Deliv 25(1):1984–1995

Chaurasiya P, Ganju E, Upmanyu N, Ray SK, Jain P (2019) Transfersomes: a novel technique for transdermal drug delivery. J Drug Deliv Ther 9(1):279–285

Chen M-H, Chiang B-H (2020) Modification of curcumin-loaded liposome with edible compounds to enhance ability of crossing blood brain barrier. Colloids Surf 599:124862. https://doi.org/10.1016/j.colsurfa.2020.124862

Chen ZP, Sun J, Chen HX, Xiao YY, Liu D, Chen J et al (2010) Comparative pharmacokinetics and bioavailability studies of quercetin, kaempferol and isorhamnetin after oral administration of Ginkgo biloba extracts, Ginkgo biloba extract phospholipid complexes and Ginkgo biloba extract solid dispersions in rats. Fitoterapia 81(8):1045–1052

Chen Y, Pan L, Jiang M, Li D, Jin L (2016) Nanostructured lipid carriers enhance the bioavailability and brain cancer inhibitory efficacy of curcumin both in vitro and in vivo. Drug Deliv 23(4): 1383–1392. https://doi.org/10.3109/10717544.2015.1049719

Chen J, Wei N, Lopez-Garcia M, Ambrose D, Lee J, Annelin C, Peterson T (2017) Development and evaluation of resveratrol, vitamin E, and epigallocatechin gallate loaded lipid nanoparticles for skin care applications. Eur J Pharm Biopharm 117:286–291

Chen B, Luo M, Liang J, Zhang C, Gao C, Wang J et al (2018) Surface modification of PGP for a neutrophil–nanoparticle co-vehicle to enhance the anti-depressant effect of baicalein. Acta Pharm Sin B 8(1):64–73

Chetoni P, Burgalassi S, Monti D, Tampucci S, Tullio V, Cuffini AM, Muntoni E, Spagnolo R, Zara GP, Cavalli R (2016) Solid lipid nanoparticles as promising tool for intraocular tobramycin delivery: pharmacokinetic studies on rabbits. Eur J Pharm Biopharm 109:214–223. https://doi.org/10.1016/j.ejpb.2016.10.006

Chirio D, Peira E, Sapino S, Dianzani C, Barge A, Muntoni E, Morel S, Gallarate M (2018) Stearoyl-chitosan coated nanoparticles obtained by microemulsion cold dilution technique. Int J Mol Sci 19(12):3833. https://doi.org/10.3390/ijms19123833

Chuan LI, Zhang J, Yu-Jiao ZU, Shu-Fang NIE, Jun CAO, Qian W et al (2015) Biocompatible and biodegradable nanoparticles for enhancement of anti-cancer activities of phytochemicals. Chin J Nat Med 13(9):641–652

Chutoprapat R, Kopongpanich P, Chan LW (2022) A mini-review on solid lipid nanoparticles and nanostructured lipid carriers: topical delivery of phytochemicals for the treatment of acne vulgaris. Molecules 27(11):3460. https://doi.org/10.3390/molecules27113460

Cipolla D, Shekunov B, Blanchard J, Hickey A (2014) Lipid-based carriers for pulmonary products: preclinical development and case studies in humans. Adv Drug Deliv Rev 75:53–80. https://doi.org/10.1016/j.addr.2014.05.001

Cirri M, Mennini N, Maestrelli F, Mura P, Ghelardini C, Mannelli LDC (2017) Development and in vivo evaluation of an innovative "Hydrochlorothiazide-in Cyclodextrins-in Solid Lipid Nanoparticles" formulation with sustained release and enhanced oral bioavailability for potential hypertension treatment in pediatrics. Int J Pharm 521(1–2):73–83

Cortés H, Hernández-Parra H, Bernal-Chávez SA, Prado-Audelo MLD, Caballero-Florán IH, Borbolla-Jiménez FV et al (2021) Non-ionic surfactants for stabilization of polymeric nanoparticles for biomedical uses. Materials 14(12):3197

da Silva Santos V, Ribeiro APB, Santana MHA (2019) Solid lipid nanoparticles as carriers for lipophilic compounds for applications in foods. Food Res Int 122:610–626

Dal Magro R, Ornaghi F, Cambianica I, Beretta S, Re F, Musicanti C, Rigolio R, Donzelli E, Canta A, Ballarini E, Cavaletti G, Gasco P, Sancini G (2017) ApoE-modified solid lipid nanoparticles: a feasible strategy to cross the blood-brain barrier. J Control Release 249:103–110. https://doi.org/10.1016/j.jconrel.2017.01.039

Dang YJ, Zhu CY (2013) Oral bioavailability of cantharidin-loaded solid lipid nanoparticles. Chin Med 8(1):1. https://doi.org/10.1186/1749-8546-8-1

Das B, Nayak AK, Mallick S (2022) Transferosomes: a novel nanovesicular approach for drug delivery. In: Systems of nanovesicular drug delivery. Academic Press, New York, pp 103–114

de Pace RC, Liu X, Sun M, Nie S, Zhang J, Cai Q, Gao W, Pan X, Fan Z, Wang S (2013) Anticancer activities of (-)-epigallocatechin-3-gallate encapsulated nanoliposomes in MCF7 breast cancer cells. J Liposome Res 23(3):187–196. https://doi.org/10.3109/08982104.2013.788023

Dhawan S, Kapil R, Singh B (2011) Formulation development and systematic optimization of solid lipid nanoparticles of quercetin for improved brain delivery. J Pharm Pharmacol 63(3):342–351. https://doi.org/10.1111/j.2042-7158.2010.01225.x

Din FU, Aman W, Ullah I, Qureshi OS, Mustapha O, Shafique S, Zeb A (2017) Effective use of nanocarriers as drug delivery systems for the treatment of selected tumors. Int J Nanomedicine 12:7291–7309. https://doi.org/10.2147/IJN.S146315

Durán N, Costa AF, Stanisic D, Bernardes JS, Tasic L (2019) Nanotoxicity and dermal application of nanostructured lipid carrier loaded with hesperidin from orange residue. J Phys Conf Ser 1323(1):012021

Esposto BS, Jauregi P, Tapia-Blácido DR, Martelli-Tosi M (2021) Liposomes vs. chitosomes: encapsulating food bioactives. Trends Food Sci Technol 108:40–48

Essendoubi M, Gobinet C, Reynaud R, Angiboust JF, Manfait M, Piot OJSR (2016) Human skin penetration of hyaluronic acid of different molecular weights as probed by Raman spectroscopy. Skin Res Technol 22(1):55–62

Fang CL, Al-Suwayeh SA, Fang JY (2013) Nanostructured lipid carriers (NLCs) for drug delivery and targeting. Recent Pat Nanotechnol 7(1):41–55

Ferlay J, Colombet M, Soerjomataram I, Mathers C, Parkin DM, Piñeros M et al (2019) Estimating the global cancer incidence and mortality in 2018: GLOBOCAN sources and methods. Int J Cancer 144(8):1941–1953

Fernandes F, Dias-Teixeira M, Delerue-Matos C, Grosso C (2021) Critical review of lipid-based nanoparticles as carriers of neuroprotective drugs and extracts. Nanomaterials (Basel, Switzerland) 11(3):563. https://doi.org/10.3390/nano11030563

Ferreira M, Silva E, Barreiros L, Segundo MA, Costa Lima SA, Reis S (2016) Methotrexate loaded lipid nanoparticles for topical management of skin-related diseases: design, characterization and skin permeation potential. Int J Pharm 512(1):14–21. https://doi.org/10.1016/j.ijpharm.2016.08.008

Ferri CP, Prince M, Brayne C, Brodaty H, Fratiglioni L, Ganguli M (2005) Global prevalence of dementia: a Delphi consensus study. Lancet 366(9503):2112–2117

Fonseca-Santos B, Gremião MP, Chorilli M (2015) Nanotechnology-based drug delivery systems for the treatment of Alzheimer's disease. Int J Nanomedicine 10:4981–5003. https://doi.org/10.2147/IJN.S87148

Freag MS, Elnaggar YS, Abdallah OY (2013) Lyophilized phytosomal nanocarriers as platforms for enhanced diosmin delivery: optimization and ex vivo permeation. Int J Nanomedicine 8:2385–2397

Gabetta B, Bombardelli E, Pifferi G (1988) U.S. patent no. 4,764,508. U.S. Patent and Trademark Office, Washington, DC

Ganesan P, Narayanasamy D (2017) Lipid nanoparticles: different preparation techniques, characterization, hurdles, and strategies for the production of solid lipid nanoparticles and nanostructured lipid carriers for oral drug delivery. Sustain Chem Pharm 6:37–56

García-Pinel B, Porras-Alcalá C, Ortega-Rodríguez A, Sarabia F, Prados J, Melguizo C, López-Romero JM (2019) Lipid-based nanoparticles: application and recent advances in cancer treatment. Nanomaterials (Basel, Switzerland) 9(4):638. https://doi.org/10.3390/nano9040638

Garg A, Bhalala K, Tomar DS, Wahajuddin (2017) In-situ single pass intestinal permeability and pharmacokinetic study of developed Lumefantrine loaded solid lipid nanoparticles. Int J Pharm 516(1–2):120–130. https://doi.org/10.1016/j.ijpharm.2016.10.064

Geetha T, Kapila M, Prakash O, Deol PK, Kakkar V, Kaur IP (2015) Sesamol-loaded solid lipid nanoparticles for treatment of skin cancer. J Drug Target 23(2):159–169. https://doi.org/10.3109/1061186X.2014.965717

Ghanbarzadeh S, Hariri R, Kouhsoltani M, Shokri J, Javadzadeh Y, Hamishehkar H (2015) Enhanced stability and dermal delivery of hydroquinone using solid lipid nanoparticles. Colloids Surf B Biointerfaces 136:1004–1010. https://doi.org/10.1016/j.colsurfb.2015.10.041

Ghasemiyeh P, Mohammadi-Samani S (2018) Solid lipid nanoparticles and nanostructured lipid carriers as novel drug delivery systems: applications, advantages and disadvantages. Res Pharm Sci 13(4):288–303. https://doi.org/10.4103/1735-5362.235156

Ghosh C, De A (2017) Basics of aging theories and disease related aging—an overview. PharmaTutor 5(2):16–23

Gonçalves LMD, Maestrelli F, Mannelli LDC, Ghelardini C, Almeida AJ, Mura P (2016) Development of solid lipid nanoparticles as carriers for improving oral bioavailability of glibenclamide. Eur J Pharm Biopharm 102:41–50

Gönüllü Ü, Üner M, Yener G, Karaman EF, Aydoğmuş Z (2015) Formulation and characterization of solid lipid nanoparticles, nanostructured lipid carriers and nanoemulsion of lornoxicam for transdermal delivery. Acta Pharm 65(1):1–13. https://doi.org/10.1515/acph-2015-0009

Göppert TM, Müller RH (2003) Plasma protein adsorption of Tween 80- and poloxamer 188-stabilized solid lipid nanoparticles. J Drug Target 11(4):225–231. https://doi.org/10.1080/10611860310001615956

Gordillo-Galeano A, Mora-Huertas CE (2018) Solid lipid nanoparticles and nanostructured lipid carriers: a review emphasizing on particle structure and drug release. Eur J Pharm Biopharm 133:285–308

Gugleva V, Ivanova N, Sotirova Y, Andonova V (2021) Dermal drug delivery of phytochemicals with phenolic structure via lipid-based nanotechnologies. Pharmaceuticals (Basel, Switzerland) 14(9):837. https://doi.org/10.3390/ph14090837

Hamishehkar H, Shokri J, Fallahi S, Jahangiri A, Ghanbarzadeh S, Kouhsoltani M (2015) Histopathological evaluation of caffeine-loaded solid lipid nanoparticles in efficient treatment of cellulite. Drug Dev Ind Pharm 41:1640–1646

Harding C, Heuser J, Stahl P (1983) Receptor-mediated endocytosis of transferrin and recycling of the transferrin receptor in rat reticulocytes. J Cell Biol 97(2):329–339. https://doi.org/10.1083/jcb.97.2.329

Harding C, Heuser J, Stahl P (1984) Endocytosis and intracellular processing of transferrin and colloidal gold-transferrin in rat reticulocytes: demonstration of a pathway for receptor shedding. Eur J Cell Biol 35(2):256–263

Hay RJ, Johns NE, Williams HC, Bolliger IW, Dellavalle RP, Margolis DJ, Marks R, Naldi L, Weinstock MA, Wulf SK, Michaud C, Murray JLC, Naghavi M (2014) The global burden of skin disease in 2010: an analysis of the prevalence and impact of skin conditions. J Invest Dermatol 134(6):1527–1534. https://doi.org/10.1038/jid.2013.446

Hazarika N, Archana M (2016) The psychosocial impact of acne vulgaris. Indian J Dermatol 61(5): 515

He X, Zhu Y, Wang M, Jing G, Zhu R, Wang S (2016) Antidepressant effects of curcumin and HU-211 coencapsulated solid lipid nanoparticles against corticosterone-induced cellular and animal models of major depression. Int J Nanomedicine 11:4975–4990. https://doi.org/10.2147/IJN.S109088

Hidalgo A, Cruz A, Pérez-Gil J (2015) Barrier or carrier? Pulmonary surfactant and drug delivery. Eur J Pharm Biopharm 95:117–127

Hosseini M, Haji-Fatahaliha M, Jadidi-Niaragh F, Majidi J, Yousefi M (2016) The use of nanoparticles as a promising therapeutic approach in cancer immunotherapy. Artif Cells Nanomed Biotechnol 44(4):1051–1061. https://doi.org/10.3109/21691401.2014.998830

Huang ZR, Hua SC, Yang YL, Fang JY (2008) Development and evaluation of lipid nanoparticles for camptothecin delivery: a comparison of solid lipid nanoparticles, nanostructured lipid carriers, and lipid emulsion. Acta Pharmacol Sin 29(9):1094–1102

Jain AK, Jain A, Garg NK, Agarwal A, Jain A, Jain SA, Tyagi RK, Jain RK, Agrawal H, Agrawal GP (2014) Adapalene loaded solid lipid nanoparticles gel: an effective approach for acne treatment. Colloids Surf B Biointerfaces 121:222–229. https://doi.org/10.1016/j.colsurfb.2014.05.041

Jenning V, Schäfer-Korting M, Gohla S (2000) Vitamin A-loaded solid lipid nanoparticles for topical use: drug release properties. J Control Release 66(2–3):115–126. https://doi.org/10.1016/s0168-3659(99)00223-0

Johnstone RM, Adam M, Hammond JR, Orr L, Turbide C (1987) Vesicle formation during reticulocyte maturation. Association of plasma membrane activities with released vesicles (exosomes). J Biol Chem 262(19):9412–9420

Jose S, Anju SS, Cinu TA, Aleykutty NA, Thomas S, Souto EB (2014) In vivo pharmacokinetics and biodistribution of resveratrol-loaded solid lipid nanoparticles for brain delivery. Int J Pharm 474(1–2):6–13. https://doi.org/10.1016/j.ijpharm.2014.08.003

Joshi MD, Müller RH (2009) Lipid nanoparticles for parenteral delivery of actives. Eur J Pharm Biopharm 71(2):161–172. https://doi.org/10.1016/j.ejpb.2008.09.003

Karewicz A, Bielska D, Loboda A, Gzyl-Malcher B, Bednar J, Jozkowicz A, Dulak J, Nowakowska M (2013) Curcumin-containing liposomes stabilized by thin layers of chitosan derivatives. Colloids Surf B Biointerfaces 109:307–316. https://doi.org/10.1016/j.colsurfb.2013.03.059

Katouzian I, Esfanjani AF, Jafari SM, Akhavan S (2017) Formulation and application of a new generation of lipid nano-carriers for the food bioactive ingredients. Trends Food Sci Technol 68: 14–25

Kaur N (2014) Transdermal drug delivery through carriers: transfersomes. PharmaTutor 2:77–85

Kaur IP, Bhandari R, Bhandari S, Kakkar V (2008) Potential of solid lipid nanoparticles in brain targeting. J Control Release 127(2):97–109. https://doi.org/10.1016/j.jconrel.2007.12.018

Kaur P, Garg T, Rath G, Murthy RS, Goyal AK (2016) Development, optimization and evaluation of surfactant-based pulmonary nanolipid carrier system of paclitaxel for the management of drug resistance lung cancer using Box-Behnken design. Drug Deliv 23(6):1912–1925. https://doi.org/10.3109/10717544.2014.993486

Kauslya A, Borawake PD, Shinde JV, Chavan RS (2021) Niosomes: a novel carrier drug delivery system. J Drug Deliv Ther 11(1):162–170

Keshavarz M, Xie K, Schaaf K, Bano D, Ehninger D (2023) Targeting the "hallmarks of aging" to slow aging and treat age-related disease: fact or fiction? Mol Psychiatry 28(1):242–255

Khan I, Saeed K, Khan I (2019) Nanoparticles: properties, applications and toxicities. Arab J Chem 12(7):908–931

Khushnud T, Mousa SA (2013) Potential role of naturally derived polyphenols and their nanotechnology delivery in cancer. Mol Biotechnol 55(1):78–86. https://doi.org/10.1007/s12033-012-9623-7

Kim WB, Jerome D, Yeung J (2017) Diagnostic et prise en charge du psoriasis. Can Fam Physician 63(4):e210–e218

Komeil IA, El-Refaie WM, Gowayed MA, El-Ganainy SO, El Achy SN, Huttunen KM, Abdallah OY (2021) Oral genistein-loaded phytosomes with enhanced hepatic uptake, residence and improved therapeutic efficacy against hepatocellular carcinoma. Int J Pharm 601:120564

Kumar R (2019) Lipid-based nanoparticles for drug-delivery systems. In: Nanocarriers for drug delivery. Elsevier, Amsterdam, pp 249–284

Kumar R, Singh M, Bala R, Seth N, Rana AC (2012) Transferosomes: a novel approach for transdermal drug delivery. Int Res J Pharm 3(1):20–24

Kumar VD, Saraswathi B, Kanth NP (2013) Transferosome: latest updates. Int J Pharm Sci Res 4(11):4125

Kumar A, Ahuja A, Ali J, Baboota S (2016) Curcumin-loaded lipid nanocarrier for improving bioavailability, stability and cytotoxicity against malignant glioma cells. Drug Deliv 23(1): 214–229

Kumar R, Dkhar DS, Kumari R, Mahapatra S, Dubey VK, Chandra P (2022) Lipid based nanocarriers: production techniques, concepts, and commercialization aspect. J Drug Deliv Sci Technol 74:103526

Kuo F, Subramanian B, Kotyla T, Wilson TA, Yoganathan S, Nicolosi RJ (2008) Nanoemulsions of an anti-oxidant synergy formulation containing gamma tocopherol have enhanced bioavailability and anti-inflammatory properties. Int J Pharm 363(1–2):206–213

Kuo YC, Lin CY, Li JS, Lou YI (2017) Wheat germ agglutinin-conjugated liposomes incorporated with cardiolipin to improve neuronal survival in Alzheimer's disease treatment. Int J Nanomedicine 12:1757–1774. https://doi.org/10.2147/IJN.S128396

Lai F, Pireddu R, Corrias F, Fadda AM, Valenti D, Pini E, Sinico C (2013) Nanosuspension improves tretinoin photostability and delivery to the skin. Int J Pharm 458(1):104–109

Lauterbach A, Müller-Goymann CC (2015) Applications and limitations of lipid nanoparticles in dermal and transdermal drug delivery via the follicular route. Eur J Pharm Biopharm 97 (Pt A):152–163. https://doi.org/10.1016/j.ejpb.2015.06.020

Li J, Chen J, Cai BC, Yang T (2013) Preparation, characterization and tissue distribution of brucine stealth liposomes with different lipid composition. Pharm Dev Technol 18(4):772–778

Li J, Guo X, Liu Z, Okeke CI, Li N, Zhao H et al (2014) Preparation and evaluation of charged solid lipid nanoparticles of tetrandrine for ocular drug delivery system: pharmacokinetics, cytotoxicity and cellular uptake studies. Drug Dev Ind Pharm 40(7):980–987
Li Z, Zhang Z, Ren Y, Wang Y, Fang J, Yue H, Ma S, Guan F (2021) Aging and age-related diseases: from mechanisms to therapeutic strategies. Biogerontology 22(2):165–187. https://doi.org/10.1007/s10522-021-09910-5
Liu L, Tang Y, Gao C, Li Y, Chen S, Xiong T, Yao P (2014) Characterization and biodistribution in vivo of quercetin-loaded cationic nanostructured lipid carriers. Colloids Surf B Biointerfaces 115:125–131
Liu W, Hou Y, Jin Y, Wang Y, Xu X, Han J (2020) Research progress on liposomes: application in food, digestion behavior and absorption mechanism. Trends Food Sci Technol 104:177–189
Liu L, Kshirsagar PG, Gautam SK, Gulati M, Wafa EI, Christiansen JC et al (2022) Nanocarriers for pancreatic cancer imaging, treatments, and immunotherapies. Theranostics 12(3):1030
Low LE, Tan LTH, Goh BH, Tey BT, Ong BH, Tang SY (2019) Magnetic cellulose nanocrystal stabilized pickering emulsions for enhanced bioactive release and human colon cancer therapy. Int J Biol Macromol 127:76–84
Luo Y, Chen D, Ren L, Zhao X, Qin J (2006) Solid lipid nanoparticles for enhancing vinpocetine's oral bioavailability. J Control Release 114(1):53–59
Luo CF, Hou N, Tian J, Yuan M, Liu SM, Xiong LG, Chen MS (2013) Metabolic profile of puerarin in rats after intragastric administration of puerarin solid lipid nanoparticles. Int J Nanomedicine 8:933–940
Luo X, Guan R, Chen X, Tao M, Ma J, Zhao J (2014) Optimization on condition of epigallocatechin-3-gallate (EGCG) nanoliposomes by response surface methodology and cellular uptake studies in Caco-2 cells. Nanoscale Res Lett 9(1):1–9
Ma H, Fan Q, Yu J, Xin J, Zhang C (2013) Anticancer activities of tanshinone microemulsion against hepatocellular carcinoma in vitro and in vivo. Mol Med Rep 7(1):59–64
Ma H, Guo D, Fan Y, Wang J, Cheng J, Zhang X (2018) Paeonol-loaded ethosomes as transdermal delivery carriers: design, preparation and evaluation. Molecules (Basel, Switzerland) 23(7): 1756. https://doi.org/10.3390/molecules23071756
Madane RG, Mahajan HS (2016) Curcumin-loaded nanostructured lipid carriers (NLCs) for nasal administration: design, characterization, and in vivo study. Drug Deliv 23(4):1326–1334. https://doi.org/10.3109/10717544.2014.975382
Makled S, Nafee N, Boraie N (2017) Nebulized solid lipid nanoparticles for the potential treatment of pulmonary hypertension via targeted delivery of phosphodiesterase-5-inhibitor. Int J Pharm 517(1–2):312–321. https://doi.org/10.1016/j.ijpharm.2016.12.026
Martins S, Sarmento B, Ferreira DC, Souto EB (2007) Lipid-based colloidal carriers for peptide and protein delivery--liposomes versus lipid nanoparticles. Int J Nanomedicine 2(4):595–607
Masyuk AI, Masyuk TV, Larusso NF (2013) Exosomes in the pathogenesis, diagnostics and therapeutics of liver diseases. J Hepatol 59(3):621–625. https://doi.org/10.1016/j.jhep.2013.03.028
Mehdi A, Lamiae B, Samira B, Ramchoun M, Abdelouahed K, Tamas F, Hicham B (2022) Pomegranate (Punica granatum L.) attenuates neuroinflammation involved in neurodegenerative diseases. Foods 11(17):2570
Mehnert W, Mäder K (2001) Solid lipid nanoparticles: production, characterization and applications. Adv Drug Deliv Rev 47(2–3):165–196. https://doi.org/10.1016/s0169-409x(01)00105-3
Mehnert W, Mäder K (2012) Solid lipid nanoparticles: production, characterization and applications. Adv Drug Deliv Rev 64:83–101
Mehta P, Bothiraja C, Mahadik K, Kadam S, Pawar A (2018) Phytoconstituent based dry powder inhalers as biomedicine for the management of pulmonary diseases. Biomed Pharmacother 108: 828–837. https://doi.org/10.1016/j.biopha.2018.09.094
Meng F, Asghar S, Gao S, Su Z, Song J, Huo M, Meng W, Ping Q, Xiao Y (2015) A novel LDL-mimic nanocarrier for the targeted delivery of curcumin into the brain to treat Alzheimer's

disease. Colloids Surf B Biointerfaces 134:88–97. https://doi.org/10.1016/j.colsurfb.2015.06.025
Meng S, Sun L, Wang L, Lin Z, Liu Z, Xi L et al (2019) Loading of water-insoluble celastrol into niosome hydrogels for improved topical permeation and anti-psoriasis activity. Colloids Surf B Biointerfaces 182:110352
Mian M, Silfvast-Kaiser A, Paek S, Kivelevitch D, Menter A (2019) A review of the most common dermatologic conditions and their debilitating psychosocial impacts. Int Arch Intern Med 3:18
Mishra DK, Shandilya R, Mishra PK (2018) Lipid based nanocarriers: a translational perspective. Nanomedicine 14(7):2023–2050. https://doi.org/10.1016/j.nano.2018.05.021
Mohite Mukesh T, Ahire Saurabh N, Shinde Supriya S, Revan K, Shubham K (2021) An overview on ethosomes: need of future
Moon B, Chang S (2022) Exosome as a delivery vehicle for cancer therapy. Cells 11(3):316. https://doi.org/10.3390/cells11030316
Moreno-Sastre M, Pastor M, Esquisabel A, Sans E, Viñas M, Fleischer A et al (2016) Pulmonary delivery of tobramycin-loaded nanostructured lipid carriers for Pseudomonas aeruginosa infections associated with cystic fibrosis. Int J Pharm 498(1–2):263–273
Mozafari MR, Reed CJ, Rostron C, Kocum C, Piskin E (2002) Construction of stable anionic liposome-plasmid particles using the heating method: a preliminary investigation. Cell Mol Biol Lett 7(3):923–927
Mozafari MR, Khosravi-Darani K, Borazan GG, Cui J, Pardakhty A, Yurdugul S (2008) Encapsulation of food ingredients using nanoliposome technology. Int J Food Prop 11(4):833–844
Müller RH, Radtke M, Wissing SA (2002a) Solid lipid nanoparticles (SLN) and nanostructured lipid carriers (NLC) in cosmetic and dermatological preparations. Adv Drug Deliv Rev 54 (Suppl 1):S131–S155. https://doi.org/10.1016/s0169-409x(02)00118-7
Müller RH, Radtke M, Wissing S (2002b) Nanostructured lipid matrices for improved microencapsulation of drugs. Int J Pharm 242(1–2):121–128
Müller RH, Runge S, Ravelli V, Mehnert W, Thünemann AF, Souto EB (2006) Oral bioavailability of cyclosporine: solid lipid nanoparticles (SLN) versus drug nanocrystals. Int J Pharm 317(1): 82–89. https://doi.org/10.1016/j.ijpharm.2006.02.045
Mura P, Maestrelli F, D'Ambrosio M, Luceri C, Cirri M (2021) Evaluation and comparison of solid lipid nanoparticles (SLNs) and nanostructured lipid carriers (NLCs) as vectors to develop hydrochlorothiazide effective and safe pediatric oral liquid formulations. Pharmaceutics 13(4): 437
Nangare S, Bhatane D, Mali R, Shitole M (2021) Development of a novel freeze-dried mulberry leaf extract-based transfersome gel. Turk J Pharm Sci 18(1):44–55. https://doi.org/10.4274/tjps.galenos.2019.98624
Naseri N, Valizadeh H, Zakeri-Milani P (2015) Solid lipid nanoparticles and nanostructured lipid carriers: structure, preparation and application. Adv Pharm Bull 5(3):305–313. https://doi.org/10.15171/apb.2015.043
Nasri S, Ebrahimi-Hosseinzadeh B, Rahaie M, Hatamian-Zarmi A, Sahraeian R (2020) Thymoquinone-loaded ethosome with breast cancer potential: optimization, in vitro and biological assessment. J Nanostruct Chem 10:19–31
Nassir AM, Ibrahim IA, Md S, Waris M, Ain MR, Ahmad I, Shahzad N (2019) Surface functionalized folate targeted oleuropein nano-liposomes for prostate tumor targeting: in vitro and in vivo activity. Life Sci 220:136–146
Natsheh H, Vettorato E, Touitou E (2019) Ethosomes for dermal administration of natural active molecules. Curr Pharm Des 25(21):2338–2348. https://doi.org/10.2174/1381612825666190716095826
Neves AR, Lúcio M, Martins S, Lima JL, Reis S (2013) Novel resveratrol nanodelivery systems based on lipid nanoparticles to enhance its oral bioavailability. Int J Nanomedicine 8:177–187. https://doi.org/10.2147/IJN.S37840

Nordin UUM, Ahmad N, Salim N, Yusof NSM (2022) Correction: lipid-based nanoparticles for psoriasis treatment: a review on conventional treatments, recent works, and future prospects. RSC Adv 12(33):21110

Nunes S, Madureira AR, Campos D, Sarmento B, Gomes AM, Pintado M, Reis F (2017) Solid lipid nanoparticles as oral delivery systems of phenolic compounds: overcoming pharmacokinetic limitations for nutraceutical applications. Crit Rev Food Sci Nutr 57(9):1863–1873

Obeid MA, Khadra I, Mullen AB, Tate RJ, Ferro VA (2017) The effects of hydration media on the characteristics of non-ionic surfactant vesicles (NISV) prepared by microfluidics. Int J Pharm 516(1–2):52–60. https://doi.org/10.1016/j.ijpharm.2016.11.015

Olbrich C, Müller RH (1999) Enzymatic degradation of SLN-effect of surfactant and surfactant mixtures. Int J Pharm 180(1):31–39. https://doi.org/10.1016/s0378-5173(98)00404-9

Opatha SAT, Titapiwatanakun V, Chutoprapat R (2020) Transfersomes: a promising nanoencapsulation technique for transdermal drug delivery. Pharmaceutics 12(9):855. https://doi.org/10.3390/pharmaceutics12090855

Pandita D, Kumar S, Poonia N, Lather V (2014) Solid lipid nanoparticles enhance oral bioavailability of resveratrol, a natural polyphenol. Food Res Int 62:1165–1174

Paranjpe M, Müller-Goymann C (2014) Nanoparticle-mediated pulmonary drug delivery: a review. Int J Mol Sci 15(4):5852–5873. https://doi.org/10.3390/ijms15045852

Paranjpe M, Finke JH, Richter C, Gothsch T, Kwade A, Büttgenbach S, Müller-Goymann CC (2014) Physicochemical characterization of sildenafil-loaded solid lipid nanoparticle dispersions (SLN) for pulmonary application. Int J Pharm 476(1–2):41–49. https://doi.org/10.1016/j.ijpharm.2014.09.031

Pardeike J, Hommoss A, Müller RH (2009) Lipid nanoparticles (SLN, NLC) in cosmetic and pharmaceutical dermal products. Int J Pharm 366(1–2):170–184. https://doi.org/10.1016/j.ijpharm.2008.10.003

Pardeike J, Weber S, Haber T, Wagner J, Zarfl HP, Plank H, Zimmer A (2011) Development of an itraconazole-loaded nanostructured lipid carrier (NLC) formulation for pulmonary application. Int J Pharm 419(1–2):329–338. https://doi.org/10.1016/j.ijpharm.2011.07.040

Patel PA, Patil SC, Kalaria DR, Kalia YN, Patravale VB (2013) Comparative in vitro and in vivo evaluation of lipid based nanocarriers of Huperzine A. Int J Pharm 446(1–2):16–23. https://doi.org/10.1016/j.ijpharm.2013.02.014

Patel D, Kesharwani R, Nair SK (2015) Solid lipid nanoparticle (SLN): a modern approach for drug delivery. Asian J Pharm Res Dev:1–17

Patidar A, Thakur DS, Kumar P, Verma J (2010) A review on novel lipid based nanocarriers. Int J Pharm Pharm Sci 2(4):30–35

Patil-Gadhe A, Kyadarkunte A, Patole M, Pokharkar V (2014) Montelukast-loaded nanostructured lipid carriers: part II pulmonary drug delivery and in vitro-in vivo aerosol performance. Eur J Pharm Biopharm 88(1):169–177. https://doi.org/10.1016/j.ejpb.2014.07.007

Patlolla RR, Chougule M, Patel AR, Jackson T, Tata PN, Singh M (2010) Formulation, characterization and pulmonary deposition of nebulized celecoxib encapsulated nanostructured lipid carriers. J Control Release 144(2):233–241. https://doi.org/10.1016/j.jconrel.2010.02.006

Patzelt A, Mak WC, Jung S, Knorr F, Meinke MC, Richter H, Rühl E, Cheung KY, Tran NBNN, Lademann J (2017) Do nanoparticles have a future in dermal drug delivery? J Control Release 246:174–182. https://doi.org/10.1016/j.jconrel.2016.09.015

Pietta P, Simonetti P, Gardana C, Brusamolino A, Morazzoni P, Bombardelli E (1998) Relationship between rate and extent of catechin absorption and plasma antioxidant status. Biochem Mol Biol Int 46(5):895–903. https://doi.org/10.1080/15216549800204442

Pinilla CMB, Lopes NA, Brandelli A (2021) Lipid-based nanostructures for the delivery of natural antimicrobials. Molecules (Basel, Switzerland) 26(12):3587. https://doi.org/10.3390/molecules26123587

Pouton CW, Porter CJ (2008) Formulation of lipid-based delivery systems for oral administration: materials, methods and strategies. Adv Drug Deliv Rev 60(6):625–637. https://doi.org/10.1016/j.addr.2007.10.010

Pradhan M, Alexander A, Singh MR, Singh D, Saraf S, Saraf S (2018) Understanding the prospective of nano-formulations towards the treatment of psoriasis. Biomed Pharmacother 107:447–463

Prasad D, Chauhan H (2014) Nanotoxicity of polymeric and solid lipid nanoparticles. In: Biointeractions of nanomaterials. CRC Press, Boca Raton, pp 141–158

Priyadarsini RV, Nagini S (2012) Cancer chemoprevention by dietary phytochemicals: promises and pitfalls. Curr Pharm Biotechnol 13(1):125–136. https://doi.org/10.2174/138920112798868610

Puglia C, Rizza L, Drechsler M, Bonina F (2010) Nanoemulsions as vehicles for topical administration of glycyrrhetic acid: characterization and in vitro and in vivo evaluation. Drug Deliv 17(3):123–129. https://doi.org/10.3109/10717540903581679

Qian ZM, Ke Y (2014) Huperzine A: is it an effective disease-modifying drug for Alzheimer's disease? Front Aging Neurosci 6:216. https://doi.org/10.3389/fnagi.2014.00216

Radhakrishnan R, Kulhari H, Pooja D, Gudem S, Bhargava S, Shukla R, Sistla R (2016) Encapsulation of biophenolic phytochemical EGCG within lipid nanoparticles enhances its stability and cytotoxicity against cancer. Chem Phys Lipids 198:51–60. https://doi.org/10.1016/j.chemphyslip.2016.05.006

Rahman M, Almalki WH, Afzal O, Alfawaz Altamimi AS, Kazmi I, Al-Abbasi FA et al (2020) Cationic solid lipid nanoparticles of resveratrol for hepatocellular carcinoma treatment: systematic optimization, in vitro characterization and preclinical investigation. Int J Nanomedicine 15:9283–9299

Rajagopal M, Paul AK, Lee MT, Joykin AR, Por CS, Mahboob T et al (2022) Phytochemicals and nano-phytopharmaceuticals use in skin, urogenital and locomotor disorders: are we there? Plants 11(9):1265

Ramteke KH, Joshi SA, Dhole SN (2012) Solid lipid nanoparticle: a review. IOSR J Pharm 2(6):34–44

Ranjan AP, Mukerjee A, Helson L, Gupta R, Vishwanatha JK (2013) Efficacy of liposomal curcumin in a human pancreatic tumor xenograft model: inhibition of tumor growth and angiogenesis. Anticancer Res 33(9):3603–3609

Rao S, Prestidge CA (2016) Polymer-lipid hybrid systems: merging the benefits of polymeric and lipid-based nanocarriers to improve oral drug delivery. Expert Opin Drug Deliv 13(5):691–707. https://doi.org/10.1517/17425247.2016.1151872

Rawat M, Singh D, Saraf S, Saraf S (2006) Nanocarriers: promising vehicle for bioactive drugs. Biol Pharm Bull 29(9):1790–1798. https://doi.org/10.1248/bpb.29.1790

Raychaudhuri SK, Maverakis E, Raychaudhuri SP (2014) Diagnosis and classification of psoriasis. Autoimmun Rev 13(4–5):490–495

Rezaei-Sadabady R, Eidi A, Zarghami N, Barzegar A (2016) Intracellular ROS protection efficiency and free radical-scavenging activity of quercetin and quercetin-encapsulated liposomes. Artif Cells Nanomed Biotechnol 44(1):128–134

Roberts MS, Mohammed Y, Pastore MN, Namjoshi S, Yousef S, Alinaghi A et al (2017) Topical and cutaneous delivery using nanosystems. J Control Release 247:86–105

Rogerson A, Cummings J, Willmott N, Florence AT (1988) The distribution of doxorubicin in mice following administration in niosomes. J Pharm Pharmacol 40(5):337–342. https://doi.org/10.1111/j.2042-7158.1988.tb05263.x

Ruz C, Alcantud JL, Vives Montero F, Duran R, Bandres-Ciga S (2020) Proteotoxicity and neurodegenerative diseases. Int J Mol Sci 21(16):5646. https://doi.org/10.3390/ijms21165646

Sachdeva AK, Misra S, Pal Kaur I, Chopra K (2015) Neuroprotective potential of sesamol and its loaded solid lipid nanoparticles in ICV-STZ-induced cognitive deficits: behavioral and biochemical evidence. Eur J Pharmacol 747:132–140. https://doi.org/10.1016/j.ejphar.2014.11.014

Sahin NO (2007) Niosomes as nanocarrier systems. In: Nanomaterials and nanosystems for biomedical applications. Springer, Dordrecht, pp 67–81

Sainsbury F, Zeng B, Middelberg AP (2014) Towards designer nanoemulsions for precision delivery of therapeutics. Curr Opin Chem Eng 4:11–17

Sala M, Elaissari A, Fessi H (2016) Advances in psoriasis physiopathology and treatments: up to date of mechanistic insights and perspectives of novel therapies based on innovative skin drug delivery systems (ISDDS). J Control Release 239:182–202. https://doi.org/10.1016/j.jconrel.2016.07.003

Sánchez-López E, Espina M, Doktorovova S, Souto EB, García ML (2017a) Lipid nanoparticles (SLN, NLC): overcoming the anatomical and physiological barriers of the eye—Part II—ocular drug-loaded lipid nanoparticles. Eur J Pharm Biopharm 110:58–69. https://doi.org/10.1016/j.ejpb.2016.10.013

Sánchez-López E, Espina M, Doktorovova S, Souto EB, García ML (2017b) Lipid nanoparticles (SLN, NLC): overcoming the anatomical and physiological barriers of the eye—part I—barriers and determining factors in ocular delivery. Eur J Pharm Biopharm 110:70–75. https://doi.org/10.1016/j.ejpb.2016.10.009

Sangsen Y, Wiwattanawongsa K, Likhitwitayawuid K, Sritularak B, Wiwattanapatapee R (2015) Modification of oral absorption of oxyresveratrol using lipid based nanoparticles. Colloids Surf B Biointerfaces 131:182–190. https://doi.org/10.1016/j.colsurfb.2015.04.055

Saporito F, Sandri G, Bonferoni MC, Rossi S, Boselli C, Icaro Cornaglia A et al (2018) Essential oil-loaded lipid nanoparticles for wound healing. Int J Nanomed 13:175–186

Sasmita AO, Wong YP, Ling APK (2018) Biomarkers and therapeutic advances in glioblastoma multiforme. Asia Pac J Clin Oncol 14(1):40–51

Sathali AH, Ekambaram P, Priyanka K (2012) Solid lipid nanoparticles: a review. Sci Rev Chem Commun 2(1):80–102

Satyam G, Shivani S, Garima G (2010) Ethosomes: a novel tool for drug delivery through the skin. J Pharm Res 3(4):688–691

Selvamuthukumar S, Velmurugan R (2012) Nanostructured lipid carriers: a potential drug carrier for cancer chemotherapy. Lipids Health Dis 11:159. https://doi.org/10.1186/1476-511X-11-159

Seyfoddin A, Shaw J, Al-Kassas R (2010) Solid lipid nanoparticles for ocular drug delivery. Drug Deliv 17(7):467–489. https://doi.org/10.3109/10717544.2010.483257

Sezer CV (2021) An in vitro assessment of the cytotoxic and apoptotic potency of silymarin and silymarin loaded solid lipid nanoparticles on lung and breast cancer cells. Pak J Zool 53(4):1–9

Shabreen MR, Sangeetha S (2020) Ethosomes: a novel drug delivery system and their therapeutic applications—a review. Res J Pharm Technol 13(4):1970–1978

Shah S, Dhawan V, Holm R, Nagarsenker MS, Perrie Y (2020) Liposomes: advancements and innovation in the manufacturing process. Adv Drug Deliv Rev 154–155:102–122. https://doi.org/10.1016/j.addr.2020.07.002

Shangguan M, Lu Y, Qi J, Han J, Tian Z, Xie Y et al (2014) Binary lipids-based nanostructured lipid carriers for improved oral bioavailability of silymarin. J Biomater Appl 28(6):887–896

Shen LN, Zhang YT, Wang Q, Xu L, Feng NP (2014) Enhanced in vitro and in vivo skin deposition of apigenin delivered using ethosomes. Int J Pharm 460(1–2):280–288. https://doi.org/10.1016/j.ijpharm.2013.11.017

Shi J, Wang Y, Luo G (2012) Ligustrazine phosphate ethosomes for treatment of Alzheimer's disease, in vitro and in animal model studies. AAPS PharmSciTech 13:485–492

Shinde RL, Devarajan PV (2017) Docosahexaenoic acid–mediated, targeted and sustained brain delivery of curcumin microemulsion. Drug Deliv 24(1):152–161

Shrestha H, Bala R, Arora S (2014a) Lipid-based drug delivery systems. J Pharm 2014:801820. https://doi.org/10.1155/2014/80182

Shrestha H, Bala R, Arora S (2014b) Lipid-based drug delivery systems. J Pharm 2014:801820. https://doi.org/10.1155/2014/801820

Siegel RL, Miller KD, Jemal A (2018) Cancer statistics, 2018. CA Cancer J Clin 68(1):7–30

Silva AC, González-Mira E, García ML, Egea MA, Fonseca J, Silva R et al (2011) Preparation, characterization and biocompatibility studies on risperidone-loaded solid lipid nanoparticles (SLN): high pressure homogenization versus ultrasound. Colloids Surf B Biointerfaces 86(1):158–165

Singh HP, Utreja P, Tiwary AK, Jain S (2009) Elastic liposomal formulation for sustained delivery of colchicine: in vitro characterization and in vivo evaluation of anti-gout activity. AAPS J 11(1):54–64. https://doi.org/10.1208/s12248-008-9078-8
Singh D, Rawat MS, Semalty A, Semalty M (2012) Quercetin-phospholipid complex: an amorphous pharmaceutical system in herbal drug delivery. Curr Drug Discov Technol 9(1):17–24
Singh S, Sharma N, Behl T, Sarkar BC, Saha HR, Garg K et al (2021) Promising strategies of colloidal drug delivery-based approaches in psoriasis management. Pharmaceutics 13(11):1978
Sivakranth M, AnjumaAra P, Krishnaveni C, Venkatesh E (2012) Ethosomes: a novel vesicular drug delivery system. Int J Adv Pharm Res 2(1):16–27
Solanki D, Kushwah L, Motiwale M, Chouhan V (2016) Transferosomes—a review. World J Pharm Pharm Sci 5(10):435–449
Song Y, Zhuang J, Guo J, Xiao Y, Ping Q (2008) Preparation and properties of a silybin-phospholipid complex. Die Pharmazie 63(1):35–42
Sonia K, Anupama D (2011) Microemulsion based transdermal drug delivery of tea tree oil. Int J Drug Dev Res 3(1):191–198
Souto EB, Nayak AP, Murthy RS (2011) Lipid nanoemulsions for anti-cancer drug therapy. Pharmazie 66(7):473–478
Sriraman SK, Torchilin VP (2014) Recent advances with liposomes as drug carriers. In: Advanced biomaterials and biodevices. Wiley, New York, pp 79–119
Sun J, Bi C, Chan HM, Sun S, Zhang Q, Zheng Y (2013) Curcumin-loaded solid lipid nanoparticles have prolonged in vitro antitumour activity, cellular uptake and improved in vivo bioavailability. Colloids Surf B Biointerfaces 111:367–375. https://doi.org/10.1016/j.colsurfb.2013.06.032
Suresh PK, Singh P, Saraf S (2013) Novel topical drug carriers as a tool for treatment of psoriasis: progress and advances. Afr J Pharm Pharmacol 7(5):138–147
Surh YJ (2003) Cancer chemoprevention with dietary phytochemicals. Nat Rev Cancer 3(10): 768–780
Tabassum H, Ahmad IZ (2018) Evaluation of the anticancer activity of sprout extract-loaded nanoemulsion of N. sativa against hepatocellular carcinoma. J Microencapsul 35(7–8):643–656
Tadros T, Izquierdo P, Esquena J, Solans C (2004) Formation and stability of nano-emulsions. Adv Colloid Interf Sci 108:303–318
Tan BJ, Liu Y, Chang KL, Lim BK, Chiu GN (2012) Perorally active nanomicellar formulation of quercetin in the treatment of lung cancer. Int J Nanomedicine 7:651–661
Tapeinos C, Battaglini M, Ciofani G (2017) Advances in the design of solid lipid nanoparticles and nanostructured lipid carriers for targeting brain diseases. J Control Release 264:306–332
Teixeira MC, Carbone C, Souto EB (2017) Beyond liposomes: recent advances on lipid based nanostructures for poorly soluble/poorly permeable drug delivery. Prog Lipid Res 68:1–11. https://doi.org/10.1016/j.plipres.2017.07.001
Thangaraj P, Junior LJQ, Ponpandian N (2022) Nanophytomedicine: an emerging platform for drug delivery. CRC Press, Boca Raton
Thi TTH, Suys EJ, Lee JS, Nguyen DH, Park KD, Truong NP (2021) Lipid-based nanoparticles in the clinic and clinical trials: from cancer nanomedicine to COVID-19 vaccines. Vaccine 9(4): 359
Thiagarajan P (2011) Nanoemulsions for drug delivery through different routes. Research in biotechnology
Tosi G, Musumeci T, Ruozi B, Carbone C, Belletti D, Pignatello R, Vandelli MA, Puglisi G (2016) The "fate" of polymeric and lipid nanoparticles for brain delivery and targeting: strategies and mechanism of blood–brain barrier crossing and trafficking into the central nervous system. J Drug Deliv Sci Technol 32:66–76
Touitou E, Dayan N, Bergelson L, Godin B, Eliaz M (2000) Ethosomes—novel vesicular carriers for enhanced delivery: characterization and skin penetration properties. J Control Release 65(3): 403–418. https://doi.org/10.1016/s0168-3659(99)00222-9
Tran TH, Ramasamy T, Truong DH, Choi HG, Yong CS, Kim JO (2014) Preparation and characterization of fenofibrate-loaded nanostructured lipid carriers for oral bioavailability

enhancement. AAPS PharmSciTech 15(6):1509–1515. https://doi.org/10.1208/s12249-014-0175-y
Tsai YJ, Chen BH (2016) Preparation of catechin extracts and nanoemulsions from green tea leaf waste and their inhibition effect on prostate cancer cell PC-3. Int J Nanomedicine 11:1907
Üner M, Yener G (2007) Importance of solid lipid nanoparticles (SLN) in various administration routes and future perspectives. Int J Nanomedicine 2(3):289–300
Vaiserman A, Koliada A, Zayachkivska A, Lushchak O (2020) Nanodelivery of natural antioxidants: an anti-aging perspective. Front Bioeng Biotechnol 7:447
Vedagiri A, Thangarajan S (2016) Mitigating effect of chrysin loaded solid lipid nanoparticles against Amyloid β25–35 induced oxidative stress in rat hippocampal region: an efficient formulation approach for Alzheimer's disease. Neuropeptides 58:111–125. https://doi.org/10.1016/j.npep.2016.03.002
Vella LJ, Sharples RA, Nisbet RM, Cappai R, Hill AF (2008) The role of exosomes in the processing of proteins associated with neurodegenerative diseases. Eur Biophys J 37:323–332
Verma P, Pathak K (2010) Therapeutic and cosmeceutical potential of ethosomes: an overview. J Adv Pharm Technol Res 1(3):274
Vijayan V, Aafreen S, Sakthivel S, Reddy KR (2013) Formulation and characterization of solid lipid nanoparticles loaded neem oil for topical treatment of acne. J Acute Dis 2(4):282–286
Walker R, Decker EA, McClements DJ (2015) Development of food-grade nanoemulsions and emulsions for delivery of omega-3 fatty acids: opportunities and obstacles in the food industry. Food Funct 6(1):41–54
Wang P, Zhang K, Zhang Q, Mei J, Chen CJ, Feng ZZ, Yu DH (2012a) Effects of quercetin on the apoptosis of the human gastric carcinoma cells. Toxicol In Vitro 26(2):221–228. https://doi.org/10.1016/j.tiv.2011.11.015
Wang G, Wang JJ, Yang GY, Du SM, Zeng N, Li DS et al (2012b) Effects of quercetin nanoliposomes on C6 glioma cells through induction of type III programmed cell death. Int J Nanomed 7:271–280
Wang S, Zhang J, Chen M, Wang Y (2013) Delivering flavonoids into solid tumors using nanotechnologies. Expert Opin Drug Deliv 10(10):1411–1428. https://doi.org/10.1517/17425247.2013.807795
Wang S, Su R, Nie S, Sun M, Zhang J, Wu D, Moustaid-Moussa N (2014) Application of nanotechnology in improving bioavailability and bioactivity of diet-derived phytochemicals. J Nutr Biochem 25(4):363–376
Wang X, Chen H, Luo Z, Fu X (2016) Preparation of starch nanoparticles in water in oil microemulsion system and their drug delivery properties. Carbohydr Polym 138:192–200. https://doi.org/10.1016/j.carbpol.2015.11.006
Weber S, Zimmer A, Pardeike J (2014) Solid lipid nanoparticles (SLN) and nanostructured lipid carriers (NLC) for pulmonary application: a review of the state of the art. Eur J Pharm Biopharm 86(1):7–22. https://doi.org/10.1016/j.ejpb.2013.08.013
Wilczewska AZ, Niemirowicz K, Markiewicz KH, Car H (2012) Nanoparticles as drug delivery systems. Pharmacol Rep 64(5):1020–1037. https://doi.org/10.1016/s1734-1140(12)70901-5
Wissing SA, Kayser O, Müller RH (2004) Solid lipid nanoparticles for parenteral drug delivery. Adv Drug Deliv Rev 56(9):1257–1272. https://doi.org/10.1016/j.addr.2003.12.002
Wong HL, Bendayan R, Rauth AM, Li Y, Wu XY (2007) Chemotherapy with anticancer drugs encapsulated in solid lipid nanoparticles. Adv Drug Deliv Rev 59(6):491–504. https://doi.org/10.1016/j.addr.2007.04.008
Xi XM, Xia SJ, Lu R (2021) Drug loading techniques for exosome-based drug delivery systems. Die Pharmazie 76(2–3):61–67
Xiang QY, Wang MT, Chen F, Gong T, Jian YL, Zhang ZR, Huang Y (2007) Lung-targeting delivery of dexamethasone acetate loaded solid lipid nanoparticles. Arch Pharm Res 30:519–525
Xiao J, Bai W (2019) Bioactive phytochemicals. Crit Rev Food Sci Nutr 59(6):827–829

Yang CR, Zhao XL, Hu HY, Li KX, Sun X, Li L, Chen DW (2010) Preparation, optimization and characteristic of huperzine a loaded nanostructured lipid carriers. Chem Pharm Bull 58(5): 656–661

Yang L, Gong L, Wang P, Zhao X, Zhao F, Zhang Z et al (2022) Recent advances in lipid nanoparticles for delivery of mRNA. Pharmaceutics 14(12):2682

Yoon G, Park JW, Yoon I (2013) Solid lipid nanoparticles (SLNs) and nanostructured lipid carriers (NLCs): recent advances in drug delivery. J Pharm Investig 43:353–362

Yuan L, Liu C, Chen Y, Zhang Z, Zhou L, Qu D (2013) Antitumor activity of tripterine via cell-penetrating peptide-coated nanostructured lipid carriers in a prostate cancer model. Int J Nanomedicine 8:4339–4350

Yue Y, Yang Y, Shi L, Wang Z (2015) Suppression of human hepatocellular cancer cell proliferation by Brucea javanica oil-loaded liposomes via induction of apoptosis. Arch Med Sci 11(4): 856–862

Yusuf M, Khan M, Khan RA, Ahmed B (2013) Preparation, characterization, in vivo and biochemical evaluation of brain targeted piperine solid lipid nanoparticles in an experimentally induced Alzheimer's disease model. J Drug Target 21(3):300–311. https://doi.org/10.3109/1061186X.2012.747529

Zhang C, Gu C, Peng F, Liu W, Wan J, Xu H, Lam C, Yang X (2013) Preparation and optimization of triptolide-loaded solid lipid nanoparticles for oral delivery with reduced gastric irritation. Molecules 18(11):13340–13356. https://doi.org/10.3390/molecules181113340

Zhang Y, Li Z, Zhang K, Yang G, Wang Z, Zhao J, Feng N (2016) Ethyl oleate-containing nanostructured lipid carriers improve oral bioavailability of trans-ferulic acid ascompared with conventional solid lipid nanoparticles. Int J Pharm 511(1):57–64

Zhu J, Huang Y, Zhang J, Feng Y, Shen L (2020) Formulation, preparation and evaluation of nanostructured lipid carrier containing naringin and coix seed oil for anti-tumor application based on "unification of medicines and excipients". Drug Des Dev Ther 14:1481–1491

Supramolecule-Mediated Delivery of Phytochemicals

6

Sunaina Chaurasiya and Hitesh Kulhari

6.1 Introduction

Supramolecules are the crucial components of in supramolecular chemistry as they can encapsulate multiple guest molecules through physical interactions such as host-guest complexation. These interactions are based on weak, noncovalent forces, including hydrogen bonding, electrostatic interactions, π-π stacking, and van der Waals forces (Ortolan et al. 2018; Español and Villamil 2019). Supramolecular compounds possess a wide range of structures, architecture such as host-guest complexes, self-assembled systems, and supramolecular polymers (Adam et al. 2019). These structures can be produced synthetically by methods like self-assembly, coordination chemistry, and templated-directed synthesis (Holliday and Mirkin 2001). By selectively picking the building blocks and their interactions, supramolecular chemistry can be used to synthesize customized system with targeted functional groups for various applications such as molecular recognition, catalysis, sensing (Kurth 2008), drug delivery system, etc. (Lu et al. 2020; Zhang et al. 2017). Molecular recognition is one of the most fundamental concepts in supramolecular chemistry. Supramolecular compounds can imitate natural processes like the interaction of an enzyme with its substrate or the recognition of ligand receptor by selectively binding to particular molecules or ions (Busschaert et al. 2015). Their ability to identify and interact with guest molecules has a great interest in various fields such as pharmaceuticals, separation methods and sensors, etc. (Huang and Anslyn 2015) Supramolecular chemistry has deepened our understanding of molecular interactions, which has also facilitated the development of novel functional materials. The ability to alter the noncovalent forces resulted in the cutting-edge breakthrough in the field of nanotechnology, materials science, molecular

S. Chaurasiya · H. Kulhari (✉)
School of Nano Sciences, Central University of Gujarat, Gandhinagar, Gujarat, India
e-mail: hitesh.kulhari@cug.ac.in

© The Author(s), under exclusive license to Springer Nature Singapore Pte Ltd. 2024

D. Pooja, H. Kulhari (eds.), *Nanotechnology Based Delivery of Phytoconstituents and Cosmeceuticals*, https://doi.org/10.1007/978-981-99-5314-1_6

electronics, and drug discovery (Chan and Yam 2022; Lepeltier et al. 2020). By using supramolecular chemistry, scientists are discovering novel approaches to design innovative and complex molecular architectures with specialized features and functionalities (Bojarska et al. 2020; Vicens and Vicens 2009). Various supramolecules, like cyclodextrins (Azzi et al. 2018), cucurbit[*n*]urils (Boraste et al. 2018), tetrapyrroles (Harit et al. 2017), resorcinarenes (Shumatbaeva et al. 2020), crown ethers (Parisi and Pappalardo 2008), phthalocyanines (Li et al. 2019), and calix[*n*]arenes are explored for the drug delivery applications due to their cyclic structure which helps to encapsulate a high amount of drug and protect the drug from the degradation. The most widely and classical macrocyclic molecules are cyclodextrin (CDs). In 1903, Schardinger reported the cyclodextrin structure (Crini 2020). In 1967, Pedersen reported the first synthesized macrocyclic compounds crown ethers (Ullah et al. 2022). In 1978, Gutsche et al. recognized that calix[*n*]arenes are composed of phenolic macrocyclic molecule (Gutsche and Stoddart 1989; Gutsche 2007). In 1981, Mock and colleagues achieved a remarkable discovery by identifying cucurbit[*n*]urils, a class of symmetrically perfect hexamers with a pumpkin-shaped structure (Freeman et al. 1981). Macrocyclic compounds with nitrogen-containing structure are extensively found in natural products and therapeutic molecules. These compounds generally have distinct physicochemical and pharmacological characteristics. The supramolecular compounds have received great attention for the characteristic properties of alternation in the size of the cavity and targeted complexation with guest molecules such as organic and inorganic small molecules and metal ions. These compounds have been extensively exploited in the domain of physical chemistry, analytical chemistry, organic synthetic chemistry, environmental sciences, and biochemistry (Fyfe and Stoddart 1997). In reported studies, it was observed that the naturally occurring macrocyclic compound has various biological activities, particularly anticancer properties. This chapter highlights the type, structure advantages, disadvantages, and properties of supramolecules. Another section discusses the work reported on the use of supramolecules for the delivery of phytochemicals.

6.2 Types and Structures of Supramolecules

Supramolecules are a subclass of macrocyclic or polyhydroxylated compounds with supramolecular characteristics. These compounds are well-known for their capability to interact noncovalently and self-assemble into a well-defined complex structure. These play a significant role in supramolecular chemistry since they have several reactive sites. Further, these molecules are less toxic and can be produced synthetically. These compounds including cyclodextrins, cucurbit[*n*]uril, crown ether, calixarenes, resorcinarenes, and pyrogallolarenes are extremely fascinating. Some examples of supramolecular macrocyclic compounds are as follows:

6.2.1 Cucurbit[*n*]urils

The macrocyclic compound **cucurbit[*n*]uril** is made up of glycoluril units connected by methylene bridges (Mock 2005; King et al. 2019). The "*n*" indicates the number of glycoluril units in the macrocycle, which can range from 5 to 8 (Xu et al. 1994). Among all cucurbit[*n*]uril, cucurbit[7]uril and cucurbit[8]uril are frequently studied (Uzunova et al. 2010). Each glycoluril unit provides two aromatic faces and two carbonyl-lined portals to form a barrel-like shape of the cucurbit[*n*]uril structure. A firm macrocycle is formed by the methylene bridge-linked glycoluril units (Madaan et al. 2014). The portals also referred to as portals 1 and 2 enable access to the core of macrocycles. The arrangement of the glycoluril units leads in a hydrophobic and electron-rich interior cavity (Isaacs 2011). The hydrophobic cavity is rich with carbonyl groups that can take part in hydrogen bonding with guest molecules (Murray et al. 2017). The size of the cavity in cucurbit [*n*]uril corresponds to the number of glycoluril units present in the macrocycle structure. As an example, cucurbit[8]uril possesses a larger cavity compared to cucurbit[7]uril as shown in Fig. 6.1. The cavity dimensions allow the cucurbit[*n*] uril to accommodate different guest molecules depending on the complementary interactions, shape, and size (Liu et al. 2005; Gao et al. 2023).

6.2.2 Crown Ethers

Crown ethers are macrocyclic compounds characterized by a pattern of repeating ether (–O–) units (Ullah et al. 2022; Nicoli et al. 2021). These are generally consisting of a ring-shaped arrangement of oxygen atoms linked together by two or more carbon atoms, which serve as bridge between the atoms of oxygen. Crown

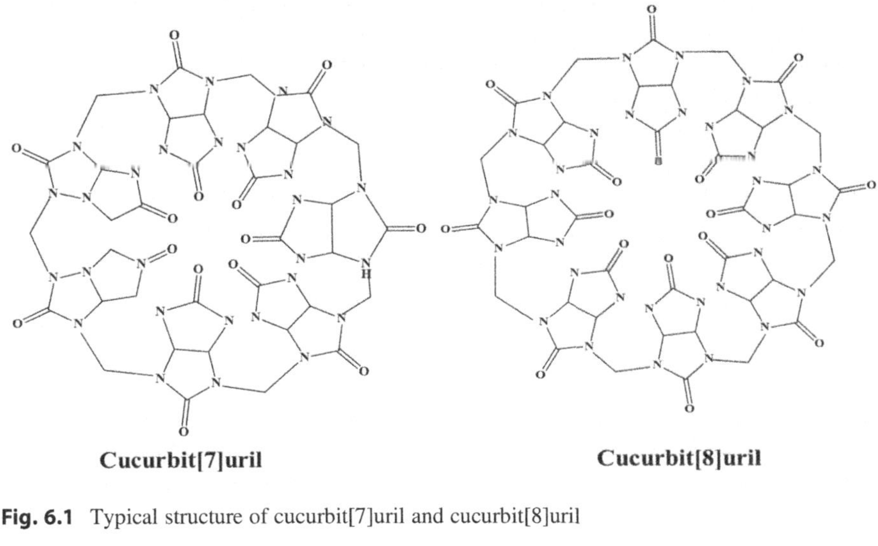

Fig. 6.1 Typical structure of cucurbit[7]uril and cucurbit[8]uril

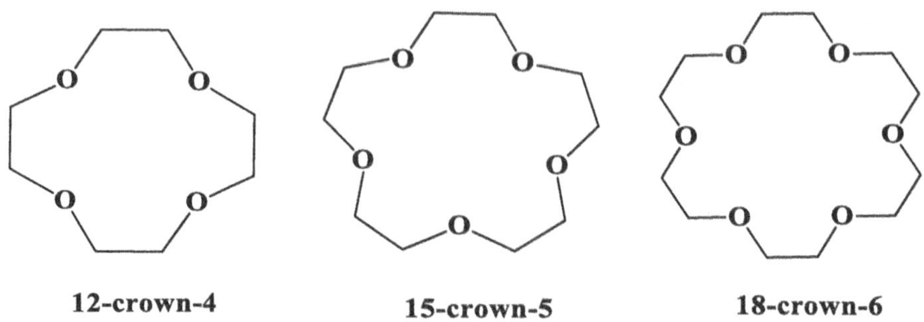

Fig. 6.2 Structure of different crown ethers

ethers are shaped like crowns and through host-guest interactions, these can selectively bind organic cations or metal ions. The number in the name denotes the number of atoms in the macrocyclic ring, comprising oxygen atoms and the linked methylene groups. The cavity size and macrocyclic ring size of the crown ether play a crucial role in determining the selectivity toward specific metal ions. For instance, 18-crown-6, which has a ring size of 12 atoms (6 oxygen atoms), has a strong affinity for the cations of alkali metals like sodium and potassium. Here are some of the main characteristics and applications of crown ethers. Crown ethers are commonly consisting of 3–8 ether oxygen (–O–) atoms in a ring structure, and the most common crown ethers are 12-crown-4, 15-crown-5, and 18-crown-6 as shown in Fig. 6.2. Crown ethers coordinate with metal cations to generate complexes. In order to form a stable complex, the oxygen atoms in the crown ether can coordinate with the metal cations. The coordination with metal cations is regulated by the electrostatic interactions and the crown ether's capacity to solvate the metal cation. Crown ethers are commonly used as phase transfer catalysts to assist molecules or ions move across indistinguishable phases (Su et al. 2020). Crown ethers can improve the solubility of these ions in organic solvents and facilitate with their transport across the phase boundary by complexing with metal cations (Jamali et al. 2017; Kunstmann-Olsen et al. 2021). Crown ethers can also form complexes with organic molecules or small ions. The complexation of these molecules can affect the solubility, stability, and reactivity of guest molecules. Crown ethers have applications in different fields. For instance, crown ethers are used in solvent extraction procedures to recover the metal ions from the mixed composition. These can help in the purification and separation of metal ions from aqueous solutions. Crown ethers can be used in sensing systems for ion sensing and detection to identify and measure metal ions or other analytes based on changes in absorbance, fluorescence, and other characteristics (Gokel et al. 2004). Crown ethers can act as ligands or co-catalysts in catalytic reactions, which may influence the rates and selectivity of reactions as well as the formation of metal complexes (Di Stefano et al. 2020; Yoo et al. 2019). Crown ethers have been investigated as a drug delivery system, facilitating the solubilization and controlled release of pharmacological substances.

6.2.3 Cyclodextrins

The cyclic oligosaccharides known as **cyclodextrins** are made up of glucose units interconnected by glycosidic bonds. α-Cyclodextrin, β-cyclodextrin, and γ-cyclodextrin are the three significant and extensively studied forms of cyclodextrins as shown in Fig. 6.3. α-Cyclodextrin consists of six glucose units, β-cyclodextrin encompasses seven glucose units, and γ-cyclodextrin comprises eight glucose units. Cyclodextrins have a doughnut-like macrocyclic structure, with a hydrophilic surface and hydrophobic core (Wüpper et al. 2021; Poulson et al. 2021). Cyclodextrins have a hydrophilic exterior with hydrophobic core, which make them more relevant for accommodating the guest molecules (Li et al. 2007; Crini et al. 2018). These compounds have been widely exploited in the food industry, analytical chemistry, and pharmaceuticals. The following are some of the salient characteristics and uses of cyclodextrins (Gonzalez Pereira et al. 2021; Paiva-Santos et al. 2022). The distinctive arrangements of glucose units within a cyclodextrin molecules result in the formation of a remarkable cylindrical-shaped cavity. The hydrophobic cavity within a cyclodextrin molecule is formed by the presence of nonpolar carbon atoms of glucose units, and hydroxyl groups on the surface of the molecule cyclodextrin contribute to its aqueous solubility by forming inclusion complex with wide range of guest molecules. The hydrophobic cavity of cyclodextrins may accommodate hydrophobic or partially hydrophobic guest molecules, leading to the development of stable inclusion complexes (Carneiro et al. 2019; Cheirsilp and Rakmai 2016). The guest compounds can be small organic molecules, pharmaceuticals, dyes, fragrances, and even larger molecules like proteins. Cyclodextrin can enhance the solubility and stability of poorly water soluble and unstable guest molecules. A cyclodextrin molecule shields the guest molecules from the surrounding medium by protecting them inside the core, and this prevents their precipitation and degradation (Tiwari et al. 2010).

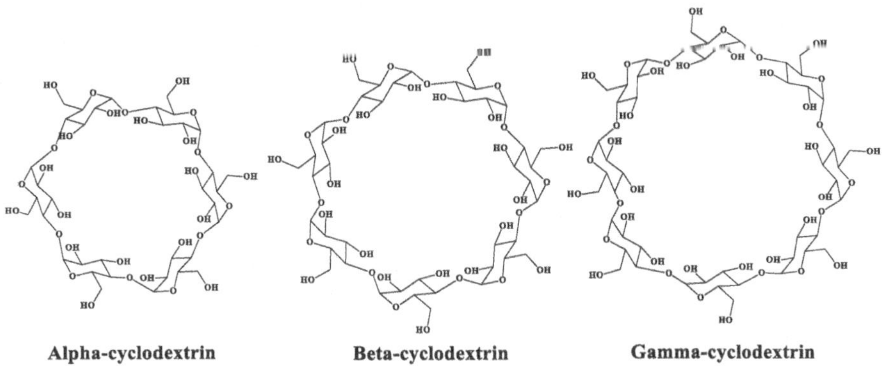

Fig. 6.3 Structure of different types of cyclodextrins

6.2.4 Calix[*n*]arenes

Calix[*n*]arenes are one of the well-known supramolecular compounds consisting of phenolic units interconnected by methylene bridges (Ortolan et al. 2018). The "*n*" in the name refers to the number of phenolic units in the macrocyclic structure, which can range from 4 to 8 or more (Fulton and Stoddart 2001; Gutsche 1998). Calix[*n*] arenes exist in various forms, but the most extensively studied forms include calix[4] arene, calix[6]arene, and calix[8]arene as shown in Fig. 6.4. Calix[*n*]arenes exhibit a characteristic structure comprising a wide-cyclic framework with phenolic group (–OH) on the surface or periphery and a hydrophobic cavity (Maier et al. 2015). The number and position of phenolic groups determine the calix[*n*]arene form. The macrocyclic structure is made up flexible by the phenolic units, which are connected by the methylene ($–CH_2–$) group (Yousaf et al. 2015; Xiao et al. 2011). Calix[*n*] arenes have a three-dimensional cavity than can encapsulate variety of biologically active molecules through the host-guest interaction including hydrogen bonding, π-π stacking, and Van der Waals (Mazzone et al. 2018; Schädel et al. 2005). In recent years, various functionalization of plain calix[*n*]arene such as sulfonation, PEGylation, acetylation, alkylation, etc. have been explored to enhance the water solubility of various biologically active molecules (Mummidivarapu et al. 2018; Makha and Raston 2001; Neri et al. 1994). It is one of the key classes in host-guest chemistry due to high water solubility, stability, less toxicity, biocompatibility, and flexible three-dimensional architecture.

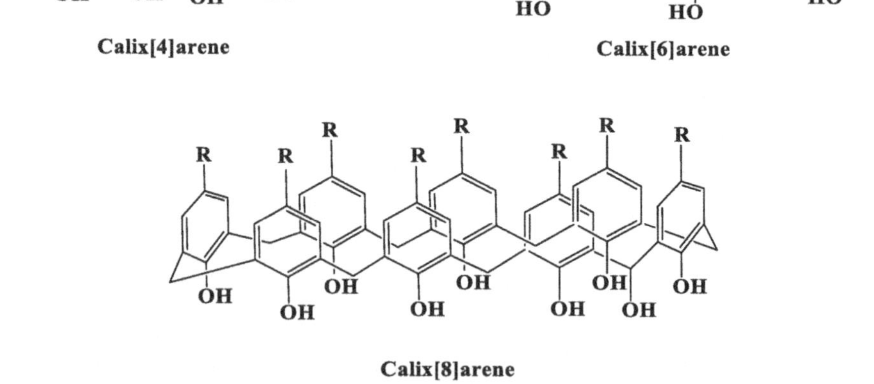

Fig. 6.4 Structure of different types of calix[*n*]arenes

6.3 Properties of Supramolecules

Supramolecular compounds have a number of unique characteristics that result from noncovalent interactions and self-assembly. Supramolecular compounds have the following important characteristics.

6.3.1 Self-Assembly

Supramolecular compounds can get self-assemble to form a well-defined structure through noncovalent interactions including hydrogen bonding, Van der Waals, π-π stacking, and electrostatic interactions. The self-assembly formation is regulated by the thermodynamics of selected molecules of interest, which can result in the development of complex structure, including capsules, fibers, cages, and gels.

6.3.2 Selective Molecular Recognition

Supramolecules commonly consist of specific binding sites or cavities that can selectively bind and recognize to guest or other. The corresponding size, shape, and functionalities of supramolecular compounds enable selective interaction with the guest molecule. This property enables the supramolecules to serve as receptors or hosts for the guest moieties, ions, or other supramolecular structures. The binding between supramolecules and guest molecules can be highly selective through the complementary interactions like hydrogen bonding, host-guest interactions, or electrostatic interactions. The use of this feature can be observed in molecular sensors, therapeutic administration systems, and separation processes.

6.3.3 Host-Guest Chemistry

Supramolecules can form complexation of host-guest, where the supramolecules encapsulate or bind to the desired guest molecules or biologically active molecules. The host guest interactions can influence the physiochemical and biological properties of both the host and the guest molecules. The field of host-guest chemistry is extensively explored due to its various applications, including drug delivery systems, sensing catalysis, and selective molecular recognition (Khan and Lee 2021; Sayed and Pal 2021).

6.3.4 Stimuli-Responsiveness

Supramolecules exhibit stimuli-responsive behavior, allowing their properties and function to be effectively alter in response to external stimuli including temperature, pH, light, and chemical signals. This dynamic property enables them to adapt and respond to their environment, which can be exploited for many applications like sensing, drug delivery, and molecular switches (Chen et al. 2020; Hatai et al. 2019).

6.3.5 Hierarchical Organization

Supramolecules have the ability to self-assemble into larger and more complex hierarchical structures like vesicles, fibers, micelles, and nanoparticles (Helttunen and Shahgaldian 2010). These hierarchical complex structures can have several distinct properties, such as controlled drug release behavior, programmable physical and optical properties, and improved stability (Datta et al. 2018).

6.3.6 Solubility

Various supramolecules are soluble in typical organic solvents, influencing their synthesis and regulation. The solubility can be customized by modifying the functional groups on the periphery of supramolecules, providing enhanced compatibility in certain applications (Yudin 2015; Zheng et al. 2012; Davis and Brewster 2004).

6.3.7 Dynamic Properties

Supramolecules are generally dynamic in nature and offer structural re-assembly or reorganization, adaptability, and responsiveness to external changes. This dynamic nature can result the reversible behavior of the noncovalent interactions which hold the supramolecular structure together (Torchi et al. 2018; Braegelman and Webber 2019).

6.3.8 Optical and Electronic Properties

Certain supramolecule compounds show remarkable optical and electronic properties. For instance, porphyrin supramolecules are well known for their absorption of vibrant dye and visible light (Mohnani and Bonifazi 2010). These properties enable them more fascinating in fields such as sensors, light-harvesting system, and photovoltaic.

6.3.9 Biological Activity

Some supramolecules have biological activity and can interact with biological targets like enzymes or receptors (Wang et al. 2021a). For instance. a very well-known supramolecule vancomycin is used as a pharmaceutical agent (antibiotic) because it can hinder the cell wall synthesis of bacteria (Flint and Davis 2022).

6.3.10 Stability

The well-defined and rigid structure of supramolecules participates in their stability. They can withstand a wide range of external conditions and keep their molecular shape intact. This stability property is beneficial in applications of host-guest chemistry and selective molecular recognition (Mukherjee et al. 2020).

6.4 Advantages and Disadvantages of Supramolecules

There are several advantages and disadvantages of supramolecular compounds.

6.4.1 Advantage of Supramolecules

1. **Self-assembly:** Supramolecular compounds can be self-assembled and facilitated by noncovalent interactions such as π-π stacking, van der Waals forces, and hydrogen bonding (Li et al. 2021). This process allows the development of complex and highly arranged structures such as capsules, cages, and helical assemblies. The self-assembly offers a bottom-up approach for the development of these systems (Gerberich et al. 2003).
2. **Reversibility:** The reversibility offers flexible and adaptive behavior, where the supramolecular compounds can react to external stimuli including temperature fluctuations, change in the pH, and the presence of guest moieties. This property is very important for drug delivery applications, molecular recognition, and sensing (Xu et al. 2015).
3. **Selectivity and specificity:** Supramolecular compounds have a high level of specificity and selectivity to specific molecules (Sommer et al. 2001). The size, shape, and functionalities of the host supramolecular compound can be modified to match certain guest molecules, enabling highly specific and efficient molecular recognition. Selectivity and specificity have high interest in fields like separation techniques and molecular sensors (Wang et al. 2021b).
4. **Nontoxicity:** Various supramolecules have been investigated for their toxicity properties and are found nontoxic or less toxic. This characteristic is an important factor for biomedical applications, as it shows low side effects during biological applications (Singh Sekhon 2015; Yui 2002; Zayed et al. 2010).
5. **Functional diversity:** Supramolecular chemistry provides a broad range of functional diversity. Supramolecular assemblies can be made from a variety of building blocks and functional groups, which enables the development of materials with distinctive optical, electrical, and catalytic functions. These properties offer a great opportunity in the field of molecular electronics, energy storage, and chemical synthesis (Lehn 2009).
6. **Targeted drug delivery:** Supramolecule compounds can be modified or functionalized to attach the targeting ligands like peptides or antibodies, facilitating targeted interaction with the specific tissues or cells. This targeted

drug delivery method elevates the efficacy of therapeutic agents while limiting off-target effects and minimizing systemic toxicity.

7. **Water solubility**: Most of the supramolecules compounds are aqueous soluble or can develop water-soluble complexes. This property is crucial for biomedical applications and pharmaceutical formulation development.
8. **Encapsulation and protection of biomolecules:** Supramolecule compounds can encapsulate and protect molecules like drugs and nucleic acid from enzymatic activity. The stability and bioavailability of the incorporated biomolecules can be improved by this protective effect, thereby increasing their efficacy during clinical applications (Ma and Zhao 2015).
9. **Low immunogenicity:** Supramolecular compounds generally show low immunogenicity, i.e., these have relatively small effects on the immune system when administered in the body. This property lowers the possibility of negative immunological reactions, making them favorable for certain applications like tissue engineering and drug delivery (Jin et al. 2020).
10. **Compatibility with biological processes:** Supramolecular compounds can interact with biological systems without disturbing important biological functions. Their architectures and features can be designed to complement and imitate the native biomolecules, enabling greater interaction and integration within biological systems.
11. **Biodegradability:** Supramolecules compounds can be conceptualized to be biodegradable so that they can be metabolized into nontoxic by-products within the biological system. This characteristic lowers the possibility of long-term toxicity or accumulation and enables their safe elimination from the body (Yui 2002; Zayed et al. 2010).

6.4.2 Disadvantages of Supramolecules

1. **Synthesis challenges:** The synthesis of the supramolecules can be difficult. To attain specific control over the supramolecular assembly and achieve a well-defined structure may be challenging due to specific 3D arrangements and multiple conformations. The optimization of reaction conditions and purification processes are required due to its complexity (Tsoucaris 2012).
2. **Precise control of self-assembly:** Supramolecule formation depends on noncovalent interactions to develop their desired structures, including hydrogen bonds, Van der Waals forces, and π-π-stacking. It can be difficult to get perfect control over these interactions in order to produce the appropriate supramolecular structure. It is important to meticulously optimize parameters such as the selection of building blocks, solvent conditions, and reaction kinetics (Sikder et al. 2021).
3. **Stereochemistry and chirality:** Supramolecules frequently exhibit chirality or contain chiral elements, which makes the synthesis process more complex. It is difficult to attain control over the chirality of the supramolecules assembly, as the desired stereochemistry requires to encode the appropriate chirality in the

building blocks and preserve it throughout the supramolecular assembly process (Adawy 2022; Huang et al. 2021).

4. **Solubility and purification:** Various supramolecules have inadequate solubility in common solvents. Therefore, their synthesis and purification are very tedious. Aggregation of the supramolecules during synthesis creates the challenges to obtain desired products in pure form. It is important to conceptualize and develop appropriate solvent media and purification technique like recrystallization and column chromatography (Fuertes-Espinosa et al. 2020).
5. **Limited scalability:** Although supramolecules have achieved great promise in the laboratory, it is difficult to scale up its industrial production. The complexity of the self-assembly process and to control over the well-maintained reaction condition may limit its synthesis at large scale and commercial feasibility (Pilgrim et al. 2022).
6. **Thermodynamic and kinetic control**: The development of supramolecular assemblies in is influenced by both kinetic and thermodynamic forces. It often becomes important to understand and manage these parameters to achieve the desired product. Thermodynamic controls stabilize the supramolecular assembly within the equilibrium conditions, while the kinetic controls focus on specific reaction time (Assaf and Nau 2019).
7. **Structural characterization:** It is very challenging to characterize the supramolecule structure due to their dynamic nature and vast size (Gasparotto et al. 2019). The available conventional analysis techniques like NMR spectroscopy and X-ray crystallography may have limitations in obtaining the full structural details of particular supramolecules. The cutting-edge characterization technique, such as mass spectroscopy (MALDI-TOF), and computation or simulation modeling are generally used to obtain detail structural information.
8. **Preorganization effects:** Many supramolecules require pre-organization steps to guide their self-assembly. The pre-organization can include the use of host-guest complex, metal ions, and other directing components. Conceptualization and execution of effective pre-organization strategies can be difficult because the pre-organized template needs to stabilize the targeted supramolecular structure and selectively influences its formation.
9. **Stability**: Supramolecular compounds, particularly those that interact noncovalently, can be less stable as compared to covalent compounds. External phenomena like temperature, pH, and solvent conditions can influence the stability of supramolecular assemblies. This limitation can prevent them from being used in certain applications that need great stability or long-term perseverance.
10. **Limited structural diversity**: Although supramolecules have a variety of structures and functions, there are yet still limitations to the arrangement and combinations of building blocks. This limitation could restrict the development of novel supramolecular structures, particularly when precise structural parameters are required.

6.5 Supramolecule-Mediated Phytochemical Delivery

Supramolecular compound-mediated phytochemical delivery refers to the use of supramolecular systems to enhance the delivery efficacy and bioavailability of phytochemicals. Various studies have explored the potential of supramolecule-mediated phytochemical delivery and have gained significant attention in several fields, like drug delivery, food industry, sensor, nutraceuticals, etc. Several significant studies have been reported in recent years, highlighting the utilization of supramolecules for delivery of phytochemicals for broad-spectrum applications. The following discussions provide detailed insights into some of these reported works.

Zeba Manzar et al. (2021) have used cucurbit[7]uril (CB7) to check the aversive nature of phytochemicals caffeine and strychnine in *Drosophila melanogaster* (Fig. 6.5). This groundbreaking study demonstrates the innovative application of CB7 as an approach of introducing unpalatable compounds into insects. This novel method extends our knowledge about insect physiology, improves pest management techniques, and is an important way to study the complicated connection between bitterness and toxicity. They have evaluated two phytochemicals with different levels of aversiveness: caffeine and strychnine. Regardless of how unpleasant each compound's taste is, they found that CB7 nanoencapsulation effectively disguises it. It is important that these phytochemicals dissociate within the insect's body and show their toxic effects. The toxicity of these compounds may be concealed if they remain attached to CB7, as is the case with neuromuscular blocking drugs, where CB7 can undo the harm imposed by such agents. Using CB7 as a delivery agent,

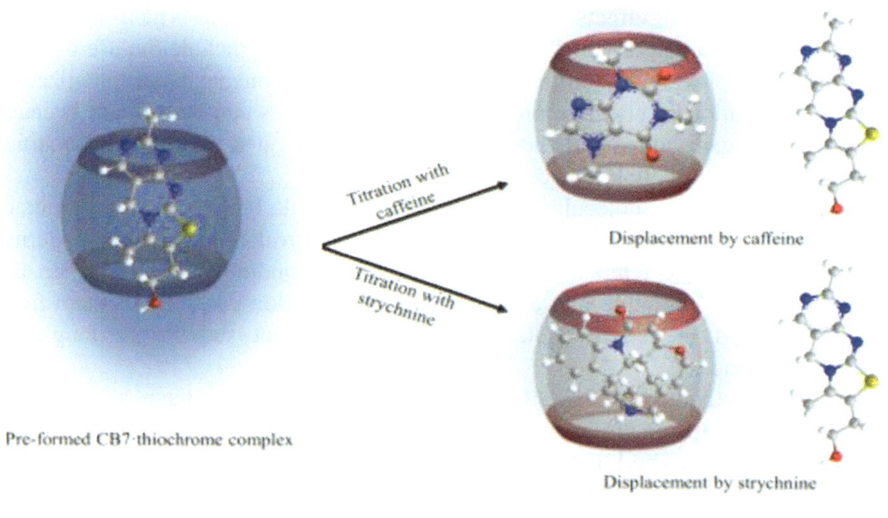

Fig. 6.5 Nanoencapsulation formation of cucurbit[7]uril (CB7): caffeine and strychnine. (Reprinted with permission from Manzar et al. 2021. Copyright © 2021, American Chemical Society)

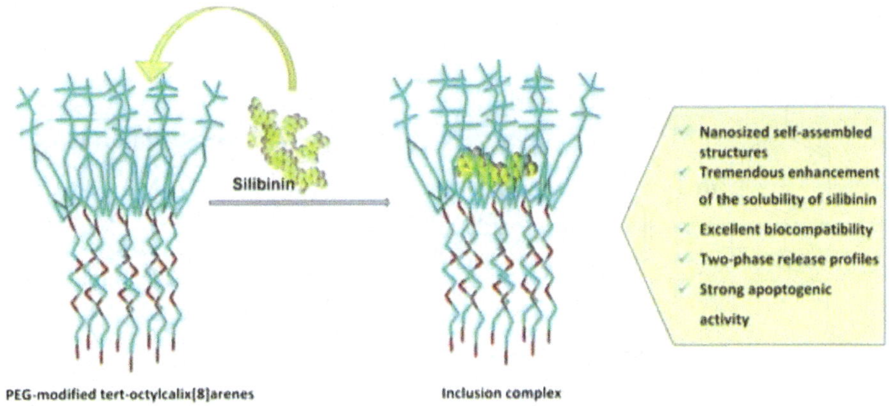

Fig. 6.6 Complex of *tert*-octylcalix[8]arenes and silibinin achieved by the "grafting from" approach. (This figure has been adapted from Budurova et al. 2021)

different unpleasant substances can be delivered to insects, allowing us to analyze the relationship between bitterness and noxiousness and open up new research opportunities to study insect physiology. Furthermore, this study may have implications for pest management. The encapsulation of these toxic compounds in CB7 can effectively suppress the pest populations (Manzar et al. 2021).

In a recent study, Desislava Budurova et al. designed a novel PEGylated *tert*-octylcalix[8]arenes that showed the exceptional potential for the delivery of silibinin, a chemoprotective, hepatoprotective, and anticancer agent (Fig. 6.6) (Pooja et al. 2014). The complex of *tert*-octylcalix[8]arenes and silibinin was achieved by the "grafting from" method. These macrocyclic molecules consisted of *tert*-octylcalix[8]arenes with PEG chains grafted onto their lower rim. The PEG chains varied in their polymerization degrees, ranging from 4 to 96. As a result, the produced amphiphilic macrocycles exhibited a hydrophobic core derived from the *tert*-octylcalix[8]arene cavity, while the surface of the molecule was composed of hydrophilic PEG chains. After reaching a critical concentration, synthesized PEGylated *tert*-octylcalix[8]arenes exhibited remarkable self-assembly in aqueous solutions and forming nanosize range structures. The solubility of silibinin was enhanced (>1700%) after the encapsulation within PEGylated *tert*-octylcalix[8]arenes. The drug release profile of silibinin was observed to occur in two distinct phases. Initially, there was faster release. Subsequently, there was a slower and sustained release from the formed complexes of silibinin and the PEGylated *tert*-octylcalix[8]arenes. The study found that the PEGylated *tert*-octylcalix[8]arenes had low inherent toxicity and did not affect the anticancer properties of silibinin. This is an important finding because it suggests that these compounds can be safely used as carriers for delivering silibinin without compromising its effectiveness against cancer (Budurova et al. 2021).

Khadija Rehman et al. have developed a novel supramolecule called benzyloxy macrocycle (BM) to improve the therapeutic effect of quercetin (QRT) (Fig. 6.7)

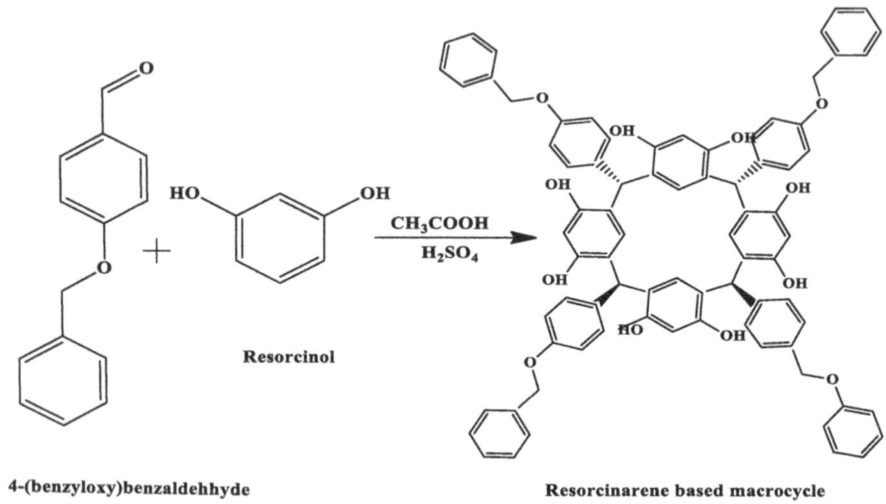

Fig. 6.7 Synthetic reaction of benzyloxy macrocycle (BM). (This figure has been adapted from Rehman et al. 2021)

(Cai et al. 2013). The developed BM was characterized by different spectroscopic techniques like ^{13}C NMR, proton NMR, FTIR, and mass spectroscopy. The obtained results showed that the developed BM was highly biocompatible, as it exhibited minimal hemolysis and cytotoxicity. The QRT-loaded BM assembly showed a small size of almost 225.5 ± 16.31 nm and 88 ± 1.52% encapsulation efficiency (EE%) of QRT. Further, the drug release pattern showed controlled and sustained release after encapsulation of QRT in BM assembly. The minimum inhibitory concentration (MIC) value of QRT drastically declined to 136 μg/mL after its the encapsulation in BM nano-assembly. Atomic force microscopy (AFM) images confirmed the antibacterial efficacy of QRT-loaded BM assembly, which exhibited the complete distortion of the morphology of the bacterial cell surface. Overall, these results indicate that the synthesized resorcinarenes macrocycle is an effective drug delivery strategy with the potential to improve QRT's therapeutic efficacy (Rehman et al. 2021).

In a study conducted by Alessia Filippone et al. for the first time, they utilized calix[4]arene supramolecule-based nano-hydrogel as a drug delivery system for skin diseases, specifically focusing on a psoriasis model induced by imiquimod (IMQ) (Fig. 6.8). A micellar system comprising choline-calix[4]arene (CALIX) amphiphile and curcumin (CUR) was developed (Granata et al. 2017). The obtained results suggest that the encapsulation of the CUR in calixarene-based nano-hydrogel keeps its anti-inflammatory activity intact. This indicates that the nano-hydrogel successfully dissolves CUR and protects it from premature deterioration. The nano-hydrogel is a promising method for delivering CUR to the skin due to its mechanical characteristics, which include skin spread ability, penetration, adhesivity, and

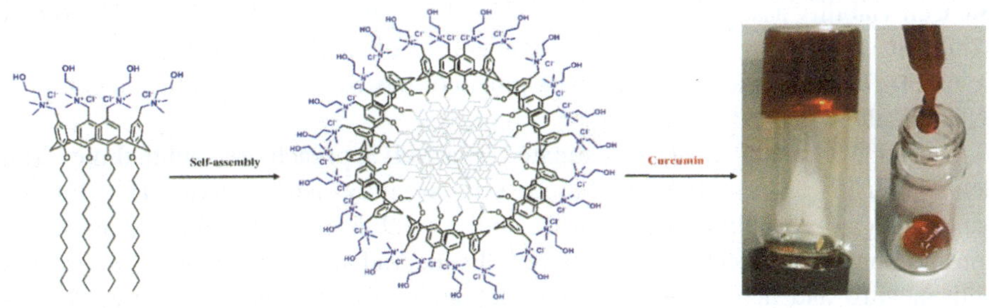

Fig. 6.8 Development of micellar choline-calix[4]arene amphiphile/curcumin (CALIX/CUR) nano-hydrogel. (This figure has been adapted from Filippone et al. 2020)

sustained drug release. The development of this calixarene-based nano-hydrogel presents a cutting-edge method of psoriasis treatment. It has the potential to increase satisfaction among patients by enhancing tolerability and effectiveness in comparison to conventional treatment approaches. This groundbreaking drug delivery method may completely change how psoriasis is treated while opening the door to new breakthroughs in the treatment of skin diseases (Filippone et al. 2020).

A notable study conducted by Loïc Leclercq et al. focused on the preparation of self-assembled Pickering emulsions using colloidal tectonics concept containing biocidal carvacrol and terpinene-4-o (phytochemicals oils) and β-cyclodextrin able to enhance the antibiofilm and antimicrobial activity of miconazoctylium bromide. In comparison to commercial cream containing miconazole nitrate, the carvacrol-containing emulsion is two times more sensitive against *S. aureus and C. albicans*, highly effective against *E. coli*, where the commercially available creams containing miconazole nitrate are completely ineffective. Additionally, these emulsions exhibit synergistic effect against fungi, additive responses against bacteria, and proficient *staphylococcal* biofilm elimination. These obtained results are related with the permeabilization of membranes, the inhibition of enzymes, and the accumulation of ROS. Finally, this system interfered with prior biofilms of MRSA (methicillin-resistant *Staphylococcus aureus*) (Leclercq et al. 2020).

Jianbin Chao et al. have reported a study with the well known phytochemical baicalein (Ba). An inclusion complex of Ba with p-sulfonatocalix[4,6,8]arenes was prepared. The prepared inclusion complex i.e. Ba-SC[4,6,8]A was characterized by fluorescence spectroscopy for titration experiment, which confirmed inclusion complex formation at a 1:1 stoichiometry-nuclear magnetic resonance spectroscopy (^{1}H NMR), and Fourier transform infrared spectroscopy (FTIR) analysis confirmed that the Ba was encapsulated inside the cavity of SC[*n*]A-atomic force microscopy (AFM) analysis observed 2D and 3D images, and it was found that the morphological structure of plain Ba and BA-SC[4,6,8]A showed distinctive differences. Additionally, the prepared illusion host-guest complex between Ba-SC[4,6,8]A complex contributed to an increase in the antioxidant activity, photostability, and thermal stability of plain Ba, which is favorable to improve the bioavailability of plain

Ba. Cell viability assay of prepared inclusion host-guest complex of Ba-SC[4,6,8]A exhibited improved cytotoxicity as compared to plain Ba, which was remarkably favorable for the application and development of BA-SC[4,6,8]A inclusion complex (Chao et al. 2019).

Giuseppe Granata et al. investigated a novel approach for ophthalmic drug delivery using a calix[4]arene-based self-assembly nanosystem. The formed nanoaggregate of calix[4]arene successfully solubilized curcumin, enhanced its anti-inflammatory effects, and buffered the degradation in ocular inflammation both in vitro and in vivo models. The prepared calix[4]arene nanoaggregate was found to be reproducible, and the resulting assembly aggregate exhibited characteristics such as clearness, low PDI, nanosized, stability, and biocompatibility. Due to these qualities, Calix-Cur supramolecular complex is a promising nanotechnological system for the ocular curcumin administration. It's significant that the polycationic calix[4]arene nano-assembly showed flexibility as a possible ocular drug carrier for both curcumin and other hydrophobic therapeutic agents. As a result, it is possible that calix[*n*]arene macrocyclic compound can be developed into a novel class of molecular scaffolding for use in a variety of ophthalmic applications (Granata et al. 2017).

In another study, Kiran Jyoti et al. have reported the development of a targeted curcumin delivery system for colon cancer treatment. They prepared a complex by encapsulating curcumin inside 2-HP-β-CD (2-hydroxypropyl-β-cyclodextrin) in a 1: 1 ratio. the curcumin-2-HP-β-CD complex was coated with chitosan microspheres, resulting in the formation of curcumin-2-HP-β-CD-CMs. The prepared formulation aimed to enable efficient curcumin delivery via the oral route of administration particularly targeted for colon cancer treatment. Several analytical, spectral, and in silico docking techniques exhibited that the curcumin was successfully incorporated in 2-HP-β-CD cavity with the 3.35×10^{-3} M stability constant. The prepared curcumin-2-HP-β-CD-CMs showed the average particle of 6.8 μm and 39.2 mV charges on surface. The microspheres' encapsulation efficiency was also approximately 79.86%. Further, a dissolution study of the prepared curcumin-2-HP-β-CD complex-CMs showed maximum curcumin release in pH of approximately 7.0–8.0 (simulated colonic fluid) with increased therapeutic index in HT-29 cell lines. In preclinical pharmacokinetic experiments, curcumin-2-HP-CD-complex-CMs consistently increased the curcumin bioavailability by 8.36-fold in comparison to curcumin suspension (Jyoti et al. 2016).

Hyeonjeong Jeon et al. have synthesized a magnetic-guided drug delivery system formed by host-guest inclusion complex of β-cyclodextrin superparamagnetic iron oxide nanoparticles (βCD-SPION) and paclitaxel (PTX) for an effective anticancer therapy. It was found that the development of nano-assemblies with SPION cluster structures is facilitated by the multivalent host-guest interactions between PTX and βCD-SPION. Due to the magnetically induced targeting effects, the generated PTX/βCD-SPION showed increased anticancer effects both in vivo and in vitro when compared to the control groups. The conceptually designed and magnetically guided drug delivery system provides the potential of enhancing drug targeting and anticancer effects. Additionally, the strong magnetism of the nano-assembly's clustered SPION provides the potential for magnetic-guided targeting even for tumors located far within the body (Jeon et al. 2016).

Adam D. Martin et al. have prepared a series of novel amphiphilic calix[4]arenes with alkyl chains on their lower rims and *p*-phosphonic acid groups on the upper rim. The prepared complex was evaluated for their effects on cell viability in PC12 rat model cells and primary mixed retinal cell cultures. The substitution patterns and the lengths of the alkyl chains or hydroxy group at the lower rim of the calixarenes vary. The obtained results showed that the compound with a hydroxy group at its lower rim was less toxic against PC12 cells compared to a compound with an alky group at its lower rim. Even at concentrations as high as 1 mg/mL, these compounds had no significant effect on the total amount of viable cells. The mixed retinal cultures, however, were more vulnerable to the toxic effects of all the calixarenes that were evaluated, demonstrating that these substances were more toxic to retinal cells than PC12 cells. When dissolved in toluene, the compound was shown to self-assemble into nanofibers and formed micelles in an aqueous solution. These structures showed potential biological applications, including the solubilization of curcumin and *p*-nitrophenol. Overall, the study demonstrated that the para-phosphonic acid calix[*n*] arenes have different effects on various cells type and emphasized the possible therapeutic applications of particular compounds on the nanoscale (Martin et al. 2012).

6.6 Conclusion

Supramolecules have shown promising potential for the efficient delivery vehicles for phytochemicals in the field of drug delivery. Supramolecular structures offer various advantages for the controlled and targeted release of phytochemicals due to their ability to produce reversible and highly selective noncovalent interactions. The encapsulation of phytochemicals within supramolecular hosts, such as cucurbiturils, cyclodextrins, and calixarenes, protects against environmental factors like enzymatic degradation, thereby enhancing solubility, bioavailability, and stability. Furthermore, self-assembled nanostructures based on supramolecular building blocks, such as micelles, liposomes, or nanoparticles, have additional advantages. These nanostructures can encapsulate a wide range of phytochemicals, within their core, protecting them from degradation and facilitating controlled release. The size, surface properties, and composition of these nanostructures can be modified to obtain prolonged circulation, targeted delivery, and enhanced cellular uptake. Additionally, by incorporating stimuli-responsive components into supramolecular systems, such as pH-, temperature-, or enzyme-sensitive moieties, phytochemicals can be released at specified times in response to particular physiological conditions. The spatiotemporal control of the release kinetics promotes the therapeutic efficacy of phytochemicals and minimizes side effects. By harnessing the supramolecular chemistry principle, researchers are developing the novel drug delivery techniques that could revolutionize the use of phytochemicals in various fields, such as medicines, nutraceuticals, and cosmetics. Further, research and development in this area holds tremendous promise for unlocking the full potential of phytochemicals and its therapeutic applications.

Acknowledgment The authors would like to thank Central University of Gujarat, Gandhinagar for the support and encouragement. S.C. acknowledges the UGC, New Delhi for a Ph.D. fellowship.

References

Adam R et al (2019) Self-assembly of catalytically active supramolecular coordination compounds within metal–organic frameworks. J Am Chem Soc 141(26):10350–10360. https://doi.org/10.1021/jacs.9b03914

Adawy A (2022) Functional chirality: from small molecules to supramolecular assemblies. Symmetry 14(2):292

Assaf KI, Nau WM (2019) Cucurbituril properties and the thermodynamic basis of host–guest binding. In: Cucurbiturils and related macrocycles, vol 28. Royal Society of Chemistry, Cambridge, p 54

Azzi J, Jraij A, Auezova L, Fourmentin S, Greige-Gerges H (2018) Novel findings for quercetin encapsulation and preservation with cyclodextrins, liposomes, and drug-in-cyclodextrin-in-liposomes. Food Hydrocoll 81:328–340. https://doi.org/10.1016/j.foodhyd.2018.03.006

Bojarska J et al (2020) A supramolecular approach to structure-based design with a focus on synthons hierarchy in ornithine-derived ligands: review, synthesis, experimental and in silico studies. Molecules 25(5):1135

Boraste DR, Chakraborty G, Ray AK, Shankarling GS, Pal H (2018) Supramolecular host-guest interaction of antibiotic drug ciprofloxacin with cucurbit[7]uril macrocycle: modulations in photophysical properties and enhanced photostability. J Photochem Photobiol A Chem 358: 26–37. https://doi.org/10.1016/j.jphotochem.2018.02.037

Braegelman AS, Webber MJ (2019) Integrating stimuli-responsive properties in host-guest supramolecular drug delivery systems. Theranostics 9(11):3017

Budurova D, Momekova D, Momekov G, Shestakova P, Penchev H, Rangelov S (2021) PEG-modified tert-octylcalix[8]arenes as drug delivery nanocarriers of silibinin. https://doi.org/10.3390/pharmaceutics13122025

Busschaert N, Caltagirone C, Van Rossom W, Gale PA (2015) Applications of supramolecular anion recognition. Chem Rev 115(15):8038–8155. https://doi.org/10.1021/acs.chemrev.5b00099

Cai X, Fang Z, Dou J, Yu A, Zhai G (2013) Bioavailability of quercetin: problems and promises. Curr Med Chem 20(20):2572–2582. https://doi.org/10.2174/09298673113209990120

Carneiro SB et al (2019) Cyclodextrin–drug inclusion complexes: in vivo and in vitro approaches. Int J Mol Sci 20(3):642. https://doi.org/10.3390/ijms20030642

Chan MH-Y, Yam VW-W (2022) Toward the design and construction of supramolecular functional molecular materials based on metal–metal interactions. J Am Chem Soc 144(50):22805–22825

Chao J, Xu M, Wang X, Guo Z (2019) Baicalein-p-sulfonatocalix[n]arenes inclusion complexes: characterization, antioxidant ability and stability. Polym Bull 76(2):989–1006. https://doi.org/10.1007/s00289-018-2415-x

Cheirsilp B, Rakmai J (2016) Inclusion complex formation of cyclodextrin with its guest and their applications. Biol Eng Med 2(1):1–6. https://doi.org/10.15761/BEM.1000108

Chen Z, Chan MH-Y, Yam VW-W (2020) Stimuli-responsive two-dimensional supramolecular polymers based on trinuclear platinum (II) scaffolds: reversible modulation of photoluminescence, cavity size, and water permeability. J Am Chem Soc 142(38):16471–16478

Crini G (2020) The contribution of Franz Schardinger to cyclodextrins: a tribute on the occasion of the centenary of his death. J Incl Phenom Macrocycl Chem 97:19–28

Crini G, Fourmentin S, Fenyvesi É, Torri G, Fourmentin M, Morin-Crini N (2018) Cyclodextrins, from molecules to applications. Environ Chem Lett 16:1361–1375. https://doi.org/10.1007/s10311-018-0763-2

Datta S, Saha ML, Stang PJ (2018) Hierarchical assemblies of supramolecular coordination complexes. Acc Chem Res 51(9):2047–2063

Davis ME, Brewster ME (2004) Cyclodextrin-based pharmaceutics: past, present and future. Nat Rev Drug Discov 3(12):1023–1035

Di Stefano S, Capocasa G, Mandolini L (2020) Supramolecular catalysts featuring crown ethers as recognition units. Eur J Org Chem 2020(23):3340–3350. https://doi.org/10.1002/ejoc.201901914

Español ES, Villamil MM (2019) Calixarenes: generalities and their role in improving the solubility, biocompatibility, stability, bioavailability, detection, and transport of biomolecules. Biomolecules 9(3). https://doi.org/10.3390/biom9030090

Filippone A et al (2020) Topical delivery of curcumin by choline-calix[4]arene-based nanohydrogel improves its therapeutic effect on a psoriasis mouse model. Int J Mol Sci 21(14):5053

Flint AJ, Davis AP (2022) Vancomycin mimicry: towards new supramolecular antibiotics. Org Biomol Chem 20(39):7694–7712

Freeman WA, Mock WL, Shih NY (1981) Cucurbituril. J Am Chem Soc 103(24):7367–7368

Fuertes-Espinosa C, Pujals M, Ribas X (2020) Supramolecular purification and regioselective functionalization of fullerenes and endohedral metallofullerenes. Chem 6(12):3219–3262

Fulton DA, Stoddart JF (2001) Neoglycoconjugates based on cyclodextrins and calixarenes. Bioconjug Chem 12(5):655–672. https://doi.org/10.1021/bc0100410

Fyfe MCT, Stoddart JF (1997) Synthetic supramolecular chemistry. Acc Chem Res 30(10): 393–401

Gao R, Ge Q, Cong H, Zhang Y, Zhao J (2023) Preparation and biomedical applications of cucurbit[n]uril-based supramolecular hydrogels. Molecules 28(8):3566. https://doi.org/10.3390/molecules28083566

Gasparotto P, Bochicchio D, Ceriotti M, Pavan GM (2019) A complete structural and dynamic characterization of supramolecular polymers via automatic learning of defects

Gerberich WW, Jungk JM, Mook WM (2003) The bottom-up approach to materials by design. In: Nano and microstructural design of advanced materials. Elsevier, Amsterdam, pp 211–220

Gokel GW, Leevy WM, Weber ME (2004) Crown ethers: sensors for ions and molecular scaffolds for materials and biological models. Chem Rev 104(5):2723–2750. https://doi.org/10.1021/cr020080k

Gonzalez Pereira A, Carpena M, García Oliveira P, Mejuto JC, Prieto MA, Simal Gandara J (2021) Main applications of cyclodextrins in the food industry as the compounds of choice to form host–guest complexes. Int J Mol Sci 22(3):1339. https://doi.org/10.3390/ijms22031339

Granata G et al (2017) Potential eye drop based on a calix[4]arene nanoassembly for curcumin delivery: enhanced drug solubility, stability, and anti-inflammatory effect. Mol Pharm 14(5): 1610–1622. https://doi.org/10.1021/acs.molpharmaceut.6b01066

Gutsche CD (1998) Calixarenes in the nanoworld. In: Calixarene revisited, vol 1, 1st edn. Royal Society of Chemistry, Cambridge

Gutsche D (2007) Calixarenes: a personal history from resinous tars to nanomaterials. In: Calixarenes in the nanoworld. Springer, Dordrecht, pp 1–19. https://doi.org/10.1007/978-1-4020-5022-4_1

Gutsche CD, Stoddart JF (1989) Monographs in supramolecular chemistry. In: Calixarenes, vol 1. Royal Society of Chemistry, Cambridge

Harit T, Bellaouchi R, Mokhtari C, El Bali B, Asehraou A, Malek F (2017) New generation of tetrapyrazolic macrocycles: synthesis and examination of their complexation properties and antibacterial activity. Tetrahedron 73(34):5138–5143. https://doi.org/10.1016/j.tet.2017.07.006

Hatai J, Hirschhäuser C, Niemeyer J, Schmuck C (2019) Multi-stimuli-responsive supramolecular polymers based on noncovalent and dynamic covalent bonds. ACS Appl Mater Interfaces 12(2): 2107–2115

Helttunen K, Shahgaldian P (2010) Self-assembly of amphiphilic calixarenes and resorcinarenes in water. New J Chem 34(12):2704–2714

Holliday BJ, Mirkin CA (2001) Strategies for the construction of supramolecular compounds through coordination chemistry. Angew Chem Int Ed 40(11):2022–2043. https://doi.org/10.1002/1521-3773(20010601)40:11<2022::AID-ANIE2022>3.0.CO;2-D

Huang F, Anslyn EV (2015) Introduction: supramolecular chemistry. Chem Rev 115(15): 6999–7000. https://doi.org/10.1021/acs.chemrev.5b00352

Huang S, Yu H, Li Q (2021) Supramolecular chirality transfer toward chiral aggregation: asymmetric hierarchical self-assembly. Adv Sci 8(8):2002132

Isaacs L (2011) The mechanism of cucurbituril formation. Israel J Chem 51(5–6):578–591. https://doi.org/10.1002/ijch.201100022

Jamali SH, Ramdin M, Becker TM, Rinwa SK, Buijs W, Vlugt TJH (2017) Thermodynamic and transport properties of crown-ethers: force field development and molecular simulations. J Phys Chem B 121(35):8367–8376. https://doi.org/10.1021/acs.jpcb.7b06547

Jeon H et al (2016) J Control Release. https://doi.org/10.1016/j.jconrel.2016.01.006

Jin H, Lin X, Gao M, Cui L, Liu Y (2020) Peptide-decorated supramolecules for subcellular targeted cancer therapy: recent advances. Front Chem 8:824

Jyoti K, Bhatia RK, Coutinho EC, Jain UK, Chandra R, Madan J (2016) Soluble curcumin amalgamated chitosan microspheres augmented drug delivery and cytotoxicity in colon cancer cells: in vitro and in vivo study. Colloids Surf B Biointerfaces 148:674–683

Khan SB, Lee S-L (2021) Supramolecular chemistry: host–guest molecular complexes. Molecules 26(13):3995

King K, Sokołowski K, Ren X, Scherman OA (2019) Functionalisation and self-assembly of nanoparticles through cucurbit[n]uril-based binding motifs. In: Cucurbiturils and related macrocycles, vol 28. Royal Society of Chemistry, Cambridge, p 362

Kunstmann-Olsen C et al (2021) Ion shuttling between emulsion droplets by crown ether modified gold nanoparticles. Nanoscale Adv 3(11):3136–3144. https://doi.org/10.1039/D1NA00009H

Kurth DG (2008) Metallo-supramolecular modules as a paradigm for materials science. Sci Technol Adv Mater 9(1):014103. https://doi.org/10.1088/1468-6996/9/1/014103

Leclercq L, Tessier J, Douyère G, Nardello-Rataj V, Schmitzer AR (2020) Phytochemical- and cyclodextrin-based pickering emulsions: natural potentiators of antibacterial, antifungal, and antibiofilm activity. ACS J Surf Colloids 36(6):4317–4323. https://doi.org/10.1021/acs.langmuir.0c00314

Lehn J-M (2009) Towards complex matter: supramolecular chemistry and self-organization. Eur Rev 17(2):263–280

Lepeltier E et al (2020) Editorial: Supramolecular nanomaterials for engineering, drug delivery, and medical applications. Front Chem 8:626468. https://doi.org/10.3389/fchem.2020.626468

Li Z et al (2007) γ-Cyclodextrin: a review on enzymatic production and applications. Appl Microbiol Biotechnol 77:245–255. https://doi.org/10.1007/s00253-007-1166-7

Li X et al (2019) Phthalocyanines as medicinal photosensitizers: developments in the last five years. Coord Chem Rev 379:147–160. https://doi.org/10.1016/j.ccr.2017.08.003

Li G-L et al (2021) Constructing π-stacked supramolecular cage based hierarchical self-assemblies via $\pi \cdots \pi$ stacking and hydrogen bonding. J Am Chem Soc 143(29):10920–10929

Liu S, Ruspic C, Mukhopadhyay P, Chakrabarti S, Zavalij PY, Isaacs L (2005) The cucurbit[n]uril family: prime components for self-sorting systems. J Am Chem Soc 127(45):15959–15967. https://doi.org/10.1021/ja055013x

Lu R, Zhang X, Cheng X, Zhang Y, Zan X, Zhang L (2020) Medical applications based on supramolecular self-assembled materials from tannic acid. Front Chem 8:583484. https://doi.org/10.3389/fchem.2020.583484

Ma X, Zhao Y (2015) Biomedical applications of supramolecular systems based on host–guest interactions. Chem Rev 115(15):7794–7839

Madaan K, Kumar S, Poonia N, Lather V, Pandita D (2014) Dendrimers in drug delivery and targeting: drug-dendrimer interactions and toxicity issues. J Pharm Bioallied Sci 6(3):139. https://doi.org/10.4103/0975-7406.130965

Maier JA et al (2015) ff14SB: improving the accuracy of protein side chain and backbone parameters from ff99SB. J Chem Theory Comput 11(8):3696–3713. Subscriber access provided by UNIV OF MISSISSIPPI. http://pubs.acs.org

Makha M, Raston CL (2001) Direct synthesis of calixarenes with extended arms: p-phenylcalix [4,5,6,8]arenes and their water-soluble sulfonated derivatives. Tetrahedron Lett 42(35): 6215–6217. https://doi.org/10.1016/S0040-4039(01)01218-7

Manzar Z, Dutta T, Koner AL, Mitra A (2021) Cucurbit[7]uril nanoencapsulation reduces the unpalatability of bitter phytochemicals. ACS Agric Sci Technol 3:117–123. https://doi.org/10.1021/acsagscitech.1c00038

Martin AD et al (2012) Synthesis and toxicology of p-phosphonic acid calixarenes and O-alkylated analogues as potential. ChemPlusChem 6009:308–313. https://doi.org/10.1002/cplu.201100081

Mazzone G, Alberto ME, Ponte F, Russo N, Toscano M (2018) Anion-π weak interactions in a heteroaromatic calixarene receptor. A theoretical investigation. Inorg Chim Acta 470:379–384

Mock WL (2005) Cucurbituril. In: Supramolecular chemistry II—host design and molecular recognition. Springer, Berlin, pp 1–24

Mohnani S, Bonifazi D (2010) Supramolecular architectures of porphyrins on surfaces: the structural evolution from 1D to 2D to 3D to devices. Coord Chem Rev 254(19–20):2342–2362

Mukherjee A, Pal DS, Kar H, Ghosh S (2020) Confined supramolecular polymers in water with exceptional stability, photoluminescence and chiroptical properties. Polym Chem 11(47): 7481–7486

Mummidivarapu VVS, Rennie ML, Doolan AM, Crowley PB (2018) Noncovalent pegylation via sulfonatocalix[4]arene—a crystallographic proof. Bioconjug Chem 29(12):3999–4003

Murray J, Kim K, Ogoshi T, Yao W, Gibb BC (2017) The aqueous supramolecular chemistry of cucurbit[n]urils, pillar[n]arenes and deep-cavity cavitands. Chem Soc Rev 46(9):2479–2496. https://doi.org/10.1039/c7cs00095b

Neri P, Battocolo E, Cunsolo F, Geraci C, Piattelli M (1994) Study on the alkylation of p-tert-butylcalix[8]arene. Partially O-alkylated calix[8]arenes. J Org Chem 59(14):3880–3889

Nicoli F, Baroncini M, Silvi S, Groppi J, Credi A (2021) Direct synthetic routes to functionalised crown ethers. Org Chem Front 8(19):5531–5549. https://doi.org/10.1039/d1qo00699a

Ortolan AO, Øestrøm I, Caramori GF, Parreira RLT, Munoz-Castro A, Bickelhaupt FM (2018) Anion recognition by organometallic calixarenes: analysis from relativistic DFT calculations. Organometallics 37(13):2167–2176. https://doi.org/10.1021/acs.organomet.8b00292

Paiva-Santos AC et al (2022) Cyclodextrins as an encapsulation molecular strategy for volatile organic compounds—pharmaceutical applications. Colloids Surf B Biointerfaces 218:112758. https://doi.org/10.1016/j.colsurfb.2022.112758

Parisi M, Pappalardo S (2008) Ten-membered rings or larger with one or more oxygen atoms. In: Comprehensive heterocyclic chemistry III, vol 14. Elsevier, Amsterdam, pp 667–750

Pilgrim BS et al (2022) MASC 2022: what challenges and opportunities do supramolecular chemists face in coming years? Cell Rep Phys Sci 3:101199

Pooja D et al (2014) Fabrication, characterization and bioevaluation of silibinin loaded chitosan nanoparticles. Int J Biol Macromol 69:267–273. https://doi.org/10.1016/j.ijbiomac.2014.05.035

Poulson BG et al (2021) Cyclodextrins: structural, chemical, and physical properties, and applications. Polysaccharides 3(1):1–31. https://doi.org/10.3390/polysaccharides3010001

Rehman K et al (2021) Synthesis of novel biocompatible resorcinarene based nanosized dendrimer-vesicles for enhanced anti-bacterial potential of quercetin. J Mol Liq 341:116921. https://doi.org/10.1016/j.molliq.2021.116921

Sayed M, Pal H (2021) An overview from simple host–guest systems to progressively complex supramolecular assemblies. Phys Chem Chem Phys 23(46):26085–26107

Schädel U, Sansone F, Casnati A, Ungaro R (2005) Synthesis of upper rim calix[4]arene divalent glycoclusters via amide bond conjugation. Tetrahedron 61(5):1149–1154

Shumatbaeva AM et al (2020) The pH-responsive calix[4]resorcinarene-mPEG conjugates bearing acylhydrazone bonds: synthesis and study of the potential as supramolecular drug delivery systems. Colloids Surf A Physicochem Eng Asp 589:124453. https://doi.org/10.1016/j.colsurfa.2020.124453

Sikder A, Xie Y, Thomas M, Derry MJ, O'Reilly RK (2021) Precise control over supramolecular nanostructures via manipulation of H-bonding in π-amphiphiles. Nanoscale 13(47):20111–20118

Singh Sekhon B (2015) Supramolecular nanomedicine—an overview. Curr Drug Targets 16(13):1407–1428

Sommer RD, Rheingold AL, Goshe AJ, Bosnich B (2001) Supramolecular chemistry: molecular recognition and self-assembly using rigid spacer-chelators bearing cofacial terpyridyl palladium (II) complexes separated by 7 Å. J Am Chem Soc 123(17):3940–3952

Su T-S et al (2020) Crown ether modulation enables over 23% efficient formamidinium-based perovskite solar cells. J Am Chem Soc 142(47):19980–19991. https://doi.org/10.1021/jacs.0c08592

Tiwari G, Tiwari R, Rai AK (2010) Cyclodextrins in delivery systems: applications. J Pharm Bioallied Sci 2(2):72. https://doi.org/10.4103/0975-7406.67003

Torchi A, Bochicchio D, Pavan GM (2018) How the dynamics of a supramolecular polymer determines its dynamic adaptivity and stimuli-responsiveness: structure–dynamics–property relationships from coarse-grained simulations. J Phys Chem B 122(14):4169–4178

Tsoucaris G (2012) Current challenges on large supramolecular assemblies, vol 519. Springer Science & Business Media, Berlin

Ullah F et al (2022) Heterocyclic crown ethers with potential biological and pharmacological properties: from synthesis to applications. Appl Sci 12(3):1102. https://doi.org/10.3390/app12031102

Uzunova VD, Cullinane C, Brix K, Nau WM, Day AI (2010) Toxicity of cucurbit[7]uril and cucurbit[8]uril: an exploratory in vitro and in vivo study. Org Biomol Chem 8(9):2037–2042

Vicens J, Vicens Q (2009) Origins and emergences of supramolecular chemistry. J Incl Phenom Macrocycl Chem 65:221–235. https://doi.org/10.1007/s10847-009-9602-1

Wang J, Li H, Xu B (2021a) Biological functions of supramolecular assemblies of small molecules in the cellular environment. RSC Chem Biol 2(2):289–305

Wang B, Chen H, Liu T, Shi S, Russell TP (2021b) Host–guest molecular recognition at liquid–liquid interfaces. Engineering 7(5):603–614

Wüpper S, Lüersen K, Rimbach G (2021) Cyclodextrins, natural compounds, and plant bioactives—a nutritional perspective. Biomolecules 11(3):401. https://doi.org/10.3390/biom11030401

Xiao K et al (2011) The effect of surface charge on in vivo biodistribution of PEG-oligocholic acid based micellar nanoparticles. Biomaterials 32(13):3435–3446

Xu S, Gantzel PK, Clark LB (1994) Glycoluril. Acta Crystallogr Sect C: Cryst Struct Commun 50(12):1988–1989

Xu J-F et al (2015) Supramolecular polymerization controlled by reversible conformational modulation. ACS Macro Lett 4(12):1410–1414

Yoo C, Dodge HM, Miller AJM (2019) Cation-controlled catalysis with crown ether-containing transition metal complexes. Chem Commun 55(35):5047–5059. https://doi.org/10.1039/C9CC00803A

Yousaf A, Hamid SA, Bunnori NM, Ishola AA (2015) Applications of calixarenes in cancer chemotherapy: facts and perspectives. Drug Des Devel Ther 9:2831–2838

Yudin AK (2015) Macrocycles: lessons from the distant past, recent developments, and future directions. Chem Sci 6(1):30–49

Yui N (2002) Supramolecular design for biological applications. CRC Press, Boca Raton

Zayed JM, Nouvel N, Rauwald U, Scherman OA (2010) Chemical complexity—supramolecular self-assembly of synthetic and biological building blocks in water. Chem Soc Rev 39(8): 2806–2816

Zhang H, Jackson JK, Chiao M (2017) Microfabricated drug delivery devices: design, fabrication, and applications. Adv Funct Mater 27(45):1703606. https://doi.org/10.1002/adfm.201703606

Zheng B, Wang F, Dong S, Huang F (2012) Supramolecular polymers constructed by crown ether-based molecular recognition. Chem Soc Rev 41(5):1621–1636

Metal/Metal Oxide Nanocarriers for the Delivery of Phytoconstituents

7

Poonam Jain and Hyuk Sang Yoo

7.1 Introduction

The broad application of various nanomaterials (NMs) in the biomedical field emphasizes many advancements such as drug delivery, diagnosis, and sensing. The top-down and bottom-up synthesis methodologies are the two fundamental approaches used for the synthesis of NMs. These methods offer a high surface area to volume ratio and can produce materials with varied dimensions (Prasher et al. 2020; Mobeen et al. 2022; Bora et al. 2022; Kulhari et al. 2016; Ray et al. 2018). Among the various NMs, metal and metal oxide nanoparticles (NPs) have gained enormous attention from the scientific community due to their unique physical and chemical properties like optical, magnetic, sensing, thermal, and magnetic resonance-based imaging via improved relaxation, tissue engineering, and therapeutic capacity and biocompatibility (Chavali and Nikolova 2019; Liu et al. 2020; Gao et al. 2021; ul Islam and Sun 2022). Owing to these properties, they have been used extensively in drug delivery to cure many diseases including cancer, antimicrobial, antiviral, neurodegenerative, bone therapy, dermal therapy, and so on (Wang et al. 2019; Zhao et al. 2021; Jain et al. 2020; Kumar et al. 2020; Kotrange et al. 2021; Yeh et al. 2020). Cancer has become one of the deadliest diseases all over the world with 8.97 million deaths per year according to the World Health Organization's (WHO) report. The most common cancer treatments are chemotherapy, radiation therapy, and surgery which are costly and have numerous side effects, leading to poor patient compliance. So, to overcome these problems

P. Jain · H. S. Yoo (✉)
Department of Medical Biomaterials Engineering, College of Biomedical Science, Kangwon National University, Chuncheon, Republic of Korea
e-mail: hsyoo@kangwon.ac.kr

© The Author(s), under exclusive license to Springer Nature Singapore Pte Ltd. 2024

D. Pooja, H. Kulhari (eds.), *Nanotechnology Based Delivery of Phytoconstituents and Cosmeceuticals*, https://doi.org/10.1007/978-981-99-5314-1_7

metal and metal oxide NPs have been used in cancer treatment extensively (Rahman et al. 2021; Chauhan et al. 2022; Pandit et al. 2022).

From ancient times, plant-based natural ingredients have been used in humans' disease treatment, wound healing through inhalation, oral administration, and topical application. Typically, phytoconstituents or phyto-drugs are plant-based active compounds that aid in curing the disease at the molecular level, even in many cases mechanism of action is still not clear. Phytoconstituents attract huge attention in drug delivery because of their several properties like high therapeutic efficacy and biocompatibility (Lin et al. 2020). The biological function of these phytoconstituents, including qualitative and quantitative studies of their positive as well as negative effects on human health, is determined by phyto- or analytical chemistry. Generally, on the basis of production via primary and secondary metabolic pathways, these phytoconstituents are categorized as: (a) an active drug component and (b) inert non-drug components. These active compounds are relatively safe and broadly satisfactory over synthetic drugs (Willenbacher et al. 2019; Gonbad et al. 2015). And so, in this chapter, we discussed the recent progress and utilization of metal and metal oxide NPs for the delivery of phytochemical drugs and their possible implication in cancer treatment.

7.2 Overview of Metal and Metal Oxide Nanoparticles and Their Unique Properties

Inorganic NPs have a distinctive structure that considerably improves physicochemical characteristics when compared to bulk form and thus serve as unique materials in drug delivery as shown in Fig. 7.1. Metal and metal oxide NPs are composed mainly of d-block elements and lanthanides of the periodic table. Most of these NPs are generated in a zero-dimensional (0D) form that shows optical and luminescent properties. Further, the focus has shifted to one-dimensional (1D) metal oxide NPs because these NPs have unusual optoelectrical properties that provide a platform for investigating size- and morphology-dependent applications. The distinctive structure, for example, 0, 1, 2, or 3D, results from the technique of synthesis, amount of chemicals used, temperature, environment, as well as some other crucial aspects.

Primary characteristic properties of metal and metal oxide NPs are as follows:

- High surface area to volume ratio
- Variable oxidation states
- Stable and long shelf-life
- Crystallinity, numerous morphology, and surface plasmon resonance
- High surface energy

Several methods have been reported to synthesize metal-based NPs including precipitation, co-precipitation method, microwave method, solvo/hydrothermal, chemical reduction methods, and phyto-mediated biological synthesis are the most

Fig. 7.1 Inorganic nanoparticles of different shapes and their unique properties

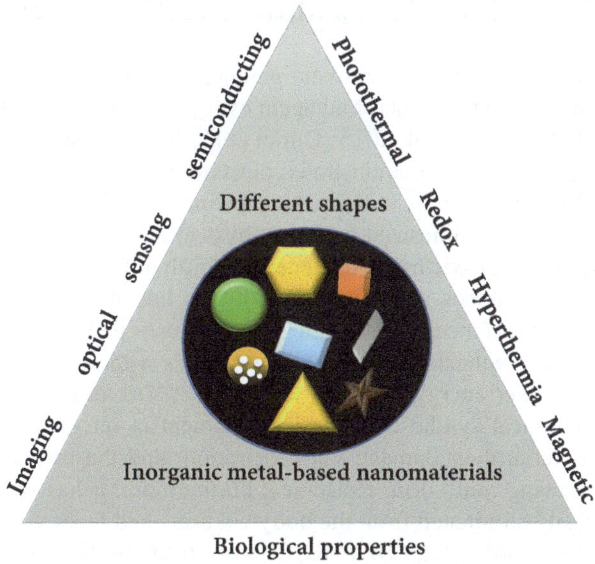

common procedures, and the obtained NPs are mainly utilized for biomedical applications. Furthermore, across a variety of environmental circumstances including temperature and pH, metal-based NPs exhibit good shelf-life as well as biological stability. Thus, some of the metallic NPs are dominating in nanotechnology mainly due to their unique physical and chemical properties for engaging in multiple applications.

7.2.1 Gold Nanoparticles (Au NPs) Properties

Au NPs, when compared to other organic and inorganic NPs, stand out due to the distinct surface plasmon resonance (SPR) property as a result of quantum confinement. Au NPs are widely used in drug delivery, diagnostic purposes, and other biomedical applications. The main functions of absorbed light to cause tumor ablation by the development of photothermal effect and imaging via light-scattering characteristics are main therapeutic application of Au NPs. Lots of studies have been performed to regulate these properties by altering the size, shape, structure, and composition of Au NPs. Also, the change in method of synthesis, in situ and ex situ surface alteration, and functionalization promote improved efficiency of traditional therapeutic drugs in vitro and in vivo biological studies. The Au NPs could also be utilized to provide on-demand release profile of therapeutic agents in response to ex situ stimuli, e.g., light. Therefore, the accomplishment of Au NPs for preclinical trials has already been established since long decades back for the treatment of biological disorders including cancer.

7.2.2 Silver Nanoparticles (Ag NPs) Properties

Silver NPs are progressively gaining attention in the biomedical field, especially as an excellent antimicrobial agent over many antibiotics (Dai et al. 2016; Mathur et al. 2018; Franci et al. 2015; Durán et al. 2016). Despite the fact that various scientists and researchers demonstrated outstanding findings, the thoroughly possible mechanism of action is still in the pipeline of discussion and arguments. Ag NPs are generally synthesized by both physical and chemical methods and considered as a costly and toxic method. Hence, the birth of biological approach for the synthesis not only produces Ag NPs of 1–100 nm but also shows eco-friendly, cost-effective method. Hence, the biological method uses parts of plants extract, bacteria, fungus, etc. to synthesize biocompatible Ag NPs without using any chemicals, temperature, and other energy (Mumtaz et al. 2022; Virk et al. 2022). The major advantage of biological synthesis is the improvement in interaction with the cell membrane of bacteria, their damage via penetration, and the formation of free radicals without showing multi-drug resistance. Furthermore, it has been reported that Ag ions are easily eliminated from the body via urine and feces without accumulating in tissues. Henceforth, Ag NPs have also been utilized in drug delivery including phytoconstituents for antibacterial and anticancer activity (Mumtaz et al. 2022).

7.2.3 Superparamagnetic Iron Oxide Nanoparticles (SPIONs) Properties

Introducing magnetically sensitive drug delivery into the diseased site is focused via a magnetic field area of interest. Among many d-block elements, SPIONs are more frequently used for magnetically sensitive drug delivery than nickel and cobalt. The chemical stability and lower harmful effect of SPIONs are the key reasons, making them the most employed materials. To regulate nucleation and growth rates, magnetic NPs have been formed through the use of a variety of chemical techniques. Subsequently, iron-based NPs have received approval for use in clinics as imaging and diagnostic tools. For instance, clinically approved SPION is ferumoxytol well-known as Fereheme™ in the USA and Rienso™ in Europe (Lazaro-Carrillo et al. 2020). In 2009, the Food and Drug Administration (FDA) approved ferumoxytol as a supplement to iron deficiency patients having anemia with chronic renal disease (Wesström 2020). In addition, several core-shell NPs have also been introduced for dual-model imaging guided photothermal effect. The SPION provides imaging and their encasement using carbon shell offers near-infrared (NIR) fluorescent properties. Therefore, SPIONs are designed for magnetic-tracked drug delivery as well as MRI contrast agents due to their photothermal characteristics having huge applications in a variety of biological fields (Neha et al. 2017).

7.2.4 Manganese Oxide Nanoparticles (MnO-Based NPs) Properties

Manganese (Mn) is the 25th transition element in the periodic table and possesses distinctive physical and chemical properties and thus has many applications, including catalytic, storage, batteries, and electronics. Among various applications, MnO-based NPs acted as excellent biocompatible metal-based NMs which have been proven to show definite application in drug delivery nanocarriers. Several other transition metals and lanthanides are being developed for imaging purposes of numerous diseases like diseased heart tissue, malignancies, and brain diseases. Currently, Mn is one of the highly demanded transition metals that are being observed and provided as magnetically active contrast agents (CAs), generally made of chelated ion manganese (Mn^{2+}). The contrast may be changed to increase the accuracy of the diagnosis. Depending on their paramagnetic susceptibilities, CAs can either rise or reduce the contrast in the tissue; these CAs are referred to as positive (bright images) or negative (dark images), respectively. As an important microelement that is actively absorbed by the small intestine, Mn acts as a necessary cofactor for numerous enzymatic activities and engages in amino acid and carbohydrate metabolism, blood coagulation, and homeostasis (Nielsen 2012; Aschner and Aschner 2005; Chen et al. 2018; Institute of Medicine 2001). Therefore, like SPIONs, Mn-based NPs are also considered for biomedical applications. Generally, Mn-based NMs with different sizes and shapes are synthesized by various methods including hydrothermal, solvothermal, chemical precipitation, and microwave synthesis methods. In recent years, chelates of Mn-based CAs have been developed such as Mn-PyC3A that showed the enhancement of blood vessels similar to that of gadolinium-based CAs (Gale and Caravan 2018). In addition to this, manganese commonly known as avasopasem of galera therapeutics showed antioxidant properties in phase II COVID 2019 infections, mucositis; esophagitis; pancreatic cancer and their phase I/II were observed for head and neck cancer (Blakaj et al. 2019; Oronsky et al. 2018). Unchelated Mn-based CAs named EVP 1001-1 is a new heart-specific MRI CA that shows excellent contrast properties directly into the myocardial site by intracellular uptake via calcium ion channels (Yuko et al. 2021). Therefore, due to their biocompatible nature and theranostic ability, MnO-based NPs have also been used for the delivery of phytoconstituents for the treatment of cancer and other diseases.

7.2.5 Mesoporous Silica Nanoparticles (MSNPs) Properties

MSNPs are a family of biocompatible nanostructures that have been gaining attention for excellent drug delivery nanocarrier (Singh et al. 2019; Kim et al. 2011). Structurally, silica NPs possess a high degree of porosity with different pore sizes and densities and are henceforth known as MSNPs. Like Au NPs and SPION, silica NPs do not exhibit imaging or magnetic properties; however, their distinctive properties such as high surface area (>900 m^2/g), large pore volume (>0.9 cm^3/g), pore diameters of 2–50 nm, high surface energy and chemical purity, and surface adsorption make them unique among other inorganic NPs (Jafari et al. 2019). Furthermore, owing to their highly porous structure, silica NPs exhibit a high payload of therapeutic agents, bioimaging dyes, quantum dots, and stimuli-responsive drug release in the biological system (Cha and Kim 2019; Tallury et al. 2008). The alteration in the particle size, porosity, enhancement of surface chemistry using biomolecules, polymers as well as doping with suitable inorganic elements allow these NPs for various biomedical applications. Due to the fact that silica NPs are among the most widely utilized NPs, the demand for these NPs has drastically expanded, reaching roughly 2.8 million metric tons in 2016 with an annual growth of 5.6%. Therefore, in order to fully satisfy the demand for applications including biomedical applications, silica NPs display various properties to manufacture MSNs, silica oxide and di-oxide NMs, and hybrid structures.

7.2.6 Zinc Oxide Nanoparticles (ZnO NPs) Properties

ZnO has been used as food additive and is considered as generally recognized as safe (GRAS) material by FDA (Narla and Lim 2020). Zinc is a trace mineral and has numerous biological functions such as the synthesis of nucleic acid, and proteins, providing bone strength, antioxidants, and defensive properties to the body (Siddiqi et al. 2018). Structurally, ZnO NPs are hydrophobic in nature, crystalline and white-colored NPs and one of the most used antibacterial and antifungal agents. Owing to their wide range of semi-conducting properties because of broad band gap (3.37 eV) and high exciton binding energy (60 meV), these NPs exhibit catalytic, optics, UV-Vis rays blocking, anti-inflammatory, and wound healing properties (Agarwal and Shanmugam 2020; Khan et al. 2021; Le et al. 2022). The intrinsic photoluminescent properties make it valuable in bio-sensing applications. Subsequently, ZnO NPs with various morphologies are extensively used in drug delivery, dentistry, cosmetics, and other clinical applications. After attempting several research experiments, the ZnO NPs are selectively confirmed in cancer treatment research because of their pH-dependent dissolution into Zn^+ ions, DNA fragmentation, reactive oxygen species generation and induce an anticancer effect (Hong et al. 2011). Thus, ZnO-based NMs with and without therapeutic agents have been shown to target solid tumors, and their stem cells, which might concurrently carry out a number of important tasks including inhibiting the growth of cancer cells.

7.2.7 Calcium Phosphate Nanoparticles (CaP NPs) Properties

CaP NPs have received substantial interest in the biomedical area in recent years owing to their extraordinary biological capabilities and are widely recognized as good drug delivery agents. Structurally, CaP NPs are similar to the natural bone composition, and synthetic CaP NPs in the form of carbonate apatite are widely employed in orthopedic implants and bone tissue engineering (Levingstone et al. 2019). In addition, its intrinsic bioactive and biodegradable properties make it an excellent adjuvant in nanomedicine applications (Craciun et al. 2017). In accordance with reported studies, CaP NPs have shown exceptional physicochemical properties such as morphology, composition, size, and surface chemistry that influence the site-specific cellular entry and treatment. CaP was discovered three decades ago for the incorporation of negatively charged nucleic acids by tuning the physical and chemical properties, specific size, and morphology (Xu et al. 2016).

Since combining calcium and phosphate does not produce particles that are of a favorable tiny size owing to spontaneous reaction, controlling their quantities and choosing the right precursors is crucial for manufacturing such particles. This is notably the difficult part of synthesizing CaP NPs. Conversely, by overcoming major difficulties in synthesis, administered CaP NPs show a variety of biological characteristics like ease of cellular entry, pH-dependent breakdown, endo-lysosomal escape, and release of therapeutic molecules like nucleic acids, protein, phytochemicals, and synthetic drugs for treating many cancers.

7.3 General Approaches for the Preparation of Metal-Based Nanoparticles

Metal and metal oxide NPs are majorly synthesized using three methods, i.e., (a) physical, (b) chemical, and (c) biological approach (Jamkhande et al. 2019; Naser et al. 2019; Shin et al. 2018; Kim et al. 2018; Vaid et al. 2020). In the case of physical method, NPs are synthesized by top-down approach in which the reduction of particle size from bulk form gives nanosized particles. The physical method for NPs generation includes different processes like ball milling, thermal evaporation, vapor phase evaporation, and lithography. In the case of chemical method, a metal precursor and a reducing or oxidizing agent react to produce NPs. Because of their low cost, simplicity of synthesis, and great productivity, chemical synthesis techniques have been shown to be the most dominant methodology in nanotechnology and nanomedicine. The chemical method of synthesis involves different techniques like chemical precipitation or co-precipitation, hydro/solvothermal, microwave synthesis method, etc. NPs with various sizes and morphologies have been produced through chemical processes. Various physical conditions, such as pH, temperature, and agitation speed, influence the outcome of these procedures. In the precipitation method, the most common reducing agents such as sodium borohydride ($NaBH_4$), and citric acid are used for the synthesis of Au and Ag NPs (Lee et al. 2022; Deshmukh et al. 2022; Santhoshkumar et al. 2017; Shiraishi et al. 2017; Paramasivam et al. 2017). Similar to this, oxidizing agents are employed to synthesize metal oxide NPs by precipitation, such as zinc, iron, and manganese oxides. Chemically produced NMs are sometimes employed for biological purposes. Owing to possible chemical toxicity, washing the manufactured NPs is a crucial step. Despite the two approaches mentioned earlier, the most widely used technique to produce different noble and metal-based NPs is now the biological approach. The use of plant extract, which contains flavonoids, alkaloids, polysaccharides, etc., that act as reducing agents is the major reason for using these in NPs synthesis. While synthesizing NPs, different strains of microorganisms, like pathogenic and non-pathogenic bacteria, fungi, yeasts, and certain algae, are also preferred because they produce enzymes that aid in particle nucleation, growth, and stabilization. Long incubation of microorganism culture media or the addition of plant extract acts as an excellent reducing agent during the synthesis process, reducing the metal precursor into zero valent NPs. This is also true for the synthesis of metal oxide NPs, while the enzymes are oxidized at room temperature. Consequently, the primary advantage of biological methods over physical and chemical ones is the provision of biocompatibility and low cost, which is also widely acknowledged in medical applications.

7.4 Alteration of the Surface of Metal-Based Nanoparticles for Phytoconstituent Delivery

While inorganic NMs offer numerous benefits, they also exhibit a number of disadvantages such as cell cytotoxicity, stability in colloidal solution, agglomeration, inadequate payload, and drug release. Most of the phytoconstituents are hydrophobic in nature and thus have very low bioavailability as well as low therapeutic index. In order to deliver these phyto-drugs to diseased sites, scientists have been working on engineering the surfaces of inorganic NPs utilizing polymers with varying chain lengths and biocompatible molecules. The surface-engineered noble NPs such as Au are mostly modified with thiol groups, amines, carboxylic acids, dendrimers, and brush-like biomolecules. Functionalization of surface-modified metal NPs further helps the drug to locate the disease site and also helps the drug to internalize and permit across the cell membrane via the receptors-ligand mechanism. Some of the natural molecules such as vitamins (folic acid and biotin), glycoprotein (transferrin), aptamers (ssDNA), and peptides (cRGD) are well-known biological ligands utilized for functionalization of metal oxide NPs that were evaluated to deliver the phytoconstituents for cancer targeting and therapy.

The effectively designed NPs can serve as a drug reservoir with ample space for drug molecules, proteins, and single- and double-stranded genetic materials. In addition, these well-designed inorganic NPs using biomolecules can stay in the blood circulation by avoiding the unwanted reticuloendothelial system, sustaining the drug release, and requiring fewer doses per day while still achieving high therapeutic indexes with the same dosage of drugs. Several phytochemical drugs such as curcumin (CUR) have been used as therapeutic adjuvants in cancer therapy. Additionally, in vitro and in vivo studies showed that NPs coated with biomolecules also enhance the post-injection imaging characteristics for several hours to days or even months, indicating the multifunctional property of inorganic NPs. Henceforth, in this chapter, the surface-modified metal NPs and their functional properties along with phyto-drugs have been described.

7.5 Role of Phytoconstituents in the Synthesis of Metal/Metal Oxide NPs

Although some of the traditional techniques may be compared to green synthesis using plant extracts, they do not give adequate information regarding the amount, stability, and reduction of ions during the formation of NPs because of which they are not ecologically benign. Researchers started utilizing more eco-friendly techniques, often known as "green synthesis," to solve this issue. Several well-established studies have shown that plant extracts have the ability to successfully convert various noble ions, such as Au^+ ions into Au NPs and Ag^+ ions into Ag NPs. Due to their facile surface-based bio-conjugation with molecular probes and outstanding plasmon-resonant characteristics, noble NPs are essential in nanotechnology and biomedicine. Some of the plants that have been used in the synthesis of NPs are *Zingiber zerumbet, Syzygium aromaticum, Peganum harmala, Fritillaria, Cinnamon zeylanicum, Coffea arabica croton, Caudatus geisel, Croton sparsiflorus, Halymenia pseudofloresii,* etc. Almost every part of the plant is utilized for the synthesis of NPs and their extract has also been found to exhibit antimicrobial, viral, fungal, inflammatory, diabetic, ROS activity, and cytotoxicity properties (Yang et al. 2019; Rajesh et al. 2018; Azizi et al. 2017; Hemmati et al. 2019; Anjum et al. 2019; Singla et al. 2021; Boomi et al. 2020; Palaniyandi et al. 2023). The green synthesized Au NPs by *Curcuma pseudomontana* using one-pot synthesis method in the neutral pH at RT is an example of green synthesis of Au NPs where the CUR present in the extract acted as an excellent reducing agent which further showed excellent antibacterial effect toward various gram-positive and gram-negative bacteria and also antioxidant and anti-inflammatory activity (Muniyappan et al. 2021). The anti-tumor effect of Au NPs was observed against HCT-116 and MCF-7 cancer cells after being chemically reduced by CUR. The NPs were found to be highly stable and aggregate resistant for up to 6 months and showed excellent anticancer properties than free CUR (Elbialy et al. 2019). Likewise, Ag and ZnO NPs were synthesized by chemical reduction method by adding freshly extracted CUR from the *Curcuma longa* in the ethanolic solution and refluxed at 60 °C. Hence, CUR not only produced the NPs with mean particle size ~20 nm but acted as a capping agent and simultaneously enhanced the antibacterial property of both the NPs toward gram-positive and gram-negative bacteria without showing any further toxic effect toward normal cell lines or Vero cells (El-Kattan et al. 2022).

7.6 Therapeutic Phytoconstituents-Loaded Metal-Based Nanoparticles

Despite the many advantages of established organic and polymer NPs, it has been shown that the inorganic NMs provide the best solution for delivering phyto-drugs because of their special features. As a result, these NPs have been loaded with a variety of drugs for inhibition of cancer proliferation and advancement over traditional therapies like radiation and surgery. These strategies of cancer treatment also control the risk of negative effects on cancer patients. Thus, herein we discussed the significant role of several phytoconstituents for many solid tumors and many other examples brought together in Table 7.1.

7.6.1 Curcumin (CUR)

For a long time, CUR has been extensively used in medicines as it has various medicinal properties. Due to this, CUR has been loading onto numerous metal and metal oxide NPs. A new approach was employed to boost the anticancer effect of CUR-loaded Au NPs by forming the strong corona of CUR and by covering it with isonicotinic acid hydrazide which showed selective toxicity against human lung squamous carcinoma and lung fibroblast cells. Further experiments of this work revealed increased cytotoxic effect was due to the generation of highly reactive oxygen species (ROS), which is the unique property of CUR that caused the morphological changes in the cancer cells and then undergo programmed cell death. Further in this study, the production of ROS by CUR was revealed by treating with free CUR against leukemia and other cells (Umapathi et al. 2020). Ag NPs were also utilized for the loading of CUR to improve its solubility and absorption into the biological cell membrane and anticancer efficacy. For example, the cytotoxic effect of CUR-loaded Ag NPs (~18 nm) on MDA-MB-435 and A549 cells was in the dose-dependent manner (Garg and Garg 2018). CUR causes DNA damage, promotes translocation of many death-related genes from cytoplasm to nucleus, encourages cell arrest at cell cycle stage G2/M, and improves ROS production. Therefore, many studies have revealed that the generation of ROS stimulates apoptosis and autophagy, thereby inhibiting cell growth, even at lower doses. The CUR-loaded SPIONs have enhanced cancer treatment at site-specific level due to which they have gained the attention of the scientific community. In one study, hyper-branched polyglycerol-coated SPIONs functionalized using folic acid (FA) showed a maximum CUR loading efficiency of 88% with improved bioavailability in cervical cancer as shown in Fig. 7.2. The impactful anticancer effect of the FA functionalized nanosystem was exhibited on HeLa cells and mouse L929 fibroblasts in a time-dependent manner up to 72 h than non-targeted one. In addition, this work provided an interesting fact about the high cytotoxicity impact of free CUR compared to CUR-loaded NPs during the early incubation hours (24 and 48) and the recovery of cells after 72 h (Ramezani Farani et al. 2022).

To enhance the bioavailability, solubility, and tissue absorption of CUR, PEG coated MSNPs were fabricated that greatly enhanced the cytotoxicity against human cervical cancer and human liver cancers. Further, the cell cycle arrest study also revealed the CUR-loaded silica NPs against HepG2 cells arrested all phases along with a high degree of cell death, while free drug only arrested early phase S followed by G2/M phase. The in vivo study also showed the significant tumor growth inhibition of approximately 80% and free CUR only was about half attributed to biocompatible MSNPs with a high holding capacity drug and due to the presence of

Table 7.1 List of various metal and metal oxide NPs, their surface modification using biomolecules, plant extract for treatment of many diseases and their significant outcomes via in vitro and in vivo studies

Nanosystems	Surface modifying agent and functionalization	Phytoconstituents	Diseases	Significant outcomes	References
Au quantum cluster (AuQC)	Gliadin	Curcumin	Human breast cancer	Higher toxicity against cancer and negligible toxicity toward normal cell	Mathew et al. (2019)
Au NPs	Gelatin	Curcumin	–	Higher encapsulation efficiency and sustained release	Khodashenas et al. (2019)
Au-based nanogels	Chitosan	Curcumin	Human breast cancer and hepatocellular cancer	Cytotoxic effect was greater than pure drug due to higher cellular uptake	Amanlou et al. (2019)
Magnetic nanogels	Hyaluronic acid	Curcumin	Bone marrow-derived mesenchymal stem cells	No morphological changes in the cells were observed even after direct exposure of formulation, and greater cell adhesion than pure drug, which showed improve the angiogenic property	Daya et al. (2022)
Superparamagnetic iron oxide NPs	–	Curcumin	*Staphylococcus aureus*	Antimicrobial properties increased under photothermal radiation, blue light radiation at 3.12 J cm^{-2}	de Santana et al. (2020)
Ag NPs	Hydroxypropyl-β-cyclodextrin complex	Curcumin	*Staphylococcus aureus*, *Pseudomonas aeruginosa*, and *Candida auris*	Excellent wound healing property, high cell-compatibility,	Gupta et al. (2020)

(continued)

Table 7.1 (continued)

Nanosystems	Surface modifying agent and functionalization	Phytoconstituents	Diseases	Significant outcomes	References
				antimicrobial effect against wound-infectious pathogen or microbes *Staphylococcus aureus, Pseudomonas aeruginosa, and Candida auris*	
ZnO NPs	Casein	Curcumin	Cervical, breast, and osteocarcinoma cancer, human keratinocytes and erythrocytes	Cytotoxic against cancer cells while non-toxic toward normal cells	Somu and Paul (2019)
ZnO NPs	Dialdehyde cellulose cross-linked chitosan hydrogels	Curcumin	*Staphylococcus aureus* and *Trichophyton rubrum*, L929 cells and A431 cells	Synergistic effect was observed on the antimicrobial activity, vitro cytotoxicity, good biocompatibility	George et al. (2019a)
MnO_2 NPs	Bovine serum albumin	Doxorubicin/ curcumin	Primary tumors CT26 cells, tumor model of BALB/c female mouse	Showed excellent dual-chemotherapeutic effect against primary tumors and combination index of 1.0–1.1, enhanced adaptive immune response by formulation than alone MnO_2 NPs	Liu et al. (2022)
Mesoporous silica NPs	Hyaluronic acid	Curcumin	Triple negative breast cancer	Cell cytotoxicity driven by the releasing reactive oxygen species, cell cycle arrest, and change of NF-κB and Bax-based	Ghosh et al. (2021)

				apoptotic pathway and decreased tumor volume	
Mesoporous silica NPs	Fucoidan	Curcumin	Colon cancer	Higher cellular uptake by endocytosis and efficient release of the drug was observed at cancerous cells than pure drug	Zhang et al. (2022)
Calcium phosphate (CP)	Hydroxyapatite	Curcumin	*Staphylococcus aureus* compared to *Pseudomonas aeruginosa*	Curcumin as a natural drug showed higher antibacterial properties than synthetic one	Lee et al. (2021)
Calcium phosphate (CP)	–	Curcumin	Lung cancer	Excellent ability to scavenge free radical and inhibit cell growth	Rao et al. (2020)
Au NPs	Gum arabic	Resveratrol	Human breast, pancreatic, and prostate	Gum arabic enhanced the protein matrix corona, enhanced the drug corona on Res-metal NPs showed superior anti-proliferative effects, due to higher cellular uptake	Thipe et al. (2019)
Au NPs	–	Resveratrol	Breast cancer	Matrix metallopeptidase-9 (MMP-9) effect, protein and messenger RNA expression were upregulated upon TPA stimulation, showed significant inhibited the activation of NF-kB and AP-1 in TPA-stimulated breast cancer cells	Kamal et al. (2018)

(continued)

Table 7.1 (continued)

Nanosystems	Surface modifying agent and functionalization	Phytoconstituents	Diseases	Significant outcomes	References
Au NPs	Polyvinylpyrrolidone	Resveratrol	Breast cancer	Showed anti-tumor effect via cell cycle arrest and reduced level of G0/G1 phase but increased level of S phase than alone drug	Lee et al. (2019)
Au NPs	–	Resveratrol	Liver cancer cells and tumor xenografts mice	Greater effect on inhibiting cell proliferation and promoted apoptosis by downregulating the pro-caspase-9, pro-caspase-3 PI3K and Akt, simultaneously upregulated the caspase-8 and bax and xenografts showed reduction of vascular endothelial growth factor (VEGF) expression in tumor tissue	Zhang et al. (2019)
Ag NPs	Leaf extract of *Aesculus hippocastanum*	Resveratrol	*Staphylococcus aureus, S. epidermidis, Listeria monocytogenes, Corynebacterium renale, Micrococcus luteus, Bacillus subtilis, B. cereus, Enterococcus faecalis, Pseudomonas aeruginosa, P. fluorescens,*	The drug-loaded NPs showed the minimum inhibitory concentration toward many pathogenic bacteria, and fungus as compared to alone antibiotics, horse chestnut leaves capped showed antioxidant properties	Küp et al. (2020)

			Escherichia coli, Enterobacter aerogenes, Klebsiella pneumonia, Proteus mirabilis, Candida albicans, C. tropicalis, C. krusei		
Ag NPs	*Rumex hymenosepalus* extracts	Resveratrol	*S. aureus, Listeria monocytogenes, Escherichia coli, Salmonella serovar typhi, Pseudomonas aeruginosa and Candida albicans and* THP-1, human leukemia monocytic cell line	Antibacterial and antifungal properties and negligible toxicity toward the THP-1 cells	Rodríguez-León et al. (2018)
Mesoporous silica NPs	Phosphonate, i.e., (3-trihydroxysilyl-propyl methyl-phosphonate amine) (added into toluene)	Resveratrol and docetaxel	Prostate cancer	Improved the in vitro cell cytotoxicity of drug in the hypoxia-inductive resistance	Chaudhary et al. (2019)
Mesoporous silica NPs	Cetrimonium bromide	Resveratrol	Human melanoma	Significantly decreased the cell proliferation than blank formulation and pure drug	Marinheiro et al. (2021)
Mesoporous silica NPs	Hyaluronic acid	Anti-miRNA21 and resveratrol	Gastric cancer cells	Targeted drug delivery (CD44 receptors) and superior anticancer efficacy, and apoptosis and necrosis and decrease in the tumor volume	Hu et al. (2019a)
Mesoporous silica NPs	Mercaptopropyl trimethoxysilane, polyethylene imine, and folic acid	Resveratrol	Gastric cancer cell lines and AGS and HGC-27	Both carrier and drug-loaded formulation were non-toxic toward normal	Lin et al. (2021)

(continued)

Table 7.1 (continued)

Nanosystems	Surface modifying agent and functionalization	Phytoconstituents	Diseases	Significant outcomes	References
			tumor-bearing mouse models	tissues of animals Significantly inhibited the growth, invasion, and movement or migration of cancer cells Greater reduction of tumor size	
Manganese-doped silica hollow mesoporous NPs	Lactoferrin	Resveratrol	Ischemic stroke	Increase the effect of superoxide dismutase and glutathione peroxidase which reduced the brain ROS and malondialdehyde level, intensification the expression of anti-inflammatory factors-10 and anti-apoptotic factors, bcl-2 in brain tissue, and decrease brain tissue-based pro-inflammatory factors-TNF-α, IL-1β, IL-6 and pro-apoptotic factors such as BAX, Cleaved caspase-3 expression	Zou et al. (2022)
Hollow manganese oxide NPs	Chitosan	Resveratrol	Spinal cord injury	Oxidative stress significantly reduced the level of ROS,	Li et al. (2022)

				malondialdehyde, superoxide dismutase, and increasing glutathione peroxidase level Immunofluorescence like (iNOS, IL-1β, and Cl caspase-3) and western blot (iNOS, cox-2, IL-1β, IL-10, Cl caspase-3, bax, and bcl-2)	
Calcium phosphate (CP)	–	Resveratrol	Skin inflammation and tumorigenesis mouse model	Significant reduction of TPA-inductive skin edema, ODC effect, and thymidine integration	Arora et al. (2021)
Superparamagnetic iron oxide NPs	Chitosan	Resveratrol	Alzheimer's disease	Improved cognitive and memory function, decreased levels of pro-inflammatory markers, and downregulated expression of NF-kB and P38	Abbas et al. (2021)
ZnO NPs	Chitosan	Resveratrol	Gestational diabetes mellitus	A-glucosidase and an amylase inhibitory effects were observed and significantly decreased the blood glucose contents in diabetic rats than normal one. Finally, decreased the amount of inflammation factors such as IL-6 and MCP-1 and ER stress	Du et al. (2020)

(continued)

Table 7.1 (continued)

Nanosystems	Surface modifying agent and functionalization	Phytoconstituents	Diseases	Significant outcomes	References
				(GRP78, p-IRE1a, p-eIF2a, and p-PERK)	
Au NPs	*p*-sulfonatocalix[4]arene	Quercetin	Colon cancer	A reduction of tumor volume in mice exhibited the alteration in the expression of more than 25 genes related to cell death	Yilmaz et al. (2019)
Au NPs	–	Quercetin	*Escherichia coli, Staphylococcus aureus, Salmonella typhimurium* and *Bacillus cereus, Aspergillus fumigatus and* fibroblasts cells	Excellent antimicrobial, antifungal, and anti-proliferative activity	Milanezi et al. (2019)
Au NPs		Quercetin	*Human lung cancer cells and normal cells*	Cytotoxic against cancer cells caused irregular morphology and enhanced oxidative stress and Au caused bioimaging	Lakshmi and Kim (2019)
Ag NPs	–	Quercetin	Colon cancer	15–90 mg/mL caused cancer cell inhibition, antioxidant effect via induction in the oxidative/nitrosative stress and lipid peroxidation	Martirosyan et al. (2016)
Ag NPs	–	Quercetin	*Staphylococcus aureus*	Quercetin acted as reducing and capping agent for NPs excellent,	Chahardoli et al. (2021)

				antibacterial, antioxidant, and anti-inflammatory properties with great hemocompatible 37.25 and 500 μg mL^{-1}	
Ag NPs	–	Quercetin and siRNA	Quercetin resistant bacteria, i.e., *Pseudomonas aeruginosa, Bacillus subtilis*	Damaged the cell wall and inhibited bacterial propagation in vitro studies as compared to ampicillin and kanamycin while showed bacteremia into the SKH1 nude mice	Sun et al. (2016)
Superparamagnetic polymeric micelles	Methoxy-poly(ethylene glycol)-*b*-oligo(ε-caprolactone), mPEG750-*b*-OCL-Bz micelles	Quercetin	Hepatocellular carcinoma cells	Cell arrest was observed at G0/G1 phase of cell cycle and potentially inhibited the cancer cell growth	Srisa-nga et al. (2019)
Iron oxide NPs		Quercetin	*Aspergillus oryzae, breast, liver, and lung cancer cells and breast cancer in induced female white albino rats*	Inhibited cell growth in a concentration manner. In vivo studies revealed the inhibition in tumor growth with enhanced radioactivity-based tumor inhibition *N*-methyl-*N*-nitrosourea	Askar et al. (2022)
Calcium phosphate	–	Quercetin	*Mouse neuroblastoma cells*	Excellent cytotoxicity, no morphological changes in the cells, potent antioxidant	Patra et al. (2017)
ZnO NPs	Phenylboronic acid conjugate	Quercetin	Breast cancer	Caused and enhanced oxidative stress and mitochondrial damage.	Sadhukhan et al. (2019)

(continued)

Table 7.1 (continued)

Nanosystems	Surface modifying agent and functionalization	Phytoconstituents	Diseases	Significant outcomes	References
				Reduced tumor growth in EAC tumor-bearing mice	
ZnO NPs	Chitosan-cellulose hydrogel	Quercetin	*Staphylococcus aureus and Trichophyton rubrum* and normal murine fibroblast cells, human skin carcinoma cell	Superior antibacterial activity over drug and biocompatible toward normal cells and cytotoxic against cancerous cells	George et al. (2019b)
Au NPs	Aptamer	Morin	SGC-7901 cells	Targeted cellular uptake and effective in vitro cytotoxic effect and decreased volume in xenograft mouse models	Ding et al. (2020)
Ag NPs	–	Morin	*Saccharomyces cerevisiae*	Significantly downregulated the hypothalamic–pituitary–gonadal axis	Arisha et al. (2019)

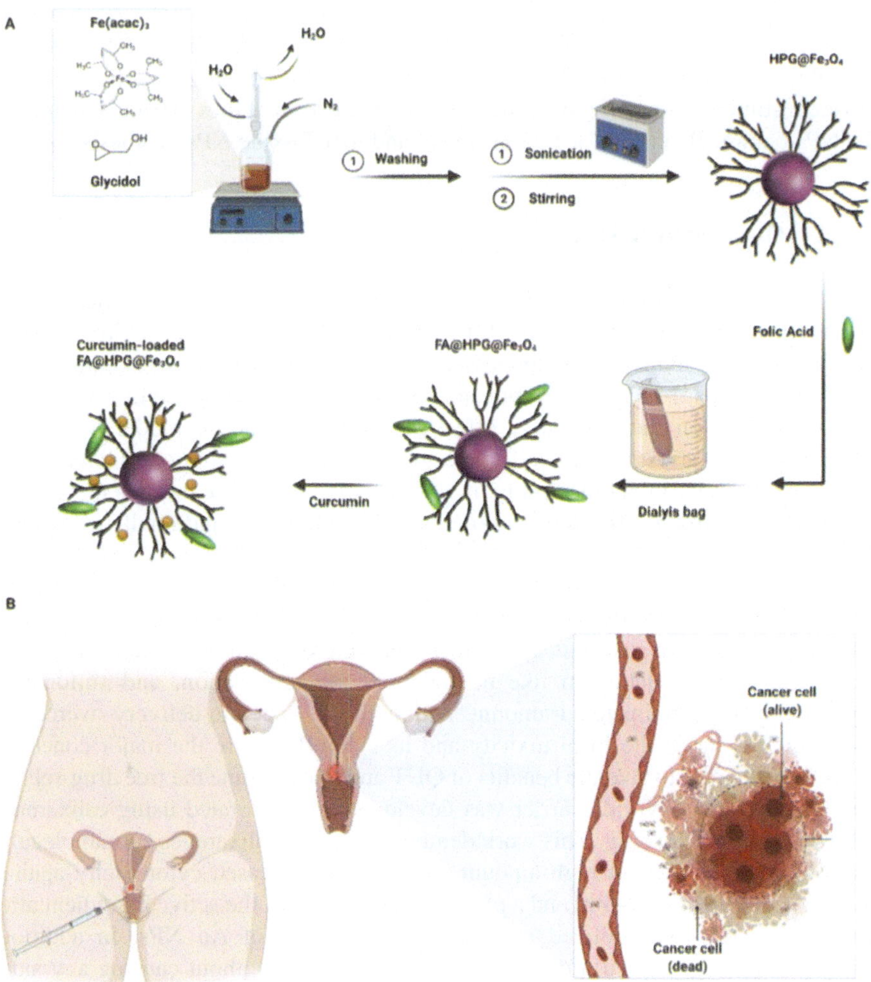

Fig. 7.2 (**a**) Synthesis of hyper-branched polyglycerol iron oxide (Hb PEG@Fe_3O_4) and folic acid FA-Hb PEG@Fe_3O_4 NPs. (**b**) Schematic representation of potential impact of folate-interaction for the treatment of cervical cancer. (This figure has been adapted from Ramezani Farani et al. 2022)

porous structure that further significantly increased the therapeutic index (Elbialy et al. 2020).

Similarly, multifunctional PEG and β-cyclodextrin functionalized ZnO NPs were synthesized to enhance the CUR loading efficiency due to its hydrophobic nature. From biological studies, it was proven that functionalized ZnO nanocarriers exhibit cell imaging property and release the drug for a longer time. The sustained release of CUR potentially decreases human metastatic breast cancer cell growth and suggested the intriguing quality of inorganic NPs as a promising agent for the delivery of other bioactive natural drugs (Anjum et al. 2021). As a result, phytoconstituents are employed to develop a variety of metal and metal oxide NPs and also loaded onto the NPs, so far, showed various biological effects. Thus, phytoconstituents not only show various biomedical activities during loading on the NPs but are also used for various metal and metal oxide NPs synthesis.

7.6.2 Quercetin (QUT)

QUT is a natural bioactive flavonoid present in many plants with antioxidant, anti-diabetic, and anticancer properties (Ebrahimpour et al. 2020). The in vitro research on QUT showed induction of apoptosis in cancer cells by downregulation of heat shock protein (hsp)-70 and prevent cancerous cell division. Further in vivo studies also shed light on the anticancer effect of QUT in a mouse model in a dose-dependent way, arresting the cell cycle in S phase, and helping to increase p53 protein expression. In another study, it was observed that QUT inhibits the activity of Bcl-xL/BcL proteins through binding its BH3 binding site and finally causes cell death.

Thus, QUT has been actively studied due to its ability in prevention of solid tumors. However, its low water solubility is the key problem that causes poor absorption rate in the gastrointestinal tract following oral administration. Moreover, quick enzymatic metabolism like methylation, glucuronidation, and sulfonation limits the actual administered amount. Although the systemic delivery overcomes this issue, yet again the local toxicity and its side effects are the major concerns. Therefore, in order to take the benefits of QUT and to overcome the free drug-related problems, Au nanohybrids carrier was developed and decorated using calixarenes followed by QUT loading. This work demonstrated that calixarenes supramolecular scaffold held the drug in high amount and showed increased cytotoxicity against SW-620 cells up to 51.9-fold and a pH-triggered release of the active ingredient after this combination was utilized to decorate the surface of Au NPs. In addition, QUT-loaded NPs showed a decrement in size of tumor without causing any side-effect to the normal tissues of tumor-bearing mice. QUT-loaded ZnO NPs demonstrated enhanced anticancer effect in metastatic ovarian cancer especially by generating oxidative stress, depolarizing the membrane potential of mitochondria and then late apoptosis via activating intrinsic signaling pathway in PA-1 cancer cells as revealed by double staining assay (Ramalingam et al. 2022). In another report, QUT was used as the model drug and the metal organic framework was made utilizing SPIONs as the metal, encapsulated dual MOF composite, and FA as the targeted ligand. The interaction of the whole nanosystem with the overexpressed folate receptor on human triple negative breast cancer cells, i.e., MDA-MB-231 resulted in excellent anticancer activity. Therefore, this investigation showed that QUT-loaded inorganic formulations were more readily taken up by cells than pure drug, and that intracellular drug administration induced substantial ROS production, nuclear fragmentation, and cell death coupled with MRI (Pandit et al. 2022).

QUT was further loaded on to the FA functionalized MSNPs to improve the bioavailability and anticancer effect in two different cancer cells lines of breast. This study evaluated the cell uptake by tagging RITC fluorescence dye to FA-MSNPs in both the cancer cells and the results showed significantly greater fluorescence in cytoplasm, especially in folate receptor containing breast cancer cells, for example, MDA-MB 231 than MCF-7 and which also exhibited nucleus uptake confirmed by DAPI staining. Further, cell arrest was observed at G1 phase in the cell cycle and subsequent decrease in number of MDA-MB 231 cells occurred at significantly lower doses of QUT that increased the Bax and caspase 3 expression associated with lowering the p-Akt expression. Ex vivo study on chick embryo model also revealed the tumor suppressive effect of QUT-loaded NPs at very low dose through relative *alu* gene expression as compared to control (Sarkar et al. 2016). In an attempt to deliver high amount of QUT, CaP NPs were used because of the highest membrane permeability, cytocompatibility, and biodegradability. This study interestingly demonstrated the fluorescence property of QUT-loaded NPs as compared to free NPs and drug without causing any toxicity and with antioxidant nature on cell viability of the hydrogen peroxide exposed to neural, N2A cells (Patra et al. 2017).

7.6.3 Resveratrol (RSV)

RSV is a phytoestrogen and has numerous medicinal properties such as antioxidant, anti-inflammatory, and anti-tumor effect against various solid tumors and metastatic cancers. Moreover, RSV has exceptional drug anti-resistant properties and causes cancer cell death (Karthikeyan et al. 2015). Au NPs loaded with RSV have been found as an excellent nanocarrier in Swiss albino mice suffering from non-small cell lung carcinoma cells and anticancer effect via cell halting in the G0/G1 phase. In another study, RSV-loaded Au NPs remarkably stopped the tumor growth of human liver cancer cells, HepG2 with enhanced apoptosis in xenograft studies than the free drug via by downregulation of the pro-caspase-9, and 3, PI3K and Akt while up-regulating caspase-8 and bax, especially through delivering the drug into mitochondrial membrane. In another study, MSNPs were synthesized to encapsulate RSV to enhance the anti-tumor activity in drug resistant prostate cancer (Chaudhary et al. 2019). Hyaluronic acid functionalized, anti miR21 and RSV-loaded MSNPs were developed to investigate the anticancer activity against gastric cancer in another study and the results showed proliferation of gastric cancer, HGC-27 and AGS cells was intensely reduced by the developed RSV-loaded MSNPs as compared to pure drug. Further in vivo study of a tumor-mouse model showed excessive decrease in size and alleviation of inflammatory cell infiltration indicated the promising RSV delivery in gastric cancers (Lin et al. 2021). For treatment of ovarian cancer, ZnO NPs were used to load RSV that caused great cell growth inhibition as demonstrated by in vitro study. The in vivo study showed antioxidant property via ROS generation and depolarization of mitochondrial membrane consistent with dysfunction of mitochondria as confirmed by DCFH assay and JC-staining methods followed by apoptosis (Khatun et al. 2016). In one report, hollow Au NPs were coated with RSV for enhanced photothermal activity and cytotoxicity against melanoma cancer. After being exposed to an 808-nm NIR laser, the NPs cause cell cycle arrest, stop cell division, and cause apoptosis in melanoma A375 cells (Wang et al. 2017).

7.6.4 (−)-Epigallocatechin-3-gallate (EGCG)

ECGC is a well-known phytochemical drug obtained from green tea and shows the anti-proliferative effect against many cancers (Almatroodi et al. 2020; Luo et al. 2017). In several reports, the in vitro and in vivo results showed the inhibition of cell growth of cancers by increasing the metabolic stress, via stopping the phosphatidylinositol-3 kinase, protein kinase B, target of rapamycin signaling cascade. These cascades also act as a hallmark of prostate cancer through hyper-activation and their inhibition is directly associated with G1 to S phase's cell cycle arrest. Numerous evidences have shown that ECGC exhibits cytotoxic properties against human non-small lung cancer. Moreover, ECGC is also associated with anti-ROS generation, antiviral, anti-inflammatory, and anti-diabetic effects (Luo et al. 2017; Shi et al. 2019; Hu et al. 2019b). Therefore, by considering the great advantages, ECGC has been used as a therapeutic agent for treatment of cancer (Yin et al. 2022). However, its oxidation in solution form, instability, and ultimately fast metabolism in vivo restricts its clinical application. Therefore, in many studies, various NPs have been used to load ECGC. For example, ECGC-loaded Au NPs were synthesized to evaluate the hemocytoxicity and in vivo anti-tumor effect in Ehrlich cell tumor-bearing mice. The results of this study showed high encapsulation and loading efficiency, aqueous stability, and non-hemotoxicity of ECGC in the Au NPs. Further, this study also demonstrated that as compared to pristine EGCG, EGCG-loaded Au NPs exhibited improved in vivo anti-tumor effect, as confirmed by reduced volume of tumor and mice body weight (Safwat et al. 2020). In another finding, silica oxide NPs were coated with chitosan and functionalized with As1411 Aptamer for targeted delivery of EGCG against ovary cancer. The Aptamer functionalization is responsible for recognition to nucleolin and subsequent cytotoxicity with late apoptosis (93%) and specially lowering in S phase and G2/M and growth in G0/G1 and arrested in G1 phase. This study also suggested the expression stages of ERK2 and hTERT have gone through downregulation in SKOV-3 cells that effectively increased its anticancer potency (Alizadeh et al. 2020). Among several proteins, bovine serum albumin (BSA) modified inorganic NPs have been found to be stable and biocompatible to many cells. The BSA stabilized SPIONs were used as nanocarrier for ECGC delivery to evaluate their cytotoxicity against lung cancer. The results of this study showed that ECGC-loaded BSA-SPIONs induced apoptosis via Nrf2/Keap1 signaling, generating ROS levels, alternating mitochondrial membrane potential, and finally apoptosis as revealed by DCFH and JC1 staining assay. The expression of Bcl-2, Bax, Bak, Bim, and Puma, members of the Bcl-2 family of pro-apoptotic proteins, again renders great efficacy against A549 cells (Velavan et al. 2018). In order to eliminate the fast oxidation of ECGC in aqueous solution, co-crystals of ZnO with EGCG NPs were prepared and found as an anticancer agent against prostate adenocarcinoma cells. The EGCG-ZnO NPs were found to be non-toxic against normal macrophage cells at the same concentration. However, some NPs internalized by endocytosis pathway, and via lysosomal disruption Zn^{+} ions and EGCG are released into the cell plasma (Samutprasert et al. 2018).

7.6.5 Morin Hydrate (MH)

MH is a yellow colored pigment and is highly soluble in methanol but poorly soluble in aqueous solution. It shows various pharmacological benefits like antioxidant, anti-inflammatory, and anticancer against many solid tumors, anti-diabetic, and also overcome MDR-related issues in cancer therapy. Several researches demonstrated the mechanism of action of MH through modulating the nuclear factor kappa-light-chain-enhancer, activating many protein signaling pathway such as mitogen-activated protein kinase, Janus kinases/signal transducer, transcription protein-based activators such as JAKs/STATs, ER and mitochondrial-mediated apoptosis (Rajput et al. 2021; Ghosh et al. 2022). However, due to hydrophobicity and non-specificity, an adequate amount of the drug does not reach to the site of action and leads to low therapeutic efficacy, poor pharmacokinetics, unwanted biodistribution and toxicity. Therefore, metal and metal oxide NPs have been developed to enhance solubility, efficient loading, and release of MH so as to achieve the potential therapeutic activity. Au NPs have been known as an excellent drug delivery nanocarrier. In one study, MH-loaded targeted Aptamer functionalized Au NPs were synthesized and were encapsulated in pH sensitive liposomes to evaluate the anticancer properties against gastric, breast, lung, cervical, and fibroblast cells. The in vitro results showed enhanced anti-tumor activity of MH-loaded NPs against SGC-7901, change in morphology but non-toxic to normal cells as confirmed by MTT assay and fluorescence microscopy. The dosage-dependent triggering of various caspase-3 and 9 caused enhanced mitochondria-mediated apoptosis due to successful delivery of MH-loaded NPs than free drug. Further, histological analysis showed significant changes in the tissue of tumor and verified the in vivo toxicity via blood biochemical assay, which showed no significant changes, and sign of inflammatory action was observed in all groups (Ding et al. 2020). In another finding, MH was loaded on to amine functionalized MSNPs for evaluating its capability against skin related diseases and suggested the potential application in cosmetic industry (Arriagada et al. 2016). With the progress of various nanosystems, multifunctional SPIONs have been gaining special attention for delivery of natural bioactive ingredients. Mannose coated SPIONs were synthesized to load MH which showed great stability, magnetic hyperthermia, and negative contrast imaging at the same time. This study suggested the possible application of the developed theranostic NPs for efficient delivery of MH for the cancer treatment via causing oxidative stress (Montiel Schneider et al. 2021).

7.6.6 Camptothecin (CPT)

CPT is an alkaloid that occurs naturally and is obtained from the plant *Camptotheca acuminata*. It was discovered in the 1960s as part of a screening process for potential anticancer drugs. It has been demonstrated via several clinical investigations that CPT works by blocking the synthesis of topoisomerase I and prevents cancer cell growth (Wen et al. 2017). Although CPT has some limitations such as limited water solubility, low biostability, the opening of its ring structure in physiological fluids renders the anti-carcinogenic property leading to low therapeutic efficacy. Despite a number of drawbacks, CPT has been utilized in cancer treatment by combining with other medications, such as irinotecan, and demonstrating the anticancerous properties in preclinical tests. Henceforth, for the better delivery of CPT, Au NPs were used as nanocarrier that showed good colloidal stability in physiological media, and different pH condition, even at room temperature (Xing et al. 2010). In another study, CPT was delivered along with the paclitaxel drug into the PVA/k-carrageenan/Au/pegylated-PU composite for lung cancer treatment via active drug delivery. The study showed the synergistic effect of drug-loaded nanocarrier on cytotoxicity of A549 cancer cells and reduction of volume of tumor in A549 tumor-bearing mice as investigated by in vitro and in vivo experiments, respectively (Irani and Nodeh 2022).

A versatile self-assembled FA functionalized polydopamine-Au NPs-zein stabilized were synthesized to enhance the loading efficiency of needle-shaped hydroxycamptothecin (HCPT) nanocrystals. The nanocarrier was found to be stable at physiological media and selectively taken by endocytosis pathway, localized into the cytoplasm via lysosomal escape, and caused superior carcinogenic effect against KB cells (overexpresses folate receptors) and HeLa cells (lower level of overexpressed folate receptors). Subsequent delivery of HCPT further successfully reached nuclear sites and induced cytotoxicity, early (40–64%) and late (~48% to 52%) apoptosis. After intravenous injection of HCPT-nanoformulation, the suppression in tumor was observed in the KB bearing mice due to efficient release of drug at tumor site. Currently, simultaneous therapeutic agent and gene delivery to tumor site have become a promising way of treatment. For example, PEGylated rod-shaped silica NPs were synthesized for targeted delivery of CPT and surviving sh-RNA against colon cancer. The synthesized nanosystem showed approximately 32% encapsulation efficiency, incorporated gene with C/P ratio of 6 of PEG coated MSNPs to DNA, and synergistic effect of iSur-pDNA on the inhibition of cancer growth, greater apoptosis via AS1411 DNA targeting-uptake approach. Further in vivo results showed higher cellular toxicity with a significant decrease in tumor volume as compared to nucleolin-negative colon cancer cells. The developed nanosystem using combined drug and gene further improved the survival rate and pharmacokinetics of mice without causing any side-effect at tissue level (Babaei et al. 2020).

7.7 Conclusion

The multifunctional metal and metal oxide NMs have been investigated for the delivery of the phytoconstituents to treat various diseases. However, drug loading and stability are major challenges in their applications. To over these, numerous surface modification approaches have been explored to improve the drug loading and prevent the drug degradation in physiological fluids. When combined with a targeting ligand, the surface-modified inorganic NPs facilitate the drug delivery to the site of action safely and facilitate the passage of molecules to the nucleus. The major benefit of the use of metal or metal oxide nanoparticles in the delivery of phytoconstituents is that these nanoparticles can also be used for the diagnosis simultaneously.

Acknowledgment H.S.Y. and P.J. acknowledge the Department of Medical Biomaterials Engineering, College of Biomedical Science, Kangwon National University, Chuncheon, Republic of Korea.

References

Abbas H, Refai H, El Sayed N, Rashed LA, Mousa MR, Zewail M (2021) Superparamagnetic iron oxide loaded chitosan coated bilosomes for magnetic nose to brain targeting of resveratrol. Int J Pharm 610:121244. https://doi.org/10.1016/j.ijpharm.2021.12124

Agarwal H, Shanmugam V (2020) A review on anti-inflammatory activity of green synthesized zinc oxide nanoparticle: mechanism-based approach. Bioorg Chem 94:103423. https://doi.org/10.1016/j.bioorg.2019.103423

Alizadeh L, Alizadeh E, Zarebkohan A, Ahmadi E, Rahmati-Yamchi M, Salehi R (2020) AS1411 aptamer-functionalized chitosan-silica nanoparticles for targeted delivery of epigallocatechin gallate to the SKOV-3 ovarian cancer cell lines. J Nanoparticle Res 22:5. https://doi.org/10.1007/s11051-019-4735-7

Almatroodi SA, Almatroudi A, Khan AA, Alhumaydhi FA, Alsahli MA, Rahmani AH (2020) Potential therapeutic targets of epigallocatechin gallate (EGCG), the most abundant catechin in green tea, and its role in the therapy of various types of cancer. Molecules 25(14):3146. https://doi.org/10.3390/molecules25143146

Amanlou N, Parsa M, Rostamizadeh K, Sadighian S, Moghaddam F (2019) Enhanced cytotoxic activity of curcumin on cancer cell lines by incorporating into gold/chitosan nanogels. Mater Chem Phys 226:151–157. https://doi.org/10.1016/j.matchemphys.2018.12.089

Anjum S, Jacob G, Gupta B (2019) Investigation of the herbal synthesis of silver nanoparticles using Cinnamon zeylanicum extract. Emergent Mater 2:113–122. https://doi.org/10.1007/s42247-019-00023-x

Anjum S, Hashim M, Malik SA, Khan M, Lorenzo JM, Abbasi BH, Hano C (2021) Recent advances in zinc oxide nanoparticles (ZnO NPs) for cancer diagnosis, target drug delivery, and treatment. Cancers (Basel) 13(18):4570. https://doi.org/10.3390/cancers13184570

Arisha AH, Ahmed MM, Kamel MA, Attia YA, Hussein MMA (2019) Morin ameliorates the testicular apoptosis, oxidative stress, and impact on blood–testis barrier induced by photo-extracellularly synthesized silver nanoparticles. Environ Sci Pollut Res 26:28749–28762. https://doi.org/10.1007/s11356-019-06066-1

Arora R, Samim M, Parkash C (2021) Evaluation of anti-inflammatory and anti-cancer activity of calcium phosphate encapsulated resveratrol in mouse skin cancer model. Biomed Pharmacol J 14:113–122. https://doi.org/10.13005/bpj/2105

Arriagada F, Correa O, Günther G, Nonell S, Mura F, Olea-Azar C, Morales J (2016) Morin flavonoid adsorbed on mesoporous silica, a novel antioxidant nanomaterial. PLoS One 11: e0164507. https://doi.org/10.1371/journal.pone.0164507

Aschner JL, Aschner M (2005) Nutritional aspects of manganese homeostasis. Mol Aspects Med 26:353–362. https://doi.org/10.1016/j.mam.2005.07.003

Askar MA, El-Nashar HAS, Al-Azzawi MA, Rahman SSA, Elshawi OE (2022) Synergistic effect of quercetin magnetite nanoparticles and targeted radiotherapy in treatment of breast cancer. Breast Cancer 16:11782234221086728. https://doi.org/10.1177/11782234221086728

Azizi M, Sedaghat S, Tahvildari K, Derakhshi P, Ghaemi A (2017) Synthesis of silver nanoparticles using Peganum harmala extract as a green route. Green Chem Lett Rev 10:420–427. https://doi.org/10.1080/17518253.2017.1395081

Babaei M, Abnous K, Taghdisi SM, Taghavi S, Saljooghi AS, Ramezani M, Alibolandi M (2020) Targeted rod-shaped mesoporous silica nanoparticles for the co-delivery of camptothecin and survivin shRNA in to colon adenocarcinoma in vitro and in vivo. Eur J Pharm Biopharm 156: 84–96. https://doi.org/10.1016/j.ejpb.2020.08.026

Blakaj A, Bonomi M, Gamez ME, Blakaj DM (2019) Oral mucositis in head and neck cancer: evidence-based management and review of clinical trial data. Oral Oncol 95:29–34. https://doi.org/10.1016/j.oraloncology.2019.05.013

Boomi P, Poorani GP, Selvam S, Palanisamy S, Jegatheeswaran S, Anand K, Balakumar C, Premkumar K, Prabu HG (2020) Green biosynthesis of gold nanoparticles using Croton sparsiflorus leaves extract and evaluation of UV protection, antibacterial and anticancer applications. Appl Organomet Chem 34:e5574. https://doi.org/10.1002/aoc.5574

Bora S, Pooja D, Kulhari H (2022) Applications of surface modified carbon nanotubes in drug delivery. In: Surface modified carbon nanotubes volume 2: industrial applications, ACS symposium series, vol 1425. American Chemical Society, Washington, DC, pp 2–19. https://doi.org/10.1021/bk-2022-1425.ch002

Cha BG, Kim J (2019) Functional mesoporous silica nanoparticles for bio-imaging applications. WIREs Nanomed Nanobiotechnol 11:e1515. https://doi.org/10.1002/wnan.1515

Chahardoli A, Hajmomeni P, Ghowsi M, Qalekhani F, Shokoohinia Y, Fattahi A (2021) Optimization of quercetin-assisted silver nanoparticles synthesis and evaluation of their hemocompatibility, antioxidant, anti-inflammatory, and antibacterial effects. Glob Challenges 5:2100075. https://doi.org/10.1002/gch2.202100075

Chaudhary Z, Subramaniam S, Khan GM, Abeer MM, Qu Z, Janjua T, Kumeria T, Batra J, Popat A (2019) Encapsulation and controlled release of resveratrol within functionalized mesoporous silica nanoparticles for prostate cancer therapy. Front Bioeng Biotechnol 7:1–9. https://doi.org/10.3389/fbioe.2019.00225

Chauhan S, Patel K, Jain P, Jangid AK, Patel S, Medicherla K, Limbad K, Mehta C, Kulhari H (2022) Matrix metalloproteinase enzyme responsive delivery of 5-fluorouracil using collagen-I peptide functionalized dendrimer-gold nanocarrier. Drug Dev Ind Pharm 48:333–342. https://doi.org/10.1080/03639045.2022.2113404

Chavali MS, Nikolova MP (2019) Metal oxide nanoparticles and their applications in nanotechnology. SN Appl Sci 1:607. https://doi.org/10.1007/s42452-019-0592-3

Chen P, Bornhorst J, Aschner M (2018) Manganese metabolism in humans. Front Biosci (Landmark Ed) 23:1655–1679. https://doi.org/10.2741/4665

Craciun I, Gunkel-Grabole G, Belluati A, Palivan CG, Meier W (2017) Expanding the potential of MRI contrast agents through multifunctional polymeric nanocarriers. Nanomedicine 12:811–817. https://doi.org/10.2217/nnm-2016-0413

Dai X, Guo Q, Zhao Y, Zhang P, Zhang T, Zhang X, Li C (2016) Functional silver nanoparticle as a benign antimicrobial agent that eradicates antibiotic-resistant bacteria and promotes wound healing. ACS Appl Mater Interfaces 8:25798–25807. https://doi.org/10.1021/acsami.6b09267

Daya R, Xu C, Nguyen N-YT, Liu HH (2022) Angiogenic hyaluronic acid hydrogels with curcumin-coated magnetic nanoparticles for tissue repair. ACS Appl Mater Interfaces 14: 11051–11067. https://doi.org/10.1021/acsami.1c19889

de Santana WMOS, Caetano BL, de Annunzio SR, Pulcinelli SH, Ménager C, Fontana CR, Santilli CV (2020) Conjugation of superparamagnetic iron oxide nanoparticles and curcumin photosensitizer to assist in photodynamic therapy. Colloids Surf B Biointerfaces 196:111297. https://doi.org/10.1016/j.colsurfb.2020.111297

Deshmukh AR, Dikshit PK, Kim BS (2022) Green in situ immobilization of gold and silver nanoparticles on bacterial nanocellulose film using Punica granatum peels extract and their application as reusable catalysts. Int J Biol Macromol 205:169–177. https://doi.org/10.1016/j.ijbiomac.2022.02.064

Ding X, Yin C, Zhang W, Sun Y, Zhang Z, Yang E, Sun D, Wang W (2020) Designing aptamer-gold nanoparticle-loaded pH-sensitive liposomes encapsulate morin for treating cancer. Nanoscale Res Lett 15:68. https://doi.org/10.1186/s11671-020-03297-x

Du S, Lv Y, Li N, Huang X, Liu X, Li H, Wang C, Jia Y-F (2020) Biological investigations on therapeutic effect of chitosan encapsulated nano resveratrol against gestational diabetes mellitus rats induced by streptozotocin. Drug Deliv 27:953–963. https://doi.org/10.1080/10717544.2020.1775722

Durán N, Durán M, de Jesus MB, Seabra AB, Fávaro WJ, Nakazato G (2016) Silver nanoparticles: a new view on mechanistic aspects on antimicrobial activity. Nanomed Nanotechnol Biol Med 12:789–799. https://doi.org/10.1016/j.nano.2015.11.016

Ebrahimpour S, Esmaeili A, Dehghanian F, Beheshti S (2020) Effects of quercetin-conjugated with superparamagnetic iron oxide nanoparticles on learning and memory improvement through targeting microRNAs/NF-κB pathway. Sci Rep 10:15070. https://doi.org/10.1038/s41598-020-71678-4

Elbialy NS, Abdelfatah EA, Khalil WA (2019) Antitumor activity of curcumin-green synthesized gold nanoparticles: in vitro study. Bionanoscience 9:813–820. https://doi.org/10.1007/s12668-019-00660-w

Elbialy NS, Aboushoushah SF, Sofi BF, Noorwali A (2020) Multifunctional curcumin-loaded mesoporous silica nanoparticles for cancer chemoprevention and therapy. Micropor Mesopor Mater 291:109540. https://doi.org/10.1016/j.micromeso.2019.06.002

El-Kattan N, Emam AN, Mansour AS, Ibrahim MA, Abd El-Razik AB, Allam KAM, Riad NY, Ibrahim SA (2022) Curcumin assisted green synthesis of silver and zinc oxide nanostructures and their antibacterial activity against some clinical pathogenic multi-drug resistant bacteria. RSC Adv 12:18022–18038. https://doi.org/10.1039/D2RA00231K

Franci G, Falanga A, Galdiero S, Palomba L, Rai M, Morelli G, Galdiero M (2015) Silver nanoparticles as potential antibacterial agents. Molecules 20:8856–8874. https://doi.org/10.3390/molecules20058856

Gale EM, Caravan P (2018) Gadolinium-free contrast agents for magnetic resonance imaging of the central nervous system. ACS Chem Neurosci 9:395–397. https://doi.org/10.1021/acschemneuro.8b00044

Gao Z, Shao S, Gao W, Tang D, Tang D, Zou S, Kim MJ, Xia X (2021) Morphology-invariant metallic nanoparticles with tunable plasmonic properties. ACS Nano 15:2428–2438. https://doi.org/10.1021/acsnano.0c06123

Garg S, Garg A (2018) Encapsulation of Curcumin in silver nanoparticle for enhancement of anticancer drug delivery. Int J Pharm Sci Res 9:1160. https://doi.org/10.13040/IJPSR.0975-8232.9(3).1160-66

George D, Maheswari PU, Sheriffa Begum KMM, Arthanareeswaran G (2019a) Biomass-derived dialdehyde cellulose cross-linked chitosan-based nanocomposite hydrogel with phytosynthesized zinc oxide nanoparticles for enhanced curcumin delivery and bioactivity. J Agric Food Chem 67:10880–10890. https://doi.org/10.1021/acs.jafc.9b01933

George D, Maheswari PU, Begum KMMS (2019b) Synergic formulation of onion peel quercetin loaded chitosan-cellulose hydrogel with green zinc oxide nanoparticles towards controlled release, biocompatibility, antimicrobial and anticancer activity. Int J Biol Macromol 132:784–794. https://doi.org/10.1016/j.ijbiomac.2019.04.008

Ghosh S, Dutta S, Sarkar A, Kundu M, Sil PC (2021) Targeted delivery of curcumin in breast cancer cells via hyaluronic acid modified mesoporous silica nanoparticle to enhance anticancer efficiency. Colloids Surf B Biointerfaces 197:111404. https://doi.org/10.1016/j.colsurfb.2020.111404

Ghosh P, Bag S, Parveen S, Subramani E, Chaudhury K, Dasgupta S (2022) Nanoencapsulation as a promising platform for the delivery of the Morin-Cu(II) complex: antibacterial and anticancer potential. ACS Omega 7:7931–7944. https://doi.org/10.1021/acsomega.1c06956

Gonbad RA, Afzan A, Karimi E, Sinniah UR, Swamy MK (2015) Phytoconstituents and antioxidant properties among commercial tea (Camellia sinensis L.) clones of Iran. Electron J Biotechnol 18:433–438. https://doi.org/10.1016/j.ejbt.2015.08.007

Gupta A, Briffa SM, Swingler S, Gibson H, Kannappan V, Adamus G, Kowalczuk M, Martin C, Radecka I (2020) Synthesis of silver nanoparticles using curcumin-cyclodextrins loaded into bacterial cellulose-based hydrogels for wound dressing applications. Biomacromolecules 21: 1802–1811. https://doi.org/10.1021/acs.biomac.9b01724

Hemmati S, Rashtiani A, Zangeneh MM, Mohammadi P, Zangeneh A, Veisi H (2019) Green synthesis and characterization of silver nanoparticles using Fritillaria flower extract and their antibacterial activity against some human pathogens. Polyhedron 158:8–14. https://doi.org/10.1016/j.poly.2018.10.049

Hong H, Shi J, Yang Y, Zhang Y, Engle JW, Nickles RJ, Wang X, Cai W (2011) Cancer targeted optical imaging with fluorescent zinc oxide nanowires. Nano Lett 11:3744–3750. https://doi.org/10.1021/nl201782m

Hu Y, Wang Z, Qiu Y, Liu Y, Ding M, Zhang Y (2019a) Anti-miRNA21 and resveratrol-loaded polysaccharide-based mesoporous silica nanoparticle for synergistic activity in gastric carcinoma. J Drug Target 27:1135–1143. https://doi.org/10.1080/1061186X.2019.1610766

Hu D, Wang G, Yu J, Zhang L, Huang Y, Wang D, Zhou H (2019b) Epigallocatechin-3-gallate modulates long non-coding RNA and mRNA expression profiles in lung cancer cells. Mol Med Rep 19:1509–1520. https://doi.org/10.1021/acs.analchem.9b03107

Institute of Medicine (2001) Dietary reference intakes for vitamin A, vitamin K, arsenic, boron, chromium, copper, iodine, iron, manganese, molybdenum, nickel, silicon, vanadium, and zinc. The National Academies Press, Washington, DC. https://doi.org/10.17226/10026

Irani M, Nodeh SM (2022) PVA/κ-carrageenan/Au/camptothecin/pegylated-polyurethane/paclitaxel nanofibers against lung cancer treatment. RSC Adv 12:16310–16318. https://doi.org/10.1039/D2RA02150A

Jafari S, Derakhshankhah H, Alaei L, Fattahi A, Varnamkhasti BS, Saboury AA (2019) Mesoporous silica nanoparticles for therapeutic/diagnostic applications. Biomed Pharmacother 109:1100–1111. https://doi.org/10.1021/ja1084095

Jain P, Bhagat S, Tunki L, Jangid AK, Singh S, Pooja D, Kulhari H (2020) Serotonin–stearic acid bioconjugate-coated completely biodegradable Mn3O4 nanocuboids for hepatocellular carcinoma targeting. ACS Appl Mater Interfaces 12:10170–10182. https://doi.org/10.1021/acsami.0c00331

Jamkhande PG, Ghule NW, Bamer AH, Kalaskar MG (2019) Metal nanoparticles synthesis: an overview on methods of preparation, advantages and disadvantages, and applications. J Drug Deliv Sci Technol 53:101174. https://doi.org/10.1016/J.JDDST.2019.101174

Kamal R, Chadha VD, Dhawan DK (2018) Physiological uptake and retention of radiolabeled resveratrol loaded gold nanoparticles (99mTc-Res-AuNP) in colon cancer tissue. Nanomed Nanotechnol Biol Med 14:1059–1071. https://doi.org/10.1016/j.nano.2018.01.008

Karthikeyan S, Hoti SL, Prasad NR (2015) Resveratrol loaded gelatin nanoparticles synergistically inhibits cell cycle progression and constitutive NF-kappaB activation, and induces apoptosis in non-small cell lung cancer cells. Biomed Pharmacother 70:274–282. https://doi.org/10.1016/j.biopha.2015.02.006

Khan AUR, Huang K, Jinzhong Z, Zhu T, Morsi Y, Aldalbahi A, El-Newehy M, Yan X, Mo X (2021) Exploration of the antibacterial and wound healing potential of a PLGA/silk fibroin based electrospun membrane loaded with zinc oxide nanoparticles. J Mater Chem B 9:1452–1465. https://doi.org/10.1039/D0TB02822C

Khatun M, Choudhury S, Liu B, Lemmens P, Pal SK, Mazumder S (2016) Resveratrol–ZnO nanohybrid enhanced anti-cancerous effect in ovarian cancer cells through ROS. RSC Adv 6: 105607–105617. https://doi.org/10.1039/C6RA16664D

Khodashenas B, Ardjmand M, Baei MS, Rad AS, Khiyavi AA (2019) Gelatin–gold nanoparticles as an ideal candidate for curcumin drug delivery: experimental and DFT studies. J Inorg Organomet Polym Mater 29:2186–2196. https://doi.org/10.1007/s10904-019-01178-0

Kim T, Momin E, Choi J et al (2011) Mesoporous silica-coated hollow manganese oxide nanoparticles as positive T1 contrast agents for labeling and MRI tracking of adipose-derived mesenchymal stem cells. J Am Chem Soc 133(9):2955–2961. https://doi.org/10.1021/ja1084095

Kim HS, Son YJ, Mao W, Leong KW, Yoo HS (2018) Atom transfer radical polymerization of multishelled cationic corona for the systemic delivery of siRNA. Nano Lett 18:314–325. https://doi.org/10.1021/acs.nanolett.7b04183

Kotrange H, Najda A, Bains A, Gruszecki R, Chawla P, Tosif MM (2021) Metal and metal oxide nanoparticle as a novel antibiotic carrier for the direct delivery of antibiotics. Int J Mol Sci. 22(17):9596. https://doi.org/10.3390/ijms22179596

Kulhari H, Pooja D, Kota R, Reddy TS, Tabor RF, Shukla R, Adams DJ, Sistla R, Bansal V (2016) Cyclic RGDfK peptide functionalized polymeric nanocarriers for targeting gemcitabine to ovarian cancer cells. Mol Pharm 13:1491–1500. https://doi.org/10.1021/acs.molpharmaceut.5b00935

Kumar A, Chaudhary RK, Singh R, Singh SP, Wang YS, Hoe YZ, Pan CT, Ling SY, Wei DQ, Kaushik AC, Dai X (2020) Nanotheranostic applications for detection and targeting neurodegenerative diseases. Front Neurosci 14:1–11. https://doi.org/10.3389/fnins.2020.00305

Küp FÖ, Çoşkunçay S, Duman F (2020) Biosynthesis of silver nanoparticles using leaf extract of Aesculus hippocastanum (horse chestnut): evaluation of their antibacterial, antioxidant and drug release system activities. Mater Sci Eng C 107:110207. https://doi.org/10.1016/j.msec.2019.110207

Lakshmi BA, Kim S (2019) Quercetin mediated gold nanoclusters explored as a dual functional nanomaterial in anticancer and bio-imaging disciplines. Colloids Surf B Biointerfaces 178:230–237. https://doi.org/10.1016/j.colsurfb.2019.02.054

Lazaro-Carrillo A, Filice M, Guillén MJ et al (2020) Tailor-made PEG coated iron oxide nanoparticles as contrast agents for long lasting magnetic resonance molecular imaging of solid cancers. Mater Sci Eng C 107:110262. https://doi.org/10.1016/j.msec.2019.110262

Le VAT, Trinh TX, Chien PN, Giang NN, Zhang X-R, Nam S-Y, Heo C-Y (2022) Evaluation of the performance of a ZnO-nanoparticle-coated hydrocolloid patch in wound healing. Polymers 14(5):919. https://doi.org/10.3390/polym14050919

Lee DG, Go EB, Lee M, Pak PJ, Kim J-S, Chung N (2019) Gold nanoparticles conjugated with resveratrol induce cell cycle arrest in MCF-7 cell lines. Appl Biol Chem 62:33. https://doi.org/10.1186/s13765-019-0440-6

Lee W-H, Rohanizadeh R, Loo C-Y (2021) In situ functionalizing calcium phosphate biomaterials with curcumin for the prevention of bacterial biofilm infections. Colloids Surf B Biointerfaces 206:111938. https://doi.org/10.1016/j.colsurfb.2021.111938

Lee MS, Park Y, Kim SR, Jung YM, Yoo HS (2022) Dual irradiation-triggered anticancer therapeutics composed of polydopamine-coated gold nanoparticles. Biomater Adv 136: 212779. https://doi.org/10.1016/j.bioadv.2022.212779

Levingstone TJ, Herbaj S, Dunne NJ (2019) Calcium phosphate nanoparticles for therapeutic applications in bone regeneration. Nanomaterials 9(11):1570. https://doi.org/10.3390/nano9111570

Li Y, Zou Z, An J, Wu Q, Tong L, Mei X, Tian H, Wu C (2022) Chitosan-modified hollow manganese dioxide nanoparticles loaded with resveratrol for the treatment of spinal cord injury. Drug Deliv 29:2498–2512. https://doi.org/10.1080/10717544.2022.2104957

Lin S-R, Chang C-H, Hsu C-F, Tsai M-J, Cheng H, Leong MK, Sung P-J, Chen J-C, Weng C-F (2020) Natural compounds as potential adjuvants to cancer therapy: preclinical evidence. Br J Pharmacol 177:1409–1423. https://doi.org/10.1111/bph.14816

Lin M, Yao W, Xiao Y, Dong Z, Huang W, Zhang F, Zhou X, Liang M (2021) Resveratrol-modified mesoporous silica nanoparticle for tumor-targeted therapy of gastric cancer. Bioengineered 12:6343–6353. https://doi.org/10.1080/21655979.2021.1971507

Liu X, Pan Y, Yang J, Gao Y, Huang T, Luan X, Wang Y, Song Y (2020) Gold nanoparticles doped metal-organic frameworks as near-infrared light-enhanced cascade nanozyme against hypoxic tumors. Nano Res 13:653–660. https://doi.org/10.1007/s12274-020-2668-1

Liu J, Li L, Zhang B, Xu ZP (2022) MnO2-shelled doxorubicin/curcumin nanoformulation for enhanced colorectal cancer chemo-immunotherapy. J Colloid Interface Sci 617:315–325. https://doi.org/10.1016/j.jcis.2022.02.132

Luo K-W, Chen W, Lung W-Y, Wei X-Y, Cheng B-H, Cai Z-M, Huang W-R (2017) EGCG inhibited bladder cancer SW780 cell proliferation and migration both in vitro and in vivo via down-regulation of NF-κB and MMP-9. J Nutr Biochem 41:56–64. https://doi.org/10.1016/j.jnutbio.2016.12.004

Marinheiro D, Ferreira BJML, Oskoei P, Oliveira H, Daniel-da-Silva AL (2021) Encapsulation and enhanced release of resveratrol from mesoporous silica nanoparticles for melanoma therapy. Materials 14(6):1382. https://doi.org/10.3390/ma14061382

Martirosyan A, Grintzalis K, Polet M, Laloux L, Schneider Y-J (2016) Tuning the inflammatory response to silver nanoparticles via quercetin in Caco-2 (co-)cultures as model of the human intestinal mucosa. Toxicol Lett 253:36–45. https://doi.org/10.1016/j.toxlet.2016.04.018

Mathew MS, Vinod K, Jayaram PS, Jayasree RS, Joseph K (2019) Improved bioavailability of curcumin in gliadin-protected gold quantum cluster for targeted delivery. ACS Omega 4:14169–14178. https://doi.org/10.1021/acsomega.9b00917

Mathur P, Jha S, Ramteke S, Jain NK (2018) Pharmaceutical aspects of silver nanoparticles. Artif Cells Nanomed Biotechnol 46:115–126. https://doi.org/10.1080/21691401.2017.1414825

Milanezi FG, Meireles LM, de Christo Scherer MM, de Oliveira JP, da Silva AR, de Araujo ML, Endringer DC, Fronza M, Guimarães MCC, Scherer R (2019) Antioxidant, antimicrobial and cytotoxic activities of gold nanoparticles capped with quercetin. Saudi Pharm J 27:968–974. https://doi.org/10.1016/j.jsps.2019.07.005

Mobeen H, Safdar M, Fatima A, Afzal S, Zaman H, Mehdi Z (2022) Emerging applications of nanotechnology in context to immunology: a comprehensive review. Front Bioeng Biotechnol 10:1–20. https://doi.org/10.3389/fbioe.2022.1024871

Montiel Schneider MG, Favatela MF, Muñoz Medina GA, Fernandez van Raap MB, Lassalle VL (2021) Multitherapy magnetic theranostic: synthesis, characterization and in vitro evaluation of their performance. Colloids Surf B Biointerfaces 198:111460. https://doi.org/10.1016/j.colsurfb.2020.111460

Mumtaz S, Ali S, Mumtaz S, Mughal TA, Tahir HM, Shakir HA (2022) Chitosan conjugated silver nanoparticles: the versatile antibacterial agents. Polym Bull 80:4719–4736. https://doi.org/10.1007/s00289-022-04321-z

Muniyappan N, Pandeeswaran M, Amalraj A (2021) Green synthesis of gold nanoparticles using Curcuma pseudomontana isolated curcumin: its characterization, antimicrobial, antioxidant and anti-inflammatory activities. Environ Chem Ecotoxicol 3:117–124. https://doi.org/10.1016/j.enceco.2021.01.002

Narla S, Lim HW (2020) Sunscreen: FDA regulation, and environmental and health impact. Photochem Photobiol Sci 19:66–70. https://doi.org/10.1039/c9pp00366e

Naser H, Alghoul MA, Hossain MK, Asim N, Abdullah MF, Ali MS, Alzubi FG, Amin N (2019) The role of laser ablation technique parameters in synthesis of nanoparticles from different target types. J Nanopart Res 21:249. https://doi.org/10.1007/s11051-019-4690-3

Neha R, Jaiswal A, Bellare J, Sahu NK (2017) Synthesis of surface grafted mesoporous magnetic nanoparticles for cancer therapy. J Nanosci Nanotechnol 17:5181–5188. https://doi.org/10.1166/jnn.2017.13853

Nielsen FH (2012) Manganese, molybdenum, boron, chromium, and other trace elements. In: Present knowledge in nutrition. Wiley, New York, pp 586–607. https://doi.org/10.1002/9781119946045.ch38

Oronsky B, Goyal S, Kim MM, Cabrales P, Lybeck M, Caroen S, Oronsky N, Burbano E, Carter C, Oronsky A (2018) A review of clinical radioprotection and chemoprotection for oral mucositis. Transl Oncol 11:771–778. https://doi.org/10.1016/j.tranon.2018.03.014

Palaniyandi T, Viswanathan S, Prabhakaran P et al (2023) Green synthesis of gold nanoparticles using Halymenia pseudofloresii extracts and their antioxidant, antimicrobial, and anti-cancer activities. Biomass Convers Biorefiner. https://doi.org/10.1007/s13399-023-03873-y

Pandit P, Bhagat S, Rananaware P, Mohanta Z, Kumar M, Tiwari V, Singh S, Brahmkhatri VP (2022) Iron oxide nanoparticle encapsulated; folic acid tethered dual metal organic framework-based nanocomposite for MRI and selective targeting of folate receptor expressing breast cancer cells. Micropor Mesopor Mater 340:112008. https://doi.org/10.1016/j.micromeso.2022.112008

Paramasivam G, Kayambu N, Rabel AM, Sundramoorthy AK, Sundaramurthy A (2017) Anisotropic noble metal nanoparticles: synthesis, surface functionalization and applications in biosensing, bioimaging, drug delivery and theranostics. Acta Biomater 49:45–65. https://doi.org/10.1016/j.actbio.2016.11.066

Patra M, Mukherjee R, Banik M, Dutta D, Begum NA, Basu T (2017) Calcium phosphate-quercetin nanocomposite (CPQN): a multi-functional nanoparticle having pH indicating, highly fluorescent and anti-oxidant properties. Colloids Surf B Biointerfaces 154:63–73. https://doi.org/10.1016/j.colsurfb.2017.03.018

Prasher P, Sharma M, Mudila H et al (2020) Emerging trends in clinical implications of bio-conjugated silver nanoparticles in drug delivery. Colloids Interface Sci Commun 35:100244. https://doi.org/10.1016/j.colcom.2020.100244

Rahman M, Alam K, Hafeez A, Ilyas R, Beg S (2021) Chapter 7—Metallic nanoparticles in drug delivery and cancer treatment. In: Beg S, Rahman M, Choudhry H, Souto EB, Ahmad FJ (eds) Nanoformulation strategies for cancer treatment (Micro and nano technologies). Elsevier, Amsterdam, pp 107–119

Rajesh KM, Ajitha B, Reddy YAK, Suneetha Y, Reddy PS (2018) Assisted green synthesis of copper nanoparticles using Syzygium aromaticum bud extract: physical, optical and antimicrobial properties. Optik (Stuttg) 154:593–600. https://doi.org/10.1080/21691401.2019.1620757

Rajput SA, Wang X, Yan H-C (2021) Morin hydrate: a comprehensive review on novel natural dietary bioactive compound with versatile biological and pharmacological potential. Biomed Pharmacother 138:111511. https://doi.org/10.1016/j.biopha.2021.111511

Ramalingam V, Muthukumar Sathya P, Srivalli T, Mohan H (2022) Synthesis of quercetin functionalized wurtzite type zinc oxide nanoparticles and their potential to regulate intrinsic apoptosis signaling pathway in human metastatic ovarian cancer. Life Sci 309:121022. https://doi.org/10.1016/j.lfs.2022.121022

Ramezani Farani M, Azarian M, Heydari Sheikh Hossein H et al (2022) Folic acid-adorned curcumin-loaded iron oxide nanoparticles for cervical cancer. ACS Appl Bio Mater 5:1305–1318. https://doi.org/10.1021/acsabm.1c01311

Rao C, Li M, Sun X, Li M, Lian X, Wang H, Jia L, Niu B, Li W (2020) Preparation and characterization of phosphate-stabilized amorphous calcium carbonate nanoparticles and their application in curcumin delivery. Mater Chem Phys 255:123552. https://doi.org/10.1016/j.matchemphys.2020.123552

Ray S, Li Z, Hsu CH, Hwang LP, Lin YC, Chou PT, Lin YY (2018) Dendrimer- and copolymer-based nanoparticles for magnetic resonance cancer theranostics. Theranostics 8:6322–6349. https://doi.org/10.7150/thno.27828

Rodríguez-León E, Íñiguez-Palomares RA, Navarro RE et al (2018) Silver nanoparticles synthesized with Rumex hymenosepalus extracts: effective broad-spectrum microbicidal agents and cytotoxicity study. Artif Cells Nanomed Biotechnol 46:1194–1206. https://doi.org/10.1080/21691401.2017.1366332

Sadhukhan P, Kundu M, Chatterjee S, Ghosh N, Manna P, Das J, Sil PC (2019) Targeted delivery of quercetin via pH-responsive zinc oxide nanoparticles for breast cancer therapy. Mater Sci Eng C 100:129–140. https://doi.org/10.1016/j.msec.2019.02.096

Safwat MA, Kandil BA, Elblbesy MA, Soliman GM, Eleraky NE (2020) Epigallocatechin-3-gallate-loaded gold nanoparticles: preparation and evaluation of anticancer efficacy in Ehrlich tumor-bearing mice. Pharmaceuticals 13(9):254. https://doi.org/10.3390/ph13090254

Samutprasert P, Chiablaem K, Teeraseranee C, Phaiyarin P, Pukfukdee P, Pienpinijtham P, Svasti J, Palaga T, Lirdprapamongkol K, Wanichwecharungruang S (2018) Epigallocatechin gallate-zinc oxide co-crystalline nanoparticles as an anticancer drug that is non-toxic to normal cells. RSC Adv 8:7369–7376. https://doi.org/10.1039/C7RA10997K

Santhoshkumar J, Rajeshkumar S, Venkat Kumar S (2017) Phyto-assisted synthesis, characterization and applications of gold nanoparticles—a review. Biochem Biophys Rep 11:46–57. https://doi.org/10.1016/j.bbrep.2017.06.004

Sarkar A, Ghosh S, Chowdhury S, Pandey B, Sil PC (2016) Targeted delivery of quercetin loaded mesoporous silica nanoparticles to the breast cancer cells. Biochim Biophys Acta 1860:2065–2075. https://doi.org/10.1016/j.bbagen.2016.07.001

Shi X, Wang S, Jasbi P, Turner C, Hrovat J, Wei Y, Liu J, Gu H (2019) Database-assisted globally optimized targeted mass spectrometry (dGOT-MS): broad and reliable metabolomics analysis with enhanced identification. Anal Chem 91:13737–13745

Shin JU, Gwon J, Lee S-Y, Yoo HS (2018) Silver-incorporated nanocellulose fibers for antibacterial hydrogels. ACS Omega 3:16150–16157. https://doi.org/10.1021/acsomega.8b02180

Shiraishi Y, Tanaka H, Sakamoto H, Ichikawa S, Hirai T (2017) Photoreductive synthesis of monodispersed Au nanoparticles with citric acid as reductant and surface stabilizing reagent. RSC Adv 7:6187–6192. https://doi.org/10.1039/C6RA27771C

Siddiqi KS, Ur Rahman A, Tajuddin HA (2018) Properties of zinc oxide nanoparticles and their activity against microbes. Nanoscale Res Lett 13:141. https://doi.org/10.1186/s11671-018-2532-3

Singh P, Srivastava S, Singh SK (2019) Nanosilica: recent progress in synthesis, functionalization, biocompatibility, and biomedical applications. ACS Biomater Sci Eng 5:4882–4898. https://doi.org/10.1021/acsbiomaterials.9b00464

Singla A, Marwaha A, Marwaha S, Chavali M (2021) Chapter 2—Green nanomaterials: pioneering approach toward hyperthermia treatment. In: Kharisov B, Kharissova OH (eds) GS of N and C. Elsevier, Amsterdam, pp 487–502. https://doi.org/10.1016/B978-0-12-822446-5.00021-6

Somu P, Paul S (2019) A biomolecule-assisted one-pot synthesis of zinc oxide nanoparticles and its bioconjugate with curcumin for potential multifaceted therapeutic applications. New J Chem 43: 11934–11948. https://doi.org/10.1039/C9NJ02501D

Srisa-nga K, Mankhetkorn S, Okonogi S, Khonkarn R (2019) Delivery of superparamagnetic polymeric micelles loaded with quercetin to hepatocellular carcinoma cells. J Pharm Sci 108: 996–1006. https://doi.org/10.1016/j.xphs.2018.08.008

Sun D, Zhang W, Li N, Zhao Z, Mou Z, Yang E, Wang W (2016) Silver nanoparticles-quercetin conjugation to siRNA against drug-resistant Bacillus subtilis for effective gene silencing: in vitro and in vivo. Mater Sci Eng C 63:522–534. https://doi.org/10.1016/j.msec.2016.03.024

Tallury P, Payton K, Santra S (2008) Silica-based multimodal/multifunctional nanoparticles for bioimaging and biosensing applications. Nanomedicine 3:579–592. https://doi.org/10.2217/17435889.3.4.579

Thipe VC, Panjtan Amiri K, Bloebaum P, Raphael Karikachery A, Khoobchandani M, Katti KK, Jurisson SS, Katti KV (2019) Development of resveratrol-conjugated gold nanoparticles: interrelationship of increased resveratrol corona on anti-tumor efficacy against breast, pancreatic and prostate cancers. Int J Nanomed 14:4413–4428. https://doi.org/10.2147/IJN.S204443

ul Islam S, Sun G (2022) Biological chemicals as sustainable materials to synthesize metal and metal oxide nanoparticles for textile surface functionalization. ACS Sustain Chem Eng 10: 10084–10104. https://doi.org/10.1021/acssuschemeng.2c03033

Umapathi A, Navya PN, Madhyastha H, Singh M, Madhyastha R, Maruyama M, Daima HK (2020) Curcumin and isonicotinic acid hydrazide functionalized gold nanoparticles for selective anti-cancer action. Colloids Surf A Physicochem Eng Asp 607:125484. https://doi.org/10.1016/j.colsurfa.2020.125484

Vaid P, Raizada P, Saini AK, Saini RV (2020) Biogenic silver, gold and copper nanoparticles—a sustainable green chemistry approach for cancer therapy. Sustain Chem Pharm 16:100247. https://doi.org/10.1016/j.scp.2020.100247

Velavan B, Divya T, Sureshkumar A, Sudhandiran G (2018) Nano-chemotherapeutic efficacy of (−) -epigallocatechin 3-gallate mediating apoptosis in A549 cells: involvement of reactive oxygen species mediated Nrf2/Keap1signaling. Biochem Biophys Res Commun 503:1723–1731. https://doi.org/10.1016/j.bbrc.2018.07.105

Virk K, Sharma K, Kapil S, Kumar V, Sharma V, Pandey S, Kumar V (2022) Synthesis of gum acacia-silver nanoparticles based hydrogel composites and their comparative anti-bacterial activity. J Polym Res 29:118. https://doi.org/10.1007/s10965-022-02978-8

Wang W, Tang Q, Yu T et al (2017) Surfactant-free preparation of Au@resveratrol hollow nanoparticles with photothermal performance and antioxidant activity. ACS Appl Mater Interfaces 9:3376–3387. https://doi.org/10.1021/acsami.6b13911

Wang W, Lu K, Yu C, Huang Q, Du Y-Z (2019) Nano-drug delivery systems in wound treatment and skin regeneration. J Nanobiotechnol 17:82. https://doi.org/10.1186/s12951-019-0514-y

Wen Y, Wang Y, Liu X, Zhang W, Xiong X, Han Z, Liang X (2017) Camptothecin-based nanodrug delivery systems. Cancer Biol Med 14:363–370. https://doi.org/10.20892/j.issn.2095-3941.2017.0099

Wesström J (2020) Safety of intravenous iron isomaltoside for iron deficiency and iron deficiency anemia in pregnancy. Arch Gynecol Obstet 301:1127–1131. https://doi.org/10.1007/s00404-020-05509-2

Willenbacher E, Khan SZ, Mujica SC, Trapani D, Hussain S, Wolf D, Willenbacher W, Spizzo G, Seeber A (2019) Curcumin: new insights into an ancient ingredient against cancer. Int J Mol Sci. 20(8):1808. https://doi.org/10.3390/ijms20081808

Xing Z, Liu Z, Zu Y, Fu Y, Zhao C, Zhao X, Meng R, Tan S (2010) Synthesis of camptothecin-loaded gold nanomaterials. Appl Surf Sci 256:3917–3920. https://doi.org/10.1016/j.apsusc.2010.01.050

Xu X, Li Z, Zhao X, Keen L, Kong X (2016) Calcium phosphate nanoparticles-based systems for siRNA delivery. Regen Biomater 3:187–195. https://doi.org/10.1093/rb/rbw010

Yang Y, Song W, Chen Z, Li Q, Liu L (2019) Ameliorative effect of synthesized silver nanoparticles by green route method from Zingiber zerumbet on mycoplasmal pneumonia in experimental mice. Artif Cells Nanomed Biotechnol 47:2146–2154

Yeh Y-C, Huang T-H, Yang S-C, Chen C-C, Fang J-Y (2020) Nano-based drug delivery or targeting to eradicate bacteria for infection mitigation: a review of recent advances. Front Chem 8:1–22. https://doi.org/10.3389/fchem.2020.00286

Yilmaz M, Karanastasis AA, Chatziathanasiadou MV et al (2019) Inclusion of quercetin in gold nanoparticles decorated with supramolecular hosts amplifies its tumor targeting properties. ACS Appl Bio Mater 2:2715–2725. https://doi.org/10.1021/acsabm.8b00748

Yin X, Huang S, Xu S, Chang L, Zhao X, Chen Z, Mei X, Gao X (2022) Preparation of pro-angiogenic, antibacterial and EGCG-modified ZnO quantum dots for treating bacterial infected wound of diabetic rats. Biomater Adv 133:112638. https://doi.org/10.3892/mmr.2019.9816

Yuko T, Santoso MR, Heidary S, Sano H, Tachibana A, Matsuura Y, Harnish P, Yang PC (2021) Dual contrast manganese-enhanced MRI and gadolinium delayed-enhanced MRI detect heterogenous myocardial viability in ischemic cardiomyopathy. JACC Cardiovasc Imaging 14: 1474–1476. https://doi.org/10.1016/j.jcmg.2020.12.025

Zhang D, Zhang J, Zeng J, Li Z, Zuo H, Huang C, Zhao X (2019) Nano-gold loaded with resveratrol enhance the anti-hepatoma effect of resveratrol in vitro and in vivo. J Biomed Nanotechnol 15: 288–300. https://doi.org/10.1166/jbn.2019.2682

Zhang X, Zhu Y, Fan L, Ling J, Yang L-Y, Wang N, Ouyang X (2022) Delivery of curcumin by fucoidan-coated mesoporous silica nanoparticles: fabrication, characterization, and in vitro release performance. Int J Biol Macromol 211:368–379. https://doi.org/10.1016/j.ijbiomac.2022.05.086

Zhao L, Liang L, Guo M, Li M, Yu X, Wang Y, Wang Y (2021) Hepatocellular carcinoma targeting and pharmacodynamics of paclitaxel nanoliposomes modified by glycyrrhetinic acid and ferric tetroxide. Curr Top Med Chem 21:1268–1284. https://doi.org/10.2174/1568026621666210621090005

Zou Z, Wen S, Li Y, An J, Wu Q, Tong L, Mei X, Tian H, Wu C (2022) Novel lactoferrin-functionalized manganese-doped silica hollow mesoporous nanoparticles loaded with resveratrol for the treatment of ischemic stroke. Mater Today Adv 15:100262. https://doi.org/10.1016/j.mtadv.2022.100262

Protein Nanocarriers for the Delivery of Phytoconstituents

8

Raghu Solanki and Sunita Patel

8.1 Introduction

Phytoconstituents are secondary metabolites derived from plants and classified on the basis of their chemical structures such as alkaloids, phenols, flavonoids, indoles, and many more. Phytoconstituents are considered an excellent source of bioactive compounds with wide pharmacological effects, including potential anticancer, antimicrobial, anti-inflammatory, neuro-protective, hepato-protective, and cardio-protective properties (Kaur et al. 2018). Despite having a lot of pharmaceutical properties, the use of phytoconstituents in pre-clinical/clinical applications is limited (Singh et al. 2019). Limited bioavailability and solubility are the major drawbacks of these phytoconstituents. Additionally, rapid metabolism, instability, hydrophobicity, and poor absorption are also associated with insufficient therapeutic efficacy of phytoconstituents (Singh et al. 2019). Therefore, the development of novel pharmaceutical formulations is an urgent need of the hour (Solanki et al. 2022). Nowadays, nano-drug delivery systems have gained extensive interest in numerous biomedical applications including diagnosis, detection, imaging, and therapy (Aqil et al. 2013). Nanotechnology is a very prominent field, which can be used to enhance the bioavailability, target specificity, and therapeutic efficacy of the phytoconstituents (Wang et al. 2014; Odeh et al. 2014; More et al. 2021).

Nanotechnology is an emerging field and promisingly it bridges with other fields of sciences including biological, biomedicinal, physical sciences, pharmaceutical, etc. Nanocarriers are synthesized from inorganic materials (metal nanoparticles, carbon nanotubes, quantum dots, etc.) or organic materials (liposomes, exosomes, dendrimers, polymers, protein-based nanoparticles, etc.) (Edis et al. 2021; Rout et al. 2018). Proteins are biodegradable, biocompatible, and nontoxic and therefore they

R. Solanki · S. Patel (✉)
School of Life Sciences, Central University of Gujarat, Gandhinagar, Gujarat, India
e-mail: sunitap@cug.ac.in

© The Author(s), under exclusive license to Springer Nature Singapore Pte Ltd. 2024

D. Pooja, H. Kulhari (eds.), *Nanotechnology Based Delivery of Phytoconstituents and Cosmeceuticals*, https://doi.org/10.1007/978-981-99-5314-1_8

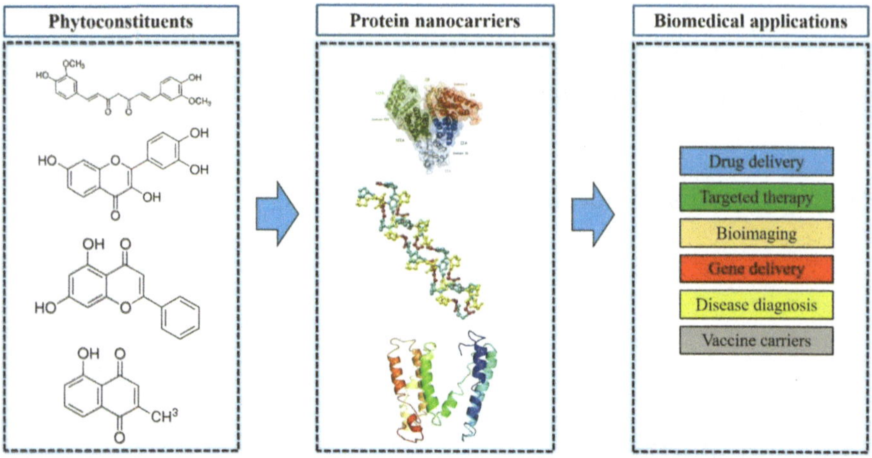

Fig. 8.1 Schematic representation of biomedical applications of phytoconstituents-loaded protein nanocarriers

are considered safe as drug delivery vehicles (Jain et al. 2018). Proteins nanocarriers are widely used as a vehicle for site-specific delivery of phytochemicals, genes, vaccines, and other biomolecules in the body. In this book chapter, we have discussed various protein-based nanocarriers including albumin, silk, collagen, gelatin, zein, and so forth for the delivery of phytoconstituents (Fig. 8.1).

Further, various approaches such as the desolvation method, emulsification method, and other methods used for the synthesis of protein nanocarriers have been discussed in depth. The administration route for any nanosystem is very important thus we have also explored the administration routes of protein nanocarriers. Finally, the biomedical applications of phytoconstituent-loaded protein nanocarriers in cancer, cardiovascular diseases, neurological disorders, and microbial diseases have been discussed.

8.2 Protein Nanocarriers for the Delivery of Phytoconstituents

Proteins are biopolymers having distinctive properties of biodegradability, biocompatibility, and non-immunogenicity, which makes them excellent nanocarriers for pharmacological applications (Hong et al. 2020). Nowadays, protein-based nanocarriers are gaining more scientific interest as nanomedicines. It triggers a plethora of signaling cascades to exhibit various functions. Further, surface modification on protein nanocarriers by attaching surfactants or ligands can increase the therapeutic effectiveness of phytoconstituents (Ling et al. 2014; Kenchegowda et al. 2021). Protein nanocarriers are nanoscale particles composed of proteins that can be used for a variety of biomedical applications, including drug delivery, imaging, and diagnostic purposes. Some of the advantages of protein nanocarriers are:

- Biocompatibility
- Biodegradability
- Non-immunogenicity
- Non-toxicity
- Ease of availability and preparation
- Low cost
- Number of functional groups can be modified for targeted delivery
- Controlled release
- High-drug loading efficiency
- Stability

Overall, protein nanocarriers have the potential to improve drug delivery and imaging efficacy and open up new diagnostic and therapeutic applications.

8.3 Types of Protein Nanocarriers Used for Delivery of Phytoconstituents

8.3.1 Albumins

Albumins are widely used nanocarriers for the delivery of phytoconstituents (Elzoghby et al. 2012). There are various sources of albumins such as bovine serum albumin (BSA), human serum albumin (HSA), egg white albumin (Ovalbumin), equine serum albumin (ESA), and rat serum albumin (RSA) (Spada et al. 2021; Majorek et al. 2012).

Majorek et al. characterized three different albumins, bovine, horse, and rabbit serum albumins, and concluded that the albumins are made up of three structurally related helical domains (I, II, and III) organized in a heart-shaped molecule in all three forms. Two subdomains (A and B) can be separated out of each domain (Fig. 8.2). Each of the three albumins contains a helical content of 74% (BSA),

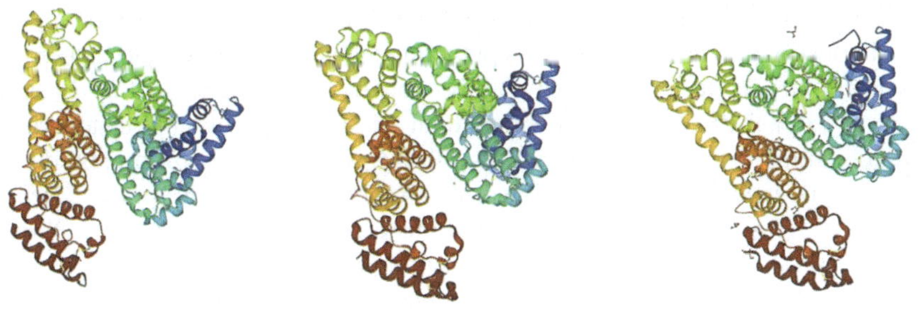

Fig. 8.2 The molecular structure illustrates the domain and secondary structural element assignment for serum albumins: HSA (PDB ID-1AO6), BSA (PDB ID-3V03), and ESA (PDB ID-4F5T)

75% (ESA), and 72% (RSA); 17 conserved disulfide bridges; and a free thiol group linked to Cys34. All three albumin structures exhibit a close resemblance to the HSA structure. Properties such as high binding efficiency and non-immunogenicity, high water solubility and selectivity make albumin nanoparticles (NPs) a suitable candidate for drug delivery systems in the biomedical field. Apart from these properties, non-immunogenicity, non-toxicity, biodegradability, and biocompatibility also support albumin as a potential nanocarrier for drug delivery (Hassanin and Elzoghby 2020). HSA and BSA are the two most commonly used albumins for drug delivery (Hassanin and Elzoghby 2020). HSA protein has a molecular weight of 66.5 kDa. It's a small globular protein made up of a single polypeptide chain of 585 amino acids arranged in 3 domains (Domain I, II, and III).

Abraxane® (albumin-bound nanoformulation of paclitaxel) is a novel FDA-approved (Food and Drug Administration, USA) albumin drug formulation, developed by Abraxis BioScience (Stinchcombe 2007). Preclinical research studies showed that Abraxane® penetrates tumor cells more deeply and shows more antitumor efficacy than the conventional paclitaxel at the same dose (Desai et al. 2006; Miele et al. 2009). A wide range of phytoconstituents such as berberine (Solanki et al. 2021a), piperlongumine (Patel et al. 2022), curcumin (Salehiabar et al. 2018), fisetin (Ghosh et al. 2016), quercetin (Antçnio et al. 2016), and chrysin (Nosrati et al. 2018a) encapsulated in albumin nanoparticles and explored its anticancer activities. To provide specific targeted delivery, the surface of albumin nanoparticles was conjugated with selectively targeted ligands such as folic acid, biotin, hyaluronic acid, etc. Reported studies suggested that surface modification of albumin nanoparticles enhanced the inhibition of cell proliferation (Nosrati et al. 2018b; Qi et al. 2014). Albumin nanoparticles get internalized by gp-60 and SPARC receptors in the tumor, which contributes toward enhanced cellular uptake. Due to their outstanding properties, albumin nanoparticles have been used in cancer therapy widely.

8.3.2 Silk Proteins

Silk is an environment-friendly and sustainable natural protein that has a long clinical track in humans due to its exceptional mechanical qualities and biocompatibility (Seib 2017). Recent research mainly focuses on its biomedical applications, and it is extensively used as a biomaterial in the form of silk nanoparticles, hydrogels, films, and others (Seib 2017). Fibroin is the natural silk protein derived from various species including silkworms (*Bombyx Mori*) and spiders (*Nephila Clavipes dragline*) widely used as a raw material in nanotechnology (Obregón et al. 2017). Silk from silkworms is a fibrous protein consisting of two main proteins: silk fibroin (SF) and sericin. SF made up of heavy chains (about 390 kDa) and light chains (about 26 kDa) are joined together by a single disulfide bond at the C-terminus of the heavy chain to form the H-L complex (Deen and Rosei 2019). Additionally, a glycoprotein of 25 kDa is also linked with the H-L complex non-covalently. Gupta et al. had fabricated curcumin (CUR)-loaded SF nanoparticles (SF NPs) with chitosan (CS) material (SFCS) (Gupta et al. 2009). The uptake and efficiency of CUR-loaded SF NPs were significantly higher in breast cancer cells. They demonstrated that SF NPs are the potential nanosystem for the treatment of breast cancer. Quercetin, another phytoconstituent, was also encapsulated in SF NPs (QSFN) by P Diez-Echave and his team, and they suggested that QSFN might be a novel drug delivery system for inflammatory bowel disease (IBD) treatment (Diez-Echave et al. 2021). Carissimi et al. extensively summarized the synthesis procedures and applications of SF NPs (Carissimi et al. 2022). The review suggests that the properties and applications of SF NPs such as biocompatibility, biodegradability, amphiphilic chemistry, and good mechanical properties make SF NPs a useful candidate for drug delivery.

8.3.3 Protamine

Protamine is an FDA-approved, nontoxic cationic peptide well known for its use as a heparin antagonist and insulin delivery system. It is an arginine-rich peptide with a molecular weight (M.W.) of 5.5–13.0 kDa and has pI (isoelectric point) of 11–12. In 1874, the first protamine called Salmine was extracted from salmon's sperm by Friedrich Miescher, and to date, various other protamines such as Clupeine, P1, and P2 are the most studied protamines (Ruseska et al. 2021). Positively charged protamine helps to protect negatively charged DNA via electrostatic interactions between the positive charge of protamines and the negative charge of DNA phosphate groups. Scientists are replicating this approach to develop revolutionary nano-drug delivery systems utilizing protamine as a nanocarrier for DNA and RNA (Ruseska et al. 2021). So far, a number of protamine products have been used in pharmaceutics, for example, in the treatment of diabetes mellitus (Owens 2011), as an anticoagulant (Boer et al. 2018; He et al. 2014), and most remarkably used as the delivery system for different bioactive molecules. Reported studies demonstrated that protamine-decorated nanoparticles enhanced the efficacy of the nanosystems (Freag et al. 2018; Han et al. 2011). Different studies demonstrated the possibility of using protamine as the low-toxic delivery vehicle that can evade the immune system, opening new opportunities for the delivery of natural bioactive ligands (Radhakrishnan et al. 2014; Lagoa et al. 2020; Elzoghby et al. 2017).

8.3.4 Gliadin

Gliadin, a natural glycoprotein (gp) with an average M.W. of 25–100 kDa, is present in wheat and other cereals (Duclairoir et al. 1998). It contributes significantly to the nutritional value and quality of wheat flour, which makes up roughly 85% of the total protein in wheat grains (Prakash et al. 2022). Gliadin structure contains disulfide linkages and hydrophobic interactions, which makes it poorly water-soluble. Gliadin's amino acids composition analysis reveals that it contains similar numbers of polar (hydrophilic) and neutral amino acids, primarily glutamine (approximately

40%), along with proline (14%) (Elzoghby et al. 2015). The structure and disulfide connections increased the ability of gliadin to encapsulate lipophilic compounds, vitamins, and enzymes, while the potent mucoadhesive properties improved its connection with the stomach lining, aiding the delivery of many pharmaceutical drugs (Prakash et al. 2022). Additionally, it has demonstrated a remarkable number of applications in pharmaceutics. Biocompatibility, hydrophobicity, and the ability of surface modifications make gliadin a potential nanocarrier in drug delivery (Arangoa et al. 2001). Biomedical application of gliadin as nanoparticles, microparticles, scaffolds, nanofibers, and films showed that they are highly stable and flexible (Alqahtani et al. 2020). Previously reported studies demonstrate that gliadin can act as a nanocarrier for the delivery of bioactive compounds like resveratrol, ascorbic acid, and many more (Voci et al. 2022; Gulfam et al. 2012; Sharif et al. 2022; Joye et al. 2015). Duclairoir et al. synthesized three different drug-encapsulated gliadin NPs and suggested that drug entrapment ability depends on the polarity of the drug and gliadin (Duclairoir et al. 1998). Gliadin NPs can be prepared by the desolvation method and electrospray method (Verma et al. 2018). NPs synthesized by using the desolvation method have low drug entrapment efficiency hence the electrospray method can be a better option for the synthesis of gliadin NPs (Sridhar and Ramakrishna 2013). Gulfam et al. prepared cyclophosphamide-loaded gliadin NPs using the electrospray technique for the treatment of breast cancer (Gulfam et al. 2012). The results implied that gliadin NPs enhanced the apoptosis-mediated anticancer activity of cyclophosphamide when encapsulated in gliadin nanocarrier. For the development of targeted drug delivery, natural proteins such as gliadin, albumin, and fibroin can be an alternative to synthetic polymers.

8.3.5 Legumin

Vegetable protein-based nanostructured vehicles have some benefits over animal proteins due to their low cost. Furthermore, they have functional groups that can be available for the surface modification of targeted NPs (Sripriyalakshmi et al. 2014). Legumin is a plant-based storage protein found in *Pisum sativum* L. and has the properties to convert into NPs after aggregation and cross-linked with different cross-linking agents (e.g., glutaraldehyde) (Mirshahi et al. 2002). Legumin is a protein of M.W. 300–400 kDa with an isoelectric point (PI) 4.8 and is rich in acidic amino acids and arginine (Mirshahi et al. 1996). There are not many studies done on legumin NPs for drug delivery, but some studies explored legumin as a nanocarrier for the transdermal delivery of therapeutics (Mirshahi et al. 1996, 2002; Belyakova et al. 1999; Irache et al. 1995). Irache et al. synthesized legumin NPs using the coacervation method (Irache et al. 1995). Synthesized NPs were in the nanoscale (250 nm) and stable at physiological pH. Legumin can be explored as a nanocarrier for the delivery of bioactive agents.

8.3.6 Collagen

Collagen is a natural protein present in connective tissues, skin, tendons, ligaments, cartilage, and bones (Sripriyalakshmi et al. 2014). Collagen is made up of three similar or dissimilar polypeptide chains, each with approximately 1000 amino acids. The particular arrangement of amino acids helps to stabilize the structure of collagen. The most prevalent repeating unit in the collagen sequence is Gly-X-Y. In every third position in the order, the smallest amino acid, glycine, is repeated, whereas proline at position X and hydroxyproline at position Y take up 35% and 10% of the non-glycine positions, respectively (Kucharz 2012). Owing to their favorable properties such as biocompatibility, biodegradability, non-immunogenicity, and biomimetic property, collagens are extensively used as a nanocarrier for drug, protein, or gene and also used as scaffold matrix in tissue engineering (Arun et al. 2021). Apart from these properties and versatile sources, the ability to self-assemble, good mechanical properties, and having a bi-functional surface make collagen a good choice of nanocarrier. Emulsification, solvent evaporation, desolvation, and coacervation are the general methods used for the formulation of collagen NPs (Saxena et al. 2005). Temperature, pH, degree of cross-linking, and amount of gelatin are the parameters that affect the preparation of NPs and their size as well as stability. A study performed by Cascone found that the gelatin matrix's degree of cross-linking is a key factor influencing the form and size of the particles (Cascone et al. 2002). It was found that raising the cross-linkers (glutaraldehyde) concentration results in a small particle size of gelatin NPs. Gelatin formulations have applications in cancer treatment, drug delivery via combination with hydrogels or liposomes, ophthalmology, and use as skin substitutes in body tissue engineering (Arun et al. 2021). Collagen is a versatile material in the biomedical field as it can be cross-linked and form gels or solids (Lo and Fauzi 2021). Apart from these benefits, the high cost of collagen limits its applications in biomedical fields. Available literature suggests that collagen NPs are emerging materials in drug delivery and tissue engineering and play a significant role in biomedicine (Lee et al. 2001; Nicklas et al. 2009; Nagarajan et al. 2014; Grigore et al. 2017).

8.3.7 Gelatin

Gelatin is a synthetic protein that is prepared from collagen by carefully hydrolyzing it (Narayanaswamy et al. 2016). The obtained gelatin can be of two types (type A and type B) depending on whether acidic or alkaline pretreatments are used during the gelatinization process (Mohiti-Asli and Loboa 2016). Glycine, proline, and 4-hydroxyproline amino acids are abundant in gelatin. It is a powder that is translucent, colorless, and almost tasteless (Deshmukh et al. 2017). Gelatin has been considered a versatile nanocarrier due to its unique chemical and physical characteristics. Surface modification of gelatin NPs with ligands such as cationized amine derivatives or polyethyl glycols (PEG) can further provide targeted delivery of bioactive molecules (Sahoo et al. 2015). Gelatin is generally regarded as safe

(GRAS) material by the FDA owing to its long history of safe use in food products, biomedicine, and cosmetics products (Kommareddy et al. 2007). Simple coacervation, nanoprecipitation, self-assembly, solvent evaporation, and emulsification are widely used preparative methods for gelatin NPs. Evidence suggests that gelatin has been widely used in drug delivery for various phytoconstituents including resveratrol (Karthikeyan et al. 2013), noscapine (Madan et al. 2011), berberine (Zhang et al. 2021a), curcumin, and piperine (Ratanavaraporn et al. 2014). These phytoconstituents-loaded gelatin NPs increased the cellular internalization of drugs at the target site thus enhancing their anticancer properties compared to the drug itself. Gelatin NPs have developed significant applications as drug carriers in the pharmaceutical field. As collagen protein is of animal origin, there are chances of occurrence of transmissible diseases like transmissible spongiform encephalopathy (TSE), thus the uses of gelatin NPs are limited. However, current studies found a solution. Studies done by Young-Wook Won and his team members suggested that recombinant human gelatin NPs can be an alternative option to human-originated gelatin to minimize safety issues (Won and Kim 2009). Overall, gelatin NPs may be promising protein-based nanocarriers for the effective, controlled, and sustainable release of phytoconstituents.

8.3.8 Casein

Casein is a major mammalian milk protein, an essential part of our daily diet and a good source of amino acids (Elzoghby et al. 2011). Casein is affordable, widely accessible, non-toxic, and highly stable and these properties help the use of casein from nutrient to nutraceutical (Gandhi and Roy 2021). Being a biodegradable and biocompatible natural food product, it is well known as GRAS protein. Moreover, its remarkable ion and small molecule binding property, stabilizing abilities, good emulsification, self-assembly characteristics, extraordinary surface-active properties, exceptional gelation, and water binding capacity facilitate its applications as a drug delivery system (Livney 2010). Casein is amphiphilic and can offer excellent drug delivery potential to phytoconstituents in the form of nanoparticles, gels, micelles, nanoemulsions, and microparticles (Horne 2002; Rehan et al. 2019). Casein contains both hydrophilic and hydrophobic groups, thus it is recognized for binding and/or encapsulating various phytoconstituents very efficiently. Casein-based NPs can be synthesized by various techniques like the emulsification method, coacervation method, cross-linking by chemical or enzymatic method, heat gelation, electrostatic complexation, and many more (Rehan et al. 2019). Several in vitro and in vivo studies suggested casein as a suitable nanocarrier for various phytoconstituents such as curcumin (Niu et al. 2020; Xu et al. 2020), quercetin (Peñalva et al. 2019), rutin (Luo et al. 2015), epigallocatechin gallate (Shukla et al. 2009), resveratrol (Cheng et al. 2020), and many more. Constructing casein-based nano-architectures as nanocarriers is one of the most effective ways or strategies to enhance anticancer activities of poor water-soluble phytoconstituents by improving the water solubility, bioavailability, and stability (Tang 2021). Mansoore Esmaili et al. demonstrated that casein nanomicelles increased the solubility of curcumin (2500-fold) resulting in enhanced cytotoxicity of bioactives after encapsulating in casein (Esmaili et al. 2011). The targeted oral delivery and targeting mechanisms for cancer treatment are still major challenges for milk proteins but advancement in nanotechnology will sort out the current problems and expand the use of milk proteins in food and biomedical sectors.

8.3.9 Zein

Zein is a prolamine protein extracted from naturally occurring corn plants, constituting 45–50% of the total protein of zein. Zein is a hydrophobic protein and contains amino acids such as proline, alanine, and glutamine. (Bhawani et al. 2019). Zein is a GRAS protein approved by the FDA as the safest biomaterial in the food, cosmetic, and pharmaceutical fields (Yu et al. 2020). Zein is widely used as a coating material for foods and pharmaceuticals as they are biocompatible and have low water uptake properties with excellent thermal resistance and good mechanical (Bhawani et al. 2019). Additionally, properties like safe material, low cost, adhesiveness, flexibility, easily modulated, and soft nature make zein a promising biomaterial for the encapsulation of hydrophobic drugs and controlled release characteristics (Labib 2018). Zein was applied in the drug/vaccine delivery and tissue engineering field to develop micro- and nanoparticles, gels, fabrics, scaffolds, micelles, and so on. Controlled delivery of phytoconstituents such as resveratrol (Huang et al. 2017; Jayan et al. 2019; Zhang et al. 2019), curcumin (Liu et al. 2020, 2022), epigallocatechin gallate and piperine (Dahiya et al. 2018), and luteolin (Shinde et al. 2019) from zein-based nanocarrier systems have been reported in the literature. Zein-based nanosystems are great in improving the solubility and oral bioavailability of poor hydrophobic drugs by protecting their degradation from the gastric environment (Paliwal and Palakurthi 2014). Additionally, the self-assembling nature of zein provides high drug encapsulation efficiency and is investigated to synthesize cutting-edge multifunctional micelles for cancer treatment and bioimaging. However, the use of zein in pharmaceuticals is still limited due to the possible immunogenic effects of zein NPs, but optimization in the dose of zein during the synthesis of formulations may minimize the immunogenic responses thus requiring more investigations in this field. Still, zein nanosystems have gained a lot of attention as protein nanocarrier-based delivery of anticancer drugs, and advancement in these nano-drug delivery systems will expand the possible applications.

8.3.10 Elastin-Like Polypeptides

Elastin is a primary protein consisting of a mixture of proteins and polysaccharides that make up the extracellular matrix (Buck and Tirrell 2012). Elastin is flexible and strong and contributes to tissue expansion and contraction. Elastin-like polypeptides (ELPs) are synthetic polypeptides motivated by human elastin. ELPs are made up of repeating amino acid sequences of Val-Pro-Gly-Xaa-Gly where Xaa is any amino acid except proline. The possible lack of toxicity, monodispersity, genetically encoded synthesis, and biocompatibility of these biopolymers make them appealing for use in biological applications. ELPs have a unique thermal property in which temperature changes induce conformational changes in individual proteins, thus it is a suitable biomaterial for stimulus (i.e., thermal) responsive applications in biomedical sectors. ELPs have been used in thermo-responsive targeted delivery of small drugs (MacKay et al. 2009; Bidwell III et al. 2007; Bidwell et al. 2007), plasmid DNA (Chen et al. 2008), proteins (Shamji et al. 2008), and peptides (Massodi et al. 2005; Bidwell and Raucher 2005) in cancer treatment. Due to beneficial properties such as biodegradability, biocompatibility, and non-toxicity, ELPs are used as a drug delivery system for the treatment of cancer, type 2 diabetes, and cardiovascular diseases (Lima et al. 2022).

8.3.11 Virus-Like Particles

Virus-like particles (VLPs) closely resemble viruses but are non-infectious as they lack viral genetic material (Nooraei et al. 2021). Infectious diseases (i.e., SARS-Cov-2) are a serious concern for human health and need innovative platforms to develop successful treatment in the form of vaccines to fight them. VLPs' nanostructures offer an alternative strategy to fight against infectious diseases and it is in progress concurrently with mRNA and viral-carrier-based vaccines (Tariq et al. 2022). VLPs are a desirable and attractive nano-system as they can be modified genetically or chemically. Additionally, VLPs are potent immune activators and suitable nanocarriers for vaccine/drug delivery as they can serve as the natural carrier for genetic materials to their host cells. VLPs provide the great benefits of morphological homogeneity, biocompatibility, and simple functionalization as an essential and growing nanocarrier platform (Ma et al. 2012). A list of protein nanocarriers used for the encapsulation of phytoconstituents is given in Table 8.1.

8.4 Methods for Preparation of Protein Nanocarriers

Protein nanocarriers can be synthesized by using different methods: chemical, physical, and self-assembly (Fig. 8.3). Emulsification and coacervation methods are chemical-based approaches. Physical methods consist of electrospray drying and nanospray drying methods. The desolvation method and self-assembly are self-assembly-based methods used for the preparation of protein nanocarriers. Each method has its advantages and disadvantages which are discussed below.

8.4.1 Emulsification

The emulsification method is frequently used for the synthesis of protein-based nanocarriers. A high-speed homogenizer or ultrasonic shear is used to emulsify a

Table 8.1 Protein nanocarriers: representing the sources, molecular weight, properties, nanoparticles preparation techniques, and the encapsulated phytoconstituents

Protein	Sources	Molecular weight	Properties	Preparation techniques	Encapsulated phytoconstituents
Albumin	Human serum; animals (i.e., cow, chicken, salmon, etc.); and plants (i.e., peanuts, sunflower, passion fruit, etc.)	~66.5 kDa	Highly water-soluble, non-immunogenic, non-toxic, biodegradable, biocompatible, high binding efficiency	Desolvation, coacervation, ultra-emulsification, and nab-technology methods	Berberine (Solanki et al. 2021a), piperlongumine (Patel et al. 2022), curcumin (Salehiabar et al. 2018; Jithan et al. 2011), fisetin (Ghosh et al. 2016), quercetin (Antçnio et al. 2016), chrysin (Nosrati et al. 2018a; Ferrado et al. 2020; Solanki et al. 2023)
Silk	Silkworms (*Bombyx mori*) and spiders (*Nephila clavipes dragline*)	Silk fibroin 60–150 kDa	Genetic control of the encoded protein sequence, slow rate of degradation in vivo, along with robust mechanical properties, biocompatible, and biodegradable	Electrospinning, self-assembly, 3D printing, and cross-linking methods	Curcumin (Gupta et al. 2009), quercetin (Diez-Echave et al. 2021)
Protamine	Salmon's sperm, mammals, recombinant protein	5.5–13.0 kDa	Small, positively charged, arginine-rich, nuclear proteins, precise nuclear localization, highly soluble in water	Ionotropic gelation, coacervation, emulsification-solvent evaporation, reverse micelle and self-assembly methods	Curcumin (Abdel-Hakeem et al. 2021), capsaicin (Cialdai et al. 2003), epigallocatechin gallate (Ding et al. 2018)
Gliadin	Wheat grains	25–100 kDa	Mucoadhesive, biodegradable, biocompatible, long-term stability, controllable size	Electrospray, desolvation, antisolvent precipitation, emulsification, coacervation, and dialysis methods	Resveratrol (Wu et al. 2020), ascorbic acid (Voci et al. 2022), γ-oryzanol (Sharif et al. 2022), retinoic acid (Ezpeleta

(continued)

Table 8.1 (continued)

Protein	Sources	Molecular weight	Properties	Preparation techniques	Encapsulated phytoconstituents
			and shape, and ability of surface modifications		et al. 1996), α-tocopherol (Duclairoir et al. 2002)
Legumin	Beans, peas, lentils, vetches, hemp, and other leguminous seeds, e.g., *Pisum sativum* L.	300–400 kDa	Low cost, biocompatibility, low toxicity, biodegradability, controllable size and shape, and ability of surface modification	Coacervation, chemical cross-linking, antisolvent precipitation, emulsification, ionotropic gelation, and spray-drying methods	Resveratrol (Fan et al. 2020)
Collagen	High-protein foods such as fish, poultry, meat, eggs, dairy, legumes, and soy	~300 kDa	Biocompatibility, biodegradability, non-immunogenicity and biomimetic property, versatility, ability to self-assemble, good mechanical properties, and have bi-functional surface	Emulsification, solvent evaporation, desolvation, self-assembly, and coacervation methods	Curcumin (Terzopoulou et al. 2020; Leng et al. 2020; Rezaii et al. 2019; Karri et al. 2016), fisetin (Cimenci et al. 2022), resveratrol (Siddiqui et al. 2021, 2022), berberine (Chiu et al. 2021), silymarin (Rathore et al. 2020)
Gelatin	Produced by the processing of animal bones, cartilage, and skin	15–400 kDa	Used as gel, thickener, foaming agent, plasticizer, and texture and binding agent. Properties such as biocompatibility, biodegradability, small size, high surface area	Simple coacervation, nanoprecipitation, self-assembly, solvent evaporation, and emulsification methods	Resveratrol (Karthikeyan et al. 2013), noscapine (Madan et al. 2011), berberine (Zhang et al. 2021a), curcumin, and piperine (Ratanavaraporn et al. 2014)
Casein	Milk and its products	20–25 kDa	Biocompatibility, biodegradability, ion and small molecule binding	Emulsification, coacervation, cross-linking by chemical or enzymatic	Curcumin (Esmaili et al. 2011; Pan et al. 2013, 2014; Barick et al. 2021),

			property, stabilizing abilities, good emulsification, self-assembly characteristics, extraordinary surface-active properties, exceptional gelation, and water binding capacity	method, heat gelation, and electrostatic complexation methods	resveratrol (Peñalva et al. 2018), epigallocatechin gallate (Malekhosseini et al. 2019; Zheng et al. 2019), genistein (Bindhya et al. 2021), quercetin (Ghayour et al. 2019), eugenol (Xue et al. 2019)
Zein	Corn plants	25–40 kDa	Biocompatible and low water uptake properties and excellent thermal resistance and good mechanical properties. It is a safe material and has low cost, adhesiveness, flexibility, easily modulated, and soft nature	Antisolvent precipitation, nanoprecipitation, heat-induced self-assembly, coacervation, electrospinning, and emulsion-solvent evaporation methods	Resveratrol (Huang et al. 2017; Jayan et al. 2019; Zhang et al. 2019), curcumin (Liu et al. 2020, 2022), epigallocatechin gallate, piperine (Dahiya et al. 2018), luteolin (Shinde et al. 2019)
Elastin-like polypetides (ELPs)	Synthetic proteins derived from human tropoelastin	16–250 kDa	Possible lack of toxicity, monodispersed, genetically encoded synthesis, and biocompatibility, biodegradability, and elastomeric characteristics	Ultrasonication method, genetic engineering methods, Gibson assembly, self-assembly, cross-linking, and coacervation methods	Curcumin (Bergonzi et al. 2020)
Virus-like particles (VLPs)	Recombinantly synthesized	Can vary depending on the specific virus and the method of production	Provides the great benefits of morphological homogeneity, biocompatibility, simple functionalization, immunogenicity, stability, and is safer	Self-assembly, recombinant engineering	

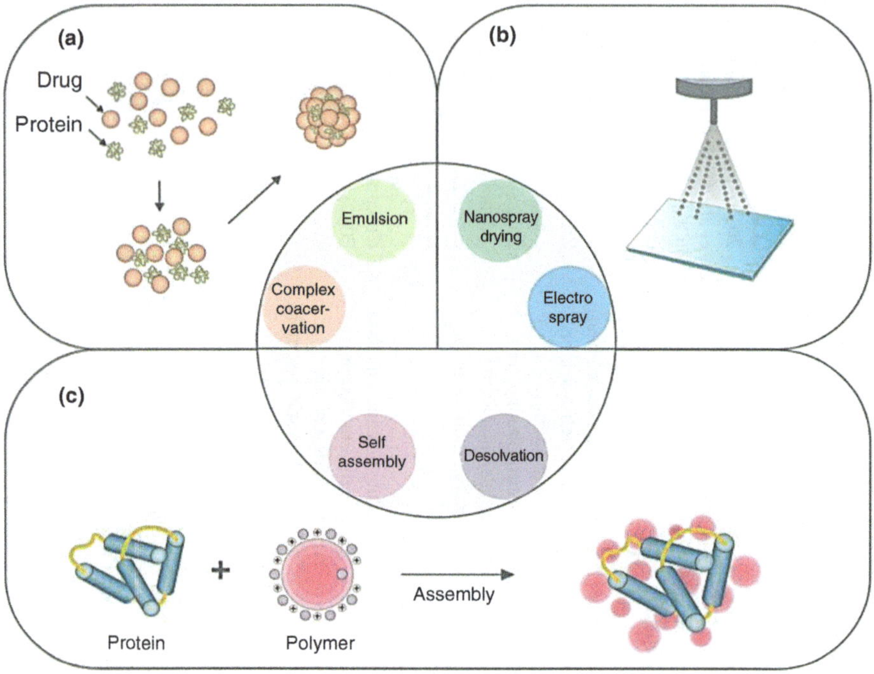

Fig. 8.3 Different methods used for the preparation of protein-based nano-drug delivery systems. Chemical method (**a**), physical method (**b**), and self-assembly (**c**). (Figure adapted from Jain et al. 2018)

protein solution in oil, and the NPs are formed at the interface of oil/water (Lohcharoenkal et al. 2014). Scheffel and his colleagues first discovered this method in 1972 for the preparation of albumin spheres, and later in 1995, Gao and his team members rediscovered this method (Verma et al. 2018). Previously reported studies suggest that albumin NPs and whey NPs are the common protein nanocarriers that have been prepared by using the emulsion-based approach (Meng et al. 2022; Mishra et al. 2006; Adjonu et al. 2014). Particle size depends on the oil-in-water ratio (w/o) and protein concentration, increasing concentration results in larger particle size. In the literature, two approaches of the emulsion method are suggested for protein NPs: single emulsion and double emulsion. The first approach is based on oil in water (o/w) (Fig. 8.4a) while the second approach depends on emulsions of water in oil in water (w/o/w) (Fig. 8.4b) (Tarhini et al. 2017).

Reported studies demonstrated that the emulsification method was employed to encapsulate various phytoconstituents such as curcumin (Rao et al. 2015; Fereydouni et al. 2021), berberine (Sharifi-Rad et al. 2021), plumbagin (Chrastina et al. 2018), and fisetin (Ragelle et al. 2012) in protein nanocarriers.

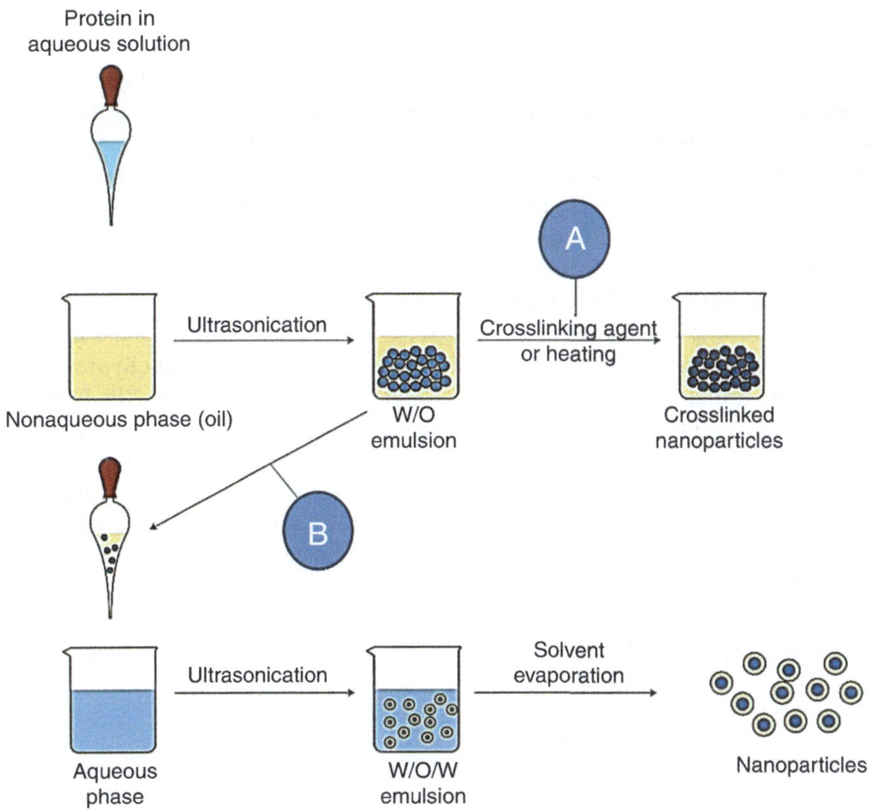

Fig. 8.4 Preparation of protein-based nanoparticles by emulsification, single emulsion (a) and double emulsion approach (b). (Figure adapted from Tarhini et al. 2017)

8.4.2 Coacervation

The coacervation method is frequently used to encapsulate proteins and hydrophilic compounds (Dubey et al. 2016). Proteins solubility is different in different solvents and the coacervation method works on this solubility approach. The solubility of proteins depends on pH, polarity, ionic strength, and the presence of electrolytes in the solvent (Tarhini et al. 2017). The size of particles and homogeneity can be monitored by altering the concentration of protein and viscosity of the non-solvent. In the coacervation method, phase separation involves forming liquid-liquid phase separation, resulting in the formation of the desired NPs as a dense phase rich in polymers at the bottom and a clear solution above (Elzoghby et al. 2015). Protein NPs including gelatin, silk, zein, and legumin were synthesized by the coacervation method (Sripriyalakshmi et al. 2014; Lohcharoenkal et al. 2014; Rahimnejad et al. 2009). Phytoconstituents-loaded protein NPs were successfully prepared using the coacervation method (Salehiabar et al. 2018; Zare-Zardini et al. 2022).

8.4.3 Electrospraying

Electrospraying is a novel electrodynamic method used for developing protein nanocarriers where protein solutions are exposed to high-voltage electric fields (Wang et al. 2019). It has been suggested that proteins with relatively large surface charges are ideally suitable for the electrospray method used to synthesize nanoparticles and nanofibers. In this method, protein and drug are dissolved in suitable solvents and filled in the capillary tube. A high electrical voltage is applied to drag the protein solution out of the capillary tube, forming a narrow jet. The electrical strength and the composition of the protein affect the fabrication of nanofibers or capsules (Tarhini et al. 2017). The distance between the capillary and the collector, where the electrospray jet travels, must be sufficient to prevent the solvent and protein particles from fusing. During the fabrication of NPs using the electrospray method, various parameters such as flow rate, electric voltage, collector distance, and parameters of the solution including protein type, concentration, viscosity, and density are optimized to synthesize desired NPs (Jafari and McClements 2017). In various studies, the electrospraying method is used for the synthesis of bioactive-loaded protein NPs (Rostamabadi et al. 2021; Solanki et al. 2021b; Süngüç 2013; Asadi et al. 2021).

Heera Jayan et al. synthesized resveratrol-loaded zein nanoparticles using the electrospraying method (Jayan et al. 2019) (Fig. 8.5). They demonstrated that

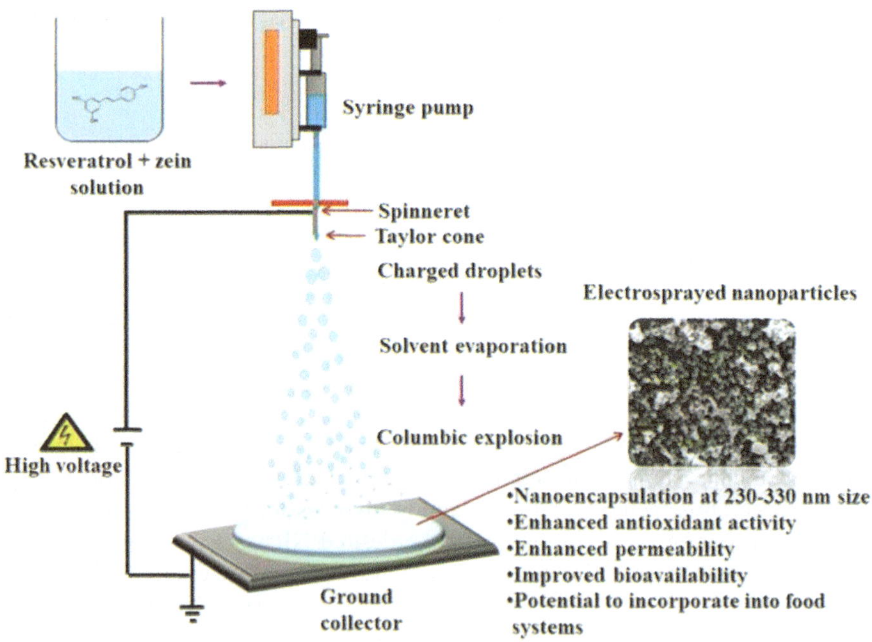

Fig. 8.5 Encapsulation of resveratrol in zein NPs using the electrospraying technique. (Figure adapted from Jayan et al. 2019)

resveratrol was protected by zein NPs in simulated stomach conditions, and it was released under simulated intestinal conditions. Additionally, resveratrol-loaded zein NPs demonstrated higher permeability (1.15-fold) in an ex vivo dynamically designed small intestine tract which is related to the improved bioavailability of resveratrol.

A key drawback of the electrospraying technology is the fact that many proteins cannot be used alone to synthesize nanofibers due to their complex structures and capability of strong interactions. To address this problem, protein solutions might be supplemented with surfactants, plasticizers, or reducing agents (Elzoghby et al. 2015). The electrospraying technology might be advantageous to other traditional methods. This method produces NPs without the use of a template or surfactant. Additionally, this technique has a high rate of self-dispersion and drug-loading efficiency.

8.4.4 Nanospray Drying

The spray drying technique gained a lot of popularity for the formulation of micro- and nano-drug delivery systems (Haggag and Faheem 2015). Conventional spray drying methods have limitations like larger particle size, high volume requirements, and low yield rate. So, to overcome these limitations the new spray drying method was developed by BUCHI Labortechnik AG in 2009 and named B-90 which is now known as the nanospray drying method (Tarhini et al. 2017). Nowadays, researchers have shifted toward nanospray drying techniques to fabricate protein-based nanocarriers such as nanoparticles, microparticles, nanoemulsions, and nanosuspensions. Several studies revealed the use of nanospray drying technology in the encapsulation of phytoconstituents-loaded protein NPs (Chang et al. 2017). Sie Huey Lee et al. prepared BSA NPs by using the nanospray drying technique. They demonstrated that prepared NPs were smooth, spherical, and nano-scale (460 nm) protein NPs and also suggested that this method is a very fast, simple, and alternative approach for protein drug delivery systems (Lee et al. 2011).

8.4.5 Desolvation

Desolvation is a thermodynamically driven, most common easy synthesis method for protein nanocarriers. In this method, NPs are formulated by adding desolvating agents such as ethanol dropwise to a protein solution under constant stirring (Langer et al. 2003). Next, a cross-linker is to be added for cross-linking of amino acid moieties. Temperature, pH, ionic strength, solvent, stirring condition, amount of protein, and amount of drug are the parameters that affect the size and shape of nanoparticles prepared using the desolvation method. The desolvation method is the most common method to design protein NPs, but the use of organic solvents and cross-linkers is a major drawback. Albumin NPs, gelatin NPs, silk NPs, whey NPs, and gliadin NPs were prepared using the desolvation method. Available literature exhibited that various phytoconstituents-loaded protein nanocarriers such as berberine, piperlongumine (Patel et al. 2022), fisetin (Ghosh et al. 2016), curcumin (Salehiabar et al. 2018), chrysin (Nosrati et al. 2018a), and quercetin (Raj et al. 2015) were synthesized using the desolvation method.

8.4.6 Self-Assembly

The self-assembly process is the arrangement of smaller subunits into a well-organized larger structure and by this procedure, proteins may be designed into nanosystems, which has distinct benefits for delivering anticancer drugs (Yadav et al. 2020; Sun et al. 2017). In this method, hydrophobic anticancer drugs are added to the aqueous protein solution to increase the hydrophobicity of the protein, which leads to the breaking of disulfide bonds and the removal of primary amines (Solanki et al. 2021b). Protein nanomicelles were generally synthesized using a self-assembly procedure and phytochemicals were encapsulated in synthesized nanomicelles. Bioactive molecules such as genistein (Jangid et al. 2022a), chrysin (Jangid et al. 2022b), and curcumin (Bai et al. 2017).

8.5 Route of Administration for Protein Nanocarriers

A location where nanocarriers are administered in the human body is defined as the route of administration, such as oral or intravenous routes. The route of administration must be the priority while synthesizing protein nanocarriers because the choice of administration route depends on the pharmacokinetics and pharmacodynamics of the drug which impacts the efficacy of drug-loaded nanosystems. Since the development of protein nanocarriers, research has focused on the administration route of protein nanocarriers to target organs/tissues and the target functionality (Escobar-Chávez et al. 2012). To obtain the maximum therapeutic efficacy of the drug delivery system, all barriers must be successfully crossed (Fig. 8.6).

From various administration routes such as oral, intravenous, and inhalation, the oral route is the standard route for administration in the human body (Bose et al. 2014). The oral route is considered a safe, convenient, and cost-effective route. Despite being the safest approach, it has various drawbacks, including poor bioavailability, rapid metabolism, and irregular absorption of the pharmaceuticals. These issues are mostly caused by the fluctuating pH conditions and the digestion of protein or peptide pharmaceuticals by proteolytic enzymes located in the intestine. Other key administration routes such as parenteral, topical, ocular, and pulmonary are therefore gaining administrative priority. Hence, while designing a nanosystem, it is crucial to know which way of administering a nanodrug would result in its highest accumulation at the target site.

Nanocarriers designed to target gastrointestinal (GI) tract–related cancer such as colon cancer and improve their accumulation at colon sites should be given orally and synthesized accordingly. Recent studies on the therapeutic efficacy of orally

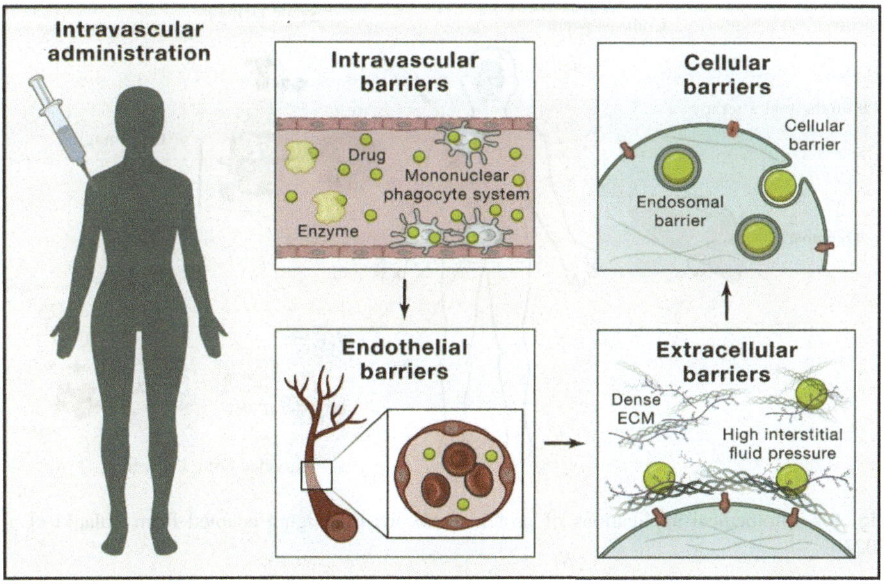

Fig. 8.6 Biological barriers for drug delivery. (Figure adapted from Zhao et al. 2020)

delivered via protein nanocarriers have shown increasing anticancer activity of phytoconstituents such as curcumin (Chang et al. 2017, 2019). Chang Chao et al. synthesized a complex nanosystem composed of proteins and polysaccharides for the oral delivery of curcumin (Chang et al. 2017). Studies demonstrated that complex nanosystems stabilized in GI conditions and also enhanced the antioxidant activity of curcumin.

8.6 Biomedical Applications of Phytoconstituents-Encapsulated Protein Nanocarriers

In recent years, protein nanocarriers have made a remarkable impact in the field of nanomedicine. They have different pharmaceutical applications such as nanocarriers, targeted proteins, fluorescent probes, and enzymes for applications in drug delivery, biosensing, and bioimaging (Ding et al. 2021). Protein nanocarriers offer a wide range of biomedical applications including targeted drug delivery, gene delivery, photothermal therapy, diagnostic, bioimaging, etc. (Fig. 8.7) (Solanki et al. 2021b).

Biocompatibility, biodegradability, ease of synthesis, abundance in natural sources, and low cost are the benefits of proteins used as nanocarriers for drug delivery. Unlike other drug delivery systems such as metallic NPs and synthetic NPs, the use of protein NPs is not limited by the issues such as potential toxicity, large particle size, accumulation, and rapid clearance from the human body (Jain et al.

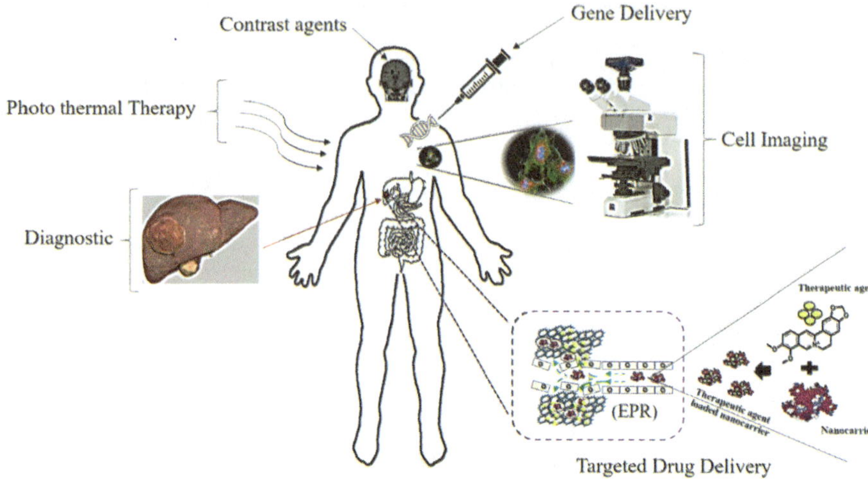

Fig. 8.7 Biomedical applications of protein nanocarriers. (Figure adapted from Solanki et al. 2021b)

2018). Furthermore, surface modification on protein NPs using peptides, other proteins, and carbohydrates offer targeted drug delivery and enhanced the functionality of drug-loaded NPs.

8.6.1 Applications in Cancer Therapy

Protein nanocarriers have proven effective in the delivery of phytoconstituents in various cancer treatments including breast, lung, brain, etc. In one study, Hae-Yong Seok et al. developed CD44-targeted hyaluronic acid cross-linked zein nanogels (HA-Zein NGs) for the delivery of curcumin (Seok et al. 2018). HA-Zein NGs were synthesized using a self-assembly procedure and hypothesized that NGs will bind to cancer cells expressing the CD44 receptor and internalized in cancer cells using a receptor-mediated endocytosis mechanism (Fig. 8.8a). Curcumin (CRC)-loaded HA-Zein NGs (30 mg/kg daily) were applied to the CT26 xenograft model and preclinical studies were performed (Fig. 8.8b). They observed that tumor volume was reduced in the CRC-loaded HA-Zein NGs treated model compared to PBS, CRC, and HA-Zein NGs treated models (Fig. 8.8b(A)). The CRC-loaded HA-Zein NGs significantly decreased the tumor volume in comparison to the control samples. Further, to confirm the in vivo apoptosis, histopathology was performed to investigate the tumor tissues of the treated and untreated groups. As shown in Fig. 8.8b(E), the results revealed a distinct apoptotic area with the highest density in the case of the CRC-loaded HA-Zein NGs treated group. While PBS, CRC, and HA-Zein NGs treated groups did not show any significant apoptotic effects. The photographs acquired after euthanasia indicated that the tumor was suppressed in CRC-loaded HA-Zein NGs compared to other groups (Fig. 8.8b(C)). During the entire treatment

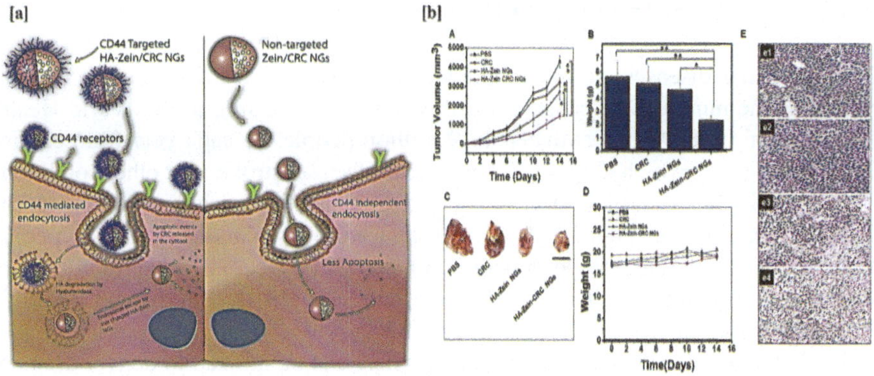

Fig. 8.8 Synthesis and in vivo evaluation of hyaluronic acid cross-linked zein nanogels (HA-Zein NGs). Schematic representation of HA-Zein NGs (**a**) and in vivo therapeutic efficacy of HA-Zein NGs (**b**). (Figure adapted from Seok et al. 2018)

period, there were no significant weight changes, indicating that these NG formulations were very safe in vivo (Fig. 8.8b(D)). The study done by Hae-Yong Seok et al. demonstrated that protein nanocarriers could serve as a novel drug delivery system for various cancer treatments.

In another study, Huaiying Zhang et al. developed curcumin-loaded zein NPs but they decorated NPs with dodecamer peptide (G23)-functionalized polydopamine (pD) (CUR-ZpD-G23 NPs) to target glioblastoma (GBM) cancer (Zhang et al. 2021b). After observing the results, they concluded that CUR-ZpD-G23 NPs have promising potential as targeted drug delivery for GBM therapy. Apart from curcumin, several phytoconstituents were encapsulated and delivered at their target sites via protein nanocarriers. Available literature suggested that berberine (Solanki et al. 2021a; Majidzadeh et al. 2020), curcumin (Salehiabar et al. 2018; Liu et al. 2020; Jithan et al. 2011; Rao et al. 2015; Zare-Zardini et al. 2022; Asadi et al. 2021; Chang et al. 2017; Seok et al. 2018; Zhang et al. 2021b), plumbagin (Chrastina et al. 2018; Rajalakshmi et al. 2018; Kamble and Shaikh 2022, b), evodiamine (Li et al. 2019; Zhang et al. 2012; Yang et al. 2022), fisetin (Ghosh et al. 2016; Ragelle et al. 2012), piperlongumine (Patel et al. 2022), luteolin (Shinde et al. 2019), epigallocatechin gallate (Dahiya et al. 2018), usnic acid (Zugic et al. 2020), etc. were encapsulated in protein nanocarriers (i.e., albumin, zein, silk, etc.) and their in vitro and/or in vivo anticancer mechanisms were explored successfully. These studies implied that phytoconstituents-loaded protein nanocarriers improved the anticancer activities of bioactive agents via targeted drug delivery and enhanced their drug efficacy with minimum side effects.

8.6.2 Cardioprotective Applications

Cardiovascular diseases are one of the leading causes of death worldwide and are ranked as the number one cause of death globally. According to the World Health Organization (WHO), an estimated 17.9 million people die each year from cardiovascular diseases, which include coronary heart disease, stroke, and other conditions that affect the heart and blood vessels (World Health Organization n.d.; Nemeroff and Goldschmidt-Clermont 2012; Kim 2021). Cardiovascular disease affects people of all ages and ethnicities and is a major public health concern. It is therefore critical to continue research and develop effective interventions to prevent and treat this disease (Chaturvedi 2003).

In cardiovascular disease, protein nanocarriers can be utilized to deliver drugs or genetic material directly to the site of the disease, thereby minimizing off-target effects and reducing systemic toxicity (Flores et al. 2019). Additionally, protein nanocarriers can enhance the stability, solubility, and pharmacokinetics of drugs/phytochemicals, and can also improve their cellular uptake and therapeutic efficacy. One potential application of phytochemical encapsulated protein nanocarriers in cardiovascular disease is in the treatment of atherosclerosis, which is a condition characterized by the buildup of plaques in the arteries. Protein nanocarriers can be engineered to target specific cells within the plaques, such as macrophages, and deliver drugs or genes that can reduce inflammation, promote plaque regression, and improve vascular function (Cicha et al. 2013). Another potential application of phytochemical encapsulated protein nanocarriers in cardiovascular disease is in the treatment of myocardial infarction (heart attack). Protein nanocarriers can be used to deliver therapeutic agents (phytoconstituents) that can promote cardiac repair and regeneration, such as growth factors, stem cells, or gene therapy vectors (Sahoo et al. 2021).

8.6.3 Neuroprotective Applications

Neurological disorders have been increasing in recent years and are highly prevalent globally. Herbal drug-loaded protein nanocarriers have the potential to play a significant role in the treatment of neurological disorders (Moradi et al. 2020). One potential application of these nanocarriers in neurological disorders is the delivery of drugs to treat brain tumors (Moradi et al. 2020). These nanocarriers can be designed to target the tumor cells specifically and deliver chemotherapy drugs directly to the tumor site, increasing drug concentration and reducing the side effects of systemic chemotherapy. Rui Yang et al. synthesized curcumin-loaded chitosan-BSA nanoparticles to improve the penetration of curcumin through the blood-brain barrier (BBB) in Alzheimer's disease (AD) treatment (Yang et al. 2018). The findings showed that prepared NPs effectively increased curcumin penetration through the BBB, promoted microglia activation, and accelerated Aβ peptide phagocytosis. NPs also inhibited the TLR4-MAPK/NF-kB signaling pathway, which further reduced M1 macrophage polarization. This study suggested that curcumin-loaded CS-BSA NPs may improve Aβ 42 phagocytosis in AD by modulating macrophage polarization. In another study, Huae Xu et al. prepared resveratrol-loaded HSA NPs and demonstrated its application in the treatment of ischemic stroke (Xu et al. 2018). Prepared drug-loaded HSA NPs showed potential in the therapy of cerebral I/R injury via enhanced neurological score, which could be due to prolonged blood circulation, localization in the I/R brain region, and a sustained release pattern of resveratrol from HSA NPs. Furthermore, protein nanocarriers can be used to deliver imaging agents to the brain for diagnostic purposes. For example, magnetic resonance imaging (MRI) contrast agents can be delivered to the brain using protein nanocarriers, allowing for the non-invasive detection of neurological disorders such as Alzheimer's disease and multiple sclerosis (Kabanov and Gendelman 2007; Ning et al. 2022).

8.6.4 Antimicrobial Applications

In recent years, encapsulation of phytoconstituents containing antimicrobial properties in protein nanoparticle systems has emerged as an innovative and promising alternative that improves antimicrobial potency and reduces undesirable drug side effects (Rudramurthy et al. 2016; Stan et al. 2021). Protein NPs can be used to develop vaccines against a variety of microbial diseases. They can be engineered to mimic the structure of the pathogen, stimulating a robust immune response and generating protective immunity. By combining the antimicrobial properties of phytochemicals with the unique properties of protein nanocarriers, researchers have developed nanoparticle-based systems that can effectively target and kill microorganisms. For example, zein NPs loaded with phytochemicals such as silymarin (Tsai et al. 2018), ellagic acid (de Souza Tavares et al. 2021), carvacrol (da Rosa et al. 2015), thymol (Li et al. 2013), anacardic acid (de Araujo et al. 2021), quercetin (Zhang et al. 2023), and so forth have been shown to have strong antimicrobial activity against bacteria and fungi. Other phytochemicals that have been loaded into nanoparticles for antimicrobial applications include essential oils, flavonoids, and alkaloids. Apart from biomedical applications, these protein nanoparticle-based systems have potential applications in agriculture and food preservation, among other areas.

8.7 Disadvantages of Protein Nanocarriers

Protein nanocarriers have many potential advantages, such as biocompatibility, low toxicity, and the ability to be designed for targeted drug delivery. However, there are some disadvantages also associated with protein nanoparticles, including:

8.7.1 Stability

Their stability can be a challenge, particularly during storage and transport. Protein NPs can aggregate or precipitate over time, which can affect their size, shape, and drug-loading capacity (Kwan et al. 2022). This can be caused by changes in temperature, pH, ionic strength, or exposure to mechanical stress (Kwan et al. 2022). Aggregation can also occur due to protein-protein interactions or interactions with other molecules present in the formulation. Another challenge is the potential for protein denaturation or degradation, which can affect the structural integrity of the nanoparticle and its drug release properties (Rampado et al. 2020). This can be caused by exposure to environmental stressors such as light, heat, or oxidation. To overcome these challenges, several strategies can be employed to improve the stability of protein nanoparticles. One approach is to optimize the formulation conditions, such as pH, temperature, and ionic strength, to minimize protein-protein interactions and prevent aggregation (Cleland et al. 1993; Frokjaer and Otzen 2005). Another approach is to use stabilizing agents such as surfactants or polymers to prevent protein aggregation and stabilize the nanoparticle structure. Additionally, advanced characterization techniques such as dynamic light scattering, zeta potential measurement, and electron microscopy can be used to monitor the stability of protein nanoparticles over time and identify any changes in their size or shape.

8.7.2 Reproducibility and Variability

Proteins are natural polymers that can be heterogeneous in size and molecular weight. This variability can result in batch-to-batch variation during industrial production, which can affect the reproducibility and consistency of the final product (Kianfar 2021). However, there are several strategies that can be employed to minimize these variations and improve the reproducibility of protein production. One strategy is to use recombinant DNA technology to produce proteins with consistent molecular weights and characteristics (Hong et al. 2020). Recombinant DNA technology involves introducing the gene that encodes a protein of interest into a host organism, such as bacteria or yeast, and allowing the host to produce the protein. By controlling the conditions of protein production, such as temperature, pH, and nutrient availability, it is possible to achieve consistent protein yields and characteristics. Overall, while the heterogeneity of proteins can present challenges in industrial production, there are strategies available to minimize these variations and improve the reproducibility of protein production.

8.7.3 Manufacturing

Protein nanocarriers can be difficult to manufacture at a large scale and can require complex processes and expensive equipment.

8.7.4 Immunogenicity

Protein nanocarriers may trigger an immune response in some individuals, which can limit their use as drug delivery vehicles (Pondman et al. 2022). Immunogenicity is the ability of a foreign substance to provoke an immune response in the body. Protein nanoparticles are considered foreign by the immune system, which can trigger an immune response against the nanoparticles themselves or the drugs they carry (Kianfar 2021). This immune response can manifest as inflammation, allergies, or even anaphylactic shock in severe cases. To minimize the risk of immunogenicity, it is important to carefully design and engineer the protein nanoparticles to avoid triggering an immune response. This can involve modifying the protein nanoparticle surface with polymers or surfactants to reduce their interaction with the immune system (Gamucci et al. 2014). Another strategy is to use natural or biocompatible materials for nanoparticle synthesis to minimize immune recognition. Furthermore, advanced characterization techniques, such as in vitro and in vivo immunogenicity assays, can be used to assess the potential immunogenicity of protein nanoparticles during the development process. This can help identify and address any potential issues before the protein nanoparticles are used in clinical trials. Overall, while immunogenicity is a challenge associated with the use of protein nanoparticles, careful design, engineering, and characterization can help minimize this risk and ensure the safety and efficacy of these drug delivery systems.

8.7.5 Chances of Transmission of Animal Diseases

The use of animal-derived materials in the production of protein nanoparticles can potentially transmit animal diseases to humans. However, the risk can be minimized through appropriate sourcing, processing, and testing of the materials. Animal diseases such as bovine spongiform encephalopathy (BSE) or "mad cow disease" can be transmitted to humans through the consumption of contaminated animal products (Kianfar 2021). In the production of protein nanoparticles, animal-derived materials may come from tissues or fluids that could potentially contain infectious agents, such as prions or viruses. To minimize the risk of transmission, animal-derived materials should be sourced from healthy animals that are free of known infectious diseases. In addition, the materials should be subjected to appropriate processing and purification methods to remove or inactivate any potential infectious agents.

Despite these limitations, protein nanocarriers continue to be a promising area of research and development for drug delivery and other biomedical applications. Advances in protein engineering and nanoparticle formulation are helping to overcome some of these limitations, including improving stability and increasing cargo capacity.

8.8 Conclusion

Research on protein nanocarriers is acknowledged to play a significant role in the cure of various life-threatening diseases because they offer safer and more convenient treatment options. Properties such as biocompatibility, biodegradability, non-toxicity, and non-immunogenic properties make them more suitable nanocarriers compared to other nanocarriers. Protein NPs including albumin, zein, silk, collagen, gliadin, legumin, protamines, casein, elastin-like polypeptides, and virus-like particles have gained a lot of attention recently due to their promising uses in bioimaging, drug/gene delivery, disease diagnosis, and vaccine development. Phytoconstituents successfully delivered using protein nanocarriers are widely explored in the treatment of various diseases and show promising potential in the nanotechnology field, particularly in drug delivery. Finally, future advancements in targeted drug delivery will expand the uses of protein nanocarriers to deliver phytoconstituents more safely with maximum health-promoting benefits.

Acknowledgments The authors thank the Central University of Gujarat (CUG), Gandhinagar, for providing the necessary facilities and support. Raghu Solanki acknowledges the Council of Scientific and Industrial Research (CSIR) for providing a Senior Research Fellowship (SRF).

Conflict of Interest The authors declare no competing interests.

References

Abdel-Hakeem MA, Mongy S, Hassan B, Tantawi OI, Badawy I (2021) Curcumin loaded chitosan-protamine nanoparticles revealed antitumor activity via suppression of NF-κB, proinflammatory cytokines and Bcl-2 gene expression in the breast cancer cells. J Pharm Sci 110:3298–3305

Adjonu R, Doran G, Torley P, Agboola S (2014) Whey protein peptides as components of nanoemulsions: a review of emulsifying and biological functionalities. J Food Eng 122:15–27

Alqahtani MS, Syed R, Alshehri M (2020) Size-dependent phagocytic uptake and immunogenicity of gliadin nanoparticles. Polymers (Basel) 12:2576

Antçnio E, Khalil NM, Mainardes RM (2016) Bovine serum albumin nanoparticles containing quercetin: characterization and antioxidant activity. J Nanosci Nanotechnol 16:1346–1353

Aqil F, Munagala R, Jeyabalan J, Vadhanam MV (2013) Bioavailability of phytochemicals and its enhancement by drug delivery systems. Cancer Lett 334:133–141

Arangoa MA, Campanero MA, Renedo MJ, Ponchel G, Irache JM (2001) Gliadin nanoparticles as carriers for the oral administration of lipophilic drugs. Relationships between bioadhesion and pharmacokinetics. Pharm Res 18:1521–1527

Arun A, Malrautu P, Laha A, Luo H, Ramakrishna S (2021) Collagen nanoparticles in drug delivery systems and tissue engineering. Appl Sci 11:11369

Asadi M, Salami M, Hajikhani M, Emam-Djomeh Z, Aghakhani A, Ghasemi A (2021) Electrospray production of curcumin-walnut protein nanoparticles. Food Biophys 16:15–26

Bai F, Diao J, Wang Y, Sun S, Zhang H, Liu Y, Wang Y, Cao J (2017) A new water-soluble nanomicelle formed through self-assembly of pectin–curcumin conjugates: preparation, characterization, and anticancer activity evaluation. J Agric Food Chem 65:6840–6847

Barick KC, Tripathi A, Dutta B, Shelar SB, Hassan PA (2021) Curcumin encapsulated casein nanoparticles: enhanced bioavailability and anticancer efficacy. J Pharm Sci 110:2114–2120

Belyakova LE, Semenova MG, Antipova AS (1999) Effect of small molecule surfactants on molecular parameters and thermodynamic properties of legumin in a bulk and at the air–water interface depending on a protein structure in an aqueous medium. Colloids Surf B Biointerfaces 12:271–285

Bergonzi C, d'Ayala GG, Elviri L, Laurienzo P, Bandiera A, Catanzano O (2020) Alginate/human elastin-like polypeptide composite films with antioxidant properties for potential wound healing application. Int J Biol Macromol 164:586–596

Bhawani SA, Moheman A, Yakout AA, Ibrahim MNM (2019) Nanostructured biopolymers for application as drug-delivery vehicles. In: Nanostructured polymer composites for biomedical applications. Elsevier, Amsterdam, pp 189–210

Bidwell GL III, Raucher D (2005) Application of thermally responsive polypeptides directed against c-Myc transcriptional function for cancer therapy. Mol Cancer Ther 4:1076–1085

Bidwell GL III, Fokt I, Priebe W, Raucher D (2007) Development of elastin-like polypeptide for thermally targeted delivery of doxorubicin. Biochem Pharmacol 73:620–631

Bidwell GL, Davis AN, Fokt I, Priebe W, Raucher D (2007) A thermally targeted elastin-like polypeptide-doxorubicin conjugate overcomes drug resistance. Investig New Drugs 25:313–326

Bindhya KP, Maheswari PU, Begum KMMS (2021) Milk protein inspired multifunctional magnetic carrier targeting progesterone receptors: improved anticancer potential of soybean-derived genistein against breast and ovarian cancers. Mater Chem Phys 272:125055

Boer C, Meesters MI, Veerhoek D, Vonk ABA (2018) Anticoagulant and side-effects of protamine in cardiac surgery: a narrative review. Br J Anaesth 120:914–927

Bose T, Latawiec D, Mondal PP, Mandal S (2014) Overview of nano-drugs characteristics for clinical application: the journey from the entry to the exit point. J Nanopart Res 16:1–25

Buck ME, Tirrell DA (2012) Artificial proteins

Carissimi G, Montalbán MG, Fuster MG, Víllora G (2022) Silk fibroin nanoparticles: synthesis and applications as drug nanocarriers. In: 21st century nanostructured materials: physics, chemistry, classification, and emerging applications in industry, biomedicine, and agriculture. IntechOpen, London, p 205

Cascone MG, Lazzeri L, Carmignani C, Zhu Z (2002) Gelatin nanoparticles produced by a simple W/O emulsion as delivery system for methotrexate. J Mater Sci Mater Med 13:523–526

Chang C, Wang T, Hu Q, Luo Y (2017) Caseinate-zein-polysaccharide complex nanoparticles as potential oral delivery vehicles for curcumin: effect of polysaccharide type and chemical cross-linking. Food Hydrocoll 72:254–262

Chang C, Meikle TG, Su Y, Wang X, Dekiwadia C, Drummond CJ, Conn CE, Yang Y (2019) Encapsulation in egg white protein nanoparticles protects anti-oxidant activity of curcumin. Food Chem 280:65–72

Chaturvedi N (2003) Ethnic differences in cardiovascular disease. Heart 89:681–686

Chen T-HH, Bae Y, Furgeson DY (2008) Intelligent biosynthetic nanobiomaterials (IBNs) for hyperthermic gene delivery. Pharm Res 25:683–691

Cheng H, Dong H, Liang L (2020) A comparison of β casein complexes and micelles as vehicles for trans-/cis-resveratrol. Food Chem 330:127209

Chiu C-F, Fu R-H, Hsu S, Yu Y-HA, Yang S-F, Tsao TC-Y, Chang K-B, Yeh C-A, Tang C-M, Huang S-C (2021) Delivery capacity and anticancer ability of the berberine-loaded gold nanoparticles to promote the apoptosis effect in breast cancer. Cancers (Basel) 13:5317

Chrastina A, Baron VT, Abedinpour P, Rondeau G, Welsh J, Borgström P (2018) Plumbagin-loaded nanoemulsion drug delivery formulation and evaluation of antiproliferative effect on prostate cancer cells. Biomed Res Int 2018:9035452

Cialdai C, Lecci A, Lazzeri M, Giuliani S, Maggi CA, Turini D (2003) Effect of protamine pretreatment on detrusor hyperactivity induced by intravesical capsaicin in anaesthetized guinea-pigs. Neurourol Urodyn 22:16

Cicha I, Lyer S, Alexiou C, Garlichs CD (2013) Nanomedicine in diagnostics and therapy of cardiovascular diseases: beyond atherosclerotic plaque imaging. Nanotechnol Rev 2:449–472

Cimenci CE, Blackburn NJR, Sedlakova V, Pupkaite J, Munoz M, Rotstein BH, Spiegel DA, Alarcon EI, Suuronen EJ (2022) Combined methylglyoxal scavenger and collagen hydrogel therapy prevents adverse remodeling and improves cardiac function post-myocardial infarction. Adv Funct Mater 32:2108630

Cleland JL, Powell MF, Shire SJ (1993) The development of stable protein formulations: a close look at protein aggregation, deamidation, and oxidation. Crit Rev Ther Drug Carrier Syst 10: 307–377

da Rosa CG, de Oliveira Brisola Maciel MV, de Carvalho SM, de Melo APZ, Jummes B, da Silva T, Martelli SM, Villetti MA, Bertoldi FC, Barreto PLM (2015) Characterization and evaluation of physicochemical and antimicrobial properties of zein nanoparticles loaded with phenolics monoterpenes. Colloids Surf A Physicochem Eng Asp 481:337–344

Dahiya S, Rani R, Dhingra D, Kumar S, Dilbaghi N (2018) Conjugation of epigallocatechin gallate and piperine into a zein nanocarrier: implication on antioxidant and anticancer potential. Adv Nat Sci Nanosci Nanotechnol 9:35011

de Araujo JTC, Martin-Pastor M, Pérez L, Pinazo A, de Sousa FFO (2021) Development of anacardic acid-loaded zein nanoparticles: physical chemical characterization, stability and antimicrobial improvement. J Mol Liq 332:115808

de Souza Tavares W, Pena GR, Martin-Pastor M, de Sousa FFO (2021) Design and characterization of ellagic acid-loaded zein nanoparticles and their effect on the antioxidant and antibacterial activities. J Mol Liq 341:116915

Deen I, Rosei F (2019) Silk fibroin-derived polypeptides additives to promote hydroxyapatite nucleation in dense collagen hydrogels. PLoS One 14:e0219429

Desai N, Trieu V, Yao Z, Louie L, Ci S, Yang A, Tao C, De T, Beals B, Dykes D (2006) Increased antitumor activity, intratumor paclitaxel concentrations, and endothelial cell transport of cremophor-free, albumin-bound paclitaxel, ABI-007, compared with cremophor-based paclitaxel. Clin Cancer Res 12:1317–1324

Deshmukh K, Ahamed MB, Deshmukh RR, Pasha SKK, Bhagat PR, Chidambaram K (2017) Biopolymer composites with high dielectric performance: interface engineering. In: Biopolymer composites in electronics. Elsevier, Amsterdam, pp 27–128

Diez-Echave P, Ruiz-Malagón AJ, Molina-Tijeras JA, Hidalgo-García L, Vezza T, Cenis-Cifuentes L, Rodríguez-Sojo MJ, Cenis JL, Rodríguez-Cabezas ME, Rodríguez-Nogales A (2021) Silk fibroin nanoparticles enhance quercetin immunomodulatory properties in DSS-induced mouse colitis. Int J Pharm 606:120935

Ding J, Liang T, Min Q, Jiang L, Zhu J-J (2018) "Stealth and fully-laden" drug carriers: self-assembled nanogels encapsulated with epigallocatechin gallate and siRNA for drug-resistant breast cancer therapy. ACS Appl Mater Interfaces 10:9938–9948

Ding S, Zhang N, Lyu Z, Zhu W, Chang Y-C, Hu X, Du D, Lin Y (2021) Protein-based nanomaterials and nanosystems for biomedical applications: a review. Mater Today 43:166–184

Dubey S, Mody N, Sharma R, Agrawal U, Vyas SP (2016) Nanobiomaterials: novel nanoplatforms for protein and peptide delivery. In: Nanobiomaterials in drug delivery. Elsevier, Amsterdam, pp 111–146

Duclairoir C, Nakache E, Marchais H, Orecchioni A-M (1998) Formation of gliadin nanoparticles: influence of the solubility parameter of the protein solvent. Colloid Polym Sci 276:321–327

Duclairoir C, Orecchioni AM, Depraetere P, Nakache E (2002) α-Tocopherol encapsulation and in vitro release from wheat gliadin nanoparticles. J Microencapsul 19:53–60

Edis Z, Wang J, Waqas MK, Ijaz M, Ijaz M (2021) Nanocarriers-mediated drug delivery systems for anticancer agents: an overview and perspectives. Int J Nanomedicine 16:1313

Elzoghby AO, El-Fotoh WSA, Elgindy NA (2011) Casein-based formulations as promising controlled release drug delivery systems. J Control Release 153:206–216

Elzoghby AO, Samy WM, Elgindy NA (2012) Albumin-based nanoparticles as potential controlled release drug delivery systems. J Control Release 157:168–182

Elzoghby AO, Elgohary MM, Kamel NM (2015) Implications of protein-and peptide-based nanoparticles as potential vehicles for anticancer drugs. Adv Protein Chem Struct Biol 98: 169–221

Elzoghby AO, Mostafa SK, Helmy MW, ElDemellawy MA, Sheweita SA (2017) Multi-reservoir phospholipid shell encapsulating protamine nanocapsules for co-delivery of letrozole and celecoxib in breast cancer therapy. Pharm Res 34:1956–1969

Escobar-Chávez JJ, Díaz-Torres R, Rodríguez-Cruz IM, Domínguez-Delgado CL, Morales RS, Ángeles-Anguiano E, Melgoza-Contreras LM (2012) Nanocarriers for transdermal drug delivery. In: Research and reports in transdermal drug delivery, vol 1. Dove Press, London, p 3

Esmaili M, Ghaffari SM, Moosavi-Movahedi Z, Atri MS, Sharifizadeh A, Farhadi M, Yousefi R, Chobert J-M, Haertlé T, Moosavi-Movahedi AA (2011) Beta casein-micelle as a nano vehicle for solubility enhancement of curcumin; food industry application. LWT-Food Sci Technol 44: 2166–2172

Ezpeleta I, Irache JM, Stainmesse S, Chabenat C, Gueguen J, Popineau Y, Orecchioni A-M (1996) Gliadin nanoparticles for the controlled release of all-trans-retinoic acid. Int J Pharm 131:191–200

Fan Y, Zeng X, Yi J, Zhang Y (2020) Fabrication of pea protein nanoparticles with calcium-induced cross-linking for the stabilization and delivery of antioxidative resveratrol. Int J Biol Macromol 152:189–198

Fereydouni N, Movaffagh J, Amiri N, Darroudi S, Gholoobi A, Goodarzi A, Hashemzadeh A, Darroudi M (2021) Synthesis of nano-fibers containing nano-curcumin in zein corn protein and its physicochemical and biological characteristics. Sci Rep 11:1–15

Ferrado JB, Perez AA, Ruiz MC, León IE, Santiago LG (2020) Chrysin-loaded bovine serum albumin particles as bioactive nanosupplements. Food Funct 11:6007–6019

Flores AM, Ye J, Jarr K-U, Hosseini-Nassab N, Smith BR, Leeper NJ (2019) Nanoparticle therapy for vascular diseases. Arterioscler Thromb Vasc Biol 39:635–646

Freag MS, Saleh WM, Abdallah OY (2018) Laminated chitosan-based composite sponges for transmucosal delivery of novel protamine-decorated tripterine phytosomes: ex-vivo mucopenetration and in-vivo pharmacokinetic assessments. Carbohydr Polym 188:108–120

Frokjaer S, Otzen DE (2005) Protein drug stability: a formulation challenge. Nat Rev Drug Discov 4:298–306

Gamucci O, Bertero A, Gagliardi M, Bardi G (2014) Biomedical nanoparticles: overview of their surface immune-compatibility. Coatings 4:139–159

Gandhi S, Roy I (2021) Drug delivery applications of casein nanostructures: a minireview. J Drug Deliv Sci Technol 66:102843

Ghayour N, Hosseini SMH, Eskandari MH, Esteghlal S, Nekoei A-R, Gahruie HH, Tatar M, Naghibalhossaini F (2019) Nanoencapsulation of quercetin and curcumin in casein-based delivery systems. Food Hydrocoll 87:394–403

Ghosh P, Roy AS, Chaudhury S, Jana SK, Chaudhury K, Dasgupta S (2016) Preparation of albumin based nanoparticles for delivery of fisetin and evaluation of its cytotoxic activity. Int J Biol Macromol 86:408–417

Grigore ME, Grumezescu AM, Holban AM, Mogoşanu GD, Andronescu E (2017) Collagen-nanoparticles composites for wound healing and infection control. Metals (Basel) 7:516

Gulfam M, Kim J, Lee JM, Ku B, Chung BH, Chung BG (2012) Anticancer drug-loaded gliadin nanoparticles induce apoptosis in breast cancer cells. Langmuir 28:8216–8223

Gupta V, Aseh A, Ríos CN, Aggarwal BB, Mathur AB (2009) Fabrication and characterization of silk fibroin-derived curcumin nanoparticles for cancer therapy. Int J Nanomedicine 4:115

Haggag YA, Faheem AM (2015) Evaluation of nano spray drying as a method for drying and formulation of therapeutic peptides and proteins. Front Pharmacol 6:140

Han R, Zhu J, Yang X, Xu H (2011) Surface modification of poly (d, l-lactic-co-glycolic acid) nanoparticles with protamine enhanced cross-presentation of encapsulated ovalbumin by bone marrow-derived dendritic cells. J Biomed Mater Res A 96:142–149

Hassanin I, Elzoghby A (2020) Albumin-based nanoparticles: a promising strategy to overcome cancer drug resistance. Cancer Drug Resistance 3:930

He H, Ye J, Liu E, Liang Q, Liu Q, Yang VC (2014) Low molecular weight protamine (LMWP): a nontoxic protamine substitute and an effective cell-penetrating peptide. J Control Release 193: 63–73

Hong S, Choi DW, Kim HN, Park CG, Lee W, Park HH (2020) Protein-based nanoparticles as drug delivery systems. Pharmaceutics 12:604

Horne DS (2002) Casein structure, self-assembly and gelation. Curr Opin Colloid Interface Sci 7: 456–461

Huang X, Dai Y, Cai J, Zhong N, Xiao H, McClements DJ, Hu K (2017) Resveratrol encapsulation in core-shell biopolymer nanoparticles: impact on antioxidant and anticancer activities. Food Hydrocoll 64:157–165

Irache JM, Bergougnoux L, Ezpeleta I, Gueguen J, Orecchioni A-M (1995) Optimization and in vitro stability of legumin nanoparticles obtained by a coacervation method. Int J Pharm 126: 103–109

Jafari SM, McClements DJ (2017) Nanotechnology approaches for increasing nutrient bioavailability. Adv Food Nutr Res 81:1–30

Jain A, Singh SK, Arya SK, Kundu SC, Kapoor S (2018) Protein nanoparticles: promising platforms for drug delivery applications. ACS Biomater Sci Eng 4:3939–3961

Jangid AK, Solanki R, Patel S, Pooja D, Kulhari H (2022a) Genistein encapsulated inulin-stearic acid bioconjugate nanoparticles: formulation development, characterization and anticancer activity. Int J Biol Macromol 206:213–221

Jangid AK, Solanki R, Patel S, Medicherla K, Pooja D, Kulhari H (2022b) Improving anticancer activity of chrysin using tumor microenvironment pH-responsive and self-assembled nanoparticles. ACS Omega 7:15919–15928

Jayan H, Leena MM, Sundari SKS, Moses JA, Anandharamakrishnan C (2019) Improvement of bioavailability for resveratrol through encapsulation in zein using electrospraying technique. J Funct Foods 57:417–424

Jithan A, Madhavi K, Madhavi M, Prabhakar K (2011) Preparation and characterization of albumin nanoparticles encapsulating curcumin intended for the treatment of breast cancer. Int J Pharm Investig 1:119

Joye IJ, Davidov-Pardo G, McClements DJ (2015) Encapsulation of resveratrol in biopolymer particles produced using liquid antisolvent precipitation. Part 2: stability and functionality. Food Hydrocoll 49:127–134

Kabanov AV, Gendelman HE (2007) Nanomedicine in the diagnosis and therapy of neurodegenerative disorders. Prog Polym Sci 32:1054–1082

Kamble PR, Shaikh KS (2022) Optimization and evaluation of self-nanoemulsifying drug delivery system for enhanced bioavailability of plumbagin. Planta Med 88:79–90

Karri VVSR, Kuppusamy G, Talluri SV, Mannemala SS, Kollipara R, Wadhwani AD, Mulukutla S, Raju KRS, Malayandi R (2016) Curcumin loaded chitosan nanoparticles impregnated into collagen-alginate scaffolds for diabetic wound healing. Int J Biol Macromol 93:1519–1529

Karthikeyan S, Prasad NR, Ganamani A, Balamurugan E (2013) Anticancer activity of resveratrol-loaded gelatin nanoparticles on NCI-H460 non-small cell lung cancer cells. Biomed Prev Nutr 3:64–73

Kaur V, Kumar M, Kumar A, Kaur K, Dhillon VS, Kaur S (2018) Pharmacotherapeutic potential of phytochemicals: implications in cancer chemoprevention and future perspectives. Biomed Pharmacother 97:564–586

Kenchegowda M, Rahamathulla M, Hani U, Begum MY, Guruswamy S, Osmani RAM, Gowrav MP, Alshehri S, Ghoneim MM, Alshlowi A (2021) Smart nanocarriers as an emerging platform for cancer therapy: a review. Molecules 27:146

Kianfar E (2021) Protein nanoparticles in drug delivery: animal protein, plant proteins and protein cages, albumin nanoparticles. J Nanobiotechnol 19:159

Kim HC (2021) Epidemiology of cardiovascular disease and its risk factors in Korea. Glob Health Med 3:134–141

Kommareddy S, Shenoy DB, Amiji MM (2007) Gelatin nanoparticles and their biofunctionalization. In: Nanotechnologies for the life sciences. Wiley, New York

Kucharz EJ (2012) The collagens: biochemistry and pathophysiology. Springer Science & Business Media, Berlin

Kwan TOC, Kolek SA, Danson AE, Reis RI, Camacho IS, Shaw Stewart PD, Moraes I (2022) Measuring protein aggregation and stability using high-throughput biophysical approaches. Front Mol Biosci 9:890862

Labib G (2018) Overview on zein protein: a promising pharmaceutical excipient in drug delivery systems and tissue engineering. Expert Opin Drug Deliv 15:65–75

Lagoa R, Silva J, Rodrigues JR, Bishayee A (2020) Advances in phytochemical delivery systems for improved anticancer activity. Biotechnol Adv 38:107382

Langer K, Balthasar S, Vogel V, Dinauer N, Von Briesen H, Schubert D (2003) Optimization of the preparation process for human serum albumin (HSA) nanoparticles. Int J Pharm 257:169–180

Lee CH, Singla A, Lee Y (2001) Biomedical applications of collagen. Int J Pharm 221:1–22

Lee SH, Heng D, Ng WK, Chan H-K, Tan RBH (2011) Nano spray drying: a novel method for preparing protein nanoparticles for protein therapy. Int J Pharm 403:192–200

Leng Q, Li Y, Pang X, Wang B, Wu Z, Lu Y, Xiong K, Zhao L, Zhou P, Fu S (2020) Curcumin nanoparticles incorporated in PVA/collagen composite films promote wound healing. Drug Deliv 27:1676–1685

Li K-K, Yin S-W, Yin Y-C, Tang C-H, Yang X-Q, Wen S-H (2013) Preparation of water-soluble antimicrobial zein nanoparticles by a modified antisolvent approach and their characterization. J Food Eng 119:343–352

Li C, Cai G, Song D, Gao R, Teng P, Zhou L, Ji Q, Sui H, Cai J, Li Q (2019) Development of EGFR-targeted evodiamine nanoparticles for the treatment of colorectal cancer. Biomater Sci 7: 3627–3639

Lima LF, Sousa MGDC, Rodrigues GR, de Oliveira KBS, Pereira AM, da Costa A, Machado R, Franco OL, Dias SC (2022) Elastin-like polypeptides in development of nanomaterials for application in the medical field. Front Nanotechnol 4:31

Ling D, Hackett MJ, Hyeon T (2014) Surface ligands in synthesis, modification, assembly and biomedical applications of nanoparticles. Nano Today 9:457–477

Liu Q, Han C, Tian Y, Liu T (2020) Fabrication of curcumin-loaded zein nanoparticles stabilized by sodium caseinate/sodium alginate: curcumin solubility, thermal properties, rheology, and stability. Process Biochem 94:30–38

Liu L, Yang S, Chen F, Cheng K-W (2022) Hyaluronic acid–zein core-shell nanoparticles improve the anticancer effect of curcumin alone or in combination with oxaliplatin against colorectal cancer via CD44-mediated cellular uptake. Molecules 27:1498

Livney YD (2010) Milk proteins as vehicles for bioactives. Curr Opin Colloid Interface Sci 15:73–83

Lo S, Fauzi MB (2021) Current update of collagen nanomaterials—fabrication, characterisation and its applications: a review. Pharmaceutics 13:316

Lohcharoenkal W, Wang L, Chen YC, Rojanasakul Y (2014) Protein nanoparticles as drug delivery carriers for cancer therapy. Biomed Res Int 2014:180549

Luo Y, Pan K, Zhong Q (2015) Casein/pectin nanocomplexes as potential oral delivery vehicles. Int J Pharm 486:59–68

Ma Y, Nolte RJM, Cornelissen JJLM (2012) Virus-based nanocarriers for drug delivery. Adv Drug Deliv Rev 64:811–825

MacKay JA, Chen M, McDaniel JR, Liu W, Simnick AJ, Chilkoti A, Conjugate SCP (2009) Self-assembling chimeric polypeptide-doxorubicin conjugate nanoparticles that abolish tumours after a single injection. Nat Mater 8:993–999

Madan J, Dhiman N, Sardana S, Aneja R, Chandra R, Katyal A (2011) Long-circulating poly (ethylene glycol)-grafted gelatin nanoparticles customized for intracellular delivery of noscapine: preparation, in-vitro characterization, structure elucidation, pharmacokinetics, and cytotoxicity analyses. Anti-Cancer Drugs 22:543–555

Majidzadeh H, Araj-Khodaei M, Ghaffari M, Torbati M, Dolatabadi JEN, Hamblin MR (2020) Nano-based delivery systems for berberine: a modern anti-cancer herbal medicine. Colloids Surf B Biointerfaces 194:111188

Majorek KA, Porebski PJ, Dayal A, Zimmerman MD, Jablonska K, Stewart AJ, Chruszcz M, Minor W (2012) Structural and immunologic characterization of bovine, horse, and rabbit serum albumins. Mol Immunol 52:174–182

Malekhosseini P, Alami M, Khomeiri M, Esteghlal S, Nekoei A, Hosseini SMH (2019) Development of casein-based nanoencapsulation systems for delivery of epigallocatechin gallate and folic acid. Food Sci Nutr 7:519–527

Massodi I, Bidwell GL III, Raucher D (2005) Evaluation of cell penetrating peptides fused to elastin-like polypeptide for drug delivery. J Control Release 108:396–408

Meng R, Zhu H, Wang Z, Hao S, Wang B (2022) Preparation of drug-loaded albumin nanoparticles and its application in cancer therapy. J Nanomater 2022:3052175

Miele E, Spinelli GP, Miele E, Tomao F, Tomao S (2009) Albumin-bound formulation of paclitaxel (Abraxane® ABI-007) in the treatment of breast cancer. Int J Nanomedicine 4:99

Mirshahi T, Irache JM, Gueguen J, Orecchioni AM (1996) Development of drug delivery systems from vegetal proteins: legumin nanoparticles. Drug Dev Ind Pharm 22:841–846

Mirshahi T, Irache JM, Nicolas C, Mirshahi M, Faure JP, Gueguen J, Hecquet C, Orecchioni AM (2002) Adaptive immune responses of legumin nanoparticles. J Drug Target 10:625–631

Mishra V, Mahor S, Rawat A, Gupta PN, Dubey P, Khatri K, Vyas SP (2006) Targeted brain delivery of AZT via transferrin anchored pegylated albumin nanoparticles. J Drug Target 14:45–53

Mohiti-Asli M, Loboa EG (2016) Nanofibrous smart bandages for wound care. In: Wound healing biomaterials. Elsevier, Amsterdam, pp 483–499

Moradi SZ, Momtaz S, Bayrami Z, Farzaei MH, Abdollahi M (2020) Nanoformulations of herbal extracts in treatment of neurodegenerative disorders. Front Bioeng Biotechnol 8:238

More MP, Pardeshi SR, Pardeshi CV, Sonawane GA, Shinde MN, Deshmukh PK, Naik JB, Kulkarni AD (2021) Recent advances in phytochemical-based nano-formulation for drug-resistant Cancer. Med Drug Discov 10:100082

Nagarajan U, Kawakami K, Zhang S, Chandrasekaran B, Nair BU (2014) Fabrication of solid collagen nanoparticles using electrospray deposition. Chem Pharm Bull(Tokyo) 62:422–428

Narayanaswamy R, Kanagesan S, Pandurangan A, Padmanabhan P (2016) Basics to different imaging techniques, different nanobiomaterials for image enhancement. In: Nanobiomaterials in medical imaging. Elsevier, Amsterdam, pp 101–129

Nemeroff CB, Goldschmidt-Clermont PJ (2012) Heartache and heartbreak—the link between depression and cardiovascular disease. Nat Rev Cardiol 9:526–539

Nicklas M, Schatton W, Heinemann S, Hanke T, Kreuter J (2009) Preparation and characterization of marine sponge collagen nanoparticles and employment for the transdermal delivery of 17β-estradiol-hemihydrate. Drug Dev Ind Pharm 35:1035–1042

Ning S, Jorfi M, Patel SR, Kim DY, Tanzi RE (2022) Neurotechnological approaches to the diagnosis and treatment of Alzheimer's disease. Front Neurosci 16:854992

Niu B, Guo J, Guo X, Sun X, Rao C, Liu C, Zhang J, Zhang C, Fan Y-Y, Li W (2020) (NaPO3) 6-assisted formation of dispersive casein-amorphous calcium phosphate nanoparticles: an excellent platform for curcumin delivery. J Drug Deliv Sci Technol 55:101412

Nooraei S, Bahrulolum H, Hoseini ZS, Katalani C, Hajizade A, Easton AJ, Ahmadian G (2021) Virus-like particles: preparation, immunogenicity and their roles as nanovaccines and drug nanocarriers. J Nanobiotechnol 19:1–27

Nosrati H, Rakhshbahar A, Salehiabar M, Afroogh S, Manjili HK, Danafar H, Davaran S (2018a) Bovine serum albumin: an efficient biomacromolecule nanocarrier for improving the therapeutic efficacy of chrysin. J Mol Liq 271:639–646

Nosrati H, Abbasi R, Charmi J, Rakhshbahar A, Aliakbarzadeh F, Danafar H, Davaran S (2018b) Folic acid conjugated bovine serum albumin: an efficient smart and tumor targeted biomacromolecule for inhibition folate receptor positive cancer cells. Int J Biol Macromol 117:1125–1132

Obregón R, Ramón-Azcón J, Ahadian S (2017) Nanofiber composites in blood vessel tissue engineering. In: Nanofiber composites for biomedical applications. Elsevier, Amsterdam, pp 483–506

Odeh F, Al-Jaber H, Khater D (2014) Nanoflora—how nanotechnology enhanced the use of active phytochemicals. In: Application of nanotechnology in drug delivery. IntechOpen, London, pp 343–368

Owens DR (2011) Insulin preparations with prolonged effect. Diabetes Technol Ther 13:S-5

Paliwal R, Palakurthi S (2014) Zein in controlled drug delivery and tissue engineering. J Control Release 189:108–122

Pan K, Zhong Q, Baek SJ (2013) Enhanced dispersibility and bioactivity of curcumin by encapsulation in casein nanocapsules. J Agric Food Chem 61:6036–6043

Pan K, Luo Y, Gan Y, Baek SJ, Zhong Q (2014) pH-driven encapsulation of curcumin in self-assembled casein nanoparticles for enhanced dispersibility and bioactivity. Soft Matter 10: 6820–6830

Patel K, Jain P, Rajput PK, Jangid AK, Solanki R, Kulhari H, Patel S (2022) Human serum albumin-based propulsive Piperlongumine-loaded nanoparticles: formulation development, characterization and anti-cancer study. Colloids Surf A Physicochem Eng Asp 652:129738

Peñalva R, Morales J, González-Navarro CJ, Larrañeta E, Quincoces G, Peñuelas I, Irache JM (2018) Increased oral bioavailability of resveratrol by its encapsulation in casein nanoparticles. Int J Mol Sci 19:2816

Peñalva R, Esparza I, Morales-Gracia J, González-Navarro CJ, Larrañeta E, Irache JM (2019) Casein nanoparticles in combination with 2-hydroxypropyl-β-cyclodextrin improves the oral bioavailability of quercetin. Int J Pharm 570:118652

Pondman K, Le Gac S, Kishore U (2022) Nanoparticle-induced immune response: health risk versus treatment opportunity? Immunobiology 228(2):152317

Prakash S, Bansal P, Kumar A, Saxena V, Kumar V, Katiyar D (2022) Overview and in-silico pharmacological profiling of gliadin: a potential biomaterial. Mater Today Proc 62:276–282

Qi L, Guo Y, Luan J, Zhang D, Zhao Z, Luan Y (2014) Folate-modified bexarotene-loaded bovine serum albumin nanoparticles as a promising tumor-targeting delivery system. J Mater Chem B 2:8361–8371

Radhakrishnan K, Gupta S, Gnanadhas DP, Ramamurthy PC, Chakravortty D, Raichur AM (2014) Protamine-capped mesoporous silica nanoparticles for biologically triggered drug release. Part Part Syst Charact 31:449–458

Ragelle H, Crauste-Manciet S, Seguin J, Brossard D, Scherman D, Arnaud P, Chabot GG (2012) Nanoemulsion formulation of fisetin improves bioavailability and antitumour activity in mice. Int J Pharm 427:452–459

Rahimnejad M, Mokhtarian N, Ghasemi M (2009) Production of protein nanoparticles for food and drug delivery system. Afr J Biotechnol 8:4738–4743

Raj LFAA, Jonisha R, Revathi B, Jayalakshmy E (2015) Preparation and characterization of BSA and chitosan nanopartices for sustainable delivery system for quercetin. J Appl Pharm Sci 5:1–5

Rajalakshmi S, Vyawahare N, Pawar A, Mahaparale P, Chellampillai B (2018) Current development in novel drug delivery systems of bioactive molecule plumbagin. Artif Cells Nanomed Biotechnol 46:209–218

Rampado R, Crotti S, Caliceti P, Pucciarelli S, Agostini M (2020) Recent advances in understanding the protein corona of nanoparticles and in the formulation of "stealthy" nanomaterials. Front Bioeng Biotechnol 8:166

Rao KM, Rao KSVK, Ramanjaneyulu G, Ha C-S (2015) Curcumin encapsulated pH sensitive gelatin based interpenetrating polymeric network nanogels for anti cancer drug delivery. Int J Pharm 478:788–795

Ratanavaraporn J, Kanokpanont S, Damrongsakkul S (2014) The development of injectable gelatin/silk fibroin microspheres for the dual delivery of curcumin and piperine. J Mater Sci Mater Med 25:401–410
Rathore P, Arora I, Rastogi S, Akhtar M, Singh S, Samim M (2020) Collagen nanoparticle-mediated brain silymarin delivery: an approach for treating cerebral ischemia and reperfusion-induced brain injury. Front Neurosci 14:538404
Rehan F, Ahemad N, Gupta M (2019) Casein nanomicelle as an emerging biomaterial—a comprehensive review. Colloids Surf B Biointerfaces 179:280–292
Rezaii M, Oryan S, Javeri A (2019) Curcumin nanoparticles incorporated collagen-chitosan scaffold promotes cutaneous wound healing through regulation of TGF-β1/Smad7 gene expression. Mater Sci Eng C 98:347–357
Rostamabadi H, Falsafi SR, Rostamabadi MM, Assadpour E, Jafari SM (2021) Electrospraying as a novel process for the synthesis of particles/nanoparticles loaded with poorly water-soluble bioactive molecules. Adv Colloid Interf Sci 290:102384
Rout GK, Shin H-S, Gouda S, Sahoo S, Das G, Fraceto LF, Patra JK (2018) Current advances in nanocarriers for biomedical research and their applications. Artif Cells Nanomed Biotechnol 46: 1053–1062
Rudramurthy GR, Swamy MK, Sinniah UR, Ghasemzadeh A (2016) Nanoparticles alternatives against drug-resistant pathogenic microbes. Molecules 21(7):836. https://doi.org/10.3390/molecules21070836. PMID 27355939
Ruseska I, Fresacher K, Petschacher C, Zimmer A (2021) Use of protamine in nanopharmaceuticals—a review. Nanomaterials 11:1508
Sahoo N, Sahoo RK, Biswas N, Guha A, Kuotsu K (2015) Recent advancement of gelatin nanoparticles in drug and vaccine delivery. Int J Biol Macromol 81:317–331
Sahoo S, Kariya T, Ishikawa K (2021) Targeted delivery of therapeutic agents to the heart. Nat Rev Cardiol 18:389–399
Salehiabar M, Nosrati H, Javani E, Aliakbarzadeh F, Kheiri Manjili H, Davaran S, Danafar H (2018) Production of biological nanoparticles from bovine serum albumin as controlled release carrier for curcumin delivery. Int J Biol Macromol 115:83–89
Saxena A, Sachin K, Bohidar HB, Verma AK (2005) Effect of molecular weight heterogeneity on drug encapsulation efficiency of gelatin nano-particles. Colloids Surf B Biointerfaces 45:42–48
Seib FP (2017) Silk nanoparticles—an emerging anticancer nanomedicine. AIMS Bioeng 42:239–258
Seok H-Y, Rejinold NS, Lekshmi KM, Cherukula K, Park I-K, Kim Y-C (2018) CD44 targeting biocompatible and biodegradable hyaluronic acid cross-linked zein nanogels for curcumin delivery to cancer cells: in vitro and in vivo evaluation. J Control Release 280:20–30
Shamji MF, Chen J, Friedman AH, Richardson WJ, Chilkoti A, Setton LA (2008) Synthesis and characterization of a thermally-responsive tumor necrosis factor antagonist. J Control Release 129:179–186
Sharif N, Golmakani M-T, Hajjari MM (2022) Integration of physicochemical, molecular dynamics, and in vitro evaluation of electrosprayed γ-oryzanol-loaded gliadin nanoparticles. Food Chem 395:133589
Sharifi-Rad A, Mehrzad J, Darroudi M, Saberi MR, Chamani J (2021) Oil-in-water nanoemulsions comprising berberine in olive oil: biological activities, binding mechanisms to human serum albumin or holo-T. J Biomol Struct Dyn 39:1029–1043
Shinde P, Agraval H, Singh A, Yadav UCS, Kumar U (2019) Synthesis of luteolin loaded zein nanoparticles for targeted cancer therapy improving bioavailability and efficacy. J Drug Deliv Sci Technol 52:369–378
Shukla A, Narayanan T, Zanchi D (2009) Structure of casein micelles and their complexation with tannins. Soft Matter 5:2884–2888
Siddiqui MA, Akhter J, Bashir DJ, Manzoor S, Rastogi S, Arora I, Aggarwal NB, Samim M (2021) Resveratrol loaded nanoparticles attenuate cognitive impairment and inflammatory markers in PTZ-induced kindled mice. Int Immunopharmacol 101:108287

Siddiqui MA, Asad M, Akhter J, Hoda U, Rastogi S, Arora I, Aggarwal NB, Samim M (2022) Resveratrol-loaded glutathione-coated collagen nanoparticles attenuate acute seizures by inhibiting HMGB1 and TLR-4 in the hippocampus of mice. ACS Chem Neurosci 13:1342–1354

Singh VK, Arora D, Ansari MI, Sharma PK (2019) Phytochemicals based chemopreventive and chemotherapeutic strategies and modern technologies to overcome limitations for better clinical applications. Phytother Res 33:3064–3089

Solanki R, Patel K, Patel S (2021a) Bovine serum albumin nanoparticles for the efficient delivery of berberine: preparation, characterization and in vitro biological studies. Colloids Surf A Physicochem Eng Asp 608:125501

Solanki R, Rostamabadi H, Patel S, Jafari SM (2021b) Anticancer nano-delivery systems based on bovine serum albumin nanoparticles: a critical review. Int J Biol Macromol 193:528–540

Solanki R, Jodha B, Prabina KE, Aggarwal N, Patel S (2022) Recent advances in phytochemical based nano-drug delivery systems to combat breast cancer: a review. J Drug Deliv Sci Technol 77:103832

Solanki R, Saini M, Mochi J, Pappachan A Patel S (2023) Synthesis, characterization, in-silico and in-vitro anticancer studies of Plumbagin encapsulated albumin nanoparticles for breast cancer treatment. J Drug Deliv Sci Technol 84:104501

Spada A, Emami J, Tuszynski JA, Lavasanifar A (2021) The uniqueness of albumin as a carrier in nanodrug delivery. Mol Pharm 18:1862–1894

Sridhar R, Ramakrishna S (2013) Electrosprayed nanoparticles for drug delivery and pharmaceutical applications. Biomatter 3:e24281

Sripriyalakshmi S, Jose P, Ravindran A, Anjali CH (2014) Recent trends in drug delivery system using protein nanoparticles. Cell Biochem Biophys 70:17–26

Stan D, Enciu A-M, Mateescu AL, Ion AC, Brezeanu AC, Stan D, Tanase C (2021) Natural compounds with antimicrobial and antiviral effect and nanocarriers used for their transportation. Front Pharmacol 12:723233

Stinchcombe TE (2007) Nanoparticle albumin-bound paclitaxel: a novel Cremphor-EL®-free formulation of paclitaxel. Nanomedicine 2(4):415–423

Sun H, Luo Q, Hou C, Liu J (2017) Nanostructures based on protein self-assembly: from hierarchical construction to bioinspired materials. Nano Today 14:16–41

Süngüç C (2013) Encapsulation of sarcopoterim spinosum extract in zein perticle by using electrospray method. Izmir Institute of Technology (Turkey), Izmir

Tang C (2021) Assembled milk protein nano-architectures as potential nanovehicles for nutraceuticals. Adv Colloid Interf Sci 292:102432

Tarhini M, Greige-Gerges H, Elaissari A (2017) Protein-based nanoparticles: from preparation to encapsulation of active molecules. Int J Pharm 522:172–197

Tariq H, Batool S, Asif S, Ali M, Abbasi BH (2022) Virus-like particles: revolutionary platforms for developing vaccines against emerging infectious diseases. Front Microbiol 12:790121

Terzopoulou Z, Michopoulou A, Palamidi A, Koliakou E, Bikiaris D (2020) Preparation and evaluation of collagen-based patches as curcumin carriers. Polymers (Basel) 12:2393

Tsai Y-H, Yang Y-N, Ho Y-C, Tsai M-L, Mi F-L (2018) Drug release and antioxidant/antibacterial activities of silymarin-zein nanoparticle/bacterial cellulose nanofiber composite films. Carbohydr Polym 180:286–296

Verma D, Gulati N, Kaul S, Mukherjee S, Nagaich U (2018) Protein based nanostructures for drug delivery. J Pharm (Cairo) 2018:9285854

Voci S, Gagliardi A, Fresta M, Cosco D (2022) Ascorbic acid-loaded gliadin nanoparticles as a novel nutraceutical formulation. Food Res Int 161:111869

Wang S, Su R, Nie S, Sun M, Zhang J, Wu D, Moustaid-Moussa N (2014) Application of nanotechnology in improving bioavailability and bioactivity of diet-derived phytochemicals. J Nutr Biochem 25:363–376

Wang J, Jansen JA, Yang F (2019) Electrospraying: possibilities and challenges of engineering carriers for biomedical applications—a mini review. Front Chem 7:258

Won Y-W, Kim Y-H (2009) Preparation and cytotoxicity comparison of type A gelatin nanoparticles with recombinant human gelatin nanoparticles. Macromol Res 17:464–468

World Health Organization (n.d.). https://www.who.int/en/news-room/fact-sheets/detail/cardiovascular-diseases-(cvds). Cardiovascular diseases (CVDs) fact sheet

Wu W, Kong X, Zhang C, Hua Y, Chen Y, Li X (2020) Fabrication and characterization of resveratrol-loaded gliadin nanoparticles stabilized by gum Arabic and chitosan hydrochloride. LWT 129:109532

Xu H, Hua Y, Zhong J, Li X, Xu W, Cai Y, Mao Y, Lu X (2018) Resveratrol delivery by albumin nanoparticles improved neurological function and neuronal damage in transient middle cerebral artery occlusion rats. Front Pharmacol 9:1403

Xu G, Li L, Bao X, Yao P (2020) Curcumin, casein and soy polysaccharide ternary complex nanoparticles for enhanced dispersibility, stability and oral bioavailability of curcumin. Food Biosci 35:100569

Xue Y, Zhou S, Fan C, Du Q, Jin P (2019) Enhanced antifungal activities of eugenol-entrapped casein nanoparticles against anthracnose in postharvest fruits. Nanomaterials 9:1777

Yadav S, Sharma AK, Kumar P (2020) Nanoscale self-assembly for therapeutic delivery. Front Bioeng Biotechnol 8:127

Yang R, Zheng Y, Wang Q, Zhao L (2018) Curcumin-loaded chitosan–bovine serum albumin nanoparticles potentially enhanced Aβ 42 phagocytosis and modulated macrophage polarization in Alzheimer's disease. Nanoscale Res Lett 13:1–9

Yang X, Leng M, Yang L, Peng Y, Wang J, Wang Q, Wu K, Zou J, Wan W, Li L (2022) Effect of evodiamine on collagen-induced platelet activation and thrombosis. Biomed Res Int 2022: 4893859

Yu X, Wu H, Hu H, Dong Z, Dang Y, Qi Q, Wang Y, Du S, Lu Y (2020) Zein nanoparticles as nontoxic delivery system for maytansine in the treatment of non-small cell lung cancer. Drug Deliv 27:100–109

Zare-Zardini H, Soltaninejad H, Ghorani-Azam A, Nafisi-Moghadam R, Haddadzadegan N, Ansari M, Saeed-Banadaki SH, Sobhan MR, Mozafari S, Zahedi M (2022) Slow release curcumin-containing soy protein nanoparticles as anticancer agents for osteosarcoma: synthesis and characterization. Prog Biomater 11:311–320

Zhang Y-T, Huang Z-B, Zhang S-J, Zhao J-H, Wang Z, Liu Y, Feng N-P (2012) In vitro cellular uptake of evodiamine and rutaecarpine using a microemulsion. Int J Nanomedicine 7:2465

Zhang F, Khan MA, Cheng H, Liang L (2019) Co-encapsulation of α-tocopherol and resveratrol within zein nanoparticles: impact on antioxidant activity and stability. J Food Eng 247:9–18

Zhang X, Miao F, Niu L, Wei Y, Hu Y, Lian X, Zhao L, Chen W, Huang D (2021a) Berberine carried gelatin/sodium alginate hydrogels with antibacterial and EDTA-induced detachment performances. Int J Biol Macromol 181:1039–1046

Zhang H, van Os WL, Tian X, Zu G, Ribovski L, Bron R, Bussmann J, Kros A, Liu Y, Zuhorn IS (2021b) Development of curcumin-loaded zein nanoparticles for transport across the blood–brain barrier and inhibition of glioblastoma cell growth. Biomater Sci 9:7092–7103

Zhang Z, Hu Y, Ji H et al (2023) Physicochemical stability, antioxidant activity, and antimicrobial activity of quercetin-loaded zein nanoparticles coated with dextrin-modified anionic polysaccharides. Food Chem 415:135736

Zhao Z, Ukidve A, Kim J, Mitragotri S (2020) Targeting strategies for tissue-specific drug delivery. Cell 181:151–167

Zheng Y, Xiao L, Yu C, Jin P, Qin D, Xu Y, Yin J, Liu Z, Du Q (2019) Enhanced antiarthritic efficacy by nanoparticles of (−)-epigallocatechin gallate–glucosamine–casein. J Agric Food Chem 67:6476–6486

Zugic A, Tadic V, Savic S (2020) Nano- and microcarriers as drug delivery systems for usnic acid: review of literature. Pharmaceutics 12:156

Dendrimers-Mediated Delivery of Phytoconstituents

9

Divya Bharti Rai, Kanakraju Medicherla, Deep Pooja, and Hitesh Kulhari

9.1 Introduction

Dendrimers are nanostructures that belong to the fourth class of polymer. They are highly branched, symmetrical macromolecules with a tree-like shape, synthesized by iterative chemical reactions that add layers of branches to a central core. The theoretical idea about dendrimers was first elucidated by Flory et al. in 1941, describing it as a three-dimensional branched molecule (Flory 1941). F. Vögtle made the initial effort to synthesize lower-generation dendron-like structures in 1978, followed by R. G. Denkewalter at Allied Corporation in 1981 (2) and D. Tomalia at Dow Chemical in 1983 (Tomalia et al. 1986, 1990). G. R. Newkome in 1985 created the first full-generation dendrimers using the divergent method (Newkome et al. 1986, 1990), also known as the inside-out method in which the dendrimer grows outward from a core (Walter and Malkoch 2012). Later, C. Hawker and J. Fréchet developed the convergent synthetic technique in 1990, as an alternate growth method for building dendrimers (Wendland and Zimmerman 1999). It is an outside-in synthesis in which monomers are covalently bonded to produce branching dendron wedges, which are then connected to a multifunctional core that makes up a dendrimer (Hawker and Frechet 1990). There are various dendrimer shapes that are classified into more than 50 families and are occasionally known as asterisks,

D. B. Rai · H. Kulhari (✉)
School of Nano Sciences, Central University of Gujarat, Gandhinagar, Gujarat, India
e-mail: hitesh.kulhari@cug.ac.in

K. Medicherla
Department of Human Genetics, College of Science and Technology, Andhra University, Visakhapatnam, Andhra Pradesh, India

D. Pooja
School of Pharmacy, National Forensic Science University, Gandhinagar, Gujarat, India

© The Author(s), under exclusive license to Springer Nature Singapore Pte Ltd. 2024

D. Pooja, H. Kulhari (eds.), *Nanotechnology Based Delivery of Phytoconstituents and Cosmeceuticals*, https://doi.org/10.1007/978-981-99-5314-1_9

arborols, cascade molecules, cauliflower polymers, starburst polymers, and star-shaped nanomolecules (Tomalia and Fréchet 2002; Newkome et al. 1985).

Structurally, dendrimers are spherical macromolecules that vary in size from 1 to 100 nm, made up of 3 distinguished regions: a central core, a hyperbranched interior with cavities, and an outer shell with exterior reactive functional groups (Al-Jamal et al. 2005). Dendrimers possess consistent structural characteristics because of the great degree of control over the synthesis of dendritic architecture (Malkoch and García-Gallego 2020). This structural uniformity of dendrimers contributes to their special physicochemical and biological features, enabling a range of applications (Klajnert-Maculewicz and Bryszewska 2001; Abbasi et al. 2014). The ability to construct dendrimers with specified features and functions is one of the advantages of its controlled synthesis process attributed to their highly defined, yet tunable structure. They are helpful for the delivery of active pharmaceutical ingredients (APIs) because they have a high surface area-to-volume ratio and a lot of reactive groups (Samad et al. 2009; Santander-Ortega et al. 2013; Svenson and Tomalia 2012; Noriega-Luna et al. 2014). Additionally, dendrimers may be functionalized with a variety of chemical groups, enabling them to be customized for certain uses (Fischer and Vögtle 1999). Many different dendrimers, including polyamidoamine (PAMAM), poly (ethylenimine) (PEI), poly (propylene imine) (PPI), poly (glycerol-*co*-succinic acid) (PGSA), poly-L-lysine (PLL), melamine, triazine, etc., have been effectively used for the transport of bioactive compounds (Gillies and Fréchet 2005).

Phytoconstituents (PCs), generally known as phytochemicals, are the newly explored class of APIs. These are naturally occurring chemical constituents found in plants and believed to have specific biological effects contributing to their medicinal effects (Nahata 2017; Khan and Gurav 2018; Ouyang et al. 2014; Chirumbolo 2012). PC includes alkaloids, flavonoids, terpenoids, phenolic acids, glycosides, polyphenols, and tannins (Shahzad et al. 2017; Jin-Jian et al. 2012; Oberlies and Kroll 2004; Stahl and Sies 2005). PC provides the advantage of a natural and holistic treatment approach, ample availability, easy accessibility, low cost, fewer side effects, and multiple biological activities—in contrast to synthetic drugs (Karimi et al. 2015; Levin 1971). But it is important to note that while PCs have been traditionally used for their potential health benefits, their efficacy and safety may vary and advanced formulations such as dendrimer-based systems of PCs are required to be developed for their consistent and efficient utilization as a medicine (Chirumbolo 2012; Pan et al. 2013; Nisar et al. 2018).

9.2 Physicochemical and Biological Principles of Dendrimers

Dendrimers are appropriate for different purposes due to their tunable physicochemical features, including size, shape, branch length, surface functionality, and production of tailored dendritic scaffolds. The biological aspects of dendrimers include their interactions with cell membranes and biological molecules (Chis et al. 2020).

9.2.1 Physicochemical Principles

Overall, the physicochemical principles of dendrimers include properties like their high surface area and solubility which are largely determined by their size, shape, and functional groups, and can be tailored through careful design and synthesis. In addition to their high surface area and solubility, dendrimers also have many other physicochemical properties that make them useful in a variety of applications (Imae 2012).

9.2.1.1 Well-Defined Topology

The stepwise growth process of dendrimers in which monomer units are added to a central core results in a well-defined structure with less size variation and a uniform chemical composition. The well-defined structure of dendrimers provides several attributes including monodispersity and stability (Hu et al. 2009).

9.2.1.2 Versatility

The highly branched structure of dendrimers and the presence of numerous terminal groups allows for functionalization with a wide range of chemical groups, making them versatile materials for a variety of applications (Simonescu 2018; Majoros et al. 2008). For instance, dendrimers can employ molecular markers to make detection and bio-distribution tracking easier by chemically incorporating APIs or ligands to target particular receptors (Menjoge et al. 2010). The great density of surface functional groups on dendrimers makes it possible to functionalize these nanocarriers with various targeting moieties, nucleic acids, and imaging agents (Gupta et al. 2020; Jang et al. 2009).

9.2.1.3 High Surface Area

One major advantage of dendrimers is their high surface area-to-volume ratio, which is a result of their highly branched structure. This high surface area allows dendrimers to have a large number of reactive groups and makes them useful for PC delivery, where high reactivity is useful (Garg et al. 2011).

9.2.1.4 Monomolecular

Dendrimers are monomolecular due to the reason that the branches are formed through a series of controlled reactions. Hence, the branches are evenly distributed and the final structure of the dendrimer molecule is homogeneous. One of the key advantages of monomolecular dendrimers is their monodispersity, which means that all the molecules in a sample have the same size and shape. The narrow size distribution of dendrimers makes it easier to control their behavior and properties (Hawker and Frechet 1990). This is in contrast to traditional polymers, which have uneven molecular weights and structures.

9.2.1.5 Generation

The generation of dendrimers refers to the number of steps in the synthesis process, which determines the size and overall structure of the dendrimer. In each generation, monomer units are added to the dendrimer, creating a sporadically branched structure with a central core and multiple branches (Munir et al. 2016). Typically, the first generation of a dendrimer is referred to as "generation 0" (G0). The simplest molecule like ammonia, which has three reaction sites, forms the core molecule. To produce the first generation, the reaction sites between the core and monomer molecules form a connection. Each unreacted end group of bound monomers provides a place for several molecules to bind together, creating higher generations. In each subsequent generation, additional monomer units are added, resulting in a dendrimer with a larger size and a more complex structure. The generation number of a dendrimer is often denoted as "G1," "G2," and so on, with "G*n*" referring to the *n*th generation. Increasing the generation number of a dendrimer typically leads to an increase in size, surface area, and branching and terminal groups available for functionalization. The application and required characteristics of a dendrimer are the factors to determine its generation number (Bryant et al. 1999). The generation number of a dendrimer must be carefully considered when designing dendrimers for specific applications. Dendrimers with 0–3 generations resemble organic molecules, have an open structure and asymmetric forms, and are tiny without much uniformity or distinctive 3D structure. Dendrimers start to develop 3D structures and become spherical after G4, where the peripheries become densely packed and combine to create a structure resembling a membrane sphere. Due to a shortage of available space, dendrimers are unable to expand after the crucial branching stage has been reached, and this phenomenon is known as the "starburst effect." Dendrimers range in diameter from 2 to 10 nm depending on the generation (Garg et al. 2011).

9.2.1.6 Solubility

Another important physicochemical property of dendrimers is their solubility. Because of their highly branched structure, dendrimers can have a range of solubilities depending on their size, shape, and functional groups. Some dendrimers are highly soluble in water, while others are more soluble in organic solvents (Gupta et al. 2006).

There are several examples of water-soluble dendrimers that have been used as solubility enhancers. One example of a water-soluble dendrimer is the PAMAM dendrimer, which has a central core and multiple layers of amine group branches emanating from the core. PAMAM dendrimers are highly soluble in water due to the presence of the amine groups, which can interact with water molecules through hydrogen bonding (Chauhan et al. 2004). Other examples of water-soluble dendrimers include PEI and PPI dendrimers (Singh and Hildgen 2006). Overall, the solubility of dendrimers in water is largely determined by the presence of hydrogen bonding groups such as oxygen and nitrogen atoms, and the solubility can be tailored through the choice of core and functional groups (Svenson and Chauhan 2008; Choudhary et al. 2017).

An example of a less water-soluble dendrimer is polystyrene (PS), in which the central core is of carbon atoms and multiple layers of branches bearing hydroxyl groups emanating from the core. PS dendrimers are less soluble in water due to the presence of carbon atoms, which do not interact with water molecules through hydrogen bonding. Instead, PS dendrimers are more soluble in organic solvents such as benzene and toluene (van Hest et al. 1979; Siyad and Kumar 2012).

9.2.1.7 High Stability

The well-defined structure of dendrimers provides stability to the polymer, making it less likely to degrade over time. Therefore, they are suitable for use in PC delivery, as they can protect PC molecules from thermal degradation during storage and administration. The stability of dendrimers is largely determined by the strength of the bonds in their structure and can be tailored through careful control of their structure and chemical composition (Wang et al. 2017; Ma et al. 2001).

9.2.2 Biological Properties

The biological properties of dendrimers refer to their effects on living organisms, including their toxicity, biocompatibility, and pharmacokinetics (Bugno et al. 2015; Cheng et al. 2011; Madaan et al. 2014). These properties are important to consider when evaluating the potential use of dendrimers in PC delivery (van der Poll et al. 2010).

9.2.2.1 Toxicity

One important physiological property of dendrimers is their toxicity. Dendrimers have been shown to have low toxicity in many studies, which makes them applicable in anticancer and antimicrobial herbal actives (HA) delivery (Madaan et al. 2014). However, the dendrimers exhibit toxicity depending on their incubation time, surface charge, concentration, and generation (Sadekar and Ghandehari 2012; Janaszewska et al. 2019).

Some studies have suggested that dendrimers may be toxic to normal cells and cause non-specific cytotoxicity (Jain et al. 2010). This has been observed in vitro (in cell cultures) and in vivo (in animal models) and may depend on the specific type and structure of the dendrimer. When injected intravenously or given orally, higher-generation dendrimers are less tolerable than the lower-generation dendrimers. Further, the cationic dendrimers like amine terminated dendrimers are more toxic than the anionic dendrimers with carboxylic acid or hydroxyl terminal groups (Jones et al. 2012; Winnicka et al. 2015). For example, dendrimers with amine-terminated functional groups, such as PAMAM and PPI dendrimers, were shown to be cytotoxic in some studies (Sadekar and Ghandehari 2012). PAMAM dendrimers were found to be cytotoxic to cells in culture, causing cell death or inhibition of cell proliferation. This cytotoxicity has been observed at high concentrations of PAMAM dendrimers (Roberts et al. 1996). When given orally, dendrimers were non-toxic at larger dosages in comparison to intravenous administration (El-Sayed et al. 2002).

9.2.2.2 Biocompatibility

Another important physiological property of dendrimers is their biocompatibility, which refers to their ability to interact with living tissue without causing adverse effects. Dendrimers have been shown to have a high degree of biocompatibility in many studies, which makes them suitable for use as efficient nanocarriers for delivering the PC (Gothwal et al. 2020).

In general, dendrimers are considered to be biocompatible materials due to their size, which is typically in the nanoscale range (2–10 nm diameter), and their well-defined structure, which provides stability to the polymer. Additionally, their ability to be functionalized with various biological molecules allows them for cellular adsorption and targeted delivery of PCs (Ciolkowski et al. 2012). However, the biocompatibility of dendrimers can vary depending on their chemical composition, size, and surface functionalization, as well as the specific biological system in which they are introduced (Mishra et al. 2009). For example, dendrimers with a high generation number and a large surface area may be more likely to elicit an immune response, while dendrimers with a low generation number and a small surface area may be more likely to penetrate cells and tissues. Studies by Florence et al. compared the absorption of dendrimers with diameters of 2.5 nm and polystyrene with 50–3000 nm diameter, to assess the effect of particle size on cellular uptake. Compared to the 50 nm polystyrene particles, dendrimer absorption in the liver, spleen, kidneys, and blood was less accumulative. Dendrimer-based formulations do not experience the problem of poor colloidal stability that is typically seen with other nano-formulations in biological contexts (Dobrovolskaia 2017; Chen et al. 2004; Do et al. 2020; Albertazzi et al. 2013).

9.2.2.3 Pharmacokinetics

In addition to their toxicity and biocompatibility, the pharmacokinetics of dendrimers is also an important physiological property to consider. Pharmacokinetics refers to the absorption, distribution, metabolism, and excretion of PC in the body. Dendrimers have been shown to have many advantages in terms of their pharmacokinetic properties, including high solubility, a high surface area-to-volume ratio, and high stability in the body. These properties make dendrimers attractive for the delivery of PC, as they can improve the bioavailability and therapeutic efficacy of PC molecules (Tunki et al. 2020; Jain and Bharatam 2014).

9.2.2.4 Encapsulation and Release

Encapsulation and release of molecules are some of the key features of dendrimers considered during the delivery of PC. Dendrimers can be used to encapsulate PCs and control their in vivo release over time. The encapsulation of PCs within dendrimers can be achieved through various methods, such as covalent conjugation, physical adsorption, or co-polymerization, depending on the application and the properties of the PC being encapsulated (Menjoge et al. 2010).

The release of PCs from dendrimers can be controlled by various mechanisms, including pH-sensitive release, redox-sensitive release, and temperature-sensitive release, usually after dendrimer swelling and chemical degradation (Kurtoglu et al.

2010). For example, dendrimers coupled to PCs through strong covalent bonds might prolong the release of PCs (Zhu et al. 2010). For example, CPT when conjugated to PAMAM through a succinic acid-glycine linker demonstrated little drug release of 4% and 6% in PBS and growth medium, respectively, at 48 h (Thiagarajan et al. 2010).

9.3 Formulation Manufacturing/Preparation Process (Step by Step)

Formulation development with dendrimers involves the design and optimization of dendrimer-based PC systems for efficient and targeted delivery of PCs (Warsi et al. 2021). This process typically involves the selection of appropriate dendrimer type, optimization of dendrimer size and surface charge, selection of appropriate PC loading method, functionalization with an additional moiety with a specific property, if needed, and the evaluation of in vitro and in vivo performance of the dendrimer-based PC formulation (Santos et al. 2019; Mittal et al. 2021). The goal of formulation development is to achieve improved pharmacokinetics and higher therapeutic effect. Ultimately, the method for improving PC pharmacokinetics and site-specific targeting using dendrimers is to alter "critical nanoscale design characteristics" which include size, morphology, surface composition, versatility, and structural and chemical properties (Chauhan and Kaul 2018; Chauhan 2020).

9.3.1 Surface Charge of Dendrimers

In particular, the numerous terminal functional groups on the surface of dendrimers such as amines, carboxyl groups, or hydroxyl groups produce a highly localized charge density, which may be a positive charge, a negative charge, or no charge. The dendrimers have a positively charged surface with amine terminal groups while dendrimers bearing carboxyl and hydroxyl end groups have a negative surface charge. This surface charge can have a significant impact on how the dendrimer interacts with the cell membrane and moves inside of cells (Dobrovolskaia et al. 2012; Shi et al. 2010; Zhang et al. 2021; Yang et al. 2012; Lombardo et al. 2016).

9.3.1.1 Cationic Dendrimers

Dendrimers with positively charged surface groups such as amine-terminated PAMAM show high permeability and are efficiently bound with the negatively charged cell membrane (Ainalem and Nylander 2011; Fox et al. 2018). It has been observed that the permeability increases with the use of cationic dendrimers (G2, G3, and G4) in place of anionic dendrimers (G 2.5 and G 3.5). However, the cytotoxicity also increases in the cell treated with cationic dendrimers (Jevprasesphant et al. 2003). Cationic dendrimers may rupture the negatively charged biological membranes and culminate in cell lysis (Ciolkowski et al. 2012; Aisina et al. 2020). Cationic dendrimers as low as 12 μg/mL concentration have shown toxicity in clinical studies (Bugno et al. 2015). Higher-generation cationic dendrimers are also reported to induce plasma protein aggregation, hemolysis, and blood coagulation (Jones et al. 2012).

9.3.1.2 Anionic Dendrimers

Anionic dendrimers have a net negative charge on their surface due to the presence of anionic end moieties, such as hydroxyl, carboxyl, sulfonic, or phosphonic acids. Anionic dendrimers imitate cellular receptors and biomolecules and are nontoxic, nonimmunogenic, and more stable than cationic dendrimers (Aisina et al. 2020). Moreover, they bind and alter the structure of cationic cell receptors, preventing them from effectively interacting with viruses. Due to these properties, anionic dendrimers may be preferred for antiviral treatment and delivering antiviral PCs (Rodríguez-Izquierdo et al. 2022).

9.3.2 Nature of Dendrimers

9.3.2.1 Hydrophilic Dendrimers

Hydrophilic dendrimers are used as solubility enhancers, especially while developing the formulation of weakly water-soluble PCs (Mittal et al. 2021). PAMAM dendrimers traded as starburst are the most commercialized dendrimers for this purpose. The possible mechanism by which PAMAM possesses high solubility is by the virtue of ionic interaction and hydrogen bonding between its terminal groups and the surrounding medium and hydrophobic interactions between the internal cavity of dendrimers and the outer aqueous medium (Lyu et al. 2019).

9.3.2.2 Amphiphilic Dendrimers

These dendrimers are primarily made by separating the two sides of the chain which can self-assemble, forming spheres and vesicles. One side possesses electron-withdrawing properties and the other portion electron-donating, which aggregates into vesicular form. These dendrimers are ideal as surfactants and solubilizers (Al-Jamal et al. 2005). Polyamidoamine organosilicon (PAMAMOS) dendrimers are the first commercialized amphiphilic dendrimers, sold as SARSOX. They are made up of hydrophilic and nucleophilic organosilicon (OS) exteriors and hydrophobic PAMAM inside. Superfect, hydraamphiphiles, and bolaamphiphiles are among the other available amphiphilic dendrimers in the market (Balagani et al. 2011).

9.3.3 Functionalization

Functionalization is the process of adding chemical moiety to introduce a desired property for a particular application (Rai et al. 2021). Dendrimer functionalization has made it possible to achieve certain goals, including decreased cytotoxicity, targeted PC administration, longer plasma residence time, and in vivo biodegradation (Shi et al. 2007a; Majoros et al. 2006; Smith et al. 2021; Kolhatkar et al. 2008). An extensive range of molecules can be used for the functionalization of plain dendrimers which could be classified into several classes (Labieniec-Watala and Watala 2015).

9.3.3.1 Polymeric Dendrimers

The polymer can be conjugated to dendrimers by various methods such as chemical modification, physical adsorption, grafting, or copolymerization and offers many benefits, including non-ionic nature, biodegradability, low toxicity, and steric properties. Moreover, the polymers attached to the dendrimer bypass the immune system, thereby extending the circulation time of dendrimers (Fox et al. 2009; Reymond and Darbre 2012; Gu et al. 2010). Polyethylene glycol (PEG) is the most commercial polymer used for the functionalization of nanoparticles including dendrimers. Jangid et al. synthesized pegylated G4 PAMAM to eliminate the dose-dependent toxicity of higher-generation amine-terminated dendrimers and used the system for the delivery of piperlongumine (PL) against colon cancer. The PEGylated dendrimers showed less hemolytic toxicity and greater cytotoxicity against HCT 116 cells in comparison to the unloaded PL (Jangid et al. 2022). According to Fox et al., PEGylation of the PLL dendrimer-camptothecin (CPT) conjugate boosted both the tumor absorption and the circulation half-life of pegylated PLL-loaded CPT as compared to the free CPT (Fox et al. 2009).

9.3.3.2 Peptide Dendrimers

Peptide dendrimers refer to dendrimers that have peptides conjugate on the surfaces or amino acids incorporated in their inner branches. These dendrimers may function as protein mimics and delivery systems (Reymond and Darbre 2012; Gu et al. 2010). A typical amino acid used in dendrimers is lysine. As antibacterial agents, dendrimers containing 4–8 residue longer chain of lysine in the core or the surface have been used. As compared to their linear polymeric counterparts, these dendrimers exhibit greater aqueous phase solubility, higher antibacterial activity, and greater proteolysis resistance (Tam et al. 2002). The scientists created two types of fourth-generation hydrophilic and amphiphilic amino acid–based dendrimers modified by arginine and lysine to improve the aqueous solubility of poorly water-soluble ellagic acid (EA). The conclusion showed that the water solubility of EA rose to 1000 times higher in arginine- and lysine-based dendrimers than the free EA (Alfei et al. 2019). An Arg-Gly-Pro-Lys peptide dendrimer was synthesized and used for improving the skin permeability of silibin (SIL). SIL-peptide dendrimer complex showed about a fourfold increase in skin permeability after topical administration on rat skin (Shetty et al. 2017).

9.3.3.3 Vitamin-Functionalized Dendrimers

Vitamins are small organic molecules that are essential for cellular processes and are involved in a variety of cell signaling pathways. The receptors for these molecules are usually overexpressed in cancer cells. Hence, vitamins can be used as targeting molecules to selectively bind to cancer cells or tissues (Gupta et al. 2010). For instance, the proliferation of cancer cells depends on vitamin B_{12} also known as folic acid (FA). Receptors having an affinity for FA are overexpressed in several disease states, including inflammation and cancer such as those of the brain, breast, kidney, lung, colon, and skin, providing a design opportunity for FA-conjugated nanoparticles for targeted treatment (Kesharwani et al. 2015a; Fatima et al. 2022). Zeynalzadeh et al. grafted FA on hydroxyl-terminated G4 PAMAM for targeted delivery of curcumin (CUR) specifically to colon and fibroblastoma cancer cells (Zeynalzadeh et al. 2023). In another study, FA-functionalized pegylated PAMAM dendrimers were used for the pH-sensitive release and targeted delivery of quercetin (QUE). The QUE-loaded PAMAM-PEG-FA released the drug faster at acidic pH 5.6 than the basic pH 7.4. In comparison to non-FA-linked nanoparticles, fluorescent microscopy results showed that the FA-labelled nanoparticles effectively accumulated in tumor tissues and displayed greater cytotoxicity in comparison to free QUE (Rezaei et al. 2019).

9.3.3.4 Glycodendrimers

Glycodendrimers are those dendrimers that have carbohydrates built into their architecture. They consist of three types: dendrimers with saccharide residues on the terminals (carbohydrate-coated dendrimers); dendrimers with carbohydrates such as cyclodextrin, dextran, chitosan, pectin, heparin, hyaluronic acid, and inulin at their branches; and dendrimers with a sugar unit within their core (Bruce Turnbull et al. 2002). The most well-known glycodendrimer is cyclodextrin-based dendrimer which can entrap a wide range of hydrophobic molecules, exhibit minimal toxicity, and does not activate the body's immune system (Namazi and Heydari 2014). Sugars are monomers or dimers of carbohydrates which are mostly used as small molecules to functionalize nanocarriers for targeting and intracellular uptake. There are sugar-specific receptors and transporters on the targeted tissues recognized by the sugar molecules decorated on dendrimers and other nanocarriers (Roy et al. 2012). Many cancer cells, bacterial, and viral cell membranes overexpress receptors like lectin and asialoglycoprotein, therefore many sugar compounds, including lactobionic acid, fucose, mannose, galactose, lactose, and sialic acid, may be employed as ligands to target specific receptors that are overexpressed in a variety of malignancies and infectious diseases. As a result, they may be used for conjugation to design a target-specific dendrimer system for delivering PCs in the case of infections, inflammatory disorders, or cancers (Toomari et al. 2015; Mousavifar and Roy 2021; Liu et al. 2012). The disaccharides D-glucuronic acid and *N*-acetyl-D-glucosamine make up the polymer of hyaluronic acid. By effectively attaching to CD44, a cell-surface glycoprotein known to be overexpressed in many malignancies including ovarian, colon, lung, and breast cancers, hyaluronic acid (HLA), a linear mucopolysaccharide, has the potential to naturally target tumors. To direct a CUR derivative against the MiaPaCa-2 pancreatic cancer cell line, hyaluronic acid (HLA)-conjugated dendrimers were employed. Better cytotoxicity and cellular uptake were shown by HLA-conjugated dendritic formulations. Endocytosis was shown by a receptor-blocking experiment to have been caused by HLA binding to CD44. It was determined that even the cationic charge of PAMAM was reduced as a result of conjugation with HLA (Kesharwani et al. 2015b). In another example, Pooja et al. used *N*-acetyl-D-glucosamine (NAG) for dual targeting of glucose transporters and lectin receptors which overexpress in lung adenocarcinoma cells. CPT-loaded G3.5 PAMAM were grafted with NAG, and biodistribution studies in tumor-bearing mice were used to determine the capacity of these NAG-PAMAM to target tumors. In comparison to the unlabeled dendrimer formulation (PAMAM-CPT), CPT-loaded NAG-PAMAM demonstrated faster and higher cellular uptake, resulting in 4.5 times increased cytotoxicity and selective accumulation of CPT in lung cancer cells (Pooja et al. 2020).

9.3.3.5 Lipidic Dendrimers

Using ligands comprised of fatty acids, such as palmitic, lauryl, and omega chains, it is possible to change the characteristics of dendrimers. This lowers their toxicity profile and increases their permeability across the biological membranes (Florence et al. 2000). For example, the PAMAM dendrimer conjugated with liposomes was used for improving the permeation of puerarin (PUE) in the cornea for preparing the ocular formulation of PUE (Liu et al. 2010). In another study, Tripathi et al. created G4 PAMAM-palmitic acid core-shell delivery carriers to increase the absorption of CUR in antistress therapy. Also, the pharmacokinetic properties were observed to be improved when the CUR with the dendrimer-lipid system was administered in Swiss albino mice as compared to a simple CUR formulation. The results of histopathological and hematological assays showed no morphological alteration proving the toxicological safety of the lipid-dendrimer core-shell system (Kececiler-Emir et al. 2021).

9.3.3.6 Antibody-Functionalized Dendrimers

When malignancies overexpress certain antigens known as tumor-marker antigens, targeted immunotherapy using monoclonal antibodies (mAbs) is used. A mAb may be used to target the PC to the tumor by being conjugated to the surface of the carrier delivering that PC. It has an advantage over other targeted ligands because the presence of endogenous moieties may inhibit other ligands via competitive inhibition while such is not the case with mAbs (Thomas et al. 2004; Wängler et al. 2008). Pegylated PAMAM dendrimers were grafted with a mAb, Margetuximab (Mab) for the targeted delivery of QUE to MDA-MB-231 breast cancer cells, and the results showed a lower IC_{50} value of QUE-loaded, Mab-grafted dendrimers (100 nM) as compared to pure QUE (200 nM) and the expression of the apoptotic Bax and Caspase9 genes rose by more than eight and five folds, respectively, after the treatment of Mab-PEG-PAMAM encapsulated CUR in MDA-MB-231 cells. Additionally, cell cycle arrest increased by more than three folds (Khakinahad et al. 2022).

9.3.3.7 Nucleic Acid–Functionalized Dendrimers

Nucleic acids, such as DNA and RNA, have specific recognition properties and can form stable base-pairing interactions with complementary nucleic acid sequences. For example, Malar et al. prepared dendrosomes for the delivery of capsaicin (CAP).

Dendrosomes are dendrimer-nucleic acid combinations enclosed inside a lipophilic shell. Dendrosomes are unique vesicular, spherical, neutral, amphipathic, and biodegradable supramolecular entities offering low cytotoxicity, reasonable transfection efficiency, and high in vivo endurance. Dendrosomes as a delivery system combine the benefits of liposomes such as increased intracellular endocytosis and reduced cytotoxicity and the advantages of dendrimers such as nucleic acid condensation and improved endosomal release, while also addressing the disadvantages of both systems such as low encapsulation efficiency of liposome and non-specific toxicity of dendrimers. Malar et al. discovered that the free CAP aggregated whereas the CAP-dendrosome displayed high dispersibility and a defined shape (Dobrovolskaia 2017). Babaei et al. also created CUR-loaded dendrosomes and compared their activity as compared to the free form. It was found that the solubility of CUR was improved in dendrosome formulation. Also, the cellular uptake and inhibitory effects of dendrosomes were improved in WEHI-164 and A431 cancer cell lines at all time points and doses of treatment. On the contrary, no inhibitory effect on normal mouse embryonic fibroblasts (MEF) was observed after dendrosome treatment (Babaei et al. 2012). CUR dendrosomes were created by Tahmasebi et al. as a non-toxic nanocarrier with a spherical form, and chemical and physical stability. They claimed that the dendrosomes had no impact on non-neoplastic cells but had an inhibitory effect on cancer cell lines of the colon, bladder, gastric, breast, liver, glioblastoma, and fibrosarcoma (Tahmasebi Birgani et al. 2015). Dendrimers have also been used to co-deliver PC and small interfering RNA (siRNA). In a study, CUR was formulated in PAMAM dendrimer to increase the solubility and bioavailability of CUR, and Bcl-2 siRNA was grafted onto the surface amine groups to create CUR-loaded Bcl-2 siRNA-PAMAM. The Bcl-2 siRNA-modified CUR-PAMAM nanoformulation exhibited a high CUR loading percentage, more cellular absorption, and a reduction in the growth of HeLa cells than PAMAM-CUR nanoformulation and free CUR (Ghaffari et al. 2019).

A new class of nucleic acid ligands has emerged known as aptamers or chemical antibodies. Aptamers are single-stranded DNA or RNA molecules that are 20–100 nucleotides long and have a high affinity and selectivity for binding to target molecules. Similar to antibodies, they can identify and bind to targets and have several benefits over antibodies (Nimjee et al. 2005). They are frequently smaller in size making their synthesis, modification, and distribution simpler and more affordable. They may be manufactured chemically, eliminating the requirement for animal hosts, and they also have decreased immunogenicity (Keefe et al. 2010). Aptamers are extremely stable because they can withstand a broad range of temperatures, ionic conditions, and pHs. They are highly reproducible and have a low probability of batch failure. They can be duplicated or altered to create a vector which allows their storage for a longer span and can be acquired again through PCR amplification. They can readily enter malignant cells due to their low molecular weight of 20 kDa (Afsana and Kesharwani 2021). In a research work, the pegylated G5 PAMAM-CPT complex was conjugated to an anti-nucleolin AS1411 aptamer specific for the nucleolin protein to target colorectal cancer. Compared to PEG-PAMAM, a regulated release of CPT occurred from Apt-PEG-PAMAM. Additionally, Apt-

PEG-PAMAM demonstrated lower hemolytic toxicity than plain PAMAM, improving the safety profile of the formulation (Alibolandi et al. 2017).

9.3.3.8 Dendrimers Functionalized with Other Small Molecules

Low molecular weight reagents such as acetic anhydride are also used for simple acetylation of dendrimers. Acetylated dendrimers are more aqueous-soluble, and this property is essential for biological applications (Majoros et al. 2003). According to Wang et al., the integration of CUR by the acetylated G5 PAMAM dendrimer improved its water solubility 200 times more than free CUR and mediated a prolonged CUR release from the CUR-loaded dendrimers. Moreover, the PAMAM dendrimer increased the anticancer activity and cell death in A549 cell lines and decreased the production of intracellular reactive oxygen radicals (Wang et al. 2013). Introducing small alkyl groups onto the surface of dendrimers was done by Soltani et al. (2017) to enable the controlled release of crocetin (CRO). They alkylated the G4 PAMAM and G4 PPI dendrimers and investigated the in vitro release behavior of CRO from those dendrimers in contrast to the release from the plain dendrimers. After 48 h, 60% and 70% of CRO were determined to be released from unmodified PAMAM and PPI while alkylated PAMAM and PPI dendrimers released only 20% and 30% of CRO, respectively. The results proved that surface-modified dendrimers are more capable of controlling the release of encapsulated PC compared to plain dendrimers (Soltani et al. 2017). Wang et al. conjugated glutathione as a small reactive oxygen species (ROS)-sensitive linker to PAMAM dendrimer for the delivery of CPT deeply into the pancreatic tumor via endocytosis and transcytosis mediated by the glutamyl transpeptidase enzyme which gets triggered by the glutathione (Wang et al. 2020) (Fig. 9.1).

9.3.4 Route of Administration

9.3.4.1 Oral Administration

The most patient-compliant delivery route of PCs is through the oral route, but the acidic gastrointestinal (GI) environment and presence of digestive enzymes confer difficulties in the oral administration of PCs, especially those with low water solubility and chemical instability (Cabrera and Gonzalez-Alvarez 2011; Leuner and Dressman 2000). The capacity of dendrimers to cross the GI epithelium has led to their potential utility for the oral delivery of PCs (D'Emanuele et al. 2004; Cheng et al. 2008). There are two main ways of dendrimers passage through the GI epithelium: transcellular and paracellular. Tight junctions regulate paracellular transport, which includes the passive diffusion of chemicals via the intercellular gaps between epithelial cells (Ensign et al. 2012). Dendrimers affect tight junction proteins and enhance the permeability of tight junctions in the GI epithelium. Low-generation dendrimers, for example, G2.5 and G3.5, up to 3 nm size may permeate cells by paracellular transport while G4 and higher-generation dendrimers could enter GI cells by adsorptive endocytosis (Florence et al. 2000). The pattern of accumulation for cationic dendrimers was distinct from that for anionic dendrimers

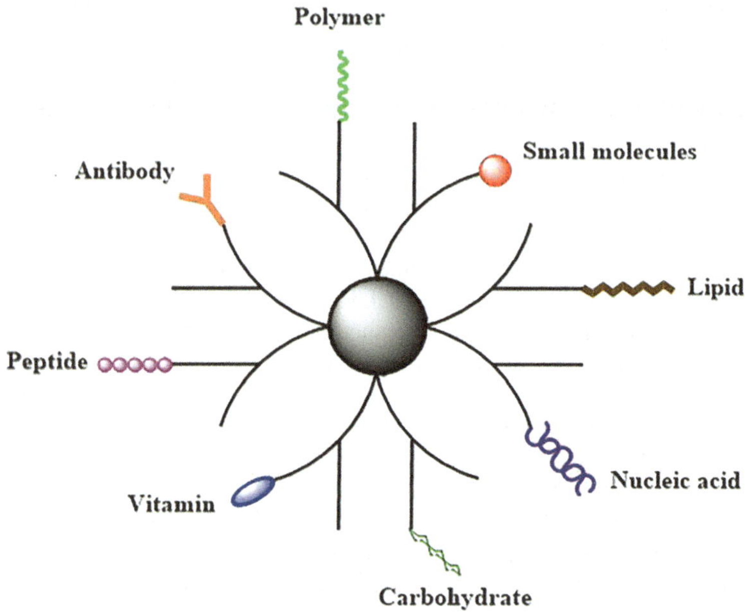

Fig. 9.1 Functionalization strategies of dendrimers for formulation development of phytoconstituents

as cationic dendrimers tend to associate with negatively charged cell membranes which slows the rate of transport for these dendrimers (Wiwattanapatapee et al. 2000). These findings suggest that the oral delivery of dendrimers is affected by their size, shape, and charge. Also, the paracellular permeability of GI epithelium increases with dendrimer concentration and incubation period (Lin et al. 2010). Gu et al. looked at the potential of PAMAM dendrimers to increase the oral bioavailability and solubility of PEU. It was observed that PEU complexed with various generations of PAMAM (G1.5, G2, G2.5, and G3) was more soluble than pure PEU. The hemolytic toxicity experiments also revealed no detectable hemolysis. Hence, a dendrimer-based formulation was found to be efficient and safe for the oral administration of PEU (Gu et al. 2013).

9.3.4.2 Intravenous Administration

The quickest and easiest way to introduce PC into systemic circulation is through the intravenous route. However, the use of intravenous administration is limited by the low water solubility of PCs, and hence developing PC formulation with dendrimers can facilitate its intravenous administration (Strickley 2004; Karandikar et al. 2017). It is vital to precisely regulate the chemistry of dendrimer molecules since the biodistribution of dendrimers after intravenous injection is governed by their structure, size, and charge (Cheng et al. 2008). After being administered intravenously, cationic dendrimers have been studied to confer general toxicity and accumulate in the liver and kidney. Coating the dendrimers with biocompatible polymers such as PEG chains, however, can overcome these problems (Santander-Ortega et al. 2013).

9.3.4.3 Topical/Cutaneous and Transdermal Administration

Topical PC delivery refers to PC distribution on the small surface of the skin, whereas transdermal PC delivery refers to PC delivery into systemic circulation through the skin. The stratum corneum is the top layer of the skin which serves as a strong barrier to the transit of PCs after topical application of PC (Dave and Venuganti 2017). Due to their extremely water-soluble and biocompatible nature, dendrimers can deliver transdermal formulations and increase PC attributes including solubility and plasma circulation time with minimum skin irritation and maximum PC loading. Dendrimers provide benefits such as improved solubilization and controlled release of PCs during their topical and transdermal delivery. For example, a dendrimer-based cream formulation of resveratrol was prepared to improve its stability and solubility, and a dendrimer-resveratrol complex demonstrated much more (2.5 times) transdermal penetration than resveratrol alone (Pentek et al. 2017). Denaturing skin keratin, changing skin lipids, and entering via aqueous skin pores are all ways that dendrimers enter the skin. Also, by increasing the partitioning and diffusion of the PC molecules into the skin, dendrimers promote PC penetration. Moreover, it is possible to alter the characteristics of dendrimers or dendrimer-PC complexes through functionalization, to increase or decrease their skin penetration as necessary (Tolia 2008).

9.3.4.4 Intranasal and Pulmonary Administration

Nasal delivery is an intriguing substitute for the very invasive parenteral method of PC administration. It prevents the metabolism of PC by the liver, degradation in the gastrointestinal tract, rapid systemic clearance, and therefore, maximizes the bioavailability of PCs. PAMAM dendrimers have been utilized to deliver PC through the intranasal route for targeting the brain. PEG-functionalized G5 PAMAM dendrimers were used to deliver poorly water-soluble paeonol (PAE) to CNS through the nasal route. Encapsulating the PAE in PEG-PAMAM dendrimers increased its solubility and release rate (Mignani et al. 2021).

9.3.4.5 Ophthalmic Administration

Administration of PCs to the posterior region of the eye containing the cornea, conjunctiva, and sclera through oral and intravenous routes is rendered by an intermediate barrier between blood capillaries and ocular tissues. Increasing the bioavailability of PCs and extending their contact time on the cornea, conjunctiva, and corneal epithelia may overcome these problems (Kalomiraki et al. 2015). Dendrimers functionalized with bioadhesive molecules like glucosamine may further lengthen the contact time of PCs in the eyes. The efficiency of ocular PC administration through dendrimers is influenced by certain factors such as dendrimer size, molecular weight, charge, and molecular shape. For instance, the findings revealed that the retention time of dendrimers with peripheral carboxylate or hydroxyl groups was more than cationic dendrimers with primary amine groups when tested in albino rabbits (Vandamme and Brobeck 2005). Yao et al. attempted dendrimer-mediated ocular delivery of PUE using different generations (G3.5, G4, G4.5, and G5) of PAMAM dendrimers. The PUE-PAMAM formulations exhibited prolonged corneal retention and slow drug release as compared to free PUE when administered to the eye of the rabbit (Yao et al. 2010).

9.3.5 Loading Methods for Phytoconstituents in a Dendrimer-Based System

There are several methods for loading PC in dendrimer-based nanocarriers, including covalent and non-covalent approaches (D'Emanuele and Attwood 2005). Covalent attachment involves the formation of a stable chemical bond between the dendrimer periphery and the PC molecules. Non-covalent attachment involves electrostatic interactions, hydrogen bonding, or hydrophobic interactions between the dendrimer and the PC (Pooja et al. 2018).

9.3.5.1 Covalent Interactions

Covalent grafting of PC to the dendrimer periphery results in more control over PC loading, increased stability, and slow PC release in systemic blood circulation. For example, Alfei et al. used two strategies for formulation preparation of gallic acid (GA) with dendrimers and observed that the percentage GA loading (68%) was higher when GA was covalently complexed with dendrimers as compared to the percentage drug loaded (50%) in the formulation in which GA was encapsulated in the dendrimers by physical interactions (Alfei et al. 2020a). Moreover, PC-covalent grafting may be used to transport the combined insertion of PCs and ligands tailored specifically to cell receptors. By selecting the right linker groups such as disulfide, hydrazone, and oxime bonds between PCs and dendrimers, it may be possible to achieve a well-localized PC release and increase sensitivity to pathological microenvironments (Jain and Bharatam 2014). For example, Yang et al. designed a pH-responsive delivery system of dendrimer to deliver CUR in osteoporosis treatment by using a pH-sensitive linker hexachlorocyclotriphosphazene (HCCP) to conjugate CUR with PAMAM dendrimers. Results of in vitro experiments showed that the CUR-conjugated HCCP-PAMAM reached lysosomes by endocytosis and released the drug after the degradation of HCCP linker at lysosomal pH (pH 5) (Alfei et al. 2020b). On the other hand, it is difficult to identify a factor, for example, pH, temperature, or a chemically active substance, that is both specific to the site of action and sufficient to separate the PC from the dendrimer surface because of which PC release is typically slow. This fact is evident in the PC release study conducted by Gupta et al. in distilled water and PBS after conjugating Berberine (BBR) with G4 PAMAM in one formulation (CBBR) and encapsulating BBR in another formulation (EBBR). After 24 h at pH 7.4, the release from CBBR and EBBR were observed to be 80% and 98.3%, respectively. Similarly, in the distilled water medium, 72.3% and 98% of BBR were released from CBBR and EBBR, respectively, which concluded that conjugated PC is released more slowly and continuously than encapsulated BBR in the G4 PAMAM cavity and hence conjugation seems to be more effective to achieve sustained release (Gupta et al. 2017). The use of linkers may also result in the production of less active PC derivatives since the possibility of PCs getting altered before performing their pharmacological action is increased in case they are covalently attached to the surface of the dendrimer and are more susceptible to environmental impact (Hu et al. 2009). Covalent conjugation requires additional chemical synthesis and purification steps, which drive up the price of the final product (Sathe and Bharatam 2022).

9.3.5.2 Non-covalent Interaction

Ionic attachments, hydrogen bonds, and hydrophobic interactions are examples of non-covalent "soft bonds" between the dendrimers and guest PCs. The PCs are bounded or encapsulated by lower energy interactions that are perfect for loading, transport, and stability of PCs against premature release at the same time PCs are chemically unaltered, and their release needs only milder conditions, such as slight changes in local pH or temperature, or through triggers like visible light. Also, a non-covalent association of PCs with dendrimers includes simple formulation processing and can increase the solubility of hydrophobic PCs. The main drawback of the non-covalent formulation approach is the comparatively less in vivo stability of PC-dendrimer complexes, which frequently may lead to an excessively fast and inconsistent PC release and in turn non-specific PC biodistribution (Fréchet 2002; Zimmerman and Lawless 2001).

9.4 Challenges for Herbal Actives

In clinical studies, PCs have been shown to have chemotherapeutic and chemopreventive properties. Yet, the limited solubility, low stability, low intracellular penetration, and non-specific activity of PCs pose significant challenges in establishing persistent administration. Pharmaceutical companies are presently using nanosized carriers for prolonged and improved delivery of phyto-derived bioactive chemicals to get over these drawbacks (Sen et al. 2011; Thillaivanan and Samraj 2014; Ekor 2014). Dendrimers are one of the widely used nanocarriers for the delivery of PCs because of their distinct features including high aqueous solubility, uniform size and structure, and tunable chemistry.

9.4.1 Aqueous Solubility

The formulation development of PCs often faces difficulties due to the limited water solubility of PC ingredients. PCs may have low water solubility due to certain reasons, for example, flavonoids, alkaloids, and terpenoids are lipophilic and tend to dissolve in nonpolar solvents like oils and fats. As a result, they dissolve poorly in water, a polar solvent. The solubility of herbal PCs had been also affected by their size, complexity of molecular structure, and crystallinity. PC molecules with larger sizes, complicated structures, or crystalline lattices are restricted from interacting with the water molecules. When given orally or via other aqueous-based methods, poor solubility may greatly influence the bioavailability and therapeutic effectiveness of these PCs.

Since dendrimers may entrap hydrophobic molecules, they are effective solubility enhancers. Most of the dendrimers themselves are highly water soluble and can be used as a water-soluble nanocarrier to entrap hydrophobic PCs (Bharathi et al. 2015; Choudhary et al. 2017). Debnath et al. investigated the water solubility of CUR derivatives and the dendrimer-CUR conjugate. Only the dendrimer-CUR combination was shown to have water solubility and hence cytotoxicity against BT549 and SKBr3 breast cancer cells rather than the free form (Debnath et al. 2013). Falconieri et al. also encapsulated CUR in G0.5 PAMAM in two different molar ratios, that is, 1:1 and 1:0.5. The solubility of CUR in formulations with a 1:1 and 1:0.5 ratio was increased by 415 and 150 times, respectively, as compared to the free CUR. Hence the results showed that the increase in the aqueous solubility of CUR encapsulated in dendrimers is directly proportional to the increasing ratio of the dendrimers (Falconieri et al. 2017). Madaan et al. used various concentrations and several generations of PAMAM and optimized the ability of dendrimers to boost the water solubility of QUE. At 4 M concentrations of G0, G1, G2, and G3 PAMAM dendrimers, the solubility of QUE was increased by 145, 161, 181, and 206 times, respectively. Because of the electrostatic interaction between the amine groups on the PAMAM surface and the hydroxyl groups of QUE, as well as the hydrogen bonds formed between internal tertiary amine groups of PAMAM and QUE molecules, it was found that solubility increased linearly along with the increasing generation number and concentration of the dendrimer. Moreover, the hydrophobic QUE is trapped within the dendrimer cavities and hence solubilized in the aqueous media (Madaan et al. 2016). Meredith et al. encapsulated CPT derivatives, that is, hydroxy-CPT (HCPT) and butylamino-CPT (BACPT), into G4.5 PGLSA dendrimer to improve their solubility in water. The encapsulated HCPT and BACPT demonstrated an increase in water solubility of 10 folds and 20 folds, respectively (Morgan et al. 2006). According to the investigation of Huang et al., the solubility of hydrophobic SIL in G2 and G3 PAMAM dendrimers increased linearly as dendrimer concentration increased because more surface amines interacted electrostatically with the phenolic hydroxyl groups of SIL and internal tertiary amines of PAMAM that formed hydrogen bonds with SIL molecules. Moreover, the hydrophobic cavities of PAMAM to encapsulate the hydrophobic SIL molecules were increased as the dendrimer concentration increased. The solubility of SIL was also affected by the pH of the PAMAM solution (pH 4.0, 7.0, 8.0, 9.0, and 10.0), and it was found to be maximum at pH 10 due to a high degree of ionization and anion production. This might be because the inner, positively charged tertiary amine groups of PAMAM engage electrostatically with the negatively charged phenolic hydroxyl group more in ionized SIL at pH values close to 10 (Huang et al. 2011).

9.4.2 Low Stability

Another frequent issue in the practical utilization of PCs is their limited stability. Many PCs have been reported to be vulnerable to environmental conditions such as oxidation, hydrolysis, temperature, pH, or light that may change their chemical makeup, reducing their effectiveness, potency, and safety.

Microencapsulation in dendrimers is one of the strategies that can provide shielding to PCs from the environment responsible for their degradation. Resveratrol (RES) is an example of photosensitive PCs that decompose when exposed to light. Yaning et al. tried to increase the storage stability of RES by preparing its formulation with dendrimers. After 1 week, it was found that the RES concentrations in the dendrimer formulation were still 92% while the pure RES form retained only 55% of the drug content. Since dendrimers have a highly branched structure and a vast number of binding sites on their surface and inside of their cavities, it has been suggested that the dendrimer matrix might have entrapped the RES within and prevented it from photodegradation (Shi et al. 2020).

9.4.3 Uncontrolled Release

It has been observed in many studies that PCs tend to release quickly and uncontrollably after in vivo administration in the absence of a formulation or delivery system. Poor effectiveness, negative effects, and difficulties in maintaining optimal therapeutic medication levels are the problems that might result from this kind of release pattern.

There are many ways to manage the release of PCs from dendrimers (Ghaffari et al. 2018). PC molecules may release slowly or quickly, depending on how strongly they bind with dendrimers (Kurtoglu et al. 2010). The SIL was formulated with PEGylated G4 PAMAM for extending the in vitro release time of SIL. The PEG chains of 0.5 and 2.0 kDa were conjugated with PAMAM, and the release studies were performed to show that 2.0 kDa PEG chains extended the release time of SIL more as compared to 0.5 kDa PEG chains. After 24 h, the complete release of free SIL was observed in vitro whereas half of the encapsulated SIL from PEGylated PAMAM was released. This may occur because dendrimers with longer PEG chains have stabilized SIL-dendrimer complexation, which may delay the release of SIL from the dendrimer (Diaz et al. 2018). Determination of the CUR release from the CUR-G0.5 PAMAM formulation prepared by Falconieri et al. (2017) also revealed a prolonged and sustained release profile (Choudhary et al. 2017). Madaan et al. investigated the in vitro QUE release from the QUE-PAMAM complex, revealing a biphasic QUE release pattern with an initial quicker release followed by a persistent release. Due to QUE's greater interaction with the inner tertiary amines of PAMAM dendrimers, a slow release of QUE was observed from the QUE-PAMAM complex (Madaan et al. 2016). CPT was covalently grafted onto the surface of the G4 PAMAM and the resulting dendrimer hydrogel enabled extended CPT diffusion at the tumor site due to ammonolysis of the ester bonds and subsequent slow self-cleavage of hydrogel cross-links (Wang et al. 2019). Pengkai et al. also used glucose-targeted pegylated PAMAM for the GSH-sensitive release of CPT in the microenvironment of liver cancer cells (Ma et al. 2018).

9.4.4 Permeability

The intracellular permeability of PCs refers to their ability to enter and penetrate inside the cells and is crucial for exerting the desired pharmacological or biological activity within the cellular environment. The intracellular permeability of PC in its native form is usually constrained due to its low solubility.

Hence many trials have been conducted to improve the permeability of different PCs by preparing their dendrimer-based formulations. For example, it was observed that at the same concentration, dendrimer-encapsulated HCPT was absorbed by cells more quickly than free HCPT. The cellular uptake and efflux assays in MCF-7 cells revealed an increase of 16-fold for cellular uptake and an increase in CPT retention inside the cell (Morgan et al. 2006). Yaning et al. conducted TEER measurements for the RES-glucan dendrimer formulation in order to assess cellular uptake and transepithelial permeability via Caco-2 cells. The concentration of RES consistently increased on the basolateral side after treatment with RES-dendrimer formulation, and RES-dendrimer was determined to have a corresponding apparent permeability compared to that of the free RES, suggesting improved paracellular transit of RES from its dendrimer complex (Shi et al. 2020).

9.4.5 Pharmacological Activity

The PCs in their native or extracted form usually have compromised and inconsistent in vivo pharmacological effects due to their low bioavailability, less adsorption, and non-specific biodistribution.

These limitations may be overcome by formulation development of the PCs using the dendrimers as an excipient for improving the physicochemical and pharmacokinetic properties of PCs such as solubility, stability, intracellular permeability, targetability, and circulation time as shown in many trials. For example, Mollazade et al. compared the cytotoxicity of G3 PAMAM-encapsulated CUR and free CUR on the T47D breast cancer cell line. Results of the MTT test showed that CUR-loaded PAMAM has a lower IC_{50} value as compared to the free CUR. Telomerase activity was also assessed using the TRAP method which showed increased inhibition of telomerase activity of the CUR-PAMAM formulation (Mollazade et al. 2013). In the investigation by Huang et al., dendrimer-SIL formulations demonstrated superior oral bioavailability over pure SIL solution after oral administration. The pharmacokinetics parameters $T_{(max)}$ and $C_{(max)}$ determined for SIL were 10 min and 134.2 ng/mL while 15 min and 182.4 ng/mL on administering SIL-PAMAM complex (Diaz et al. 2018). Sharma et al. (2011) used a surface-modified dendrimer for the delivery of GA. GA conjugated with G4 PAMAM tested for cytotoxicity against the MCF-7 cell line. PAMAM-GA were shown to have IC_{50} values of 36.2 μg/mL, lower than that of GA (81 μg/mL) (Sharma et al. 2011). Moreover, Soltani et al. showed that whereas CRO alone exhibited lesser cytotoxicity, complexation with alkylated G4 PAMAM and G4 PPI increased the cytotoxicity of CRO (Soltani et al. 2017). The MCF-7 cells were shown to be resistant to the anticancer effects of CAP in the cell survival experiment carried out by Meredith et al. On the contrary, the G4.5 PGLSA-encapsulated HCPT showed an IC_{50} value of 32.3 nmol/L, which is 3.5-fold better than the IC_{50} of free HCPT dissolved in dimethyl sulfoxide (DMSO) and the insensitive MCF-7 cell line showed the increased cell death (Morgan et al. 2006). Gupta et al. used the MTT test on the cell lines MCF-7 and MDAMB-468 to assess the cytotoxicity of BBR-G4 PAMAM. The IC_{50} values for CBBR, EBBR, and free BBR against the MCF-7 cell line were determined to be 4.03, 4.97, and 6.9 mg/mL, respectively, demonstrating an improvement in BBR cytotoxicity upon conjugation and encapsulation. Similar to this, in MDA-MB-468 cell lines, CBBR (2.79 mg/mL) and EBBR (3.23 mg/mL) had lower IC_{50} values than BBR (4.17 mg/mL). The greater intracellular absorption and penetration of dendrimer-encapsulated BBR caused an increase in the cytotoxicity of EBBR (Gupta et al. 2017).

9.4.6 Target Specificity

Targeted delivery of PCs is crucial to enhance its therapeutic effectiveness, lowering the PC payload and its adverse effects. PCs can be delivered to a particular location, such as an infected organ, tissue, or cell, by employing nanocarriers such as dendrimers as a targeted drug delivery (TDD) system.

Dendrimers are a perfect structure for creating a TDD system to get over biological barriers and deliver a PC to or close to the target spot because of their polyvalent surface which provides tunable end groups for the target-specific modification. For example, Gallien et al. complexed CUR in Cyt-modified PAMAM dendrimers to deliver the whole payload of CUR to glioblastoma cells. It was examined using the cell viability assay that Cys-PAMAM encapsulated CUR increased cell death in glioblastoma cell lines while unaffecting the normal cells. On the contrary, the unencapsulated CUR was ineffective in killing the glioblastoma cells, and non-modified dendrimer increased the mortality of both glioblastoma and healthy cells (Gallien et al. 2021). Similarly, Shi et al. used CUR-Cyt PAMAM for the treatment of a brain disorder called amyloid fibrosis. It was observed that dendrimer-conjugated CUR specifically eliminated neuronal tumor cells as compared to free CUR while being non-cytotoxic to normal cells (Shi et al. 2007b).

Herbal actives	Type of dendrimer	Disease	Remark	References
Curcumin	G3 PAMAM	Breast cancer	• Higher cytotoxicity	Mollazade et al. (2013)
			• Higher telomerase inhibition	
	Dendrosome	Fibrosarcoma	• Improved solubility	Babaei et al. (2012)
			• Increased anticancer activity	
			• Specific cytotoxicity to cancer cells	
	Dendrosome	Colon, bladder, gastric, breast, liver, glioblastoma, fibrosarcoma	• Improved stability	Tahmasebi Birgani et al. (2015)
			• Enhanced anticancer activity	
	G4 PAMAM	Breast cancer	• Enhanced solubility in different solvents	Debnath et al. (2013)
			• Higher cellular uptake	
	Cystamine-G4 PAMAM	Glioblastoma	• Improved toxicity in cancer cells	Gallien et al. (2021)
			• Non-toxic to normal cells	
	Cystamine-G4 PAMAM	Amyloid fibrosis	• Neurotumor-specific cytotoxicity	Shi et al. (2007b)
	G0.5 PAMAM	Breast cancer	• Improved solubility up to 415 times	Falconieri et al. (2017)
			• Prolonged release	
	Dendrosome	Glioma, fibroblast	• Improved cytotoxicity in a time- and concentration-dependent manner	Tahmasebi Mirgani et al. (2014)
	G3 PAMAM	Glioma	• Increased solubility	Gamage et al. (2016)
			• Tumor-specific accumulation	
	Acetyl terminated G5 PAMAM	Lung cancer	• Increased solubility	Wang et al. (2013)
			• Sustained drug release	
			• Improved anticancer activity	
	Citric acid-G5 PAMAM	Breast cancer	• pH-mediated drug release	Nosrati et al. (2018)

	Triphenylphosphonium-G4 PAMAM	Liver cancer	• Increased solubility and stability • Mitochondrial-specific targeting	Kianamiri et al. (2020)
	Bcl-2 siRNA-G4 PAMAM	Cervical cancer	• Improved solubility and bioavailability • Enhanced cellular uptake and antiproliferative activity	Nimjee et al. (2005)
	Hexachlorocyclotriphosphazene-PAMAM	Osteoporosis	• Improved water solubility • pH-responsive drug release	Yang et al. (2023)
	Palmitic acid-G4 PAMAM	Acute stress	• Improved pharmacokinetic properties	Tripathi et al. (2020)
Camptothecin	G4.5 PGLSA	Breast, lung, colorectal cancer, and glioblastoma	• Enhanced aqueous solubility • Enhanced cellular uptake • Improved cytotoxicity	Morgan et al. (2006)
	G4.5 PAMAM	Glioma	• Sustained release • 185 folds improvement in cytotoxicity	Zolotarskaya et al. (2015)
	Pegylated PLL	Colon cancer	• Prolonged circulation half-life • Increased cellular uptake	Fox et al. (2009)
	G4 PAMAM	Lung cancer	• Slow and controlled drug release • Increased cytotoxicity	Oledzka et al. (2023)
	Succinic acid-glycine linker-G4 PAMAM	Colorectal cancer	• Slow drug release • Increased cytotoxicity	Thiagarajan et al. (2010)
	Glutathione-PAMAM	Pancreatic cancer	• Enzyme-mediated uptake • Deeper penetration into tumor tissue	Wang et al. (2020)
	PEG-diacrylate-G3 PAMAM	Head and neck tumor	• Prolonged drug release through the self-cleaving mechanism	Wang et al. (2019)
	Glucose–PEG–PAMAM-Cy7	Liver cancer	• Specific targeting	Ma et al. (2018)

(continued)

Herbal actives	Type of dendrimer	Disease	Remark	References
			• Tumor microenvironment-responsive drug release	
Quercetin	G0 to G3 amine-terminated PAMAM	–	• Enhanced aqueous solubility	Madaan et al. (2016)
			• Biphasic release and sustained release	
	Resorcinarene dendrimer	–	• Improved antibacterial activity against multidrug-resistant *S. aureus*	Rehman et al. (2021)
	G4 PAMAM	Brain tumor	• Enhanced water solubility at pH 6	Kim et al. (2023)
			• High antioxidant and neuroprotective activity	
	Margetuximab-pegylated G4 PAMAM	Breast cancer	• Targeted delivery	Khakinahad et al. (2022)
			• Increased cytotoxicity	
	Folic acid-pegylated PAMAM	–	• Release in acidic pH	Rezaei et al. (2019)
			• Accumulation in tumor	
Gallic acid	G4 PAMAM	Breast cancer	• 2.2 folds improved cytotoxicity	Sharma et al. (2011)
			• Four times improved selectivity	
	G5 PAMAM	Glioblastoma	• High cytotoxicity to chemoresistant cells	Alfei et al. (2020a)
	Hydroxyl-terminated G0 and G1 PAMAM	Alzheimer's disease	• Increased interaction with the affected tissue	Araújo et al. (2021)
	G5 dendrimers	–	• Four folds higher antioxidant activity	Alfei et al. (2020b)
	G4 PAMAM	Liver cancer	• Sustained drug release	Abdou and Masoud (2018)
			• Improved cytotoxicity	
Resveratrol	Glucan-based dendrimer	Skin cancer	• Prolonged storage stability	Shi et al. (2020)

			• Improved permeability and cellular uptake • Improved antioxidant activity	
	G4 PAMAM	Skin aging	• Improved solubility, stability, and drug loading • Improved transdermal permeation	Pentek et al. (2017)
	Silica PAMAM hybrid dendrimer	Breast cancer	• Higher cytotoxicity	Kececiler-Emir et al. (2021)
	Octenylsuccinate hydroxypropyl phytoglycogen dendrimer	–	• Improved solubility and dissolution	Xie and Yao (2018)
Silybin	PEGylated G4 PAMAM	Renal cancer	• Five folds increment in aqueous solubility • Sustained release • Maximum drug loading	Diaz et al. (2018)
	G2, G3 PAMAM	–	• Enhanced aqueous solubility • Controlled release • Improved bioavailability	Huang et al. (2011)
	Arginine peptide dendrimer	Skin	• Enhanced permeation	Shetty et al. (2017)
Puerarin	G3.5, G4, G4.5, and G5 PAMAM	Eye	• Slow drug release • Extended residence time in the eye	Yao et al. (2010)
	Liposome-G2/G3 PAMAM	Eye	• Improved corneal permeation	Liu et al. 2010
	G1.5, G2, G2.5, and G3 PAMAM	–	• Increased solubility • No hemolytic toxicity • Oral formulation	Gu et al. (2013)
Berberine	G4 PAMAM	Breast cancer	• Slow and sustained release • Improved bioavailability • Higher cytotoxicity	Gupta et al. (2017)
	Pegylated PAMAM	Breast cancer	• Prolonged drug release	Yadav et al. (2023)

(continued)

Herbal actives	Type of dendrimer	Disease	Remark	References
			• Higher cellular uptake and cytotoxicity	
Baicalin	Folic acid-G3/G4/G5/G6 PAMAM	Lung cancer	• Sustained release in acidic pH	Lv et al. (2017)
			• Specific targeting and increased cellular uptake	
Crocetin	G4 PAMAM, G4 PPI	Breast cancer	• Sustained drug release by alkylated dendrimer	Soltani et al. (2017)
			• Improved cytotoxicity due to alkylated dendrimer	
Ellagic acid	Lysine modified-G4 dendrimer, arginine modified-G3 dendrimer	–	• Improved solubility up to 300 times	Alfei et al. (2019)
			• Higher drug loading	
Daidzein	G3 PAMAM and G4 PPI	Breast and lung cancer	• Improved solubility and stability	Zhao et al. (2011)
			• Prolonged drug release	
			• Increased antioxidant activity and cytotoxicity	
Capsaicin	Dendrimer	Breast cancer, liver cancer	• Improved stability	Malar and Bavanilathamuthiah (2015)
			• Improved cytotoxicity	
Anthocyanin	Silica-PAMAM	Neuroblastoma	• Prolonged drug release	Yesil-Celiktas et al. (2017)
			• Improved anti-proliferative effect	
Sinomenine	Hydroxyl-terminated G4 PAMAM	Traumatic brain injury	• Targeted delivery and rapid cellular uptake	Sharma et al. (2020)

9.5 Clinical Trials and Marketed Products

Dendrimers were used extensively in many commercialized products with various applications to improve the desired properties of guest molecules. During the last 10 years, dendrimer-based nanomedicine has made significant advancements. Several dendrimer-based formulations and products are still in the early phases of research or are undergoing clinical trials, but a dozen have already acquired FDA clearance and are on the market. Several authorized dendrimer-based medications are based on traditional pharmaceutical ingredients.

For instance, PAMAM dendrimers used in personal care and cosmetics have been investigated by the cosmetics company Revlon. To stabilize and control the adverse effect of the acidity of salicylic acid, Revlon marketed a dendrimer-based cosmetic formulation of salicylic acid in which free acid groups of salicylic acid were bounded by the surface amino groups of the PAMAM dendrimer. Salicylic acid is a hydroxy group containing natural compound and used as a keratolytic agent. Due to its anti-inflammatory and topical antibacterial effect, it is applied to remove dead skin cells (Wolf and Snyder 1995).

Vitamins, which are necessary components to the skin and may be given via cosmetic creams, have been formulated with PAMAM dendrimers for improving the skin permeability of vitamins. The studies described the PAMAM dendrimer as a carrier for vitamins A (trans-retinal) and B6 (pyridoxal and pyridoxal phosphate) and have enhanced their bioavailability and skin penetration in vitro model (Filipowicz and Wołowiec 2012). Also, PAMAM dendrimers of various generations have been used to improve the solubility of vitamin B2 (riboflavin). Water-soluble PAMAM dendrimers G2 and G3 were shown to work best as riboflavin permeation enhancers and to be helpful in cosmetic and dermatologic emulsions (Filipowicz and Wołowiec 2011).

Pentek et al. have developed dendrimer-resveratrol formulations to improve the solubility and stability of RES in topical anti-aging cream. The research found that RES-dendrimer formulations have boosted the solubility, stability, and transdermal penetration of RES in dendrimer architecture. Additionally, dendrimers have been shown to increase the effectiveness of RES loading in aqueous solution, and the produced formulations were safe to apply to the skin since they were water-based (Pentek et al. 2017).

9.6 Conclusion and Prospects

The PCs are explored in various medical trials as they offer several benefits such as natural and holistic treatment approach, availability, accessibility, low cost, fewer side effects, and multiple biological activities. But their translation from lab to market is hindered by many limitations of the PCs such as their poor physicochemical properties and non-specificity. Dendrimers are suitable carriers for delivering such PCs effectively since the molecular weight, chemical structure, and surface activity of dendrimers may be accurately regulated and their attributes can be adjusted for formulation development with different biological, chemical, and biopharmaceutical purpose interests. Dendrimers have been reported to improve the solubility, cellular permeability, stability, bioactivity, and targetability of many PCs.

Despite these benefits, it is still difficult to produce dendrimer-based products on a large scale and testing them through clinical trials before finally commercializing them is time-consuming. However, if these drawbacks are overcome, dendrimers are possible candidates for developing the commercial formulations of PCs. In summary, dendrimers have a promising future in PC-based medicine.

Acknowledgments The authors acknowledge the Central University of Gujarat, Gandhinagar for providing the necessary facilities and support. DBR acknowledges the University Grant Commission, New Delhi, India, for a Ph.D. fellowship. D.P. acknowledges the National Forensic Science University, Gujarat for providing the necessary facilities and support.

References

Abbasi E, Aval SF, Akbarzadeh A, Milani M, Nasrabadi HT, Joo SW, Hanifehpour Y, Nejati-Koshki K, Pashaei-Asl R (2014) Dendrimers: synthesis, applications, and properties. Nanoscale Res Lett 9:247. https://doi.org/10.1186/1556-276X-9-247

Abdou EM, Masoud MM (2018) Gallic acid-PAMAM and gallic acid-phospholipid conjugates, physicochemical characterization and in vivo evaluation. Pharm Dev Technol 23(1):55–66

Afsana S, Kesharwani P (2021) An insight into aptamer engineered dendrimer for cancer therapy. Eur Polym J 159:110746. https://doi.org/10.1016/j.eurpolymj.2021.110746

Ainalem M-L, Nylander T (2011) DNA condensation using cationic dendrimers—morphology and supramolecular structure of formed aggregates. Soft Matter 7:4577–4594. https://doi.org/10.1039/C0SM01171A

Aisina R, Mukhametova L, Ivanova E (2020) Influence cationic and anionic PAMAM dendrimers of low generation on selected hemostatic parameters in vitro. Mater Sci Eng: C 109:110605. https://doi.org/10.1016/j.msec.2019.110605

Albertazzi L, Gherardini L, Brondi M, Sulis Sato S, Bifone A, Pizzorusso T, Ratto GM, Bardi G (2013) In vivo distribution and toxicity of PAMAM dendrimers in the central nervous system depend on their surface chemistry. Mol Pharm 10:249–260. https://doi.org/10.1021/mp300391v

Alfei S, Turrini F, Catena S, Zunin P, Parodi B, Zuccari G, Pittaluga AM, Boggia R (2019) Preparation of ellagic acid micro and nano formulations with amazingly increased water solubility by its entrapment in pectin or non-PAMAM dendrimers suitable for clinical applications. New J Chem 43:2438–2448. https://doi.org/10.1039/C8NJ05657A

Alfei S et al (2020a) Dendrimer nanodevices and gallic acid as novel strategies to fight chemoresistance in neuroblastoma cells. Nanomaterials 10(6):1243

Alfei S, Catena S, Turrini F (2020b) Biodegradable and biocompatible spherical dendrimer nanoparticles with a gallic acid shell and a double-acting strong antioxidant activity as potential device to fight diseases from "oxidative stress". Drug Deliv Transl Res 10:259–270. https://doi.org/10.1007/s13346-019-00681-8

Alibolandi M, Taghdisi SM, Ramezani P, Hosseini F, Amel S, Abnous K, Ramezani M (2017) Smart AS1411-aptamer conjugated pegylated PAMAM dendrimer for the superior delivery of camptothecin to colon adenocarcinoma in vitro and in vivo. Int J Pharm 519:352–364. https://doi.org/10.1016/j.ijpharm.2017.01.044

Al-Jamal KT, Ramaswamy C, Florence AT (2005) Supramolecular structures from dendrons and dendrimers. Adv Drug Deliv Rev 57:2238–2270. https://doi.org/10.1016/j.addr.2005.09.015

Araújo AR, Correa J, Dominguez-Arca V, Reis RL, Fernandez-Megia E, Pires RA (2021) Functional gallic acid-based dendrimers as synthetic nanotools to remodel amyloid-β-42 into

noncytotoxic forms. ACS Appl Mater Interfaces 13:59673–59682. https://doi.org/10.1021/acsami.1c17823
Babaei E, Sadeghizadeh M, Hassan ZM, Feizi MAH, Najafi F, Hashemi SM (2012) Dendrosomal curcumin significantly suppresses cancer cell proliferation in vitro and in vivo. Int Immunopharmacol 12(1):226–234
Balagani PK, Chandiran S, Bhavya B, Sravanthi U, Manubolu S (2011) Dendrimer: a complete drug carrier. Int J Pharm Rev Res 1:25
Bharathi R, Latha S, Selvamani P (2015) Dendrimers as solubility enhancers—a review. Int J Innov Pharm Sci Res 3(4):332–339
Bruce Turnbull W, Kalovidouris SA, Fraser J, Stoddart. (2002) Large oligosaccharide-based glycodendrimers. Chemistry 8(13):2988–3000
Bryant LH Jr, Brechbiel MW, Wu C, Bulte JWM, Herynek V, Frank JA (1999) Synthesis and relaxometry of high-generation (G = 5, 7, 9, and 10) PAMAM dendrimer-DOTA-gadolinium chelates. J Magn Reson Imaging 9:348–352. https://doi.org/10.1002/(SICI)1522-2586(199902)9:2<348::AID-JMRI30>3.0.CO;2-J
Bugno J, Hsu H-J, Hong S (2015) Tweaking dendrimers and dendritic nanoparticles for controlled nano-bio interactions. J Drug Target 23(7–8):642–650
Cabrera MA, Gonzalez-Alvarez I (2011) Oral bioavailability: basic principles, advanced concepts, and applications. In: Oral bioavailability. Wiley, New York, pp 519–534. https://doi.org/10.1002/9781118067598.ch33
Chauhan A (2020) Chapter 2—Dendrimers in drug delivery and the role of "critical nanoscale design parameters" (CNDPs). In: Chauhan A, Kulhari H (eds) Pharmaceutical applications of dendrimers. Elsevier, Amsterdam, pp 49–56. https://doi.org/10.1016/B978-0-12-814527-2.00002-0
Chauhan AS, Kaul M (2018) Engineering of "critical nanoscale design parameters" (CNDPs) in PAMAM dendrimer nanoparticles for drug delivery applications. J Nanopart Res 20:226. https://doi.org/10.1007/s11051-018-4318-z
Chauhan AS, Jain NK, Diwan PV, Khopade AJ (2004) Solubility enhancement of indomethacin with poly(amidoamine) dendrimers and targeting to inflammatory regions of arthritic rats. J Drug Target 12:575–583. https://doi.org/10.1080/10611860400010655
Chen H-T, Neerman MF, Parrish AR, Simanek EE (2004) Cytotoxicity, hemolysis, and acute in vivo toxicity of dendrimers based on melamine, candidate vehicles for drug delivery. J Am Chem Soc 126:10044–10048. https://doi.org/10.1021/ja048548j
Cheng Y, Xu Z, Ma M, Xu T (2008) Dendrimers as drug carriers: applications in different routes of drug administration. J Pharm Sci 97:123–143. https://doi.org/10.1002/jps.21079
Cheng Y, Zhao L, Li Y, Xu T (2011) Design of biocompatible dendrimers for cancer diagnosis and therapy: current status and future perspectives. Chem Soc Rev 40(5):2673–2703. https://doi.org/10.1039/c0cs00097c
Chirumbolo S (2012) Plant phytochemicals as new potential drugs for immune disorders and cancer therapy: really a promising path? J Sci Food Agric 92:1573–1577. https://doi.org/10.1002/jsfa.5670
Chis AA et al (2020) Applications and limitations of dendrimers in biomedicine. *Molecules 25*(17): 3982
Choudhary S, Gupta L, Rani S, Dave K, Gupta U (2017) Impact of dendrimers on solubility of hydrophobic drug molecules. Front Pharmacol 8:261. https://doi.org/10.3389/fphar.2017.00261
Ciolkowski M, Petersen JF, Ficker M, Janaszewska A, Christensen JB, Klajnert B, Bryszewska M (2012) Surface modification of PAMAM dendrimer improves its biocompatibility. Nanomedicine 8(6):815–817
D'Emanuele A, Attwood D (2005) Dendrimer–drug interactions. Adv Drug Deliv Rev 57:2147–2162. https://doi.org/10.1016/j.addr.2005.09.012
D'Emanuele A, Jevprasesphant R, Penny J, Attwood D (2004) The use of a dendrimer-propranolol prodrug to bypass efflux transporters and enhance oral bioavailability. J Control Release 95: 447–453. https://doi.org/10.1016/j.jconrel.2003.12.006

Dave K, Venuganti VV (2017) Dendritic polymers for dermal drug delivery. Ther Deliv 8:1077–1096. https://doi.org/10.4155/tde-2017-0091

Debnath S, Saloum D, Dolai S, Sun C, Averick S, Raja K, Fata JE (2013) Dendrimer-curcumin conjugate: a water soluble and effective cytotoxic agent against breast cancer cell lines. Anticancer Agents Med Chem 13:1531–1539. https://doi.org/10.2174/18715206113139990139

Diaz C, Guzmán J, Jiménez VA, Alderete JB (2018) Partially PEGylated PAMAM dendrimers as solubility enhancers of silybin. Pharm Dev Technol 23:689–696. https://doi.org/10.1080/10837450.2017.1315134

Do VMH, Bach LG, Tran D-HN, Du Cao V, Nguyen TNQ, Hoang DT, Ngo VC, Nguyen DH, Thi TTH (2020) Effective elimination of charge-associated toxicity of low generation polyamidoamine dendrimer eases drug delivery of oxaliplatin, biotechnology and bioprocess. Engineering 25:224–234. https://doi.org/10.1007/s12257-019-0047-0

Dobrovolskaia MA (2017) Dendrimers effects on the immune system: insights into toxicity and therapeutic utility. Curr Pharm Des 23(21):3134–3141

Dobrovolskaia MA, Patri AK, Simak J, Hall JB, Semberova J, De Paoli SH, Lacerda SE, McNeil. (2012) Nanoparticle size and surface charge determine effects of PAMAM dendrimers on human platelets in vitro. Mol Pharm 9(3):382–393

Ekor M (2014) The growing use of herbal medicines-issues relating to adverse reactions and challenges in monitoring safety. Front Pharmacol 4:177

El-Sayed M, Ginski M, Rhodes C, Ghandehari H (2002) Transepithelial transport of poly (amidoamine) dendrimers across Caco-2 cell monolayers. J Control Release 81(3):355–365

Ensign LM, Cone R, Hanes J (2012) Oral drug delivery with polymeric nanoparticles: the gastrointestinal mucus barriers. Adv Drug Deliv Rev 64:557–570. https://doi.org/10.1016/j.addr.2011.12.009

Falconieri MC, Adamo M, Monasterolo C, Bergonzi MC, Coronnello M, Bilia AR (2017) New dendrimer-based nanoparticles enhance curcumin solubility. Planta Med 83:420–425. https://doi.org/10.1055/s-0042-103161

Fatima M, Sheikh A, Hasan N, Sahebkar A, Riadi Y, Kesharwani P (2022) Folic acid conjugated poly(amidoamine) dendrimer as a smart nanocarriers for tracing, imaging, and treating cancers over-expressing folate receptors. Eur Polym J 170:111156. https://doi.org/10.1016/j.eurpolymj.2022.111156

Filipowicz A, Wołowiec S (2011) Solubility and in vitro transdermal diffusion of riboflavin assisted by PAMAM dendrimers. Int J Pharm 408(1–2):152–156

Filipowicz A, Wołowiec S (2012) Bioconjugates of PAMAM dendrimers with trans-retinal, pyridoxal, and pyridoxal phosphate. Int J Nanomed 7:4819–4828

Fischer M, Vögtle F (1999) Dendrimers: from design to application—a progress report. Angew Chem Int Ed 38:884–905. https://doi.org/10.1002/(SICI)1521-3773(19990401)38:7<884::AID-ANIE884>3.0.CO;2-K

Florence AT, Sakthivel T, Toth I (2000) Oral uptake and translocation of a polylysine dendrimer with a lipid surface. J Control Release 65:253–259. https://doi.org/10.1016/S0168-3659(99)00237-0

Flory PJ (1941) Molecular size distribution in three dimensional polymers. I. Gelation1. J Am Chem Soc 63:3083–3090. https://doi.org/10.1021/ja01856a061

Fox ME, Guillaudeu S, Fréchet JMJ, Jerger K, Macaraeg N, Szoka FC (2009) Synthesis and in vivo antitumor efficacy of PEGylated poly(L-lysine) dendrimer-camptothecin conjugates. Mol Pharm 6(5):1562–1572

Fox LJ, Richardson RM, Briscoe WH (2018) PAMAM dendrimer—cell membrane interactions. Adv Colloid Interface Sci 257:1–18. https://doi.org/10.1016/j.cis.2018.06.005

Fréchet JMJ (2002) Dendrimers and supramolecular chemistry. Proc Natl Acad Sci 99:4782. https://doi.org/10.1073/pnas.082013899

Gallien J et al (2021) Curcumin loaded dendrimers specifically reduce viability of glioblastoma cell lines. Molecules 26(19):6050

Gamage NH, Jing L, Worsham MJ, Ali MM (2016) Targeted theranostic approach for glioma using dendrimer-based curcumin nanoparticle. J Nanomed Nanotechnol 7(4):393

Garg T, Singh O, Arora S, Murthy R (2011) Dendrimer—a novel scaffold for drug delivery. Int J Pharm Sci Rev Res 7:211–220

Ghaffari M, Dehghan G, Abedi-Gaballu F, Kashanian S, Baradaran B, Dolatabadi JEN, Losic D (2018) Surface functionalized dendrimers as controlled-release delivery nanosystems for tumor targeting. Eur J Pharm Sci 122:311–330. https://doi.org/10.1016/j.ejps.2018.07.020

Ghaffari M, Dehghan G, Baradaran B, Zarebkohan A, Mansoori B, Soleymani J, Dolatabadi JEN, Hamblin M (2019) Co-delivery of curcumin and Bcl-2 siRNA by PAMAM dendrimers for enhancement of the therapeutic efficacy in HeLa cancer cells. Colloids Surf B Biointerfaces 188: 110762. https://doi.org/10.1016/j.colsurfb.2019.110762

Gillies ER, Fréchet JMJ (2005) Dendrimers and dendritic polymers in drug delivery. Drug Discov Today 10:35–43. https://doi.org/10.1016/S1359-6446(04)03276-3

Gothwal A, Malik S, Gupta U, Jain NK (2020) Chapter 11—Toxicity and biocompatibility aspects of dendrimers. In: Chauhan A, Kulhari H (eds) Pharmaceutical applications of dendrimers. Elsevier, Amsterdam, pp 251–274. https://doi.org/10.1016/B978-0-12-814527-2.00011-1

Gu Z, Luo K, She W, Wu Y, He B (2010) New-generation biomedical materials: peptide dendrimers and their application in biomedicine. Sci China Chem 53:458–478

Gu L, Wu Zh, Qi X, He H, Ma X, Chou X, Wen X, Zhang M, Jiao F (2013) Polyamidomine dendrimers: an excellent drug carrier for improving the solubility and bioavailability of puerarin. Pharm Dev Technol 18:1051–1057. https://doi.org/10.3109/10837450.2011.653822

Gupta U, Agashe HB, Asthana A, Jain NK (2006) Dendrimers: novel polymeric nanoarchitectures for solubility enhancement. Biomacromolecules 7:649–658. https://doi.org/10.1021/bm050802s

Gupta U, Dwivedi SKD, Bid HK, Konwar R, Jain NK (2010) Ligand anchored dendrimers based nanoconstructs for effective targeting to cancer cells. Int J Pharm 393:186–197. https://doi.org/10.1016/j.ijpharm.2010.04.002

Gupta L, Sharma AK, Gothwal A, Khan MS, Khinchi MP, Qayum A, Singh SK, Gupta U (2017) Dendrimer encapsulated and conjugated delivery of berberine: a novel approach mitigating toxicity and improving in vivo pharmacokinetics. Int J Pharm 528:88–99. https://doi.org/10.1016/j.ijpharm.2017.04.073

Gupta N, Rai D, Kulhari H (2020) Dendrimers for diagnostic applications. In: Pharmaceutical applications of dendrimers. Elsevier, Amsterdam, pp 291–324. https://doi.org/10.1016/B978-0-12-814527-2.00013-5

Hawker CJ, Frechet JMJ (1990) Control of surface functionality in the synthesis of dendritic macromolecules using the convergent-growth approach. Macromolecules 23:4726–4729. https://doi.org/10.1021/ma00223a036

Hu J, Cheng Y, Wu Q, Zhao L, Xu T (2009) Host−guest chemistry of dendrimer-drug complexes. 2. Effects of molecular properties of guests and surface functionalities of dendrimers. J Phys Chem B 113:10650–10659. https://doi.org/10.1021/jp9047055

Huang X, Wu Z, Gao W, Chen Q, Yu B (2011) Polyamidoamine dendrimers as potential drug carriers for enhanced aqueous solubility and oral bioavailability of silybin. Drug Dev Ind Pharm 37:419–427. https://doi.org/10.3109/03639045.2010.518150

Imae T (2012) Physicochemical properties of dendrimers and dendrimer complexes. In: Dendrimer-based drug delivery systems. Wiley, New York, pp 55–92. https://doi.org/10.1002/9781118275238.ch2

Jain V, Bharatam P (2014) Pharmacoinformatic approaches to understand complexation of dendrimeric nanoparticles with drugs. Nanoscale 6(5):2476–2501. https://doi.org/10.1039/c3nr05400d

Jain K, Kesharwani P, Gupta U, Jain NK (2010) Dendrimer toxicity-let's meet the challenge. Int J Pharm 394(1–2):122–142

Janaszewska A, Lazniewska J, Trzepiński P, Marcinkowska M, Klajnert-Maculewicz B (2019) Cytotoxicity of dendrimers. Biomolecules. 9(8):330

Jang W-D, Kamruzzaman Selim KM, Lee C-H, Kang I-K (2009) Bioinspired application of dendrimers: from bio-mimicry to biomedical applications. Prog Polym Sci 34:1–23. https://doi.org/10.1016/j.progpolymsci.2008.08.003

Jangid AK, Patel K, Joshi U, Patel S, Singh A, Pooja D, Saharan VA, Kulhari H (2022) PEGylated G4 dendrimers as a promising nanocarrier for piperlongumine delivery: synthesis, characterization, and anticancer activity. Eur Polym J 179:111547. https://doi.org/10.1016/j.eurpolymj.2022.111547

Jevprasesphant R, Penny J, Attwood D, McKeown NB, D'Emanuele A (2003) Engineering of dendrimer surfaces to enhance transepithelial transport and reduce cytotoxicity. Pharm Res 20:1543–1550. https://doi.org/10.1023/A:1026166729873

Jin-Jian L, Bao J-L, Chen X-P, Huang M, Wang Y-T (2012) Alkaloids isolated from natural herbs as the anticancer agents. Evid Based Complement Alternat Med 2012:485042

Jones CF, Campbell RA, Brooks AE, Assemi S, Tadjiki S, Thiagarajan G, Mulcock C, Weyrich AS, Brooks BD, Ghandehari H, Grainger DW (2012) Cationic PAMAM dendrimers aggressively initiate blood clot formation. ACS Nano 6:9900–9910. https://doi.org/10.1021/nn303472r

Kalomiraki M, Thermos K, Chaniotakis NA (2015) Dendrimers as tunable vectors of drug delivery systems and biomedical and ocular applications. Int J Nanomedicine 11:1–12

Karandikar S, Mirani A, Waybhase V, Patravale VB, Patankar S (2017) Chapter 10—Nanovaccines for oral delivery-formulation strategies and challenges. In: Andronescu E, Grumezescu AM (eds) Nanostructures for oral medicine. Elsevier, Amsterdam, pp 263–293. https://doi.org/10.1016/B978-0-323-47720-8.00011-0

Karimi A, Majlesi M, Rafieian-Kopaei M (2015) Herbal versus synthetic drugs; beliefs and facts. J Nephropharmacol 4:27–30. https://www.ncbi.nlm.nih.gov/pubmed/28197471

Kececiler-Emir C, Ilhan-Ayisigi E, Celen-Erden C, Nalbantsoy A, Yesil-Celiktas O (2021) Synthesis of resveratrol loaded hybrid silica-PAMAM dendrimer nanoparticles with emphases on inducible nitric oxide synthase and cytotoxicity. Plant Foods Hum Nutr 76:219–225. https://doi.org/10.1007/s11130-021-00897-5

Keefe A, Pai S, Ellington A (2010) Aptamers as therapeutics. Nat Rev Drug Discov 9:537–550. https://doi.org/10.1038/nrd3141

Kesharwani P, Tekade RK, Jain NK (2015a) Generation dependent safety and efficacy of folic acid conjugated dendrimer based anticancer drug formulations. Pharm Res 32:1438–1450. https://doi.org/10.1007/s11095-014-1549-2

Kesharwani P, Xie L, Banerjee S, Mao G, Padhye S, Sarkar FH, Iyer AK (2015b) Hyaluronic acid-conjugated polyamidoamine dendrimers for targeted delivery of 3,4-difluorobenzylidene curcumin to CD44 overexpressing pancreatic cancer cells. Colloids Surf B Biointerfaces 136:413–423

Khakinahad Y, Sohrabi S, Razi S, Narmani A, Khaleghi S, Asadiyun M, Jafari H, Mohammadnejad J (2022) Margetuximab conjugated-PEG-PAMAM G4 nano-complex: a smart nano-device for suppression of breast cancer. Biomed Eng Lett 12:317–329. https://doi.org/10.1007/s13534-022-00225-z

Khan T, Gurav P (2018) Phyto-nanotechnology: enhancing delivery of plant based anti-cancer drugs. Front Pharmacol 8:1002. https://doi.org/10.3389/fphar.2017.01002

Kianamiri S, Dinari A, Sadeghizadeh M, Rezaei M, Daraei B, Bahsoun NE-H, Nomani A (2020) Mitochondria-targeted polyamidoamine dendrimer–curcumin construct for hepatocellular cancer treatment. Mol Pharm 17:4483–4498. https://doi.org/10.1021/acs.molpharmaceut.0c00566

Kim HT, Yoo M, Yang E-J, Song K-S, Park EJ, Na DH (2023) The importance of pH for the formation of stable and active quercetin–polyamidoamine dendrimer complex. Bull Korean Chem Soc 44:363–369. https://doi.org/10.1002/bkcs.12669

Klajnert-Maculewicz B, Bryszewska M (2001) Dendrimers: properties and applications. Acta Biochim Pol 48(1):199–208

Kolhatkar R, Sweet D, Ghandehari H (2008) Functionalized dendrimers as nanoscale drug carriers. In: Torchilin V (ed) Multifunctional pharmaceutical nanocarriers. Springer, New York, pp 201–232. https://doi.org/10.1007/978-0-387-76554-9_7

Kurtoglu YE, Mishra MK, Kannan S, Kannan RM (2010) Drug release characteristics of PAMAM dendrimer–drug conjugates with different linkers. Int J Pharm 384:189–194. https://doi.org/10.1016/j.ijpharm.2009.10.017

Labieniec-Watala M, Watala C (2015) PAMAM dendrimers: destined for success or doomed to fail? Plain and modified PAMAM dendrimers in the context of biomedical applications. J Pharm Sci 104:2–14. https://doi.org/10.1002/jps.24222

Leuner C, Dressman J (2000) Improving drug solubility for oral delivery using solid dispersions. Eur J Pharm Biopharm 50:47–60. https://doi.org/10.1016/S0939-6411(00)00076-X

Levin DA (1971) Plant phenolics: an ecological perspective. Am Nat 105:157–181. https://doi.org/10.1086/282712

Lin Y-L, Khanafer K, El-Sayed MEH (2010) Quantitative evaluation of the effect of poly (amidoamine) dendrimers on the porosity of epithelial monolayers. Nanoscale 2(5):755–762

Liu Y, Sun K, Yao W, Liang N, Mu H, Liang R, Yao C (2010) [Corneal penetration of PAMAM dendrimers-coated puerarin liposomes]. Zhongguo Zhong Yao Za Zhi 35(1):30–34

Liu J, Gray W, Davis M, Luo Y (2012) Peptide- and saccharide-conjugated dendrimers for targeted drug delivery: a concise review, interface. Focus 2:307–324. https://doi.org/10.1098/rsfs.2012.0009

Lombardo D, Calandra P, Bellocco E, Laganà G, Barreca D, Magazù S, Wanderlingh U, Kiselev MA (2016) Effect of anionic and cationic polyamidoamine (PAMAM) dendrimers on a model lipid membrane. Biochim Biophys Acta 1858:2769–2777. https://doi.org/10.1016/j.bbamem.2016.08.001

Lv T, Yu T, Fang Y, Zhang S, Jiang M, Zhang H, Zhang Y, Li Z, Chen H, Gao Y (2017) Role of generation on folic acid-modified poly(amidoamine) dendrimers for targeted delivery of baicalin to cancer cells. Mater Sci Eng C Mater Biol Appl 75:182–190

Lyu Z, Ding L, Huang AYT, Kao C-L, Peng L (2019) Poly(amidoamine) dendrimers: covalent and supramolecular synthesis. Mater Today Chem 13:34–48. https://doi.org/10.1016/j.mtchem.2019.04.004

Ma H, Chen B, Sassa T, Dalton LR, Jen AK-Y (2001) Highly efficient and thermally stable nonlinear optical dendrimer for electrooptics. J Am Chem Soc 123:986–987. https://doi.org/10.1021/ja003407c

Ma P, Sun Y, Chen J, Li H, Zhu H, Gao X, Bi X, Zhang Y (2018) Enhanced anti-hepatocarcinoma efficacy by GLUT1 targeting and cellular microenvironment-responsive PAMAM–camptothecin conjugate. Drug Deliv 25:153–165. https://doi.org/10.1080/10717544.2017.1419511

Madaan K, Kumar S, Poonia N, Lather V, Pandita D (2014) Dendrimers in drug delivery and targeting: drug-dendrimer interactions and toxicity issues. J Pharm Bioallied Sci 6:139–150. https://doi.org/10.4103/0975-7406.130965

Madaan K, Lather V, Pandita D (2016) Evaluation of polyamidoamine dendrimers as potential carriers for quercetin, a versatile flavonoid. Drug Deliv 23:254–262. https://doi.org/10.3109/10717544.2014.910564

Majoros IJ, Keszler B, Woehler S, Bull T, Baker JR (2003) Acetylation of poly(amidoamine) dendrimers. Macromolecules 36:5526–5529. https://doi.org/10.1021/ma021540e

Majoros IJ, Myc A, Thomas T, Mehta CB, Baker JR (2006) PAMAM dendrimer-based multifunctional conjugate for cancer therapy: synthesis, characterization, and functionality. Biomacromolecules 7:572–579. https://doi.org/10.1021/bm0506142

Majoros IJ, Williams CR, Baker JR Jr (2008) Current dendrimer applications in cancer diagnosis and therapy. Curr Top Med Chem 8:1165–1179. https://doi.org/10.2174/156802608785849049

Malar C, Bavanilathamuthiah (2015) Dendrosomal capsaicin nanoformulation for the invitro anticancer effect on HEp 2 and MCF-7 cell lines. Int J Appl Bioeng 9(2):30–35. https://doi.org/10.18000/ijabeg.10133

Malkoch M, García-Gallego S (2020) Chapter 1: Introduction to dendrimers and other dendritic polymers. In: Dendrimer chemistry: synthetic approaches towards complex architectures. The Royal Society of Chemistry, London, pp 1–20. https://doi.org/10.1039/9781788012904-00001

Menjoge AR, Kannan RM, Tomalia DA (2010) Dendrimer-based drug and imaging conjugates: design considerations for nanomedical applications. Drug Discov Today 15:171–185. https://doi.org/10.1016/j.drudis.2010.01.009

Mignani S, Shi X, Karpus A, Majoral J-P (2021) Non-invasive intranasal administration route directly to the brain using dendrimer nanoplatforms—an opportunity to develop new CNS drugs. Eur J Med Chem 209:112905

Mishra V, Gupta U, Jain NK (2009) Surface-engineered dendrimers: a solution for toxicity issues. J Biomater Sci Polym Ed 20:141–166. https://doi.org/10.1163/156856208X386246

Mittal P, Saharan A, Verma R, Altalbawy FMA, Alfaidi MA, Batiha GE-S, Akter W, Gautam RK, Uddin MS, Rahman MS (2021) Dendrimers: a new race of pharmaceutical nanocarriers. Biomed Res Int 2021:8844030. https://doi.org/10.1155/2021/8844030

Mollazade M, Nejati-Koshki K, Akbarzadeh A, Zarghami N, Nasiri M, Jahanban-Esfahlan R, Alibakhshi A (2013) PAMAM dendrimers augment inhibitory effects of curcumin on cancer cell proliferation: possible inhibition of telomerase. Asian Pac J Cancer Prev 14:6925–6928. https://doi.org/10.7314/APJCP.2013.14.11.6925

Morgan MT, Nakanishi Y, Kroll DJ, Griset AP, Carnahan MA, Wathier M, Oberlies NH, Manikumar G, Wani MC, Grinstaff MW (2006) Dendrimer-encapsulated camptothecins: increased solubility, cellular uptake, and cellular retention affords enhanced anticancer activity in vitro. Cancer Res 66:11913–11921. https://doi.org/10.1158/0008-5472.CAN-06-2066

Mousavifar L, Roy R (2021) Design, synthetic strategies, and therapeutic applications of heterofunctional glycodendrimers. Molecules 26:2428. https://doi.org/10.3390/molecules26092428

Munir M, Hanif M, Ranjha N (2016) Dendrimers and their applications: a review article. Pak J Pharm Res 2:55. https://doi.org/10.22200/pjpr.2016155-66

Nahata A (2017) Anticancer agents: a review of relevant information on important herbal drugs. Int J Clin Pharmacol Toxicol 6(2):250–255. https://doi.org/10.19070/2167-910X-1700042

Namazi H, Heydari A (2014) Synthesis of β-cyclodextrin-based dendrimer as a novel encapsulation agent. Polym Int 63:1447–1455. https://doi.org/10.1002/pi.4637

Newkome GR, Yao Z, Baker GR, Gupta VK, Micelles. (1985) Part 1. Cascade molecules: a new approach to micelles. A [27]-arborol. J Org Chem 50:2003–2004. https://doi.org/10.1021/jo00211a052

Newkome GR, Baker GR, Saunders MJ, Russo PS, Gupta VK, Yao Z, Miller JE, Bouillion K (1986) Two-directional cascade molecules: synthesis and characterization of [9]-n-[9] arborols. J Chem Soc Chem Commun:752–753. https://doi.org/10.1039/C39860000752

Newkome GR, Baker GR, Arai S, Saunders MJ, Russo PS, Theriot KJ, Moorefield CN, Rogers LE, Miller JE (1990) Cascade molecules. Part 6. Synthesis and characterization of two-directional cascade molecules and formation of aqueous gels. J Am Chem Soc 112:8458–8465. https://doi.org/10.1021/ja00179a034

Nimjee S, Rusconi C, Sullenger B (2005) APTAMERS: an emerging class of therapeutics. Annu Rev Med 56:555–583. https://doi.org/10.1146/annurev.med.56.062904.144915

Nisar B, Sultan A, Rubab SL (2018) Comparison of medicinally important natural products versus synthetic drugs—a short commentary. Nat Prod Chem Res. 6:2. https://doi.org/10.4172/2329-6836.1000308

Noriega-Luna B et al (2014) Applications of dendrimers in drug delivery agents, diagnosis, therapy, and detection. J Nanomater 2014:507273

Nosrati H, Adibtabar M, Sharafi A, Danafar H, Kheiri MH (2018) PAMAM-modified citric acid-coated magnetic nanoparticles as pH sensitive biocompatible carrier against human breast cancer cells. Drug Dev Ind Pharm 44(8):1377–1384

Oberlies NH, Kroll DJ (2004) Camptothecin and taxol: historic achievements in natural products research. J Nat Prod 67:129–135. https://doi.org/10.1021/np030498t

Oledzka E, Paśnik K, Domańska I, Zielińska-Pisklak M, Piotrowska U, Sobczak M, Szeleszczuk Ł, Laskowska A (2023) Poly(amidoamine) dendrimer-camptothecin complex: from synthesis to in vitro cancer cell line studies. Molecules 28(6):2696

Ouyang L, Luo Y, Tian M, Zhang S-Y, Lu R, Wang J-H, Kasimu R, Li X (2014) Plant natural products: from traditional compounds to new emerging drugs in cancer therapy. Cell Prolif 47: 506–515. https://doi.org/10.1111/cpr.12143

Pan S-Y, Zhou S-F, Gao S-H, Yu Z-L, Zhang S-F, Tang M-K, Sun J-N, Ma D-L, Han Y-F, Fong W-F, Ko K-M (2013) New perspectives on how to discover drugs from herbal medicines: CAM's outstanding contribution to modern therapeutics. Evid Based Complement Alternat Med 2013:627375. https://doi.org/10.1155/2013/627375

Pentek T, Newenhouse E, O'Brien B, Chauhan AS (2017) Development of a topical resveratrol formulation for commercial applications using dendrimer nanotechnology. Molecules 22:137. https://doi.org/10.3390/molecules22010137

Pooja D, Sistla R, Kulhari H (2018) Chapter 7—Dendrimer-drug conjugates: synthesis strategies, stability and application in anticancer drug delivery. In: Grumezescu AM (ed) Design of nanostructures for theranostics applications. William Andrew Publishing, New York, pp 277–303. https://doi.org/10.1016/B978-0-12-813669-0.00007-5

Pooja D, Srinivasa Reddy T, Kulhari H, Kadari A, Adams DJ, Bansal V, Sistla R (2020) N-acetyl-d-glucosamine-conjugated PAMAM dendrimers as dual receptor-targeting nanocarriers for anti-cancer drug delivery. Eur J Pharm Biopharm 154:377–386. https://doi.org/10.1016/j.ejpb.2020.07.020

Rai D, Kulhari D, Kulhari H (2021) Functionalisation of dendrimers. In: Dendrimers in nanomedicine. CRC Press, Boca Raton, pp 123–145. https://doi.org/10.1201/9781003029915-7

Rehman K, Ali I, El-Haj BM, Kanwal T, Maharjan R, Saifullah S, Imran M, Shafiullah, Simjee SU, Shah MR (2021) Synthesis of novel biocompatible resorcinarene based nanosized dendrimer-vesicles for enhanced anti-bacterial potential of quercetin. J Mol Liq 341:116921. https://doi.org/10.1016/j.molliq.2021.116921

Reymond J-L, Darbre T (2012) Peptide and glycopeptide dendrimer apple trees as enzyme models and for biomedical applications. Org Biomol Chem 10:1483–1492. https://doi.org/10.1039/C2OB06938E

Rezaei SJT, Malekzadeh AM, Ramazani A, Niknejad H (2019) pH-sensitive magnetite nanoparticles modified with hyperbranched polymers and folic acid for targeted imaging and therapy. Curr Drug Deliv 16(9):839–848

Roberts JC, Bhalgat MK, Zera RT (1996) Preliminary biological evaluation of polyamidoamine (PAMAM) starburst dendrimers. J Biomed Mater Res 30(1):53–65

Rodríguez-Izquierdo I, Sepúlveda-Crespo D, Lasso JM, Resino S, Muñoz-Fernández MÁ (2022) Baseline and time-updated factors in preclinical development of anionic dendrimers as successful anti-HIV-1 vaginal microbicides. WIREs Nanomed Nanobiotechnol 14:e1774. https://doi.org/10.1002/wnan.1774

Roy R, Shiao T, Rittenhouse-Olson K (2012) Glycodendrimers: versatile tools for nanotechnology. Braz J Pharm Sci 49:85–108. https://doi.org/10.1590/S1984-82502013000700008

Sadekar S, Ghandehari H (2012) Transepithelial transport and toxicity of PAMAM dendrimers-implications for oral drug delivery. Adv Drug Deliv Rev 64(6):571–588

Samad A, Alam M, Saxena K (2009) Dendrimers: a class of polymers in the nanotechnology for the delivery of active pharmaceuticals. Curr Pharm Des 15:2958–2969. https://doi.org/10.2174/138161209789058200

Santander-Ortega MJ, Chooi KW, Schätzlein AG, Uchegbu IF (2013) Dendrimers for drug delivery. In: Roberts GCK (ed) Encyclopedia of biophysics. Springer, Berlin, pp 441–451. https://doi.org/10.1007/978-3-642-16712-6_710

Santos A, Veiga F, Figueiras A (2019) Dendrimers as pharmaceutical excipients: synthesis, properties, toxicity and biomedical applications. Materials (Basel) 13(1):65

Sathe RY, Bharatam PV (2022) Drug-dendrimer complexes and conjugates: detailed furtherance through theory and experiments. Adv Colloid Interface Sci 303:102639. https://doi.org/10.1016/j.cis.2022.102639

Sen S, Chakraborty R, De B (2011) Challenges and opportunities in the advancement of herbal medicine: India's position and role in a global context. J Herb Med 1:67–75. https://doi.org/10.1016/j.hermed.2011.11.001

Shahzad N, Khan W, Shadab MD, Ali A, Saluja SS, Sharma S, Al-Allaf FA, Abduljaleel Z, Ibrahim IAA, Abdel-Wahab AF, Afify MA, Al-Ghamdi SS (2017) Phytosterols as a natural anticancer agent: current status and future perspective. Biomed Pharmacother 88:786–794. https://doi.org/10.1016/j.biopha.2017.01.068

Sharma A, Gautam S, Gupta A (2011) Surface modified dendrimers: synthesis and characterization for cancer targeted drug delivery. Bioorg Med Chem 19(11):3341–3346. https://doi.org/10.1016/j.bmc.2011.04.046

Sharma R, Kambhampati SP, Zhang Z, Sharma A, Chen S, Duh EI, Kannan S, Tso MOM, Kannan RM (2020) Dendrimer mediated targeted delivery of sinomenine for the treatment of acute neuroinflammation in traumatic brain injury. J Control Release 323:361–375. https://doi.org/10.1016/j.jconrel.2020.04.036

Shetty PK, Manikkath J, Tupally K, Kokil G, Hegde AR, Raut SY, Parekh HS, Mutalik S (2017) Skin delivery of EGCG and silibinin-potential of peptide dendrimers for enhanced skin permeation and deposition. AAPS PharmSciTech 18(6):2346–2357

Shi X, Wang S, Sun H, Baker JR (2007a) Improved biocompatibility of surface functionalized dendrimer-entrapped gold nanoparticles. Soft Matter 3:71–74. https://doi.org/10.1039/B612972B

Shi W, Dolai S, Rizk S, Hussain A, Tariq H, Averick S, L'Amoreaux W, El Idrissi A, Banerjee P, Raja K (2007b) Synthesis of monofunctional curcumin derivatives, clicked curcumin dimer, and a PAMAM dendrimer curcumin conjugate for therapeutic applications. Org Lett 9:5461–5464. https://doi.org/10.1021/ol702370m

Shi X, Lee I, Chen X, Shen M, Xiao S, Zhu M, Baker JR, Wang SH (2010) Influence of dendrimer surface charge on the bioactivity of 2-methoxyestradiol complexed with dendrimers. Soft Matter 6(11):2539–2545

Shi Y, Ye F, Lu K, Hui Q, Miao M (2020) Characterizations and bioavailability of dendrimer-like glucan nanoparticulate system containing resveratrol. J Agric Food Chem 68:6420–6429. https://doi.org/10.1021/acs.jafc.0c01315

Simonescu CM (2018) Introductory chapter: Dendrimers as nanoengineered materials and their applications. In: Simonescu CM (ed) Dendrimers. IntechOpen, Rijeka. https://doi.org/10.5772/intechopen.75844

Singh R, Hildgen P (2006) Synthesis and evaluation of novel dendrimers with a hydrophilic interior as nanocarriers for drug delivery. Bioconjug Chem 17:29–41. https://doi.org/10.1021/bc050184c

Siyad MA, Kumar GSV (2012) Poly(ethylene glycol) grafted polystyrene dendrimer resins: novel class of supports for solid phase peptide synthesis. Polymer 53:4076–4090. https://doi.org/10.1016/j.polymer.2012.07.011

Smith RJ, Gorman C, Menegatti S (2021) Synthesis, structure, and function of internally functionalized dendrimers. J Polym Sci 59:10–28. https://doi.org/10.1002/pol.20200721

Soltani F, Ramezani M, Farzad SA, Mokhtarzadeh A, Hashemi M (2017) Comparison study of the effect of alkyl-modified and unmodified PAMAM and PPI dendrimers on solubility and antitumor activity of crocetin. Artif Cells Nanomed Biotechnol 45(7):1356–1362

Stahl W, Sies H (2005) Bioactivity and protective effects of natural carotenoids. Biochim Biophys Acta 1740:101–107. https://doi.org/10.1016/j.bbadis.2004.12.006

Strickley R (2004) Solubilizing excipients in oral and injectable formulations. Pharm Res 21(2): 201–230. https://doi.org/10.1023/B:PHAM.0000016235.32639.23

Svenson S, Chauhan AS (2008) Dendrimers for enhanced drug solubilization. Nanomedicine 3: 679–702. https://doi.org/10.2217/17435889.3.5.679

Svenson S, Tomalia DA (2012) Dendrimers in biomedical applications-reflections on the field. Adv Drug Deliv Rev 64:102–115. https://doi.org/10.1016/j.addr.2012.09.030

Tahmasebi Birgani M, Erfani V, Babaei E, Najafi F, Zamani M, Shariati M, Nazem S, Farhangi B, Motahari P, Sadeghizade M (2015) Dendrosomal nano-curcumin; the novel formulation to improve the anticancer properties of curcumin. Prog Biol Sci 5:143–158

Tahmasebi Mirgani M, Isacchi B, Sadeghizadeh M, Marra F, Bilia AR, Mowla SJ, Najafi F, Babaei E (2014) Dendrosomal curcumin nanoformulation downregulates pluripotency genes via miR-145 activation in U87MG glioblastoma cells. Int J Nanomedicine 9:403–417. https://doi.org/10.2147/IJN.S48136

Tam JP, Lu Y-A, Yang J-L (2002) Antimicrobial dendrimeric peptides. Eur J Biochem 269(3): 923–932

Thiagarajan G, Ray A, Malugin A, Ghandehari H (2010) PAMAM-camptothecin conjugate inhibits proliferation and induces nuclear fragmentation in colorectal carcinoma cells. Pharm Res 27(11):2307–2316

Thillaivanan S, Samraj K (2014) Challenges, constraints and opportunities in herbal medicines—a review. Int J Herb Med 2(1):21–24

Thomas TP, Patri AK, Myc A, Myaing MT, Ye JY, Norris TB, Baker JR (2004) In vitro targeting of synthesized antibody-conjugated dendrimer nanoparticles. Biomacromolecules 5:2269–2274. https://doi.org/10.1021/bm049704h

Tolia G (2008) The role of dendrimers in topical drug delivery. Pharm Technol 32:88–98

Tomalia DA, Fréchet JMJ (2002) Discovery of dendrimers and dendritic polymers: a brief historical perspective. J Polym Sci A Polym Chem 40:2719–2728. https://doi.org/10.1002/pola.10301

Tomalia DA, Baker H, Dewald J, Hall M, Kallos G, Martin S, Roeck J, Ryder J, Smith P (1986) Dendritic macromolecules: synthesis of starburst dendrimers. Macromolecules 19:2466–2468. https://doi.org/10.1021/ma00163a029

Tomalia DA, Naylor AM, Goddard WA III (1990) Starburst dendrimers: molecular-level control of size, shape, surface chemistry, topology, and flexibility from atoms to macroscopic matter. Angew Chem Int Ed Engl 29:138–175. https://doi.org/10.1002/anie.199001381

Toomari Y, Namazi H, Akbar EA (2015) Synthesis of the dendritic type β-cyclodextrin on primary face via click reaction applicable as drug nanocarrier. Carbohydr Polym 132:205–213

Tripathi PK et al (2020) Curcumin loaded poly(amidoamine) dendrimer-plamitic acid core-shell nanoparticles as anti-stress therapeutics. Drug Dev Ind Pharm 46(3):412–426

Tunki L, Kulhari H, Bhargava SK, Pooja D (2020) Chapter 4—Pharmacokinetic considerations in design of dendrimer-based nanomedicines. In: Chauhan A, Kulhari H (eds) Pharmaceutical applications of dendrimers. Elsevier, Amsterdam, pp 93–106. https://doi.org/10.1016/B978-0-12-814527-2.00004-4

van der Poll DG, Kieler-Ferguson HM, Floyd WC, Guillaudeu SJ, Jerger K, Szoka FC, Fréchet JM (2010) Design, synthesis, and biological evaluation of a robust, biodegradable dendrimer. Bioconjug Chem 21:764–773. https://doi.org/10.1021/bc900553n

van Hest JCM, Delnoye DAP, Baars MWPL, van Genderen MHP, Meijer EW (1979) Polystyrene-dendrimer amphiphilic block copolymers with a generation-dependent aggregation. Science 268(1995):1592–1595. https://doi.org/10.1126/science.268.5217.1592

Vandamme TF, Brobeck L (2005) Poly(amidoamine) dendrimers as ophthalmic vehicles for ocular delivery of pilocarpine nitrate and tropicamide. J Control Release 102(1):23–38

Walter MV, Malkoch M (2012) Simplifying the synthesis of dendrimers: accelerated approaches. Chem Soc Rev 41:4593–4609. https://doi.org/10.1039/C2CS35062A

Wang L, Xiaoping X, Zhang Y, Zhang Y, Zhu Y, Shi J, Sun Y, Huang Q (2013) Encapsulation of curcumin within poly(amidoamine) dendrimers for delivery to cancer cells. J Mater Sci Mater Med 24(9):2137–2144

Wang L, Yang Q, Cui Y, Gao D, Kang J, Sun H, Zhu L, Chen S (2017) Highly stable and biocompatible dendrimer-encapsulated gold nanoparticle catalysts for the reduction of 4-nitrophenol. New J Chem 41:8399–8406. https://doi.org/10.1039/C7NJ01567D

Wang J, He H, Cooper RC, Gui Q, Yang H (2019) Drug-conjugated dendrimer hydrogel enables sustained drug release via a self-cleaving mechanism. Mol Pharm 16:1874–1880. https://doi.org/10.1021/acs.molpharmaceut.8b01207

Wang G, Zhou Z, Zhao Z, Li Q, Wu Y, Yan S, Shen Y, Huang P (2020) Enzyme-triggered transcytosis of dendrimer–drug conjugate for deep penetration into pancreatic tumors. ACS Nano 14:4890–4904. https://doi.org/10.1021/acsnano.0c00974

Wängler C, Moldenhauer G, Eisenhut M, Haberkorn U, Mier W (2008) Antibody−dendrimer conjugates: the number, not the size of the dendrimers, determines the immunoreactivity. Bioconjug Chem 19:813–820. https://doi.org/10.1021/bc700308q

Warsi M, Rahman M, Yusuf M, Ali A, Muheem A, Amin S, Ahmad J (2021) Drug delivery potential of dendrimeric formulation. In: Dendrimers in nanomedicine. CRC Press, Boca Raton, pp 255–275. https://doi.org/10.1201/9781003029915-14

Wendland MS, Zimmerman SC (1999) Synthesis of cored dendrimers. J Am Chem Soc 121:1389–1390. https://doi.org/10.1021/ja983097m

Winnicka K, Wroblewska M, Sosnowska K, Car H, Kasacka I (2015) Evaluation of cationic polyamidoamine dendrimers' dermal toxicity in the rat skin model. Drug Des Devel Ther 9: 1367–1377

Wiwattanapatapee R, Carreño-Gómez B, Malik N, Duncan R (2000) Anionic PAMAM dendrimers rapidly cross adult rat intestine in vitro: a potential oral delivery system? Pharm Res 17:991–998. https://doi.org/10.1023/A:1007587523543

Wolf BA, Snyder F (1995) Cosmetic compositions having keratolytic and anti-acne activity

Xie Y, Yao Y (2018) Octenylsuccinate hydroxypropyl phytoglycogen, a dendrimer-like biopolymer, solubilizes poorly water-soluble active pharmaceutical ingredients. Carbohydr Polym 180: 29–37

Yadav D, Semwal BC, Dewangan HK (2023) Grafting, characterization and enhancement of therapeutic activity of berberine loaded PEGylated PAMAM dendrimer for cancerous cell. J Biomater Sci Polym Ed 34(8):1053–1066

Yang Y, Sunoqrot S, Stowell C, Ji J, Lee C-W, Kim JW, Khan SA, Hong S (2012) Effect of size, surface charge, and hydrophobicity of poly(amidoamine) dendrimers on their skin penetration. Biomacromolecules 13:2154–2162. https://doi.org/10.1021/bm300545b

Yang X, Kuang Z, Yang X, Hu X, Luo P, Lai Q, Zhang B, Zhang X, Wei Y (2023) Facile synthesis of curcumin-containing poly(amidoamine) dendrimers as pH-responsive delivery system for osteoporosis treatment. Colloids Surf B Biointerfaces 222:113029. https://doi.org/10.1016/j.colsurfb.2022.113029

Yao W, Sun K, Hongjie M, Liang N, Liu Y, Yao C, Liang R, Wang A (2010) Preparation and characterization of puerarin-dendrimer complexes as an ocular drug delivery system. Drug Dev Ind Pharm 36(9):1027–1035

Yesil-Celiktas O, Cansu Pala E, Cetin-Uyanikgil O, Sevimli-Gur C (2017) Synthesis of silica-PAMAM dendrimer nanoparticles as promising carriers in neuro blastoma cells. Anal Biochem 519:1–7

Zeynalzadeh S, Dehghani E, Hassani A, Baradar Khoshfetrat A, Salami-Kalajahi M (2023) Effect of curcumin-loaded poly(amidoamine) dendrimer on cancer cell lines: a comparison between physical loading and chemical conjugation of drug. Polym Bull. https://doi.org/10.1007/s00289-023-04783-9

Zhang J, Li M, Wang M, Xu H, Wang Z, Li Y, Ding B, Gao J (2021) Effects of the surface charge of polyamidoamine dendrimers on cellular exocytosis and the exocytosis mechanism in multidrug-resistant breast cancer cells. J Nanobiotechnol 19:135. https://doi.org/10.1186/s12951-021-00881-w

Zhao C, Wang Y, Su Y, Zhang H, Ding L, Yan X, Zhao D, Shao N, Ye X, Cheng Y (2011) Inclusion complexes of isoflavones with two commercially available dendrimers: solubility, stability, structures, release behaviors, cytotoxicity, and anti-oxidant activities. Int J Pharm 421: 301–309. https://doi.org/10.1016/j.ijpharm.2011.09.044

Zhu S, Hong M, Tang G, Qian L, Lin J, Jiang Y, Pei Y (2010) Partly PEGylated polyamidoamine dendrimer for tumor-selective targeting of doxorubicin: the effects of PEGylation degree and drug conjugation style. Biomaterials 31:1360–1371. https://doi.org/10.1016/j.biomaterials.2009.10.044

Zimmerman S, Lawless L (2001) Supramolecular chemistry of dendrimers. In: Dendrimers. Springer, Berlin, pp 95–120. https://doi.org/10.1007/3-540-45003-3_3

Zolotarskaya OY, Xu L, Valerie K, Yang H (2015) Click synthesis of a polyamidoamine dendrimer-based camptothecin prodrug. RSC Adv 5:58600–58608. https://doi.org/10.1039/C5RA07987J

Nanocarriers for the Delivery of Cosmeceuticals

10

Shalini Shukla, Akshada Mhaske, and Rahul Shukla

10.1 Introduction

The US FDA has defined cosmetics as "substances intended for application to the human body aimed at cleansing, beautifying, promoting attractiveness or altering the appearance without affecting the body physiology or functions." These are the most well-known customer goods in the international business, making this an alluring niche for different industries. Pollution, unpredictably changing climate conditions, and changing lifestyles have all contributed to an increase in demand and supply for cosmetics. The global cosmetics industry, which was worth $532.43 billion USD in 2017, is projected to rise up to 7.14% compound annual growth rate (CAGR) to $805.61 billion USD by the year 2023. A recent study predicts that the Middle East and African regions would experience the highest CAGR of 21% in the market for beauty products (Gautam et al. 2011). Cosmetics that contain pharmacological active agents are known as "cosmeceuticals." Cosmeceuticals are getting positive responses from the cosmetic industry. Cosmeceuticals are frequently used in a variety of situations to treat or avoid wrinkles, black spots, dry skin, uneven skin tone, photo aging, hair damage, and hyperpigmentation. In the market for personal care items, cosmetics has recently been the sector that is growing the fastest and is providing the highest scope for medical professionals to cure skin-related illnesses in patients (Erdo et al. 2016).

Currently, the cosmetics sector is investigating the use of nanotechnology for a variety of potential purposes. Christian Dior first coined the term "nanocosmetics" in 1986 to describe cosmetic products using nanoparticles. These products gained attention after L'Oréal's 2005 discovery of the economic advantages of

S. Shukla · A. Mhaske · R. Shukla (✉)
Department of Pharmaceutics, National Institute of Pharmaceutical Education and Research-Raebareli, Lucknow, UP, India
e-mail: rahul.shukla@niperraebareli.edu.in

© The Author(s), under exclusive license to Springer Nature Singapore Pte Ltd. 2024

D. Pooja, H. Kulhari (eds.), *Nanotechnology Based Delivery of Phytoconstituents and Cosmeceuticals*, https://doi.org/10.1007/978-981-99-5314-1_10

nanocosmetics (Oberdörster et al. 2005). Many skin care products now contain nanomaterials to benefit from the special characteristics of matter at the nanoscale. Nanotechnology in cosmeceuticals has quickly proliferated and become commercialized, achieving significant scientific and financial goals (Nagaich 2016). Among the benefits of cosmeceuticals based on nanotechnology are the prolonged duration of action, enhanced aesthetic appeal, and increased bioavailability. These products differ from conventional cosmeceuticals in numerous aspects, such as their microscopic size along with high surface-to-volume ratio, which render them useful additives in cosmetics. Additionally, nanomaterials use in cosmetic compositions enhances the cosmeceuticals' appearance, coverage, and skin adherence rather than altering their characteristics. Manufacturers of cosmetics use nanosized components to enhance a range of features, including UV protection, skin penetration, color, finish quality, aroma release, anti-aging impact, and many more. Although they offer many advantages, they also have drawbacks in terms of stability, toxicity, scalability, expense, etc. Additionally, there are still questions about the toxicity and safety characteristics of nanomaterials (Santos et al. 2019). This chapter highlights the role of nanotechnology in cosmetic preparations along with its merits and demerits. The regulatory guidelines concerned with the use of nano-cosmeceuticals have also been explored and the future potential of nanomaterials in cosmetic preparations has been discussed.

10.2 Current Status of Cosmeceuticals

10.2.1 Natural Cosmetics

Since earlier times, plant-obtained bioactive substances have been utilized by people for various cosmetic therapies for the skin, hair, lips, and nails that have positive effects against hair fall, inflammation, psoriasis, lip care, and ultraviolet (UV) damage. UV light tans the skin and kills endothelium cells, thereby inhibiting immune system function. By using natural phyto-active substances, these harmful consequences can be avoided. Some phytocompounds used in cosmeceuticals are curcumin, vitamin E, vitamin C, resveratrol, aloe vera, quercetin, natural green tea, lycopene, and papain. These compounds not only have aesthetic effects but also possess anti-inflammatory, antibacterial, anti-carcinogenic, and antifungal activity. Turmeric, a plant that is indigenous to Asia, contains the active ingredient curcumin with numerous applications for beauty, health, and cooking. Due to its numerous benefits, including, moisture locking, anti-aging, and antioxidant activity, women frequently use turmeric as tuber powder or crude extract for skin care in their everyday routine. Turmeric powder paste has been used to treat inflammation and skin damage, and as an antibacterial (Meghea 2008). Vitamin E is a natural fat-soluble and heat-stable compound. It is widely utilized in cosmetics owing to its skin-shielding abilities such as anti-aging, improved skin hydration, and skin disorder prevention capabilities. There are scientific reports on the benefits of vitamin E protection against sunlight. Vitamin E application before UV exposure significantly induces early skin reactions, including edema and erythema, sunburn, lipid peroxidation, immunosuppression, DNA adduct formation, ultraviolet A (UVA)-initiated attachment of photosensitizers, and chemiluminescence, according to a number of topical studies. Topical vitamin E formulations significantly reduced the occurrence of skin tumors and chronic skin reactions brought on by persistent ultraviolet B (UVB) or ultraviolet A irradiation, including skin wrinkles (Kotyla et al. 2008). Vitamin C is present in large quantities in plants, and it possesses dual functions in both elegance and wellness, including collagen formation. It gets degraded in severe conditions and UV light. A consistent supply of vitamin C compounds raises the aesthetic quality of skin. It is essential for the production of collagen, the fibrous substance that accounts for roughly 95% of the dermal matrix and whose loss with photoaging is primarily responsible for skin laxity and drooping. Collagen is the major constituent of the dermal matrix (Teeranachaideekul et al. 2007).

Resveratrol is a polyphenolic compound offering several benefits such as quenching free radicals and also protects the skin from damaging environmental factors. Polyphenolic substances prevent UV-induced radiation and oxidative processes. They are frequently utilized in functional foods for good health and skin care and provide a higher level of skin UV-B radiation protection than other antioxidants (Ido et al. 2015). It also has anti-angiogenic, antimicrobial, antioxidant, anti-inflammatory, and antiproliferative properties. SkinCeuticals resveratrol BE® serum helps in reducing signs of aging and provides firmness and radiance to the skin.

The aloe is a succulent plant that consists of about 75 distinct nutrients, including vitamins, minerals, enzymes, carbohydrates, anthraquinones or phenolic compounds, saponins, lignin, sterols, salicylic acid, and amino acids. The herb is currently most often used in the fields of wound healing, skin care, and cosmetics. Numerous biological properties have been attributed to this plant, particularly its polysaccharides, including antiviral, laxative, antibacterial, radiation protection, immunostimulant, and anti-inflammation. It has been demonstrated to trigger macrophages and promote wound healing (Ali et al. 2014). Various aloe vera cosmetic products are currently present in the market such as Patanjali's aloe vera gel® and cream®, Khadi's® aloe vera moisturizer®, Himalayas's aloe vera body lotion®, etc. Quercetin is another phyto-active flavonoid component with effectiveness in a variety of areas, including skin care and beauty through antioxidant properties; as a result, it improves skin appearance and health care. Structurally, quercetin has abundant –OH groups due to which it demonstrates greater skin protection ability and enhances elegance when compared to other flavonoids. This prompts numerous researchers to examine the various efficacies of quercetin in cosmetics. Products with quercetin include Paula's antiwrinkle cream with SPF 30® and Sesderma factor G renew serum® (Ali et al. 2015). Green tea extract contains biologically active substances like catechin, epigallocatechin, epigallecto-3-catechin, and epicatechin, which are potent antioxidants being utilized in many skin care and cosmetic procedures. The potential of these compounds on an individual basis varies depending on how they are used in the creation of cosmeceutical products. Various skin care benefits of green tea include anti-aging and protection against UV-induced photoaging. Wow's green tea face serum® and Lotus green tea night gel and day cream® are some of the green tea cosmetics available in the market (Ahmad and Mukhtar 2001). Natural phyto-bioactive lycopene has numerous uses in skin therapy because it has anti-aging and antioxidant properties. Lycopene has limited use in skin care due to its hydrophobic nature and poor skin distribution. In addition to being lipophilic, lycopene easily degrades when exposed to sunlight and oxygen, thus limiting its application in cosmeceuticals. Some examples of lycopene cosmetics include Goree's beauty cream®, Lycopene's crema rinnovante®, and Cosmesis' lycopene cream® (Ganesan and Choi 2016). The dried latex made from papaya fruit (*Carica papaya L.*) contains papain. It has a proteolytic enzyme, which breaks down proteins, as well as other potential ingredients. In the food, pharmaceutical, and textile industries, papain is frequently utilized. It is a protease that is frequently utilized in applications involving food preparation. The healing of wounds has also been aided by it. It is employed in cosmetics to exfoliate keratosis skin. Some examples include Puresense's papaya day cream® (Starley et al. 1999).

10.2.2 Synthetic Cosmetics

Synthetic cosmetics are products produced artificially through chemical reactions. There are a variety of skin care cosmeceuticals available in stores that are used by people to get healthy, beautiful, and smooth skin. Although these products are effective not all of these are safe. Cosmetics can improve skin texture, complexion, and appearance, but the harmful chemicals present in them can lead to adverse skin conditions. Various studies have reported the presence of carcinogenic compounds in some products. Ammonium lauryl sulfate (ALS) functions as an anionic surfactant and reduces liquid surface tension. This feature makes it useful in cosmetic products including shampoos, cleaning agents, and bathing agents. It has a relatively low acute toxicity and can irritate the eyes. Many cosmetics contain talc as a key ingredient. Hydrated magnesium silicate makes up the chemical composition of talc. Talc is a key component of face powders, concealers, and other products. It is also used in dusting powders, chalk, deodorants, and soaps. However, recent studies have pointed out that prolonged usage of talc can develop ovarian cancer (Pure Chemicals Co 2022).

Cosmetics typically contain anti-bacterial preservatives to protect against microbial proliferation and decomposition, which can lead to a lowering of the longevity of the product. Formaldehyde (FA) is one of the best options for preserving cosmetics due to its fungicidal and bactericidal properties. It is one of the most often used ingredients in nail polishes, nail polish cleaners, and other cosmetic and skin care essentials. The amount of formaldehyde in various products such as shampoos and conditioners, hair straightening solutions, skin moisturizers, cream cleansers, and toothpaste is very low. This is because of their tendency to cause irritation or contact allergies in various parts of the body as well as certain types of malignancies, asthma, and developmental and reproductive damage on long-term exposure (Lv et al. 2015). Perfumes frequently use alcohol as a component. Fatty alcohols, which are components of detergent and serve as cleaning agents, are less dense alcohols that remain after fats and oils are removed. In order to prevent moisture loss, glycols are utilized as compounds. Isopropanol is an alcohol constituent in many cosmeceuticals. It is generally used to reduce liquid thickness. Mineral waxes are most often used in cosmetics and are important ingredients in products including paraffin oil, paraffin wax, and petrolatum (Vaseline) (Pure Chemicals Co 2022). Numerous products, including cosmetics, employ titanium dioxide (TiO_2). The sole type of TiO_2 utilized as an ultraviolet (UV) filter in sunscreens, as well as lip balms, some day creams, and foundations, is in nanoparticle form (nano-TiO_2). Although its effectiveness as a UV filter in preventing skin cancer and sunburns is established, certain questions have been raised about its safety. It is believed to be able to pass through skin, mucous membranes, and gastrointestinal and respiratory barriers; disperse throughout the body; and ultimately pose a risk to users due to its small size (Dréno et al. 2019).

10.3 Current Market Status of Cosmeceuticals

The growing consumer emphasis on grooming and personal appearance is driving the cosmetics sector. Rapid urbanization, the introduction of new products, and rising per capita incomes are all factors contributing to the industry's expansion. Upsurges for natural and organic products, appealing marketing tactics, and cutting-edge packaging designs are all predicted to boost market expansion over the stimulated time span. Beginning in the year 2020, the COVID-19 pandemic had a profound effect on the global cosmetics sector. During the pandemic, the European cosmetic business also faced a decline in the overall demand for cosmeceuticals. The Asia-Pacific region was severely impacted by the COVID-19 pandemic. However, the crisis caused significant harm to many businesses, altered customer attitudes and beliefs, and is likely to fuel new trends like e-commerce and the usage of DIY kits for hair coloring and nail care, among other things. In 2020, the global market for cosmetics reached a value of USD 476 billion, owing to an increase in consumer attention to grooming and beauty. The market is anticipated to continue increasing in the forecast period of 2022–2027, growing at a CAGR of 6%, aided by the expanding use of the internet, and the simplicity and convenience of online purchasing. By 2026, the industry is anticipated to grow to 675 billion USD.

In India, the cosmetics and personal care market is the market with the fastest growth rate, which offers significant opportunities in relation to foreign businesses. India's personal care products and cosmetics have continued to expand quickly. There is now more shelf space available for international cosmetics in India's shops and retail establishments. In the past few years, the Indian market of cosmetics and beauty essentials has been steadily expanding. Body care, face care, hair care, hand care, and color cosmetics make up the majority of its five main categories. An estimated $8 billion is spent on beauty and personal care (BPC) in the Indian market. Along with the country's GDP development, per capita expenditure in India on personal care and cosmetics is rising. Various global brands, including Revlon (the first foreign cosmetics brand to enter India in the middle of the 1990s), Calvin Klein®, Avon®, Burberrys®, Max Mara®, Max Factor®, Body Shop®, Estee Lauder®, Maybelline New York®, MAC, Bobbi Brown, Christian Dior, L'Oréal, and many more, are flourishing businesses in India (Expert Market Research 2022).

10.4 Challenges Associated with Cosmeceuticals

The personal care industry is growing tremendously, with cosmetics being thought to be the area of the industry that will grow the fastest, when compared to medications for the therapy of skin conditions like photoaging, wrinkles, xeroderma, hair damage, hyperpigmentation, etc. Synthetic cosmetics improve the skin's appearance, but some carry harmful chemicals which are toxic and can accumulate in the body for a much longer time, causing irritation to the skin. Some of these chemicals are even carcinogenic which can have detrimental effects on the body. The use of herbal products in cosmetics has shown immense benefits. The polyphenol content present in herbal products has skin protection ability and can prevent wrinkles, aging, and UV radiance without being harsh on the skin (Raj et al. 2012). Despite the immense advantages of cosmeceuticals and the significant advancements created by the cosmetic industries for the progress as well as integration of novel, efficient active ingredients in their products, the greater size and low penetration ability of the skin are considered to be significant factors that limit the permeation and absorption of these active constituents inside the deeper skin layers. Moreover, the extraction process of a crude compound from natural sources is also a difficult and time-consuming task as it requires a series of processes that need to be completed to get the final crude drugs. The extraction requires trained personnel who are able to extract the substances efficiently. The extraction method is referred to as the "sample preparation technique." It is the most crucial part of extraction and is mostly done by non-trained personnel, which can cause inappropriate ongoing procedures in later studies (Oliveira et al. 2022). The bioactive compounds are present along with other compounds and require efficient recovery. Other challenges involved are the requirement of costly and high-quality solvents, low extraction selectivity, evaporation of large amounts of solvent, and chances of thermal decomposition of thermolabile compounds. Improper drying leads to unintentional adulteration (Pauwels and Rogiers 2010). Moreover, the formulation of natural cosmetics is difficult as the herbal ingredients are more prone to contamination compared to synthetic molecules. Herbals also require preservatives to prevent their contamination and denaturation. Further, there is less information associated with socioeconomic advantages that can be obtained from the commercial utilization of herbals in cosmetics (Parveen et al. 2015). Other than the knowledge of these herbal products for local healthcare needs, there is minimal information on their market strength. A key challenge is to evaluate the toxicological, epidemiological, and other data and the verification of herbal compounds used. Some herbs such as turmeric and quercetin are unstable in environmental conditions (Chanchal and Swarnlata

2008). The lipophilic nature of some herbs such as lycopene causes its degradation on exposure to light and air. Rural people use these herbs directly which can cause irritation, itching, and inflammation on the skin. Some of these products may be harmful and toxic in direct use. In order to address these shortcomings of conventional cosmeceutical products, the use of nanotechnology is expanding in the cosmeceuticals area and occupies a remarkable position (Saggar et al. 2022). Nano-delivery devices penetrate the skin and increase the solubility of the phyto-bioactive substances. Effective nano-delivery techniques were used to improve the penetration of active phyto-based components, which enhanced the effectiveness of cosmeceuticals. Nanotechnology also improves the therapeutic potential of dual-role cosmeceuticals because phyto-based bioactive compounds treat the majority of skin ailments (Das et al. 2022).

10.5 Need for the Assistance of Nanotechnology for the Development of Cosmeceuticals

Nanotechnology is the most promising advanced technology of the twenty-first century. It is considered an innovative strategy in the cosmetics sector. Moreover, it has been described as an inventive branch of research that encompasses the development and evaluation of particles in the size range between 10 nm and 1000 nm (Google Books 2022). Nanocosmeceuticals are defined as a cosmetic composition that uses nanosizing as a method of delivery strategy to achieve improvement in the performance of bioactive components (Kaul et al. 2018). This technique makes it possible to create cosmetic chemicals in tiny nanoparticles with active substances that can easily permeate into the skin, heal, and enhance product performance. The quality and performance of several cosmetic products have been improved by incorporating bioactive ingredients with nanocarriers, such as liposomes, niosomes, nano-capsules micelles, and dendrimers (Aziz et al. 2019; Lu et al. 2015). Using this method, cosmetics can be created with a long-lasting fragrance, improved sunlight protection, and featuring enhanced anti-aging effects.

Nanocosmeceuticals have a variety of advantages. They specifically release drugs from carriers, and the release can be affected by a wide range of parameters, such as drug content, polymer and additives ratio, and production process, as well as physical or chemical interactions between the constituents. In order to treat hair loss and prevent gray hair, they are utilized in hair care products including Identik Masque Floral® Repair, Origem® hair recycling shampoo, and Nirvel® hair fall rescue shampoo. They also make the perfumes last longer such as Allure Eau® Parfum spray and Chanel's Allure® Parfum. By enhancing UV protection, they improve sunscreen effectiveness and the efficacy of products. The small particle sizes of these nanocosmeceuticals enhance the surface area, which facilitates the delivery of active constituents into the skin. Skin moisture is raised and penetration is improved by occlusion. Cosmeceuticals are more efficacious than traditional cosmetics and possess good sensory properties as well as a high entrapment efficiency. Most of the nanocarriers may carry both hydrophilic and lipophilic drugs.

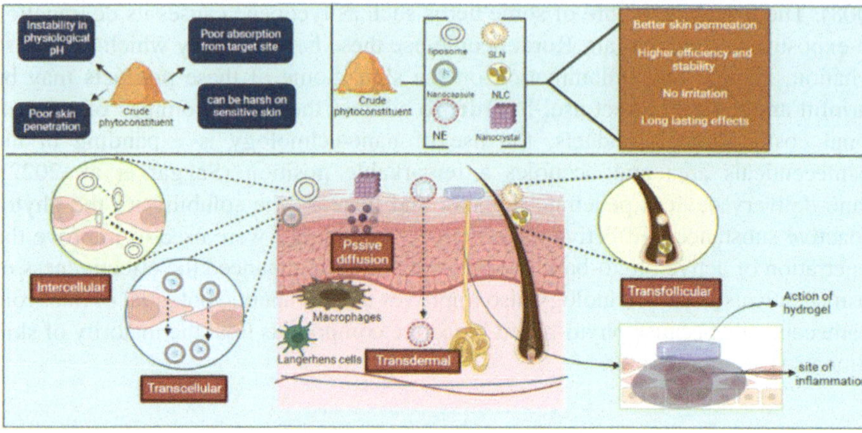

Fig. 10.1 Demerits of crude phytoconstituents and the role of novel technologies in targeted and controlled delivery to the site of action on the skin

Nanomaterials are frequently utilized to create anti-aging creams, shampoos, conditioners, hair serums, skin-whitening creams, and moisturizing creams. In proportion to the rise in customer demand for the efficacy of skin care goods, it is tougher to differentiate between cosmetic and pharmaceutical active components for topical use. Because the effectiveness of penetration depends on numerous factors such as molecular size, the degree of ionization, and lipophilicity, a specific amount of active ingredient permeation into the skin coupled with nanotechnology is required to achieve the high effectiveness of cosmetics. A lipid-based nanoparticle formulation would likely be the appropriate method to ensure an efficient topical delivery system of cosmetic active compounds given the barrier properties of the skin (Souto and Müller 2008; Müller et al. 2002).

Nanocosmeceuticals possess few disadvantages as nanoparticles can damage DNA, proteins, and membranes as well as produce a significant number of oxygen species, oxidative stress, inflammation, and other negative effects. There are a few nanomaterials, such as carbon nanotubes, copper nanoparticles, silver nanoparticles, and carbon-based fullerenes, that could harm human organs and cells. The prime sunscreen ingredient titanium dioxide has been reported to cause harmful effects on the DNA, RNA, and lipids present in cells. For the licensing and monitoring of nanocosmeceuticals, the regulatory agencies did not apply rigorous examination. Environmental harm could also be caused by nanocosmeceuticals. Nanocosmeceuticals are approved without the need for clinical trials, which raises questions about potential side effects (Kaul et al. 2018). Figure 10.1 depicts the limitations associated with phytoconstituents and the role of novel technologies in targeted and controlled delivery to the intended site.

10.6 Phytoconstituents-Based Nanocosmeceuticals

Phytoconstituents are bioactive compounds synthesized in plants via primary and secondary metabolic pathways and are categorized as active drug components and inert non-drug compounds. These have therapeutic properties, including moisturizing, reviving, anti-aging, UV protection, and the prevention of skin-related disorders. Plant-based bioactive compounds are highly desirable active ingredients in the cosmeceuticals business. Instead of synthetic chemical substances that irritate the skin, natural bioactive plant compounds are preferred since they are environmentally beneficial and suitable for a variety of skin types. There are several different types of nanocarrier systems that are created to deliver plant-based bioactive substances in cosmetics products. Various phytocompounds-based nanocarriers along with their applications are shown in Table 10.1.

10.6.1 Phytosomes

Phytosomes are products, specifically for plant phenolics, that aid in increasing the bioavailability of phyto-active compounds in the skin as well as for different therapeutic applications. The physical stability of biologically active compounds is improved by this method (Google Books 2022). For their beneficial effects on health and beauty, phytosomes are frequently employed in the cosmeceutical industry. They can be combined with bioactive ingredients or crude extracts derived from plants. *Glycyrrhiza glabra* and *Citrus auranticum* extracts have been recently used for phytosomes production as well as for studies on skin aging (Pandita and Sharma 2013). The bioavailability of these substances for use in pharmaceutical products and cosmetics is improved by phytosome nanosizing. In addition, gold nanoparticles with plant extracts rich in quercetin were integrated into phytosomes to increase the effectiveness of quercetin (Gunasekaran et al. 2014).

10.6.2 Liposomes

Liposomes are useful for transporting phyto-bioactive ingredients in cosmetics. Liposome-containing products improve the barrier function of skin and aesthetics. The majority of nanoliposomes are spherical, have one or more lamellae, and are only a few nanometers in size. Phyto-based chemicals can be delivered to the dermal layer with high efficiency by liposomes (SCMSJournal.com 2022). Nanoliposomes are utilized in cosmeceutical products including both synthetic and phyto-bioactive ingredients, such as anti-aging creams, sunblock creams, hair creams, and skin moisturizers. The main benefits of nanoliposomes include their special size, the capacity to encapsulate bioactive substances that are both hydrophilic and hydrophobic, and the potential of UV-absorbing lipids to shield the skin and bioactive compounds from deterioration (Kaur and Agrawal 2008). In studies on nanoliposomes containing phyto-bioactive chemicals, curcumin was found to have

Table 10.1 Various phytocompounds-based nanocarriers

S. No.	Phytocompounds	Nanodelivery methods	Applications	References
1.	Rosemary extract	Solid lipid nanoparticles (SLNs) and nanostructured lipid carriers (NLCs)	In skin care—antioxidant	Lacatusu et al. (2010)
2.	Safflower extract	NLCs	Hair care	Kumar et al. (2011)
3.	Aloe vera extract	Liposome	Skin care	Takahashi et al. (2009)
4.	Resveratrol	Niosomes	Skin care	Pando et al. (2015)
5.	Resveratrol and curcumin	Lipid care nanocapsules	Skin care	Friedrich et al. (2015)
6.	Lycopene	Transfersomes, ethosomes	Skin care	Ascenso et al. (2015)
7.	d-Limonene	Nanoemulsions	Skin care	Lu et al. (2014)
8.	Ganoderma triterpenoids	NLC gel	Enhancing skin appearance	Shen et al. (2015)
9.1	Hinokitiol	Poly(epsilon-caprolacton) nanocapsules	Hair care	Hwang and Kim (2008)
10.	Rice bran oil	Nanoemulsion	Anti-aging and moisturizers	Bernardi et al. (2011)
11.	Rice bran and raspberry seed oil	Lipid nanocarriers	Sunscreen	Niculae et al. (2014)
12.	Lavender extracts	Polymeric poly(lactic-co-glycolic) acid (PLGA) nanoparticle	Anti-aging	Pereira et al. (2015)

improved stability and sustained transdermal activity. Recently, 80 nm-sized curcumin nanoliposomes were developed and results reveled skin protective effects (Chen et al. 2012). Curcumin's bioavailability in the skin and body is increased when combined with hyaluronic acid, which opens up a variety of new nanotechnologies for skin-based nanomedicine and cosmetics. Liposomal vitamin C supplements are also available in the market. Despite the fact that numerous phytocompound-encapsulated liposome techniques are employed in cosmetics and therapeutics, the development of nanosized liposomes with active phyto-derived compounds is still in its early stages (Manca et al. 2015).

10.6.3 Niosomes

Niosomes are widely employed in the cosmeceutical industry because they have many advantages over liposomes, including greater stability, enhanced skin permeation, minimal toxicity, and greater bioactive substance protection (Manca et al. 2015). The efficacy of niosomes depends on their size, which ranges from nanometers to micrometers. Niosomes are unilamellar, smaller structures with a size of less than 100 nm. Niosomes are widely utilized to provide antioxidants, including ascorbic acid, ellagic acid, and resveratrol, via skin and are used to treat skin-related disorders. Regardless of the fact that synthetic compounds are often utilized in cosmeceutical creams and lotions, a new area in the development of nanoniosomes is the use of bioactive substances originating from plants (Baccarin et al. 2015). By employing phyto-derived antioxidants including resveratrol, alpha-tocopherol, and curcumin to make a niosome that was a little bigger and ranged in size from 471 to 565 nm, Tavano et al. reported improvements in skin permeability and antioxidant activity (Ascenso et al. 2015). This study established the enhanced skin safety of using phyto-derived compounds in cosmeceutical therapies without reducing their bioactivity. A change in production methods and a reduction in size to the nanoscale, therefore, may increase their activity with a variety of applications in the cosmeceutical industries. Recently developed phytocompounds with better skin protection properties include curcumin, which was created using nanoniosome methods (Singh et al. 2015).

10.6.4 Next-Generation Liposomes

Ethosomes, hyalurosomes, and glycerosomes are modified liposomes that were given their names depending on the phospholipids they contained, such as ethanol, sodium hyaluronate, and glycerol. They are specifically employed in cosmeceutical sectors for therapy and aesthetic treatments and modified for improved distribution of the active chemicals to the skin (Shukla et al. 2010). Additionally, nanosized vesicles with phytocompound inclusion exhibited improved, protective and cosmetic effects. Recently, quercetin glycerosomes with a unilamellar structure and a size between 80 and 110 nm were produced and showed improved skin protection action. To produce antioxidant skin creams in the future, this can be a useful method (Hua 2015). In order to make ethosome nanovesicles in the size range of 128 nm, polyphenolic extracts of *Fraxinus angustifolia* leaf and bark were employed. These nanovesicles showed better skin protection and wound healing. Similar to ethosomes, 100 nm-sized hyalurosomes containing licorice extract were formulated with an aim to improve skin appearance. The same research group created nanohyalurosomes of curcumin, improved skin texture with minimum particle diameter of 112 nm (Moulaoui et al. 2015).

10.6.5 Fullerenes

Another cutting-edge technology is fullerene, which is made of C60, or carbon atoms with an even number and dimension of 1 nm. Being an antioxidant, it prevents premature aging of the skin. Vitamins are effectively delivered by fullerene to improve the skin (Ito et al. 2010). A fullerene nanocapsule was recently developed that contained vitamin E and ascorbic acid. Curcumin and other bioactive chemicals were employed in fullerene-based photodynamic therapy to treat diseases of the skin. However, more research needs to be done on the use of fullerenes and diverse phyto-derived bioactive chemicals in cosmeceutical-based skin therapy (Yin and Hamblin 2015; Zhang et al. 2015).

10.6.6 Carbon Nanotubes

The cosmeceutical industry, specifically the skin care industry, is currently using carbon nanotubes as an innovative technology. Carbon nanotubes act as an antioxidant, at a diameter of 100 nm. In addition to their use in the cosmeceutical industry, they work well as phytocompound delivery systems in biomedical applications. A single-wall carbon nanotube was conjugated to curcumin for efficient administration. The significance of hyaluronic acids in the carbon nanotube-based drug delivery system was recently reviewed and additional clinical studies are required to determine their biosafety (Tripodo et al. 2015).

10.6.7 Solid Lipid Nanoparticles

One of the nanotechnologies that is widely used in both pharmaceuticals and cosmeceuticals is solid lipid nanoparticles (SLNs). Many cosmeceutical firms make cosmetics with the use of this technology to improve skin appearance. This nanoparticle system was created by combining solid lipids like stearic acid, cetyl, and palmitic acid with microemulsion systems, as well as phytocompounds like flavonoids, phenolic acids, and other bioactive constituents. In comparison to other conventional and nano delivery techniques, SLNs are more stable, produce more, and may be stored for a long time (Meghea 2008). They also increase the effectiveness of cosmeceuticals by improving qualities such as increased hydration and therapeutic advantages, reduced systemic circulation absorption, and delayed and regulated release in the skin and hair (Zamarioli et al. 2015). Curcumin SLNs were also developed of size 210 nm, and this formulation showed increased efficacy in the hydrogels that were prepared for the treatment of inflammation in porcine skin. The addition of quercetin, however, caused certain solid lipid nanoparticles to exhibit larger particle sizes, with an average of 574 nm. Silica can be used to minimize this size, bringing the average size down to 483 nm, which will increase quercetin's skin penetration while also increasing its photostability and skin-protecting effects. In a different study, solid lipid nanoparticles containing caffeine were formulated. The particles were 182 nm in diameter, and they provided better bioactive component penetration into the skin while providing more skin protection. Additionally, skin penetration was improved by resveratrol-loaded 180 nm solid lipid nanoparticles. It also demonstrated greater skin penetration with improved protection. SLNs loaded with lutein and available in nano-sized range. Protecting against UV rays and maintaining antioxidant activity enhance the appearance of healthy skin. SLNs formed with phyto-bioactive components are thought to promote the effectiveness of cosmeceuticals and their health advantages, according to past research investigations. They have the potential to develop the nanocosmeceuticals industry with future production of phyto SLNs (Mitri et al. 2011).

10.6.8 Nanostructured Lipid Carriers

Nanostructured lipid carriers (NLCs) are an important delivery method utilized in cosmeceutical preparations to more effectively distribute both natural and artificial bioactive substances. This method differs from solid lipid nanocarriers by using liquid lipids such as oleic acid and has additional advantages including greater carrier stability and bulk production than other nanotechnologies. Additionally, nanostructured lipid carriers come in a variety of shapes, including imperfect types with different lipid structures, formless types with a lipid that prevents crystallization, and numerous types with lipid oil that contains bioactive compounds, which limits the expulsion of bioactive compounds, a major disadvantage in solid lipid nanoparticles. The creation of nanostructured lipid carriers makes significant use of a variety of phyto-bioactive compounds, that are particularly effective in topical treatments, consequently enhancing the beauty and protecting the skin. Recently, Okonogi and Riangjanapatee formulated an NLC for lycopene delivery through topical route in the size range from 150 to 160 nm (Bose and Michniak-Kohn 2013). Similar to this, in pig ear skin research, a lutein-NLC with particle diameters between 163 and 350 nm was developed and tested for UV prevention. These findings confirm that different nanocarrier types have different impacts on lutein nanoparticle production and skin protection. SLNs show greater lutein photostability, followed by NLCs and nanoemulsions. An example of a marketed NLC formulation is the cutanova nanorepair cream (Jain et al. 2015). This quercetin product improves skin appearance by reducing oxidative stress. Similar to this, an NLC with enhanced photostability and water stability was created utilizing phenylethyl resorcinol with 218 nm-sized particles. In the near future, this substance might be utilized as a bleaching agent. Resveratrol was also produced using lipid nanocarrier systems, and it outperformed solid resveratrol-loaded lipid nanoparticles in terms of skin defense and antioxidant activities. Additionally, E-resveratrol offers significant skin protection from UV radiation by way of a lipid-based nanoparticle system. In some studies, bioactive substances like squalene help drugs reach the hair follicles. For instance, a patient with alopecia areata was treated with a squalene-enriched NLC, which has particles with a size between 208 and 265 nm. These latest investigations show that phyto-bioactive chemicals can be incorporated into nanostructured lipid carrier systems, enhancing cosmetics and personal care (Sharma et al. 2021).

10.6.9 Nanoemulsions

Currently, single or multiple nanoemulsions with a size range between 10 nm and 300 nm are employed in a variety of cosmetic products for their unique applications with increased activity. One of the main benefits of employing nanoemulsions in cosmetics is that they have a special size that can improve the bioactivity of components that are both hydrophilic and lipophilic. However, given that emulsions become unstable when stored, this approach has disadvantages when producing cosmeceuticals (Ahmad et al. 2017). Additionally, the emulsion core components in the nanosized emulsions determine whether a single or several nanoemulsions can be generated. Multiple emulsions are preferable to single emulsions due to their greater applicability and can help keep the cost of cosmeceuticals down. Nanoemulsions contain a number of bioactive substances originating from plants, including flavonoids and polyphenols. These substances increase cosmeceutical value along with reducing toxicity. The ethanolic extract of *Achyrocline satureioides* was used recently to make quercetin-rich nanoemulsions of 200–300 nm size, which demonstrated improved antioxidant activity in experiments on pig skin and greater skin retention of bioactive chemicals (McClements 2012). In a different study, multiple nanoemulsions with the particle size of 173–208 nm were created by co-loading the soy bioactive component genistein with tocomin, and the results showed better skin protection from UV ray damage. The creation of sunscreen lotion will benefit from this procedure. Similar to this, retinyl palmitate was employed to create nanoemulsions with a size range of 275 nm or smaller that had improved skin penetration and higher inner skin protection (Handa et al. 2021). Additionally, self-nanoemulsifying curcumin nanoemulsions with particle sizes of 85 nm were created, and they showed improved transdermal penetration without degrading (Jain et al. 2011). This can be of great significance to the skin care industry to create innovative skin care emulsions. When compared to alternative lipid carriers, lutein-encapsulated nanoemulsions were created in the size range of 150–350 nm, which efficiently stimulated the release of active ingredients. Nanoemulsions have a great role in therapies for enhancing skin appearance. Though nanoemulsions have better skin penetration and protection effects, their stability is yet an unsolvable problem with the cosmetics manufacturing process using these delivery methods (Bidone et al. 2015).

10.6.10 Nanospheres

The spherical particles known as nanospheres have a core-shell configuration. The size has a diameter that varies between 10 and 200 nm. Nanospheres are utilized in skin protection products in the cosmetics industry to deliver the beneficial effects of the active substances more accurately and effectively to the skin's affected area. The use of nanospheres in the emerging cosmetics industry is growing, particularly in skin care products including anti-aging, moisturizing, and anti-acne creams (Subedi et al. 2016).

10.6.11 Dendrimers

Dendrimers are multivalent nanoparticles with highly branching, monomolecular, globular, and micellar nanostructures and whose theoretical synthesis permits monodisperse molecules. A dendrimer usually consists of a core into which one or more molecules can be incorporated. The number of branches determines the generation of dendrimers. If there are only one or two series, the dendrimer is of the first generation; otherwise, it belongs to the second generation. The size of dendrimers ranges from 2 nm to 200 nm. Its additional qualities, such as monodispersity, polyvalence, and stability, make it suitable carrier system for topical delivery. Several cosmetic products, such as shampoos, sunscreen, hair styling products, and acne treatments, include this ingredient carrier for precise and selective drug delivery. Drugs are integrated into the dendrimers' inner core for controlled release, and they are also adhered to the surface (Klajnert and Bryszewska 2001).

The usage of dendrimers as cosmeceuticals is growing in the fields of hair, skin, and nail care. Dendrimers are a novel class of macromolecular architecture that are useful. Various cosmeceutical companies like The Dow Company, Unilever, L'Oréal, and Wella have been granted several patents for the application of dendrimers in cosmetics (Tiwari and Talreja 2020).

10.6.12 Gold Nanoparticles

Gold nanoparticles or nanogold have a size range from 5 nm to 400 nm. The assembly of gold nanoparticles with interparticle interactions plays a significant role in the determination of their properties. Numerous morphologies, including nanosphere, nanorod, nanostar, nanoshell, nanocube, nanocluster, branching, and nanotriangles, can be seen in gold nanoparticles. The size, shape, dielectric characteristics, and environmental factors all have a significant impact on the resonance frequency of gold nanoparticles. They are also accessible in conjugated and unconjugated forms. After staining on membranes, nanogold is extremely durable in liquid or dried form and does not fade. Because of their diminutive size, large surface area, rounded form, crystallinity, and higher drug-loading capacity, they can quickly enter the target cell (Khan et al. 2014).

Gold nanoparticles have been investigated as a desirable material in the cosmetics industry because of their potent antifungal and antibacterial properties. Gold nanoparticles are used by cosmetic businesses like L'Oréal and L'Core Paris to create creams and lotions that are more effective. The primary benefits of nanogold in beauty care include promoting faster blood flow, reducing inflammation and infection, improving the skin's firmness and elasticity, slowing down the aging process, and reviving skin metabolism (Yeh et al. 2012).

10.6.13 Nanocrystals

These are made from the assembly groups made up of thousands of molecules that range in size from 10 nm to 400 nm and encapsulate the therapeutic moiety with the least usage of other excipints. Nanocrystals are mostly employed for the administration of poorly soluble drugs. They encapsulate bioactive compounds and increase the dissolution rate. Studies show that rutin nanocrystals demonstrate more action in comparison to standard rutin glycoside. Another study conducted by Kopke et al. on SymUrban, a pollution-reduction agent, depicted that the solubility and permeation of SymUrban were improved in its nanocrystal form. The dermal product bioavailability of the active ingredient in SymUrban was increased significantly and the nanocrystal was found to be a convenient delivery method for the material (Sakamoto et al. 2007).

10.6.14 Cubosomes

Cubosomes are fluid crystalline nanoparticles consisting of surfactant and water in an appropriate proportion combined in a nanostructure. Cubosomes are generally made of monoglyceride glycerol monoolein. Cubsomes are utilized for different skin care preparations as well as also used as an antiperspirant. Numerous studies and cosmetic companies are attempting to employ cubosomes to absorb contaminants from cosmeceutical formulations. They also play a role as a stabilizer in o/w emulsions (Lohani et al. 2014). In one of the investigations conducted by Khan et al., erythromycin cubosomes were formulated that exhibited improved activity and effectiveness in the prevention as well as treatment of acne and worked in a sustained release manner. In addition, El-Komy et al. prepared a cubosomal gel for topical use that consists of alpha-lipoic acid and found it to be an effective and safe option for curing skin aging problems (El-Komy et al. 2017).

10.6.15 Micellar Nanoparticles

These are one of the most commonly used nanotechnological products in the cosmetic industry. They provide a strong and adaptable framework for incorporating several lipophilic active agents with a variety of physicochemical properties in cosmetic compositions. The essential characteristics that make them more effective than other nanocarriers are smaller-sized particles, greater encapsulation, effectiveness, and affordable manufacturing costs. They are generally used in skin washing solutions in place of standard cleansers to effectively remove oil and debris. The topical administration of systemic drugs is a desirable application for micellar

Table 10.2 Nanocosmeceuticals available in the market

Nanoformulation	Key features	Commercially available product	Type of cosmeceutical	References
Liposomes	Ability to incorporate both hydrophilic and lipophilic substances	Dermosome—microfluidics Capture Totale—Dior	Moisturizer Antiwrinkle cream	Fytianos et al. (2020)
Niosomes	Functional groups are present on the hydrophilic head, which make surface development and alteration easier	Lancome®—L'Oréal, Paris	Anti-aging cream	Kusuma Priya et al. (2020)
Ethosomes	Higher amount of ethanol present	Supravir Cream—Trima, Israel	Moisturizer	Sudhakar et al. (2012)
Solid lipid nanoparticles	Consists of solid lipids and higher drug loading	Chanel Allure	Perfume and cream	Fytianos et al. (2020)
Nanostructured lipid carriers	Matrix contains a blend of solid and liquid lipids	Dr. Rimpler—Cutanova	Face spa cream	Fytianos et al. (2020)
Dendrimers	Maintenance of the stability of the drug in cosmetic formulations Increase in shelf life	Topical resveratrol formulation	Sunscreen	Fytianos et al. (2020)
Nanoemulsions	No issues of creaming, flocculation, and coalescence	Cosmeceutical Vitamin A, D, E, K—Vita lipid Nano emulsion multi-peptide moisturizer—Hanacure	Body lotions, skin creams, balsams, salves, and gels Moisturizer	Blanco-Padilla et al. (2014)
Nanocrystals	100% drug loading and increase in drug solubility	Nano-In Hand and Nail Moisturizing Serum and Foot Moisturizing Serum—NanoInfinity Nanotech Nano Whitening Toothpaste—Whitewash	Moisturizer Toothpaste	Gigliobianco et al. (2018)

nanoparticle-based emulsions. However, facial cleansers containing micellar nanotechnology are popular and effective way of removing makeup and skin cleansing (Dhapte-Pawar et al. 2020; Lee et al. 2010). Various nanocosmeceuticals in the market are shown in Table 10.2.

10.7 Classes of Nanocosmeceuticals

10.7.1 Skin Care

The area of the personal care business that is expected to develop the fastest is cosmetics. The nail-care, hair, lip, and skin care industries all incorporate a variety of nanocosmeceuticals. Cosmetics are widely used for skin care for the betterment of skin function and texture by promoting collagen formation and fighting off the damaging effects of free radicals. By preserving the keratin's healthy structural integrity, they improve the health of the skin. The most efficacious minerals for skin protection in sunscreen lotions are zinc oxide and titanium dioxide nanoparticles, which penetrate the skin's deeper layers and make the product less oily, odorous, and transparent. Additionally, SLNs, nanoemulsions, and liposomes are often used in moisturizers because of their capacity to create thin humectant film layers and hold onto moisture for extended time periods. Products marketed as anti-aging nanocosmeceuticals that incorporate nanocarriers show benefits including collagen regrowth, skin renewal, and protect the skin against environmental factor.

Shampoos, conditioners, hair growth accelerators, coloring agents, and styling treatments all fall under the category of hair nanocosmeceuticals. The intrinsic qualities and distinct size of nanoparticles enable them to target the hair follicle and boost the amount of active substance (Kaul et al. 2018).

10.7.2 Hair Care

Nanocosmeceuticals comprise of shampoos, hair growth regulators, coloring, and hair treatments. The intrinsic qualities and distinct size of nanoparticles enable them to reach the hair follicle and increase the quantity of the active ingredient. Nanoparticles dispersed in shampoos trap moisture within the cuticles by maximizing local interaction with scalp and hair follicles, thus creating a protective coating. Cosmeceutical conditioning chemicals are designed to improve hair detangling by enhancing hair softness, bounce, silkiness, and gloss. The main function of new carriers such as liposomes, nanospheres, and nanoemulsions is to heal damaged tissues, restore texture and shine, and make hair less brittle and greasy (Verma et al. 2022; Hu et al. 2012).

10.7.3 Lip Care

Lip gloss, lip balm, lipstick, and lip volumizers are among the nanocosmeceuticals' lip care products. In order to smoothen the lips by minimizing transepidermal water loss and to prevent pigments from migrating from the lips and fading, several nanoparticles may be incorporated into lip glosses and lipsticks. Liposome-based lip volumizer fills in lip contour wrinkles while also improving lip volume, hydrating, and outlining the lips (SESDERMA 2022).

10.7.4 Nail Care

Products for maintaining nails that are based on nanocosmeceuticals are far superior to those that are not. The nanotechnology-based nail paints have advantages such as increased toughness, chip resistance, quick drying, durability, and ease of application because of their elastic properties (Kushwaha et al. 2020). Modern techniques like amalgamating silver and metal oxide nanoparticles have been found to have antifungal properties and are therefore used in the treatment of toenails with fungal infections (Pereira et al. 2014).

Various nanocosmeceuticals available in the market are shown in Table 10.2.

10.8 Conclusion

Researchers are looking for novel nanocosmeceutical-based treatments due to the growing demand for the use of phyto-based nanosized biologically active compounds in the creation of nanocosmeceuticals for skin care-based treatments such as anti-aging, UV protection, moisturizing, whitening effects, and skin care health. Nanosized cosmeceuticals that contain bioactive components derived from plants are more stable and protect more of these ingredients in the skin, which will enhance the skin's appearance over time. The sizes of compounds at the nanoscale vary. The capacity of bioactive chemicals to penetrate the skin is constrained as the nano range increases, which restricts cosmetic and therapeutic effects. The kind of delivery technology used for the creation of nanocosmeceuticals depends on the solubility of phyto-based bioactive chemicals and how they are used in cosmeceuticals. Targeting site release from the delivery agents requires additional study of their mechanisms. The advancement of cosmeceutical research using nanoscale phyto-active compounds and skin wellness is consequently done.

Acknowledgments The authors acknowledge the Department of Pharmaceuticals, Ministry of Chemical and Fertilizers, Govt. of India for support. The NIPER-R communication number for the chapter is NIPER-R/392.

References

Ahmad N, Mukhtar H (2001) Cutaneous photochemoprotection by green tea: a brief review. Skin Pharmacol Physiol 14(2):69–76. Available at https://www.karger.com/Article/FullText/56336

Ahmad U, Akhtar J, Singh SP, Badruddeen B, Ahmad FJ, Siddiqui S et al (2017) Silymarin nanoemulsion against human hepatocellular carcinoma: development and optimization. Artif Cells Nanomed Biotechnol 46(2):231–241. https://doi.org/10.1080/21691401.2017.1324465

Ali SW, Purwar R, Joshi M, Rajendran S (2014) Antibacterial properties of Aloe vera gel-finished cotton fabric. Cellulose 21(3):2063–2072

Ali H, Dixit S, Ali H, Dixit S (2015) Quercetin attenuates the development of 7, 12-dimethyl benz (a) anthracene (DMBA) and croton oil-induced skin cancer in mice. J Biomed Res 29(2): 139–144. https://doi.org/10.7555/JBR.29.20130025

Ascenso A, Raposo S, Batista C, Cardoso P, Mendes T, Praça FG et al (2015) Development, characterization, and skin delivery studies of related ultradeformable vesicles: transfersomes, ethosomes, and transethosomes. Int J Nanomedicine 10(1):5837–5851. Available at https://www.dovepress.com/development-characterization-and-skin-delivery-studies-of-related-ultr-peer-reviewed-fulltext-article-IJN

Aziz ZAA, Mohd-Nasir H, Ahmad A, Siti SH, Peng WL, Chuo SC et al (2019) Role of nanotechnology for design and development of cosmeceutical: application in makeup and skin care. Front Chem 7:739

Baccarin T, Mitjans M, Ramos D, Lemos-Senna E, Vinardell MP (2015) Photoprotection by Punica granatum seed oil nanoemulsion entrapping polyphenol-rich ethyl acetate fraction against UVB-induced DNA damage in human keratinocyte (HaCaT) cell line. J Photochem Photobiol B 153:127–136

Bernardi DS, Pereira TA, Maciel NR, Bortoloto J, Viera GS, Oliveira GC et al (2011) Formation and stability of oil-in-water nanoemulsions containing rice bran oil: in vitro and in vivo assessments. J Nanobiotechnol 9(1):1–9. https://doi.org/10.1186/1477-3155-9-44

Bidone J, Argenta DF, Kratz J, Pettenuzzo LF, Horn AP, Koester LS et al (2015) Antiherpes activity and skin/mucosa distribution of flavonoids from achyrocline satureioides extract incorporated into topical nanoemulsions. Biomed Res Int 2015:238010. Available at https://www.researchgate.net/publication/277920645_Antiherpes_Activity_and_SkinMucosa_Distribution_of_Flavonoids_from_Achyrocline_satureioides_Extract_Incorporated_into_Topical_Nanoemulsions

Blanco-Padilla A, Soto KM, Hernández Iturriaga M, Mendoza S (2014) Food antimicrobials nanocarriers. Sci World J 2014:837215

Bose S, Michniak-Kohn B (2013) Preparation and characterization of lipid based nanosystems for topical delivery of quercetin. Eur J Pharm Sci 48(3):442–452

Chanchal D, Swarnlata S (2008) Novel approaches in herbal cosmetics. J Cosmet Dermatol 7(2): 89–95

Chen H, Wu J, Sun M, Guo C, Yu A, Cao F et al (2012) N-trimethyl chitosan chloride-coated liposomes for the oral delivery of curcumin. J Liposome Res 22(2):100–109

Das S, Das S, Das MK (2022) Nanocosmeceuticals: concept, opportunities, and challenges. In: Nanocosmeceuticals. Elsevier, Amsterdam, pp 31–69

Dhapte-Pawar V, Kadam S, Saptarsi S, Kenjale PP (2020) Nanocosmeceuticals: facets and aspects. Future Sci OA 6(10):613. Available at https://www.researchgate.net/publication/343483440_Nanocosmeceuticals_facets_and_aspects

Dréno B et al (2019) Safety of titanium dioxide nanoparticles in cosmetics. J Eur Acad Dermatol Venereol 33(Suppl 7):34–46. https://doi.org/10.1111/jdv.15943

El-Komy M, Shalaby S, Hegazy R, Abdel Hay R, Sherif S, Bendas E (2017) Assessment of cubosomal alpha lipoic acid gel efficacy for the aging face: a single-blinded, placebo-controlled, right-left comparative clinical study. J Cosmet Dermatol 16(3):358–363. https://doi.org/10.1111/jocd.12298

Erdo F, Hashimoto N, Karvaly G, Nakamichi N, Kato Y (2016) Critical evaluation and methodological positioning of the transdermal microdialysis technique. A review. J Control Release 233: 147–161

Expert Market Research (2022) Cosmetics market size, trends, industry share, analysis 2023-2028. Available at https://www.expertmarketresearch.com/reports/cosmetics-market

Friedrich RB, Kann B, Coradini K, Offerhaus HL, Beck RCR, Windbergs M (2015) Skin penetration behavior of lipid-core nanocapsules for simultaneous delivery of resveratrol and curcumin. Eur J Pharm Sci 78:204–213

Fytianos G, Rahdar A, Kyzas GZ (2020) Nanomaterials in cosmetics: recent updates. Nanomaterials 10(5):979. Available at www.mdpi.com/journal/nanomaterials

Ganesan P, Choi DK (2016) Current application of phytocompound-based nanocosmeceuticals for beauty and skin therapy. Int J Nanomedicine 11:1987–2007. Available at https://www.

dovepress.com/current-application-of-phytocompound-based-nanocosmeceuticals-for-beau-peer-reviewed-fulltext-article-IJN
Gautam A, Singh D, Vijayaraghavan R (2011) Dermal exposure of nanoparticles: an understanding. J Cell Tissue Res 11(1):2703–2708
Gigliobianco MR, Casadidio C, Censi R, di Martino P (2018) Nanocrystals of poorly soluble drugs: drug bioavailability and physicochemical stability. Pharmaceutics 10(3):134. Available at https://www.mdpi.com/1999-4923/10/3/134/htm
Google Books (2022) Role of novel drug delivery vehicles in nanobiomedicine. Available at https://books.google.co.in/books?hl=en&lr=&id=GkP8DwAAQBAJ&oi=fnd&pg=PR11&dq=Dr.+Rahul+Shukla+&ots=ttKoeFuT9h&sig=RWQ_eL5sI3N-eQINQjOnokIyKIE&redir_esc=y#v=onepage&q=Dr.%20Rahul%20Shukla&f=false
Gunasekaran T, Haile T, Nigusse T, Dhanaraju MD (2014) Nanotechnology: an effective tool for enhancing bioavailability and bioactivity of phytomedicine. Asian Pac J Trop Biomed 4:S1–S7
Handa M, Tiwari S, Yadav AK, Almalki WH, Alghamdi S, Alharbi KS et al (2021) Therapeutic potential of nanoemulsions as feasible wagons for targeting Alzheimer's disease. Drug Discov Today 26(12):2881–2888
Hu Z, Liao M, Chen Y, Cai Y, Meng L, Liu Y et al (2012) A novel preparation method for silicone oil nanoemulsions and its application for coating hair with silicone. Int J Nanomedicine 7:5719–5724. Available at https://www.dovepress.com/a-novel-preparation-method-for-silicone-oil-nanoemulsions-and-its-appl-peer-reviewed-fulltext-article-IJN
Hua S (2015) Lipid-based nano-delivery systems for skin delivery of drugs and bioactives. Front Pharmacol 6:219
Hwang SL, Kim JC (2008) In vivo hair growth promotion effects of cosmetic preparations containing hinokitiol-loaded poly(ε-caprolacton) nanocapsules. J Microencapsul 25(5):351–356. https://doi.org/10.1080/02652040802000557
Ido Y, Duranton A, Lan F, Weikel KA, Breton L, Ruderman NB (2015) Resveratrol prevents oxidative stress-induced senescence and proliferative dysfunction by activating the AMPK-FOXO3 cascade in cultured primary human keratinocytes. PLoS One 10(2):e0115341. https://doi.org/10.1371/journal.pone.0115341
Ito Y, Warner JH, Brown R, Zaka M, Pfeiffer R, Aono T et al (2010) Controlling intermolecular spin interactions of La@C82 in empty fullerene matrices. Phys Chem Chem Phys 12(7):1618–1623. Available at https://www.researchgate.net/publication/41401776_Controlling_intermolecular_spin_interactions_of_LaC82_in_empty_fullerene_matrices
Jain V, Singodia D, Gupta GK, Garg D, Keshava GBS, Shukla R et al (2011) Ciprofloxacin surf-plexes in sub-micron emulsions: a novel approach to improve payload efficiency and antimicrobial efficacy. Int J Pharm 409(1–2):237–244
Jain A, Garg NK, Jain A, Kesharwani P, Jain AK, Nirbhavane P et al (2015) A synergistic approach of adapalene-loaded nanostructured lipid carriers, and vitamin C co-administration for treating acne. Drug Dev Ind Pharm 42(6):897–905. https://doi.org/10.3109/03639045.2015.1104343
Kaul S, Gulati N, Verma D, Mukherjee S, Nagaich U (2018) Role of nanotechnology in cosmeceuticals: a review of recent advances. J Pharm 2018:3420204. https://doi.org/10.1155/2018/3420204
Kaur I, Agrawal R (2008) Nanotechnology: a new paradigm in cosmeceuticals. Recent Pat Drug Deliv Formul 1(2):171–182
Khan AK, Rashid R, Murtaza G, Zahra A (2014) Gold nanoparticles: synthesis and applications in drug delivery. Trop J Pharm Res 13(7):1169–1177
Klajnert B, Bryszewska M (2001) Dendrimers: properties and applications. Acta Biochim Pol 48(1):199–208
Kotyla T, Kuo F, Moolchandani V, Wilson T, Nicolosi R (2008) Increased bioavailability of a transdermal application of a nano-sized emulsion preparation. Int J Pharm 347(1–2):144–148
Kumar N, Tharatha S, Chaiyasut C (2011) Development and validation of simple isocratic high performance liquid chromatography-ultraviolet (HPLCUV) method for determination of safflower yellow in Carthamus tinctorius (L.)-loaded nanostructured lipid carriers (NLC). Afr J

Pharm Pharmacol 5(20):2335–2341. Available at https://www.researchgate.net/publication/287031997_Development_and_validation_of_simple_isocratic_high_performance_liquid_chromatography-ultraviolet_HPLCUV_method_for_determination_of_safflower_yellow_in_Carthamus_tinctorius_L-loaded_nanostructured_li

Kushwaha N, Minocha N, Kumar N (2020) Use of nanotechnology in cosmeceuticals: a review. Int J Pharm Sci Invent 9:43–51

Kusuma Priya MD, Kumar V, Damini VK, Eswar K, Reddy KR, Brito Raj S et al (2020) Somes: a review on composition, formulation methods and evaluations of different types of "somes" drug delivery system. Int J Appl Pharm 12(6):7–18. Available at https://innovareacademics.in/journals/index.php/ijap/article/view/38996/23896

Lacatusu I, Badea N, Murariu A, Nichita C, Bojin D, Meghea A (2010) Antioxidant capacity of lipid nanoparticles loaded with rosemary extract. Mol Cryst Liq Cryst 523:260–272. https://doi.org/10.1080/15421401003719886

Lee R, Shenoy D, Sheel R (2010) Chapter 2 micellar nanoparticles. In: Applications for topical and passive transdermal drug delivery. Elsevier, Amsterdam

Lohani A, Verma A, Joshi H, Yadav N, Karki N (2014) Nanotechnology-based cosmeceuticals. ISRN Dermatol 2014:1–14

Lu WC, Chiang BH, Huang DW, Li PH (2014) Skin permeation of d-limonene-based nanoemulsions as a transdermal carrier prepared by ultrasonic emulsification. Ultrason Sonochem 21(2):826–832

Lu PJ, Huang SC, Chen YP, Chiueh LC, Shih DYC (2015) Analysis of titanium dioxide and zinc oxide nanoparticles in cosmetics. J Food Drug Anal 23(3):587–594

Lv C, Hou J, Xie W, Cheng H (2015) Investigation on formaldehyde release from preservatives in cosmetics. Int J Cosmet Sci 37(5):474–478. https://doi.org/10.1111/ics.12212

Manca ML, Castangia I, Zaru M, Nácher A, Valenti D, Fernàndez-Busquets X et al (2015) Development of curcumin loaded sodium hyaluronate immobilized vesicles (hyalurosomes) and their potential on skin inflammation and wound restoring. Biomaterials 71:100–109

McClements DJ (2012) Advances in fabrication of emulsions with enhanced functionality using structural design principles. Curr Opin Colloid Interface Sci 5(17):235–245. Available at https://www.infona.pl//resource/bwmeta1.element.elsevier-c1e84752-388b-38c2-9a1c-e08956b17a3b

Meghea A (2008) Pharmaceuticals and cosmeceuticals based on soft nanotechnology techniques with antioxidative, immunostimulative and other therapeutic activities. Recent Pat Nanotechnol 2(2):137–145

Mitri K, Shegokar R, Gohla S, Anselmi C, Müller RH (2011) Lipid nanocarriers for dermal delivery of lutein: preparation, characterization, stability and performance. Int J Pharm 414(1–2):267–275

Moulaoui K, Caddeo C, Manca ML, Castangia I, Valenti D, Escribano E et al (2015) Identification and nanoentrapment of polyphenolic phytocomplex from Fraxinus angustifolia: in vitro and in vivo wound healing potential. Eur J Med Chem 89:179–188

Müller RH, Radtke M, Wissing SA (2002) Solid lipid nanoparticles (SLN) and nanostructured lipid carriers (NLC) in cosmetic and dermatological preparations. Adv Drug Deliv Rev 54(suppl):S131–S155

Nagaich U (2016) Nanocosmeceuticals: a boon to personal care products. J Adv Pharm Technol Res 7(1):1. Available at https://www.japtr.org/article.asp?issn=2231-4040;year=2016;volume=7;issue=1;spage=1;epage=1;aulast=Nagaich

Niculae G, Lacatusu I, Badea N, Stan R, Vasile BS, Meghea A (2014) Rice bran and raspberry seed oil-based nanocarriers with self-antioxidative properties as safe photoprotective formulations. Photochem Photobiol Sci 13(4):703–716

Oberdörster G, Oberdörster E, Oberdörster J (2005) Nanotoxicology: an emerging discipline evolving from studies of ultrafine particles. Environ Health Perspect 113(7):823–839

Oliveira C, Coelho C, Teixeira JA, Ferreira-Santos P, Botelho CM (2022) Nanocarriers as active ingredients enhancers in the cosmetic industry-The European and North America Regulation Challenges. Molecules 27:1669. Available at https://www.mdpi.com/1420-3049/27/5/1669/htm

Pandita A, Sharma P (2013) Pharmacosomes: an emerging novel vesicular drug delivery system for poorly soluble synthetic and herbal drugs. ISRN Pharm 2013:1–10

Pando D, Matos M, Gutiérrez G, Pazos C (2015) Formulation of resveratrol entrapped niosomes for topical use. Colloids Surf B Biointerfaces 128:398–404

Parveen A, Parveen B, Parveen R, Ahmad S (2015) Challenges and guidelines for clinical trial of herbal drugs. J Pharm Bioallied Sci 7(4):329

Pauwels M, Rogiers V (2010) Human health safety evaluation of cosmetics in the EU: a legally imposed challenge to science. Toxicol Appl Pharmacol 243(2):260–274

Pereira L, Dias N, Carvalho J, Fernandes S, Santos C, Lima N (2014) Synthesis, characterization and antifungal activity of chemically and fungal-produced silver nanoparticles against Trichophyton rubrum. J Appl Microbiol 117(6):1601–1613. https://doi.org/10.1111/jam.12652

Pereira F, Baptista R, Ladeiras D, Madureira AM, Teixeira G, Rosado C et al (2015) Production and characterization of nanoparticles containing methanol extracts of Portuguese Lavenders. Measurement 74:170–177

Pure Chemicals Co (2022) Commonly used chemicals in cosmetics. Available at https://www.pure-chemical.com/blog/commonly-used-chemicals-in-cosmetics/

Raj S, Jose S, Sumod US, Sabitha M (2012) Nanotechnology in cosmetics: opportunities and challenges. J Pharm Bioallied Sci 4(3):186

Saggar S, Mir PA, Kumar N, Chawla A, Uppal J, Shilpa S et al (2022) Traditional and herbal medicines: opportunities and challenges. Pharm Res 14(2):107–114

Sakamoto J, Annapragada A, Decuzzi P, Ferrari M (2007) Antibiological barrier nanovector technology for cancer applications. Expert Opin Drug Deliv 4(4):359–369

Santos AC, Morais F, Simões A, Pereira I, Sequeira JAD, Pereira-Silva M et al (2019) Nanotechnology for the development of new cosmetic formulations. Expert Opin Drug Deliv 16(4): 313–330. https://doi.org/10.1080/17425247.2019.1585426

SCMSJournal.com (2022) Nanotechnology use with cosmeceuticals. Available at https://www.scmsjournal.com/article/buy_now/?id=158

SESDERMA (2022) FILLDERMA lips. Available at https://www.sesderma.com/int_en/FILLDERMA-Lips-40004133.html

Sharma S, Sathasivam T, Rawat P, Pushpamalar J (2021) Lycopene-loaded nanostructured lipid carrier from carboxymethyl oil palm empty fruit bunch cellulose for topical administration. Carbohydr Polym Technol Appl 2:100049

Shen CY, Dai L, de Shen B, Zhou X, Bai JX, Xu H et al (2015) Nanostructured lipid carrier based topical gel of ganoderma triterpenoids for frostbite treatment. Chin J Nat Med 13(6):454–460

Shukla P, Gupta G, Singodia D, Shukla R, Verma AK, Dwivedi P et al (2010) Emerging trend in nano-engineered polyelectrolyte-based surrogate carriers for delivery of bioactives. Expert Opin Drug Deliv 7(9):993–1011. https://doi.org/10.1517/17425247.2010.510830

Singh D, Pradhan M, Nag M, Singh MR (2015) Vesicular system: versatile carrier for transdermal delivery of bioactives. Artif Cells Nanomed Biotechnol 43(4):282–290. https://doi.org/10.3109/21691401.2014.883401

Souto EB, Müller RH (2008) Cosmetic features and applications of lipid nanoparticles (SLN®, NLC®). Int J Cosmet Sci 30(3):157–165. https://doi.org/10.1111/j.1468-2494.2008.00433.x

Starley IF, Mohammed P, Schneider G, Bickler SW (1999) The treatment of paediatric burns using topical papaya. Burns 25(7):636–639

Subedi G, Shrestha AK, Shakya S (2016) Study of effect of different factors in formulation of micro and nanospheres with solvent evaporation technique. Open Pharm Sci J 3(1):182–195

Sudhakar CK, Upadhyay N, Jain S, Charyulu RN (2012) Ethosomes as non-invasive loom for transdermal drug delivery system. Nanomed Drug Deliv 1:1–16. Available at https://www.taylorfrancis.com/chapters/edit/10.1201/b13119-5/ethosomes-non-invasive-loom-transdermal-drug-delivery-system

Takahashi M, Kitamoto D, Asikin Y, Takara K, Wada K (2009) Liposomes encapsulating aloe vera leaf gel extract significantly enhance proliferation and collagen synthesis in human skin cell lines. J Oleo Sci 58(12):643–650

Teeranachaideekul V, Müller RH, Junyaprasert VB (2007) Encapsulation of ascorbyl palmitate in nanostructured lipid carriers (NLC)-Effects of formulation parameters on physicochemical stability. Int J Pharm 340(1–2):198–206

Tiwari S, Talreja S (2020) A concept of nanotechnology in cosmetics: a complete overview. Adalya J 9(11):14–23

Tripodo G, Trapani A, Torre ML, Giammona G, Trapani G, Mandracchia D (2015) Hyaluronic acid and its derivatives in drug delivery and imaging: Recent advances and challenges. Eur J Pharm Biopharm 97:400–416

Verma A, Jain A, Hurkat P, Jain SK (2022) Transfollicular drug delivery: current status of the problem. Res Rep Transderm Drug Deliv 5:1–17. Available at https://www.researchgate.net/publication/340814556_Transfollicular_Drug_Delivery_Current_Status_Of_The_Problem_Review_Article_in_Russian

Yeh YC, Creran B, Rotello VM (2012) Gold nanoparticles: preparation, properties, and applications in bionanotechnology. Nanoscale 4(6):1871–1880

Yin R, Hamblin M (2015) Antimicrobial photosensitizers: drug discovery under the spotlight. Curr Med Chem 22(18):2159–2185

Zamarioli CM, Martins RM, Carvalho EC, Freitas LAP (2015) Nanoparticles containing curcuminoids (Curcuma longa): development of topical delivery formulation. Rev Bras 25: 53–60. https://doi.org/10.1016/j.bjp.2014.11.010

Zhang Z, Chen X, Rao W, Long F, Yan L, Yin Y (2015) Preparation of novel curcumin-imprinted polymers based on magnetic multi-walled carbon nanotubes for the rapid extraction of curcumin from ginger powder and kiwi fruit root. J Sep Sci 38(1):108–114. https://doi.org/10.1002/jssc.201400814

Toxicity Issues of Nanoparticles in the Delivery of Phytoconstituents and Cosmeceuticals

11

Mounisha Bandakinda, Ankit Kumar, and Awanish Mishra

11.1 Introduction

Plants are essential sources of food, medicine, and supplemental health products, including their bioactive components. The plant is protected from infections, predation, or infestations through microorganisms, predators, or phytoconstituents, which are non-nutrient active plant chemical compounds or bioactive substances (Sharma et al. 2022). The five basic classes of phytoconstituents include phenolics, carotenoids, alkaloids, organosulfur, and molecules containing nitrogen (Liu 2004). Phytoconstituents have anticancer, antioxidant, anti-inflammatory, immunomodulatory, and other beneficial properties (Cencic and Chingwaru 2010). Nanoscale-sized particles (1–1000 nm), also known as nanoparticles (NPs), have gained popularity as efficient tools with several uses in the biological and non-biological fields of drug administration, diagnostics, cosmetics, and others. These recent advances raise concerns regarding the safety of nanoparticles (Najahi-Missaoui et al. 2020). In the field of therapeutics and drug delivery, nanoparticles have been developed into scaffolds for tissue engineering as well as drug delivery platforms for the treatment of a variety of disorders. Nanoparticles' small size makes it possible for them to enter cells and organelles, providing novel methods like targeted drug delivery (Singh and Lillard Jr. 2009). To recognize and bind to certain receptors on the target cells, NP surfaces can be conjugated with ligands or antibodies (Ahmad et al. 2006). For greater biocompatibility and biodegradability, NPs are frequently coated with a variety of substances (Singh 2010). In cosmetic formulations, phytoconstituents are becoming more and more popular as

M. Bandakinda · A. Kumar · A. Mishra (✉)
Department of Pharmacology and Toxicology, National Institute of Pharmaceutical Education and Research (NIPER), Guwahati, Assam, India
e-mail: awanish@niperguwahati.in

© The Author(s), under exclusive license to Springer Nature Singapore Pte Ltd. 2024

D. Pooja, H. Kulhari (eds.), *Nanotechnology Based Delivery of Phytoconstituents and Cosmeceuticals*, https://doi.org/10.1007/978-981-99-5314-1_11

ingredients since they protect the skin from external and internal aggressors and treat a variety of skin disorders.

Cosmeceuticals are a combination of cosmetic products that contain pharmaceutical active ingredients that have therapeutic values and display beautifying and cleaning effects in addition to therapeutic effectiveness when applied to various regions of the body (Ganesan and Choi 2016). Nanocosmeceuticals are defined as cosmetics having medicinally beneficial biologically active ingredients that were created utilizing nanotechnology. Currently, nanotechnology is being successfully used to increase the effectiveness of cosmeceuticals. These uses include therapy for ageing, dry skin, wrinkles, and hair damage (Gupta et al. 2022). Growing consumer interest in natural and herbal cosmetics has allowed businesses to design and create new cosmetics with lower toxicity concerns. Some hurdles in making cosmeceutical products are less solubility, stability, and skin permeability of the herbal and natural pharmaceutical actives. Hence, nanotechnology using nanocarriers and nanoparticles is showing promising tools to enhance personal care effects along with therapeutic effectiveness. As nanoparticles show certain toxicity issues in the cosmeceuticals industry, guidelines and standardized criteria must be established to assess the safety, efficacy, and toxicity of nano cosmeceutical products (Johnson et al. 2022).

11.2 Nanomaterials Used in the Delivery of Phytoconstituents

The nanomaterials systems used in the delivery of phytoconstituents have been extensively explored for a wide variety of diseases (mentioned in Table 11.1). The nanotechnology approach appears effective in overcoming several translational challenges available with phytoconstituents, such as poor aqueous solubility,

Table 11.1 Nanomaterials used in the delivery of phytoconstituents for the treatment of diseases

Sl. no.	Phytoconstituents	Nanomaterials used	Disease	References
1.	Apigenin	APG-PLGA nanoparticles	Skin cancer	Kim et al. (2020)
2.	Artemisinin	MPEG5000-PLLA3200 amphiphilic block copolymeric micelles	Breast cancer	Ghafarifar et al. (2020)
3.	Berberine	PLGA nanoparticles	Hepatocellular carcinoma (HCC)	Majidzadeh et al. (2020)
4.	Combretastatin-A4	PEG-PLA and Arg-Gly-Asp peptides	Advanced anaplastic thyroid cancer, pathologic myopia, and polypoidal choroidal vasculopathy	Barnes et al. (2020)
5.	Lapachone	Polymer micelles of PEG-PLA	Colon cancer	Gomes et al. (2021)

membrane permeability, lower metabolic stability, and higher clearance rate. Thus, it helps in improving the bioavailability and overall efficacy of phytoconstituents (Alam et al. 2021; Pietroiusti et al. 2017).

11.3 Nanocarrier Delivery Systems Suitable for Phytoconstituents

Traditional drug delivery already uses nanotechnology, which provides the opportunity to create tools specifically suited for improving the therapeutic efficacy of naturally occurring bioactive chemicals (Squillaro et al. 2018). Nanocarriers are alternatives to conventional methods of medication formulation that allow for increased drug bioavailability, site-specific targeted delivery, and a reduction in hazardous side effects (Aqil et al. 2013; Huang et al. 2010). The use of nano vectors for drug delivery has many benefits over conventional formulations, including more effective water-insoluble drug delivery at high doses, drug protection from adverse environments (such as the digestive tract's acidic pH), and targeted and controlled delivery to achieve precise intake to a particular tissue over a predetermined period (Sun et al. 2014).

Different carrier systems have been used as nanocarriers for passive or active targeting. Surface modification of metal particles, such as iron oxide or gold nanosized assemblies, can function as drug carriers (Huang et al. 2012). The interaction with the body environment is heavily influenced by the nanocarrier's surface characteristics (Verma and Stellacci 2010). Nano vectors, for instance, with positive or negative surface charges have better reticuloendothelial clearance. By extending circulation durations and preventing nanoparticle loss, steric stabilization using biological or synthetic macromolecules, such as opsonin or polyethylene glycol (PEG), is beneficial (Pasche et al. 2005). To combat adverse effects and reduce systemic drug administration, targeting methods are used (Singh and Lillard Jr. 2009). Passive targeting uses the "enhanced permeability and retention effect" of damaged blood vessels and involves altering the physiochemical properties, pH, or hydrophobicity of NP. In contrast, active targeting uses biomarkers to get to the target areas, such as mutant genes, RNAs, proteins, lipids, carbohydrates, and relatively small metabolite molecules (Bertrand et al. 2014).

Due to their diverse properties, including antioxidant, anticancer, and antibacterial properties, natural polyphenols have gained significant interest in medicine. Over the past few years, natural polyphenol-based drug delivery methods have gained popularity as a study area (Li et al. 2016). Some examples of suitable nanocarrier systems have been summarized in Table 11.2.

Table 11.2 Nano drug delivery systems used in the delivery of phytoconstituents

Phytoconstituent	Nano vectors	Type of delivery system	Experimental model	References
Quercetin	NPs	NP of soya lecithin, Tween 80, and PEG	In vivo (rats)	Li et al. (2009)
		Eudragit-polyvinyl alcohol quercetin- loaded NP	In vitro	Wu et al. (2008)
		Quercetin PLGA NP	In vivo (rats)	Chakraborty et al. (2012)
	Nano-capsules	Lipid -coated NC	In vitro	Barras et al. (2009)
		Quercetin-PLGA NC		

11.4 Types of Nanocarriers for the Delivery of Cosmeceuticals

Based on the materials used in the formulation of nano cosmeceuticals, four main groups of nanocarriers are employed in the delivery of cosmeceuticals. These include lipid-based nanocarriers, metal-based nanocarriers, polymeric-based nanocarriers, and miscellaneous nanocarriers.

11.4.1 Lipid-Based Nanocarriers

Lipid-based nanocarriers may entangle hydrophilic and hydrophobic medicines, are biocompatible and biodegradable, and offer appealing properties as a means of delivering pharmaceutical actives to the target site. Excellent heat stability, ease of formulation, low manufacturing costs, superior loading volume, and widespread commercial manufacturing are all positive sides of lipid-based nanocarriers (García-Pinel et al. 2019).

11.4.1.1 Liposomes

Liposomes are most frequently used in cosmeceutical formulations. Most of these nanoparticles are composed of phospholipids, and because of their amphipathic properties, they are arranged in a bilayer configuration and form vesicles in the presence of water. These vesicular nanoparticles range from a few nanometres to a few micrometres and can encapsulate both hydrophobic as well as hydrophilic active ingredients (Yingchoncharoen et al. 2016). Some of the major reasons which make liposomes an excellent choice for cosmeceuticals are their safety, biocompatibility, enhanced dermal permeability, and biphasic encapsulating capacity. Antioxidants such as lycopene and vitamins like vitamins A and E have been encapsulated with liposomes. Liposome-based nano cosmeceuticals are widely used in the preparation of moisturizers, sun screens, hair care products, and antiwrinkle products (Monteiro et al. 2015).

11.4.1.2 Niosomes

Vesicles that can enclose both hydrophilic and hydrophobic actives can be multilamellar or unilamellar, made up of non-ionic surfactants like tweens, alkyl amide, spans, and alkyl ether which have self-assembly properties (Kaul et al. 2018). These are superior to liposomes as these vesicles have high loading capacity, low cost, higher stability, and enhanced skin penetration capacity. Enhanced skin penetration to the deeper layer of skin enables increased bioavailability of the active components, required for collagen synthesis and skin tightening. In 1987, niosome was first developed by L'Oréal under the commercial name Lancôme. Currently, several niosome-based cosmeceutical products exist commercially for the remedy of ageing, skin whitening, hair damage, and skin dryness (Naseri et al. 2015). Ellagic acid, a putative skin-whitening agent with antioxidant properties, exhibits limited water solubility. Thus, several carbon-based solvents were utilized to prepare ellagic acid niosomes employing Span 60 as well as Tween 60 (Junyaprasert et al. 2012).

11.4.1.3 Ethosomes, Transferosomes, and Cubosomes

These three innovative vesicular nanocarriers were developed due to some topical disadvantages of niosomes and liposomes, such as leakage of the active components due to the fusion of the vesicles into large-sized vesicles. Ethosomes are elastic vesicles made up of ethanol and phospholipids and have less aggregating and enhanced skin permeation properties. Transferosomes with a size of 500 nm consist of an aqueous core, a lipid bilayer, and a blend of surfactants, which function as edge activators to allow transferosomes to pass through the stratum corneum layer of skin (Santos et al. 2019). Cubosomes are hexagonal nanostructures which contains biphasic liquid crystalline cubic structures and are known for biocompatibility and suitability for nanocarrier mediated drug delivery process. The cubosomes offers large internal surface area for superior loading volumes for both hydrophilic and hydrophobic cosmeceutical products for targeted and controlled release. These nanosystems are applied in the formulation of hair care and beauty products (Santos et al. 2019; Lohani et al. 2014).

11.4.1.4 Nanoemulsions

Nanoemulsions are structures with a precise mix of oil, water, and surfactant. These nanostructures are between 50 and 200 nm in size. Nanoemulsions have advantages like better hydration of the skin, high permeation to the skin layer, less sedimentation, and less toxicity. These nanosystems are currently being used in the formulation of skin care, hair care, deodorants, and other cosmetic products.

11.4.1.5 Solid Lipid Nanoparticles and Nanostructured Lipid Carriers

Both solid lipid nanoparticles (SLNs) and nanostructured lipid carriers (NLCs) are excellent colloidal delivery carriers for low water-soluble, lipophilic, and hydrophilic cosmetic ingredients. The size range of SLNs and NLCs is 50–1000 nm and 10–1000 nm, respectively (Pardeike et al. 2009). SLN are biocompatible and biodegradable (hence less toxicity) and show enhanced skin permeation. NLCs are a better form of SLNs due to less systemic toxicity, high drug loading capacity, less spillage of actives during storage, due to its amorphous type (Chauhan et al. 2020). Both SLNs and NLCs are widely utilized in cosmeceuticals for controlled and sustained release, and enhanced skin permeation of active ingredients of cosmetic products. SLNs are applied in the delivery of fragrance of perfumes. NLCs are commercially used in antiwrinkle, skin moisturizing, anti-ageing, and skin tone products (Naseri et al. 2015).

11.4.2 Polymeric Nanocarriers

Natural or synthetic-based colloidal nanocarriers having a particle size ranging from 1 to 1000 nm have the capacity to encapsulate both hydrophilic and hydrophobic cosmetic active ingredients (Zielińska et al. 2020). Nanosphere and nanocapsules are the two widely used polymeric nanocarriers for the encapsulation and delivery of active ingredients. Nanospheres are entirely a solid mass of polymers over which surface modification like conjugation of active ingredients of cosmetic products can be done. Nanocapsules have a hollow core, can encapsulate unstable cosmetic actives, and provide more targeted delivery and enhanced bioavailability (Hickey et al. 2015; Kreuter 2007). Non-biocompatibility, high toxicity, and production cost are some drawbacks of these nanocarriers (Masood 2016). The polymeric nanocarriers based skin care products (like antiaging, moisturizing and antiacne creams) are in high demand (Lu et al. 2011).

11.4.3 Inorganic-Based Nanocarriers

Inorganic nanoparticles are made up of metal as well as metal oxides with varying sizes, ranging from 10 to 100 nm. These nanoparticles are excellent in surface modification (conjugating a variety of substances to the surface of metals) for the targeted transport of the active pharmaceutical ingredients. Apart from this, these nanosystems are biocompatible and easy to manufacture. The most widely used metal nanoparticles in cosmeceutical products include gold (Au), titanium oxide (TiO_2), silver (Ag), and zinc oxide (ZnO). However, due to the rising issue of cytotoxicity to good cells such as keratinocytes and fibroblasts, the use of these nanoparticles is very limited.

11.4.3.1 Gold Nanoparticles

Gold nanoparticles are one of the ideal metal-based nanocarriers in many aspects like high loading capacity, biocompatible, inert, stable, nonbleaching, small (5–400 nm), various changeable shapes and sizes, and various surface modifications with active ingredients for targeted drug delivery. In addition, nanogold has antibacterial, anti-inflammatory, and antifungal properties. Nanogolds are widely applied in the production of several categories of cosmeceutical products such as skin care and beauty care (Abbasi et al. 2016).

11.4.3.2 Silver Nanoparticles

Silver nanoparticles also have antifungal, anti-inflammatory, and antibacterial properties. Silver nanoparticles exerts its antimicrobial activity by damaging the microbial cell membrane and increasing the reactive oxygen species inside the cells. Silver nanoparticles are exercised in the commercial production of skin care, nail polish, oral care, and hair care products (Ong and Nyam 2022).

11.4.3.3 Titanium Oxide

Titanium dioxide nanoparticles (TiO_2) are metallic colourless nanoparticles with a size range of 1–150 nm. They are used in sunscreens for protecting the skin from ultraviolet (UV) light and shielding it from the dangerous consequences of UVA and UVB exposure. Previously TiO_2 was used in lip balms and foundations in the microcrystalline form, but due to raised safety concerns related to data on its systemic toxicity and photocytotoxicity to the skin, its use in cosmetic and cosmeceutical products has become limited (Dréno et al. 2019).

11.4.4 Miscellaneous Nanocarriers

11.4.4.1 Dendrimers

Dendrimers are novel nanocarriers having properties like being very small in size (size range from 2 to 20 nm), with compact molecular structure, highly branched surface which favors attachment of multiple functional groups (Abbasi et al. 2014). These nanostructures have positive aspects in cosmeceuticals due to enhanced stability and controlled release of the API. Dendrimers are used in cosmeceuticals to formulate hair care products like shampoo, skin care products like antiacne treatments, and nail care products by many companies like Unilever and L'Oréal (Heegaard et al. 2010).

11.4.4.2 Carbon Nanotubes and Fullerenes

Carbon nanotubes are cylindrical with hollow cores, which is composed of rolled-up graphene sheets and size ranging from 0.7 mm to 50 nm. Carbon nanotubes are novel nanostructures and are very light and hence have been used in hair colourants (Teixeira-Santos et al. 2020). Fullerenes are another allotrope of carbons, having the shape of the carbon skeleton in the spherical or cylindrical form. Fullerenes are biologically active nanostructures with antioxidant, anti-bacterial, and anti-viral activity and have high demand in cosmeceuticals. Low aqueous solubility is one of the limitations but encapsulation in more liposomes can be employed to enhance solubility. Fullerenes have been implicated in the commercial production of skin care products (Lens 2009).

11.4.4.3 Polymerosomes

Polymerosomes are vesicular structures with hydrophilic core and lipophilic lipid bilayer shell, made up of block copolymer. The size ranges from a few nanometres to micrometres and can entrap both hydrophilic (inside core) and hydrophobic (inside lipid bilayer) actives like liposomes. These nanostructures are better than liposomes for providing better stability of the active ingredients as polymerosomes have more rigid surroundings around the core (Kim et al. 2011). These nanosystems can be applied in formulating cosmeceuticals having sensitive components like enzymes, antioxidants, vitamins, and other pharmaceutical actives (Zhang and Zhang 2017).

11.5 Toxicological Facets of Nanosystems in Drug Delivery in Phytoconstituents

The major focus of nanoformulations utilized for delivery of the API is on the medication's reduced toxicity, and the carrier's potential toxicity is not considered. Particularly, likely residual effects of such a treatment might cover up potential harmful local and/or systemic reactions. So, it is very essential to bear in mind the potentially detrimental effects of nanocarriers used in nano drug delivery systems (Pandey et al. 2020).

The most crucial in vitro testing method for predicting GI tract interactions with drug delivery systems is cell culture models, which give insightful information about the molecular processes underlying NPs' absorption and toxicity. Additionally, cell culture–based methods will be required to select the best formulations for future in vivo experiments. Because it is important to assess how NPs affect the GI tract's cell viability, various culture techniques have been proposed. The cell-based Caco-2 systems are the most often used cell culture models to analyse the molecular mechanisms of toxicity on a molecular basis (Song et al. 2020).

11.6 Toxicological Facets of Nanosystems in Drug Delivery in Cosmeceuticals

The use of nanoparticles in the production of cosmeceutical products is increasing day by day because of the many positive aspects of these nanoparticles in cosmeceuticals, such as providing API to enhance skin penetration, better stability, controlled and targeted release, enhanced solubility, and therapeutic effectiveness (Ahmadian et al. 2020). However, toxicity aspects of these nanosystems are also arising and need to be regulated on the safety of cosmeceutical products related to consumers and environment on short-term and long-term exposure. It has been reported that the toxicity of nanoparticles gets enhanced as the particle size is reduced (Kaul et al. 2018). Moreover, particle size, surface nature like charge and other modifications, solubility, and property of aggregation influence the toxicity profile of these nanocosmeceutical products (Raj et al. 2012). Dermal, ingestion, and respiratory inhalation are the three main routes that toxicity from nano cosmeceutical compounds may manifest (Yah et al. 2012). We will see them out one by one.

11.6.1 Dermal

The dermal route is considered a direct route for toxicity for nano cosmeceuticals as nanoparticles of lesser size may penetrate the skin more deeply and can reach the deeper cells such as fibroblasts via transcellular, intracellular, or trans follicular pathways and can damage these cells. Moreover, smaller sizes (less than 10 nm) may get access to the systemic circulation and lymph nodes too. Use of metal-based nanoparticles (such as ultrafine TiO_2 and ZnO) exhibits dermal toxicity due to UV exposure induced reactive oxygen species generation culminating to cell damage and other dermal manifestations (Smijs and Pavel 2011). Fullerenes, having antimicrobial and enhanced skin permeation capacity, are also thought to have toxicity issues as their skin penetration capability is much higher than other nanosystems, and safety data is needed for proper use in the cosmeceutical preparation (McShan and Yu 2014). Apart from this, carbon nanotubes with surface modification can enter deep skin layers and cause toxic effects such as gene mutation and DNA damage to cells such as fibroblasts and keratinocytes (Shan et al. 2014). On the other side, dendrimers having branches attached to the core may interact with the skin cells and can damage the physiological and biological activity of the dermal cells, leading to health hazards.

11.6.2 Ingestion

Unintentional ingestion of cosmeceutical products like lip balms, lipsticks, and face powders may occur which may lead to hazardous health effects (Hoet et al. 2004). After ingestion, these nanoparticles may build up in major organs like the liver, heart, kidney, and spleen which may lead to organ toxicity, and consequently, organ damage may be manifested. Studies have revealed that silver nanoparticles have neurotoxicity and cytotoxicity properties on rat neuronal cells and germ line cells of mice, respectively. In addition, another previous study has revealed that gold nanoparticles after ingestion into mice even at low concentrations caused reduced body weight along with RBC count. Hence, more safety assessment of these cosmeceutical products is much needed for further use by consumers and keeping in mind the safety of the environment (Kaul et al. 2018).

11.6.3 Inhalation

Upon inhalation, nanoparticles present in cosmeceutical products reach the lungs and brain via the nasal route and systemic circulation, which can cause hazardous health effects. Therefore, the use of cosmeceutical products like deodorants and powders is becoming a major concern in the well-being of consumers, workers, and environmental protection personnel. Gold nanoparticles of different sizes have been found to accumulate in the liver. Further, carbon nanotubes have been shown to cause granulomatous lung lesions and inflammation upon chronic exposure

(Kobayashi et al. 2017). Ultrafine TiO_2 has also shown pulmonary damage (Warheit et al. 2007). Moreover, complete DNA damage at a smaller size range (20 nm) of the TiO_2 has been studied. Along with this, silicon dioxide has also shown toxicological effects at a much smaller size range, which is about 1–5 nm (Kaul et al. 2018).

11.7 Current Safety Regulations and Future Perspectives for Cosmeceuticals

Since cosmeceuticals are neither pure cosmetics nor drugs, we can consider this as the borderline between cosmetics and pharmaceutical products. Till now the FDA has not applied any stringent regulation for these products and manufacturers are taking advantage of this by trying to escape from clinical trials and giving safety data. However, many countries are thinking of establishing safety regulations for cosmeceuticals. Because of their toxicity, the Scientific Committee on Consumer Products (SCCP) is concerned about the use of insoluble nanoparticles in cosmetics that are employed for topical use. The European Union (EU) has recently brought one regulation into force, that is, to list all nanomaterials used in cosmeceutical products. Japan calls cosmeceuticals a "quasi-drug", and all active pharmaceutical ingredients of pharmaceutical products should be preapproved. In Australia, these borderline products are known as therapeutic goods and API should be preapproved before being incorporated into cosmetic products. In China, all foreign cosmetic products for special use must undergo safety and toxicity studies before marketing. In India, cosmetics are regulated under the Drugs and Cosmetics Act of 1940 and Rules 1945, and the new cosmetic rules, 2020. For import of the cosmetics, an import licence and registration is required and all activities of cosmetics are controlled by the Central Drugs Standard Control Organisation (CDSCO).

These nanomaterial-based cosmeceuticals have many benefits, but due to the rising safety concerns, manufacturers need to provide safety data for the proper use of nano cosmeceuticals by the consumers and should label the products clearly by mentioning all nano-based ingredients. Companies should make the product by keeping in mind the possible health hazards, to reduce the chances of toxicity, and if possible, a safety study should be done for suspected products.

References

Abbasi E, Aval SF, Akbarzadeh A, Milani M, Nasrabadi HT, Joo SW, Hanifehpour Y, Nejati-Koshki K, Pashaei-Asl R (2014) Dendrimers: synthesis, applications, and properties. Nanoscale Res Lett 9(1):247. https://doi.org/10.1186/1556-276X-9-247. PMID: 24994950; PMCID: PMC4074873.

Abbasi E, Milani M, Fekri Aval S, Kouhi M, Akbarzadeh A, Tayefi Nasrabadi H, Nikasa P, Joo SW, Hanifehpour Y, Nejati-Koshki K, Samiei M (2016) Silver nanoparticles: Synthesis methods, bio-applications and properties. Crit Rev Microbiol 42(2):173–180. https://doi.org/10.3109/1040841X.2014.912200. Epub 2014 Jun 17. PMID: 24937409.

Ahmad Z, Pandey R, Sharma S, Khuller GK (2006) Alginate nanoparticles as antituberculosis drug carriers: formulation development, pharmacokinetics and therapeutic potential. Indian J Chest Dis Allied Sci 48(3):171–176. PMID: 18610673.

Ahmadian E, Samiei M, Hasanzadeh A, Kavetskyy T, Jafari S, Alipour M, Salatin S, Rameshrad M, Sharifi S, Eftekhari A, Hasanzadeh M (2020) Monitoring of drug resistance towards reducing the toxicity of pharmaceutical compounds: past, present and future. J Pharm Biomed Anal 186: 113265. https://doi.org/10.1016/j.jpba.2020.113265. Epub 2020 Mar 19. PMID: 32283481.

Alam MB, Naznin M, Islam S, Alshammari FH, Choi HJ, Song BR, Kim S, Lee SH (2021) High resolution mass spectroscopy-based secondary metabolite profiling of *Nymphaea nouchali* (Burm. f) stem attenuates oxidative stress via regulation of MAPK/Nrf2/HO-1/ROS pathway. Antioxidants 10(5):719. https://doi.org/10.3390/antiox10050719. PMID: 34063678; PMCID: PMC8147620.

Aqil F, Munagala R, Jeyabalan J, Vadhanam MV (2013) Bioavailability of phytochemicals and its enhancement by drug delivery systems. Cancer Lett 334(1):133–141. https://doi.org/10.1016/j.canlet.2013.02.032. Epub 2013 Feb 19. PMID: 23435377; PMCID: PMC3815990.

Barnes NG, Parker AW, Ahmed Mal Ullah AA, Ragazzon PA, Hadfield JA (2020) A 2-step synthesis of Combretastatin A-4 and derivatives as potent tubulin assembly inhibitors. Bioorg Med Chem 28(19):115684. https://doi.org/10.1016/j.bmc.2020.115684. Epub 2020 Aug 5. PMID: 32912434.

Barras A, Mezzetti A, Richard A, Lazzaroni S, Roux S, Melnyk P, Betbeder D, Monfilliette-Dupont N (2009) Formulation and characterization of polyphenol-loaded lipid nanocapsules. Int J Pharm 379(2):270–277. https://doi.org/10.1016/j.ijpharm.2009.05.054. Epub 2009 Jun 6. PMID: 19501139.

Bertrand N, Wu J, Xu X, Kamaly N, Farokhzad OC (2014) Cancer nanotechnology: the impact of passive and active targeting in the era of modern cancer biology. Adv Drug Deliv Rev 66:2–25. https://doi.org/10.1016/j.addr.2013.11.009. Epub 2013 Nov 22. PMID: 24270007; PMCID: PMC4219254.

Cencic A, Chingwaru W (2010) The role of functional foods, nutraceuticals, and food supplements in intestinal health. Nutrients 2(6):611–625. https://doi.org/10.3390/nu2060611. Epub 2010 Jun 1. PMID: 22254045; PMCID: PMC3257668.

Chakraborty S, Stalin S, Das N, Choudhury ST, Ghosh S, Swarnakar S (2012) The use of nano-quercetin to arrest mitochondrial damage and MMP-9 upregulation during prevention of gastric inflammation induced by ethanol in rat. Biomaterials 33(10):2991–3001. https://doi.org/10.1016/j.biomaterials.2011.12.037. Epub 2012 Jan 16. PMID: 22257724.

Chauhan I, Yasir M, Verma M, Singh AP (2020) Nanostructured lipid carriers: a ground breaking approach for transdermal drug delivery. Adv Pharm Bull 10(2):150–165. https://doi.org/10.34172/apb.2020.021. Epub 2020 Feb 18. PMID: 32373485; PMCID: PMC7191226.

Dréno B, Alexis A, Chuberre B, Marinovich M (2019) Safety of titanium dioxide nanoparticles in cosmetics. J Eur Acad Dermatol Venereol 33(Suppl 7):34–46. https://doi.org/10.1111/jdv.15943. PMID: 31588611.

Ganesan P, Choi DK (2016) Current application of phytocompound-based nanocosmeceuticals for beauty and skin therapy. Int J Nanomedicine 11:1987–2007. https://doi.org/10.2147/IJN.S104701. PMID: 27274231; PMCID: PMC4869672.

García-Pinel B, Porras-Alcalá C, Ortega-Rodríguez A, Sarabia F, Prados J, Melguizo C, López-Romero JM (2019) Lipid-based nanoparticles: application and recent advances in cancer treatment. Nano 9(4):638. https://doi.org/10.3390/nano9040638. PMID: 31010180; PMCID: PMC6523119.

Ghafarifar F, Molaie S, Abazari R, Hasan ZM, Foroutan M (2020) Fe3O4@Bio-MOF nanoparticles combined with artemisinin, Glucantime®, or shark cartilage extract on Iranian strain of *Leishmania major* (MRHO/IR/75/ER): an in-vitro and in-vivo study. Iran J Parasitol 15(4):537–548. https://doi.org/10.18502/ijpa.v15i4.4859. PMID: 33884011; PMCID: PMC8039492.

Gomes CL, de Albuquerque Wanderley Sales V, Gomes de Melo C, Ferreira da Silva RM, Vicente Nishimura RH, Rolim LA, Rolim Neto PJ (2021) Beta-lapachone: natural occurrence, physicochemical properties, biological activities, toxicity and synthesis. Phytochemistry 186: 112713. https://doi.org/10.1016/j.phytochem.2021.112713. Epub 2021 Mar 2. PMID: 33667813.

Gupta V, Mohapatra S, Mishra H, Farooq U, Kumar K, Ansari MJ, Aldawsari MF, Alalaiwe AS, Mirza MA, Iqbal Z (2022) Nanotechnology in cosmetics and cosmeceuticals-a review of latest advancements. Gels 8(3):173. https://doi.org/10.3390/gels8030173. PMID: 35323286; PMCID: PMC8951203.

Heegaard PM, Boas U, Sorensen NS (2010) Dendrimers for vaccine and immunostimulatory uses. A review. Bioconjug Chem 21(3):405–418. https://doi.org/10.1021/bc900290d. Epub 2009 Nov 4. PMID: 19886668.

Hickey JW, Santos JL, Williford JM, Mao HQ (2015) Control of polymeric nanoparticle size to improve therapeutic delivery. J Control Release 219:536–547. https://doi.org/10.1016/j.jconrel.2015.10.006. Epub 2015 Oct 9. PMID: 26450667; PMCID: PMC4656075.

Hoet PH, Brüske-Hohlfeld I, Salata OV (2004) Nanoparticles - known and unknown health risks. J Nanobiotechnol 2(1):12. https://doi.org/10.1186/1477-3155-2-12. PMID: 15588280; PMCID: PMC544578.

Huang Q, Yu H, Ru Q (2010) Bioavailability and delivery of nutraceuticals using nanotechnology. J Food Sci 75(1):R50–R57. https://doi.org/10.1111/j.1750-3841.2009.01457.x. PMID: 20492195.

Huang K, Ma H, Liu J, Huo S, Kumar A, Wei T, Zhang X, Jin S, Gan Y, Wang PC, He S, Zhang X, Liang XJ (2012) Size-dependent localization and penetration of ultrasmall gold nanoparticles in cancer cells, multicellular spheroids, and tumors in vivo. ACS Nano 6(5):4483–4493. https://doi.org/10.1021/nn301282m. Epub 2012 May 4. PMID: 22540892; PMCID: PMC3370420.

Johnson PI, Favela K, Jarin J, Le AM, Clark PY, Fu L, Gillis AD, Morga N, Nguyen C, Harley KG (2022) Chemicals of concern in personal care products used by women of color in three communities of California. J Expo Sci Environ Epidemiol 32(6):864–876. https://doi.org/10.1038/s41370-022-00485-y. Epub 2022 Nov 2. PMID: 36323919; PMCID: PMC9628299.

Junyaprasert VB, Singhsa P, Suksiriworapong J, Chantasart D (2012) Physicochemical properties and skin permeation of Span 60/Tween 60 niosomes of ellagic acid. Int J Pharm 423(2): 303–311. https://doi.org/10.1016/j.ijpharm.2011.11.032. Epub 2011 Dec 2. PMID: 22155414.

Kaul S, Gulati N, Verma D, Mukherjee S, Nagaich U (2018) Role of nanotechnology in cosmeceuticals: a review of recent advances. J Pharm 2018:3420204. https://doi.org/10.1155/2018/3420204. PMID: 29785318; PMCID: PMC5892223.

Kim SH, Shum HC, Kim JW, Cho JC, Weitz DA (2011) Multiple polymersomes for programmed release of multiple components. J Am Chem Soc 133(38):15165–15171. https://doi.org/10.1021/ja205687k. Epub 2011 Aug 30. PMID: 21838246.

Kim JS, Jeong E, Jo SM, Park J, Kim JY (2020) Comparative study of the effects of light controlled germination conditions on saponarin content in barley sprouts and lipid accumulation suppression in HepG2 hepatocyte and 3T3-L1 adipocyte cells using barley sprout extracts. Molecules 25(22):5349. https://doi.org/10.3390/molecules25225349. PMID: 33207773; PMCID: PMC7697669.

Kobayashi N, Izumi H, Morimoto Y (2017) Review of toxicity studies of carbon nanotubes. J Occup Health 59(5):394–407. https://doi.org/10.1539/joh.17-0089-RA. Epub 2017 Aug 8. PMID: 28794394; PMCID: PMC5635148.

Kreuter J (2007) Nanoparticles–a historical perspective. Int J Pharm 331(1):1–10. https://doi.org/10.1016/j.ijpharm.2006.10.021. Epub 2006 Oct 21. PMID: 17110063.

Lens M (2009) Use of fullerenes in cosmetics. Recent Pat Biotechnol 3(2):118–123. https://doi.org/10.2174/187220809788700166. PMID: 19519567.

Li H, Zhao X, Ma Y, Zhai G, Li L, Lou H (2009) Enhancement of gastrointestinal absorption of quercetin by solid lipid nanoparticles. J Control Release 133(3):238–244. https://doi.org/10.1016/j.jconrel.2008.10.002. Epub 2008 Oct 12. PMID: 18951932.

Li J, Wu S, Wu C, Qiu L, Zhu G, Cui C, Liu Y, Hou W, Wang Y, Zhang L, Teng IT, Yang HH, Tan W (2016) Versatile surface engineering of porous nanomaterials with bioinspired polyphenol coatings for targeted and controlled drug delivery. Nanoscale 8(16):8600–8606. https://doi.org/10.1039/c6nr00600k. PMID: 27050780.

Liu RH (2004) Potential synergy of phytochemicals in cancer prevention: mechanism of action. J Nutr 134(12 Suppl):3479S–3485S. https://doi.org/10.1093/jn/134.12.3479S. PMID: 15570057.

Lohani A, Verma A, Joshi H, Yadav N, Karki N (2014) Nanotechnology-based cosmeceuticals. ISRN Dermatol 2014:843687. https://doi.org/10.1155/2014/843687. PMID: 24963412; PMCID: PMC4055113.

Lu XY, Wu DC, Li ZJ, Chen GQ (2011) Polymer nanoparticles. Prog Mol Biol Transl Sci 104:299–323. https://doi.org/10.1016/B978-0-12-416020-0.00007-3. PMID: 22093222.

Majidzadeh H, Araj-Khodaei M, Ghaffari M, Torbati M, Ezzati Nazhad Dolatabadi J, Hamblin MR (2020) Nano-based delivery systems for berberine: a modern anti-cancer herbal medicine. Colloids Surf B Biointerfaces 194:111188. https://doi.org/10.1016/j.colsurfb.2020.111188. Epub 2020 Jun 8. PMID: 32540763.

Masood F (2016) Polymeric nanoparticles for targeted drug delivery system for cancer therapy. Mater Sci Eng C Mater Biol Appl 60:569–578. https://doi.org/10.1016/j.msec.2015.11.067. Epub 2015 Nov 27. PMID: 26706565.

McShan D, Yu H (2014) DNA damage in human skin keratinocytes caused by multiwalled carbon nanotubes with carboxylate functionalization. Toxicol Ind Health 30(6):489–498. https://doi.org/10.1177/0748233712459914. Epub 2012 Sep 25. PMID: 23012341; PMCID: PMC3912206.

Monteiro N, Martins A, Ribeiro D, Faria S, Fonseca NA, Moreira JN, Reis RL, Neves NM (2015) On the use of dexamethasone-loaded liposomes to induce the osteogenic differentiation of human mesenchymal stem cells. J Tissue Eng Regen Med 9(9):1056–1066. https://doi.org/10.1002/term.1817. Epub 2013 Oct 7. PMID: 24123949.

Najahi-Missaoui W, Arnold RD, Cummings BS (2020) Safe nanoparticles: are we there yet? Int J Mol Sci 22(1):385. https://doi.org/10.3390/ijms22010385. PMID: 33396561; PMCID: PMC7794803.

Naseri N, Valizadeh H, Zakeri-Milani P (2015) Solid lipid nanoparticles and nanostructured lipid carriers: structure, preparation and application. Adv Pharm Bull 5(3):305–313. https://doi.org/10.15171/apb.2015.043. Epub 2015 Sep 19. PMID: 26504751; PMCID: PMC4616893.

Ong WTJ, Nyam KL (2022) Evaluation of silver nanoparticles in cosmeceutical and potential biosafety complications. Saudi J Biol Sci 29(4):2085–2094. https://doi.org/10.1016/j.sjbs.2022.01.035. Epub 2022 Jan 21. PMID: 35531241; PMCID: PMC9073040.

Pandey P, Gulati N, Makhija M, Purohit D, Dureja H (2020) Nanoemulsion: a novel drug delivery approach for enhancement of bioavailability. Recent Pat Nanotechnol 14(4):276–293. https://doi.org/10.2174/1872210514666200604145755. PMID: 32496999.

Pardeike J, Hommoss A, Müller RH (2009) Lipid nanoparticles (SLN, NLC) in cosmetic and pharmaceutical dermal products. Int J Pharm 366(1-2):170–184. https://doi.org/10.1016/j.ijpharm.2008.10.003. Epub 2008 Oct 17. PMID: 18992314.

Pasche S, Vörös J, Griesser HJ, Spencer ND, Textor M (2005) Effects of ionic strength and surface charge on protein adsorption at PEGylated surfaces. J Phys Chem B 109(37):17545–17552. https://doi.org/10.1021/jp050431+. PMID: 16853244.

Pietroiusti A, Bergamaschi E, Campagna M, Campagnolo L, De Palma G, Iavicoli S, Leso V, Magrini A, Miragoli M, Pedata P, Palombi L, Iavicoli I (2017) The unrecognized occupational relevance of the interaction between engineered nanomaterials and the gastro-intestinal tract: a consensus paper from a multidisciplinary working group. Part Fibre Toxicol 14(1):47. https://doi.org/10.1186/s12989-017-0226-0. PMID: 29178961; PMCID: PMC5702111.

Raj S, Jose S, Sumod US, Sabitha M (2012) Nanotechnology in cosmetics: opportunities and challenges. J Pharm Bioallied Sci 4(3):186–193. https://doi.org/10.4103/0975-7406.99016. PMID: 22923959; PMCID: PMC3425166.

Santos AC, Rodrigues D, Sequeira JAD, Pereira I, Simões A, Costa D, Peixoto D, Costa G, Veiga F (2019) Nanotechnological breakthroughs in the development of topical phytocompounds-based formulations. Int J Pharm 572:118787. https://doi.org/10.1016/j.ijpharm.2019.118787. Epub 2019 Oct 31. PMID: 31678376.

Sharma S, Choudhary M, Budhwar V (2022) Role of bioactive phytoconstituents as modulators of hepatic carbohydrates metabolising enzymes: a target specific approach to treat diabetes mellitus. Curr Diabetes Rev 18(9):e100222201016. https://doi.org/10.2174/1573399818666220210140745. PMID: 35142270.

Singh S (2010) Nanomedicine-nanoscale drugs and delivery systems. J Nanosci Nanotechnol 10(12):7906–7918. https://doi.org/10.1166/jnn.2010.3617. PMID: 21121278.

Singh R, Lillard JW Jr (2009) Nanoparticle-based targeted drug delivery. Exp Mol Pathol 86(3): 215–223. https://doi.org/10.1016/j.yexmp.2008.12.004. Epub 2009 Jan 7. PMID: 19186176; PMCID: PMC3249419.

Smijs TG, Pavel S (2011) Titanium dioxide and zinc oxide nanoparticles in sunscreens: focus on their safety and effectiveness. Nanotechnol Sci Appl 4:95–112. https://doi.org/10.2147/NSA.S19419. PMID: 24198489; PMCID: PMC3781714.

Song P, Xiao S, Zhang Y, Xie J, Cui X (2020) Mechanism of the intestinal absorption of six flavonoids from *Zizyphi Spinosi* semen across Caco-2 cell monolayer model. Curr Drug Metab 21(8):633–645. https://doi.org/10.2174/1389200221666200714100455. PMID: 32664838.

Squillaro T, Cimini A, Peluso G, Giordano A, Melone MAB (2018) Nano-delivery systems for encapsulation of dietary polyphenols: an experimental approach for neurodegenerative diseases and brain tumors. Biochem Pharmacol 154:303–317. https://doi.org/10.1016/j.bcp.2018.05.016. Epub 2018 May 24. PMID: 29803506.

Sun T, Zhang YS, Pang B, Hyun DC, Yang M, Xia Y (2014) Engineered nanoparticles for drug delivery in cancer therapy. Angew Chem Int Ed Engl 53(46):12320–12364. https://doi.org/10.1002/anie.201403036. Epub 2014 Oct 7. PMID: 25294565.

Teixeira-Santos R, Gomes M, Gomes LC, Mergulhão FJ (2020) Antimicrobial and anti-adhesive properties of carbon nanotube-based surfaces for medical applications: a systematic review. iScience 24(1):102001. https://doi.org/10.1016/j.isci.2020.102001. PMID: 33490909; PMCID: PMC7809508.

Verma A, Stellacci F (2010) Effect of surface properties on nanoparticle-cell interactions. Small 6(1):12–21. https://doi.org/10.1002/smll.200901158. PMID: 19844908.

Warheit DB, Webb TR, Reed KL, Frerichs S, Sayes CM (2007) Pulmonary toxicity study in rats with three forms of ultrafine-TiO2 particles: differential responses related to surface properties. Toxicology 230(1):90–104. https://doi.org/10.1016/j.tox.2006.11.002. Epub 2006 Nov 10. PMID: 17196727.

Wu TH, Yen FL, Lin LT, Tsai TR, Lin CC, Cham TM (2008) Preparation, physicochemical characterization, and antioxidant effects of quercetin nanoparticles. Int J Pharm 346(1-2): 160–168. https://doi.org/10.1016/j.ijpharm.2007.06.036. Epub 2007 Jun 28. PMID: 17689897.

Yah CS, Simate GS, Iyuke SE (2012) Nanoparticles toxicity and their routes of exposures. Pak J Pharm Sci 25(2):477–491. PMID: 22459480.

Yingchoncharoen P, Kalinowski DS, Richardson DR (2016) Lipid-based drug delivery systems in cancer therapy: what is available and what is yet to come. Pharmacol Rev 68(3):701–787. https://doi.org/10.1124/pr.115.012070. PMID: 27363439; PMCID: PMC4931871.

Zhang XY, Zhang PY (2017) Polymersomes in nanomedicine - a review. Curr Med Chem 13(2): 124–129. https://doi.org/10.2174/1573413712666161018144519. Epub 2017 Apr. PMCID: PMC5374840.

Zielińska A, Carreiró F, Oliveira AM, Neves A, Pires B, Venkatesh DN, Durazzo A, Lucarini M, Eder P, Silva AM, Santini A, Souto EB (2020) Polymeric nanoparticles: production, characterization, toxicology and ecotoxicology. Molecules 25(16):3731. https://doi.org/10.3390/molecules25163731. PMID: 32824172; PMCID: PMC7464532.

Regulatory Aspects for Clinical Applications of Nanophytomedicines

12

Shalini Shukla, Akshada Mhaske, and Rahul Shukla

12.1 Regulatory Guidelines

The regulation of nanoparticulate systems has confronted multiple challenges in recent years. In order to compare the variations in the safety of nanoformulation in cosmetics, a number of studies are being performed.

12.1.1 Guidelines for Nanopharmaceuticals

Under the leadership of the International Council for Harmonisation of Technical Requirements for Pharmaceuticals for Human Use (ICH), the International Pharmaceutical Regulators Programme (IPRP) investigates and resolves new patterns of concern in pharmaceutical regulation. The topics of nanomaterials in pharmaceutical medicine, combination and borderline medications, as well as innovation and assessment methods, are all discussed. Those who have signed the IPRP nanomedicines steering committee agreement include representatives from governments and organizations across the Americas, Asia, Europe, and Oceania. Its objectives include reaching out to groups of creators of nanopharmaceuticals as well as other stakeholders, regulation, harmonization, coordination with foreign authorities on the training providers, and non-confidential information sharing (IPRP 2023). Moreover, it should be remembered that the laws that govern certain industries and nations determine what needs to be reported to authorities about nanomaterials and in what metrics (e.g. based on particle size or weight). For instance, according to EU REACH specifications, particle size distribution because

S. Shukla · A. Mhaske · R. Shukla (✉)
Department of Pharmaceutics, National Institute of Pharmaceutical Education and Research-Raebareli, Lucknow, UP, India
e-mail: rahul.shukla@niperraebareli.edu.in

© The Author(s), under exclusive license to Springer Nature Singapore Pte Ltd. 2024

D. Pooja, H. Kulhari (eds.), *Nanotechnology Based Delivery of Phytoconstituents and Cosmeceuticals*, https://doi.org/10.1007/978-981-99-5314-1_12

of the quantity of particles is required for the evaluation of nanoforms, whereas the weight percentage of nanopharmaceuticals determines whether a material must be reported to the US EPA under the Toxic Substances Control Act. When a nanomaterial must be reported to the European Commission's Cosmetics Products Notification Portal (CPNP), both measurements are occasionally necessary (Federal Register 2023).

In terms of quality standards of herbal nanoformulation, there are no regulations that expressly address nano phytomedicines. In regions like the European Union (EU), Africa, Korea, and the Philippines, the management of the product quality system, regulations for pre-market approval, post-market inspection are crucial for conventional herbal medicines. Due to the recent increase in the use of plant-based nanoformulations, their advantages over traditional herbal formulations, and the safety considerations, there is a pressing requirement for regulations and monitoring bodies that can regulate and examine the quality of nano phytopharmaceuticals through every step, beginning with manufacturing and ending with clinical uses and post-clinical assessments (Kavya Teja et al. 2021).

12.1.2 Guidelines for Nanocosmeceuticals

Several regulation papers can be used to identify and analyse the safety aspects of nanomaterials in cosmetics by the cosmetics industry as well as other interested parties (academicians, researchers, etc.). There are summaries of the legal implications of nanoparticles or the usage of nanomaterials in the literature, particularly in cosmetics.

12.1.2.1 Food and Drug Administration

The FDA's current viewpoint regarding the safety evaluation of nanostructures in cosmetic goods is laid forth in this document as guidance for the industry and other stakeholders. The guideline materials provided by the FDA do not establish any legally binding obligations. Instead, guidelines should only be taken as recommendations until clear regulatory or legislative requirements are stated. The OECD framework for Manufactured Nanomaterials has published "Preliminary Review of OECD Test Guidelines for their Applicability to Manufactured Nanomaterials" and "Currently Available Methods for Characterization of Nanomaterials" in this arena.

In conclusion, data requirements and testing procedures should be assessed for cosmetic products with new or changed attributes in order to address any special qualities and functions of the nanomaterials employed in cosmetics. By comparing each ingredient's physical, chemical, and toxicological endpoints to the anticipated exposure caused by the intended application of the completed product, the safety of a cosmetic product can be assessed. This recommendation suggests that the manufacturer should meet with the FDA to discuss the test procedures and information required to confirm the product's safety, including short-term toxicity and other long-term toxicity data, as appropriate, if the manufacturer wishes to use a nanomaterial in a cosmetic product that is either new or an altered version of an already marketed ingredient (FDA 2023a).

12.1.2.2 The International Cooperation on Cosmetics Regulation

Discussions about cosmetics and pharmaceuticals with cosmetic-like properties during the fourth annual meeting of the International Cooperation on Cosmetics Regulation (ICCR-4) in Canada in July 2010 led to the creation of a Joint Industry/Regulator Working Group (WG) for nanomaterial safety. The goal of this Joint WG was to determine whether current safety precautions might be applied to nanoparticles that were pertinent to activities in the cosmetic industry. The main objectives were to perform an investigation of current safety procedures, identify any particular consumer safety concerns that have to be taken into consideration when assessing nanomaterials in cosmetics, and provide a report for discussion by the ICCR members (FDA 2023b).

12.1.3 Guidelines for Nanopharmaceuticals and Nanocosmeceuticals Across Different Countries

According to their respective regulations, many regulatory organizations from nations all over the world oversee the efficacy and safety of cosmetic products. Prior to marketing approval by the makers, the safety of the finished product is guaranteed in a few countries. The formulation's whole list of ingredients, as well as the specified limits for cosmetic and cosmeceutical substances and goods, must be listed on the label. The mentioned limits must also adhere to the defined limits. Various regulatory guidelines issued by the FDA as well as in other countries for the safe use of nanopharmaceuticals and nanocosmeceuticals have been discussed below (Tripathy and Dureja 2015).

12.1.3.1 Food and Drug Administration

The Food and Drug Administration (FDA) is in charge of a wide range of cosmetic items thanks to the Federal Food, Drug, and Cosmetic Act. The FDA does not actively oversee cosmetic items. Cosmetics do not now require any previous clearance, but safety measures must be followed to ensure safety and efficacy of cosmetics. Cosmetics safety must be the responsibility of the producer. To regulate the items containing nanoparticles, the FDA established an internal task group on nanotechnology in 2006. This action was taken to accelerate the production of novel, safe, and efficient nanomaterials for the pharmaceutical and cosmetic industries (FDA 2022a). FDA recommended changes in 2007, and many of them have since been put into effect. Additional changes are also on the way. Three regulations concerning the safety of nanoparticulate systems were released by the FDA in 2014. According to dimension-dependent phenomena and particle size, the first set of instructions explains how to spot FDA-regulated products that contain nanoparticles. The safety of nanoparticles in cosmetics was the subject of the second recommendation. The FDA does not mandate that the completed product label include information about the nanoparticles utilized in its formulation. The FDA collaborated with the Personal Care Products Council (DC, USA) to draught cosmetic rules, and the Voluntary Cosmetic Registration Programme relies on voluntary reporting of the components and adverse reactions. The FDA informs producers about the issues surrounding nanoparticles in order to continuously increase the safety of cosmetic products. With this approach, manufacturers may limit the usage of harmful nanocosmeceuticals (FDA 2022b).

12.1.3.2 The European Union

According to the EU's regulations, nanomaterials are described as intentionally created insoluble or bio-permanent substances with at least one or more exterior domains or an interior structure in the range of 1–100 nm. The word "nano" specifically must be used to identify all nanomaterials. Before nanoparticle-based goods and nanocosmeceuticals are introduced to the market, the manufacturers must notify of the product's specifications, toxicity, safety profile, and unfavourable consequences six months in advance. The cosmeceuticals, sunscreens, colourants, and antiaging treatments made using nanotechnology need premarket approval. Nanomaterials are covered under European Commission Regulation No. 1907/2006 on the Registration, Evaluation, Authorisation, and Restriction of Chemical Substances and the Scientific Committee on Consumer Product Safety (Kumud and Sanju 2018; Ali and Sinha 2014).

12.1.3.3 Australia

In Australia, cosmetic regulations are governed by the Therapeutics Goods Administration (TGA). Industrial Chemicals Notification and Assessment Act of 1989 governs all components (including natural goods) and is enforced by the National Industrial Chemicals Notification and Assessment Scheme (NICNAS). To protect employees, customers, and the environment, the Australian government conducts risk assessments for all cosmetic items made or imported into Australia (Therapeutic Goods Administration (TGA) 2022).

12.1.3.4 India

The Nano Science and Technology Initiative, funded by the Indian government, provides well-organized setups in numerous universities, academic institutions, national research centres, start-ups, and R&D agencies. The Council of Scientific and Industrial Research, Indian Council of Medical Research, Department of Science and Technology, and Department of Biotechnology are the top-performing institutions in the national health research systems (New Delhi, India). In India's capital city New Delhi, the Ministry of Health and Family Welfare is essential in the prevention and management of illnesses. Additionally, the Nanotechnology Sectional Committee actively promotes the standardization of nanomaterials and is made up of experts affiliated with numerous research organizations and institutes (FDA 2022c).

12.2 Conclusion

The nanocosmeceutical industries are rapidly rising worldwide. Nanotechnology-assisted cosmeceuticals exhibit their role in the therapy of several skin disorders. This technology is highly suggested for providing safe and cost-effective products to consumers. There is increased demand for nano-cosmeceuticals with phyto-based bioactive compounds for skincare due to their antiaging, sunblock, moisturizing, and whitening effects together with providing healthy skin. The nanosized particles are stable and retain large quantities of bioactive compounds in cutaneous tissue, and that too for a longer duration. However, due to a lack of particular guidelines, the commercialization and growth of these cosmeceuticals have increased to a great extent and the industries are gaining huge profits. Although the commercial value of these nanomaterial goods is increasing significantly, there is a great deal of controversy around their safety and toxicity profile in humans, necessitating further research. In order to ensure the safety of the use of cosmetic products, the cosmetic regulation should give a precise list of references and the substances that generate unanticipated environmental consequences for all customers of beauty products, including general and professional users.

Acknowledgements The authors acknowledge the Department of Pharmaceuticals, Ministry of Chemical and Fertilizers, Govt. of India for support. The NIPER-R communication number for the chapter is NIPER-R/437.

References

Ali A, Sinha K (2014) Prospects of nanotechnology development in the health sector in India. Int J Health Sci 2(2):2372–5079

FDA (2022a) Cosmetics & U.S. Law. Available at https://www.fda.gov/cosmetics/cosmetics-laws-regulations/cosmetics-us-law

FDA (2022b) Consumer updates. Available at https://www.fda.gov/consumers/consumer-updates

FDA (2022c) Prohibited & restricted ingredients in cosmetics. Available at https://www.fda.gov/cosmetics/cosmetics-laws-regulations/prohibited-restricted-ingredients-cosmetics

FDA (2023a) Guidance for industry: safety of nanomaterials in cosmetic products. Available at https://www.fda.gov/regulatory-information/search-fda-guidance-documents/guidance-industry-safety-nanomaterials-cosmetic-products

FDA (2023b) International cooperation on cosmetics regulation (ICCR). Available at https://www.fda.gov/cosmetics/cosmetics-international-activities/international-cooperation-cosmetics-regulation-iccr

Federal Register (2023) Chemical substances when manufactured or processed as nanoscale materials; TSCA reporting and recordkeeping requirements. Available at https://www.federalregister.gov/documents/2017/01/12/2017-00052/chemical-substances-when-manufactured-or-processed-as-nanoscale-materials-tsca-reporting-and

IPRP (2023) International pharmaceutical regulators programme. Available at https://www.iprp.global/home

Kavya Teja P, Mithiya J, Kate AS, Bairwa K, Chauthe SK (2021) Herbal nanomedicines: recent advancements, challenges, opportunities and regulatory overview. Phytomedicine 96:153890. https://doi.org/10.1016/j.phymed.2021.153890

Kumud M, Sanju N (2018) Nanotechnology driven cosmetic products: commercial and regulatory milestones. Appl Clin Res Clin Trials Regul Affairs 5(2):112–121

Therapeutic Goods Administration (TGA) (2022) Cosmetics. Available at https://www.tga.gov.au/cosmetics

Tripathy S, Dureja H (2015) Cosmetics: regulatory scenario in USA, EU and India. J Pharm Technol Res Manage 3(2):127–139. Available at https://www.researchgate.net/publication/305868114_Cosmetics_Regulatory_Scenario_in_USA_EU_and_India

SPRINGER NATURE

GPSR Compliance

The European Union's (EU) General Product Safety Regulation (GPSR) is a set of rules that requires consumer products to be safe and our obligations to ensure this.

If you have any concerns about our products, you can contact us on ProductSafety@springernature.com

In case Publisher is established outside the EU, the EU authorized representative is:

Springer Nature Customer Service Center GmbH
Europaplatz 3
69115 Heidelberg, Germany

The manufacturer's authorised representative in the EU is Springer Nature Customer Service Centre GmbH, Europaplatz 3, 69115 Heidelberg, Germany. If you have any concerns regarding our products, please contact ProductSafety@springernature.com

Printed and bound by CPI Group (UK) Ltd, Croydon, CR0 4YY
30/09/2026
02225322-0012